Biology of Insect-Induced Galls

Biology of Insect-Induced Galls

Edited by

JOSEPH D. SHORTHOUSE
Department of Biology
Laurentian University
Sudbury, Ontario, Canada

ODETTE ROHFRITSCH
Laboratoire de Cécidologie
Institut de Botanique
Université Louis Pasteur
Strasbourg, France

New York Oxford
OXFORD UNIVERSITY PRESS
1992

Oxford University Press

Oxford New York Toronto
Delhi Bombay Calcutta Madras Karachi
Petaling Jaya Singapore Hong Kong Tokyo
Nairobi Dar es Salaam Cape Town
Melbourne Auckland

and associated companies in
Berlin Ibadan

Published by Oxford University Press, Inc.,
200 Madison Avenue, New York, New York 10016

Oxford is a registered trademark of Oxford University Press

Library of Congress Cataloging-in-Publication Data
Biology of insect-induced galls/
edited by Joseph D. Shorthouse,
Odette Rohfritsch.
p. cm.
Includes bibliographical references and index.
ISBN 978-0-19-506716-3
1. Galls (Botany) 2. Gall insects.
3. Insect-plant relationships.
I. Shorthouse, Joseph D. II. Rohfritsch, Odette.
SB767.B58 1992 632'.2—dc20 91-10014

Preface

Plant galls have fascinated and bewildered naturalists from the time of Hippocrates (ca. 350 B.C.). Although early naturalists knew that insects were found within galls, the debate over the biological nature of galls continues to this day. However, it is now generally accepted that the atypical growths we call galls are a physiological response or defensive reaction by plants to certain insects that feed or deposit their eggs on or within plant tissues. The attacking insect is not only highly specific in selecting its host, organ, and tissue, but also has control over the form, structure, color, and physiology of the gall that the plant produces. The insect accomplishes these changes by imparting highly specific wound-like stimuli when the host tissues are at a particular stage of development. Thus, to understand galls, one must be aware of the life strategies of both the host plant and insect. All benefits in the association appear to go to the insect as the plant is manipulated into providing the attackers with shelter and high quality food.

We suspect that interest in galls has increased over the past few years as a result of several recent books and review articles on various aspects of gall biology. Most important among the books are Anatomie des Galles by J. Meyer and H. J. Maresquelle, Biology of Gall Insects edited by T. N. Ananthakrishnan, Plant Galls and Gall Inducers by J. Meyer, and The Plant-Feeding Gall Midges of North America by R. J. Gagné. However, each of these books concentrates on one particular aspect of gall biology. This book, in contrast, attempts to take an overview of cecidology, the study of plant galls, and emphasizes the recent advances in its fundamental principles. Furthermore, we have strived to illustrate the multidisciplinary nature of gall studies. To this end, we have assembled a series of contributions written by plant anatomists and physiologists, as well as entomologists. Each chapter is written by a specialist(s) in one of the fields of gall biology. Following an introduction to the science by the eminent gall biologist M. S. Mani, and a survey of gall-inducing insects and their galls, the evolution of the gall-inducing guild and fossil galls is discussed. Next is a series of contributions on the anatomy and developmental morphology of galls, including a review of the specialized gall cells referred to as "nutritive cells" so important to both gall development and the nutrition of gall insects. Following contributions on the physiology of both galled plants and gall induction are a series of contributions on the ecology of gall-inhabiting insects.

Cecidology is a discipline that draws entomologists and plant anatomists and physiologists together to study these complex plant–insect systems. In many instances, insect taxonomists must make the first contribution as they are required to identify the gall inducer and other inhabitants before botanists and insect ecologists can begin their studies. We suspect that few plant morphologists or physiologists are familiar with the intricacies and research potential of gall systems. Indeed, most textbooks in plant morphology and physiology make at best only passing mention of galls and none, to our knowledge, illustrates the structures of even one of the many thousand of galls found in the world. This is unfortunate since each gall represents a "new" plant organ in which control of developmental and physiological processes has passed from the plant to a specialized herbivore. Galls not only illustrate the plasticity of developing plant tissues, they show us how, with the right stimulus, plants can be reprogrammed to produce atypical structures and to accumulate substances not normally associated with the attacked organ. More than one student of gall studies has recognized gall inducers as the genetic engineers of the insect world.

Recently, it is the insect ecologists who have gained the most from the contribution of gall insects to their discipline. Several laboratories around the world have recognized that gall insects provide ideal systems in which to increase our understanding of herbivore–plant relationships, community ecology, and life history evolution. Galls of most insects are prominent and easily located and their internal communities provide unlimited opportunities for studying competition, seasonal survivorship, niche partitioning, zoogeography, links between first and third trophic levels, etc.

This volume would not have been possible without the cooperation and encouragement of many people. We are indebted to the chapter authors who amended their contributions in light of many fruitful exchanges. We thank many colleagues in our respective institutions and reviewers elsewhere for their help in improving the various contributions. In particular we thank M. Wehr of the Institut de Botanique in Strasbourg for her help in the preparation of the manuscript and M. Roche, a photographer with Laurentian University in Sudbury, for her help with many of the plates.

Sudbury, Ontario J. D. S.
January 1991 O. R.

Contents

Contributors

Taracad N. Ananthakrishnan
Entomology Research Institute,
Loyola College,
Madras, 600 034, India

Ming Anthony
Laboratoire de Cécidologie,
Institut de Botanique,
28 rue Goethe,
67083 Strasbourg Cedex, France

M. Linda Birch
20 Nethercott Road,
Tackley, Oxon,
0X5 3AW, England

J. Wayne Brewer
Department of Entomology,
Auburn University,
Auburn, Alabama, 36849-5413

Roberte Bronner
Laboratoire de Cécidologie,
Institut de Botanique,
28 rue Goethe,
67083 Strasbourg Cedex, France

Françoise Dreger-Jauffret
Laboratorie de Cécidologie,
Institut de Botanique,
28 rue Goethe,
67083 Strasbourg Cedex, France

Gordon W. Frankie
Department of Entomological Sciences,
University of California,
Berkeley, California, 94720

Eric E. Grissell
Systematic Entomology Laboratory,
NHB 168, USDA, ARS,
c/o U.S. National Museum,
Washington, D.C., 20560

Koji Hori
Laboratory of Entomology,
Obihiro University,
Obihiro, Hokkaido, Japan

Hiram G. Larew
Office of the Science Advisor,
Room 320, 8A-18,
U.S. Agency for International Development,
Washington, D.C., 20593-1818

Mahadev S. Mani
12 Second Street,
Padmanabha Nagar,
Adyar, Madras 600 020, India

David L. Morgan
Department of Agronomy,
Horticulture and Entomology,
Texas Tech University,
Lubbock, Texas, 79409

Peter W. Price
Department of Biological Sciences,
Northern Arizona University,
Box 5640
Flagstaff, Arizona, 86011

Lucien A. Rey
Laboratoire de Cytopathologie Végétale,
U.E.R. des Sciences de la Nature,
2 rue de la Houssinière,
44072 Nantes Cedex, France

Odette Rohfritsch
Laboratoire de Cécidologie,
Institut de Botanique,
28, rue Goethe,
67083 Strasbourg Cedex, France

J. (Hans) C. Roskam
Evolutionary Biology Section,
University of Leiden,
2313 ZT., Leiden,
Schelpenkade 14a, The Netherlands

Joseph D. Shorthouse
Department of Biology,
Laurentian University,
Sudbury, Ontario,
P3E 2C6, Canada

Marcela Skuhravá
140 00 Praha 4,
Bitovská 1227, Čzechoslovakia

Václav Skuhravý
Institute of Entomology,
České Budějovice,
370 05 Branišovská 31
Čzechoslovakia

Evelyne Westphal
Laboratoire de Cécidologie,
Insitut de Botanique,
28, rue Goethe,
67083 Strasbourg Cedex, France

Thomas G. Whitham
Department of Biological Sciences,
Box 5640, Northern Arizona University,
Flagstaff, Arizona, 86011

Naan A. Wiebes-Rijks
Division of Systematics
and Evolutionary Biology,
University of Leiden,
Kaiserstraat 63, RA Leiden,
The Netherlands

Biology of Insect-Induced Galls

1
Introduction to Cecidology

M. S. MANI

Cecidology, the study of plant galls, is one of the most interdisciplinary pursuits in the realm of natural sciences. This is so because galls represent a complex series of interactions between the tissues of a plant and another living organism. An understanding of these interactions involves studies of gall structure and development, etiology of gall formation, cytology, biochemistry, pathology, and physiology of gall tissues. Also included are studies on the life cycles, behavior, host specificity, nutritional physiology, and trophic relationships of the complex assemblages of organisms associated with galls. Studies on the nature, significance, and evolution of the gall-inducing guild are also basic to the field.

Although plant galls were known and used extensively in medicine, industry, and even as human food for over 1000 years in China, India, and many parts of Europe, the first scientific studies of galls were those of Malpighi of Italy in the seventeenth century (Figure 1-1 of a cynipid leaf gall was taken from Malpighi's 1679 work). Since then thousands of galls and causative organisms have been described from around the world. Galls also came to occupy an important place in plant pathology. However, it was Trotter of Italy (see references in Mani 1964), working early in the present century, who was most responsible for the development of cecidology as a major field of biological research.

Trotter founded and nourished the international journal *Marcellia,* which was devoted exclusively to cecidology. In the early 1950s Jean Meyer of France became the editor and continued in that position until 1976. More recently the journal was produced in India. Cecidological literature has appeared in many other scientific journals throughout the world but *Marcellia* was perhaps the most important single source of information on the topic.

We owe our knowledge of galls to the immense labors of many cecidologists and among their most important works are Buhr (1965), Cook (1902–1904), Cosens (1912), Docters van Leeuwen-Reijnvaan and Docters van Leeuwen (1926), Felt (1940), Fockeu (1896), Houard (1908–1913, 1922–1923, 1933), Küster (1911, 1925, 1930), Magnus (1914), Meyer (1987), Maresquelle and Meyer (1965), Meyer and Maresquelle (1983), Molliard (1913), Nalepa (1929), Němec (1913), Réaumur (1737–1742), Ross (1932), Ross and Hedicke (1927), Rübsaamen and Hedicke (1923–1925), Weidel (1911), and Zweigelt (1931).

As the science of cecidology developed we can see a gradual shift in emphasis and a series of conceptual changes. First, plant galls were studied mainly as curious objects of natural history. Later came an increasing emphasis on pathological aspects. It was only in recent times that serious attention came to be paid to the morphogenetic problems of galls, the origin and evolution of the complex interspecific interrelationships, and the nature and factors underlying neoplasms [any new growth, usually applied to a tumor (Bloch 1965)]. Nearly every modern scientific technique is now employed in gall studies.

The concept of what constitutes a gall has drastically changed since Malpighi. We now recognize as galls a bewildering variety of plant structures and growth forms, ranging from the

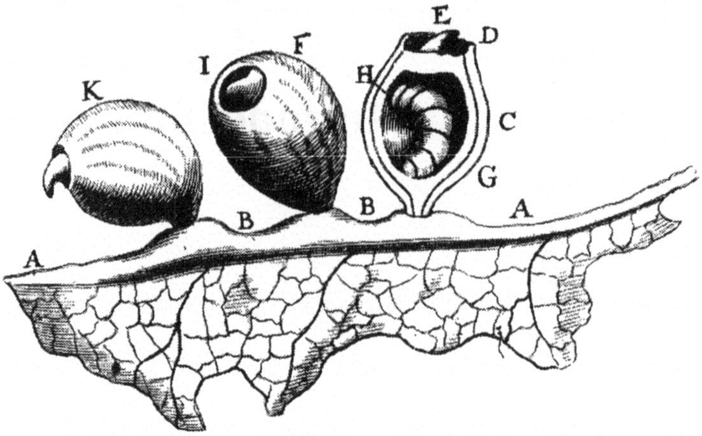

Figure 1-1. *Gallen von Andricus gallae-urnaeformis nach Malpighi 1679* [Early drawing of a cynipid leaf gall taken from Malpighi's 1679 work], in K. Böhner 1933.

nearly normal plant organs to extremely bizarre and highly complex growth abnormalities (Mani 1964 and Chapter 2). Galls arise on all plant organs; some even are found on other galls. Many galls superficially resemble normal plant parts such as berries, drupes, nuts, and other fruit types and even have been described as such in many herbaria. Other galls show a striking resemblance to sea urchins, spiny or hairy balls as well as grotesquely developed horns or clubs. Some are like exquisitely shaped vases, cups, discs, or fantastic polyps and flower-like growths. A number of interesting galls look so much like fungi that for years they were misidentified by plant taxonomists. For example, the gall of *Asteromyia* on the leaves of *Solidago* was described as the fungus *Rhytisma.*

A whole plant organ may turn into a gall or the gall may be a strictly localized histological abnormality. Most galls are usually recognized by a more or less pronounced swelling of some sort whereas others are simple overgrowths of plant tissue. However, most galls involve complex developmental inhibition, differentiation, growth, and suppression of tissues and parts (Maresquelle and Meyer 1965). Besides these processes, there is often overdevelopment or out-of-place development of tissues and also activation of tissues that have long ceased growth and development. Some galls are simply swollen masses of nearly undifferentiated parenchyma cells, but others exhibit a surprising degree of differentiation of highly specialized cells and tissues, often characteristic only of galls. Thus galls, like other regular plant organs, have their own characteristic anatomy and histology (Küster 1911; Meyer and Maresquelle 1983).

Galls are induced by a great variety of organisms including viruses, bacteria, fungi, nematodes, mites, and insects. The association of the gall and gall-inducing organisms was probably recognized in the earliest cecidological studies. However, it was not until relatively recent times that gall development and growth were correlated with the feeding activities and nutritional requirements of the gall-inducing organism.

We now know that galls are essentially the product of an active growth reaction of the plant in response to the feeding activities of the causative organism. It has also been recognized that the development of a gall is contingent on the continued presence and feeding activity of the gall inducer (Rohfritsch 1971).

The intriguing problem of how the feeding activities of a foreign organism induces the formation of a gall has naturally engaged the attention of workers from the very beginning. However, the classical idea of a specific gall-inducing secretion produced by the inducer is now virtually discarded. It was conclusively demonstrated by Anders (1960–1961) that specific cecidotoxins per se are not secreted by gall insects. The physiological changes that induce hypertrophy, hyperplasy and other growth and development movements involved in gall formation are not produced directly by the cecidozoa, but by the plant cells themselves. Thus cecidogenesis is a highly specific and specialized plant reaction to wounding and feeding by the cecidozoa. By reacting in this characteristic manner the plant seems to show atypical morphogenetic behavior (Sinnot 1962; Bloch 1965).

Normal plant morphogenesis is believed to be governed by morphogenetic fields (Wardlaw 1968; Steeves and Sussex 1989). The operation of a field generates gradients of cell activation around the active center (the apex) and inhibits similar growth elsewhere. The inhibition generated by the central cells of the apex provides the mechanism by which a meristem is integrated. There is an intensity gradient from the center to the margins in the central field with outgrowths occurring at the periphery. Inhibitory fields occur around the peripheral outgrowths operating in conjunction with the central field and partially explain the regular patterns in normal morphogenesis.

All normal development and growth of plants are controlled by a series of rigidly integrated and interdependent phasic processes in which each event evokes or abolishes other events. Development also is determined by genetic and physiological mechanisms. As well, environmental conditions modulate the rate at which the basic program is followed. In this way, the developmental processes of cell division, enlargement, and differentiation can be both quantitatively and qualitatively altered in response to changes in the environment. Thus environmental stimuli are transduced in developmental processes and confer to the plants their high degree of developmental plasticity. Gall-inducing organisms are one of these environmental factors and they modify normal plant growth by disturbing the normal morphogenetic fields.

Gall inducers create a new activated region accompanied by the appearance of a new gradient field, the center of which is localized in the cells nearest to the gall-inducing organism. These cells exhibit changes in pH, polarity, nuclear and nucleolar hypertrophy, excess free amino acids and sugars, and the presence of hydrolytic enzymes such as amylase and protease. They also acquire dominance over the other cells of the gall, especially as far as cell division, cell hypertrophy, and growth polarity are concerned. However, these cells also are the center of cytological and biochemical gradients (Chapter 8).

The cecidogenetic field that develops in galls does not totally obliterate the normal morphogenetic field, but the interaction of the two results in the altered pattern of development that leads to the formation of a gall. Thus gall growth consists of both inhibition near the insect and accelerated growth further away. The normal morphogenetic field interacts with, limits, and localizes the cecidogenetic field so that the departure from the normal morphogenetic control is neither total nor even permanent. The cecidogenetic field operates only so long as the gall-inducing organism is active, primarily because the plant reacts only when the alien stimulus is operative. Most plant galls are thus recognized as limited neoplasms, in contrast to the unlimited growth of cancerous tumors or of crown gall tissues.

The mutual interaction, interdependence, and control of the normal and cecidogenetic fields are strong evidence of concomitant adjustments and coevolution of gall-inducing organisms and the gall-bearing plants. Therefore we have a remarkable case of interactions and far-reaching mutual adjustments in nutrition and general biology, no less remarkable than the coevolution of flowers and cross-pollinating insects.

The growth and eventual enclosing of the developing cecidozoa and its virtual encapsulation involve a remarkable avoidance reaction by the plant tissue within the immediate vicinity of the cecidozoan. This encapsulation leads to the cecidozoan becoming completely imbedded within the plant tissues and as a result there must be a series of changes in the gall tissue that facilitate the escape of the adult insect at the appropriate season (Zweigelt 1931; Lalonde and Shorthouse 1982).

The failure to fully appreciate the importance of the above processes partially explains the continued dominance of narrow teleological considerations in cecidology. Instead of the gall being formed by the plant solely for the benefit of the gall insect, as was preached by classical cecidology, galls are evidence of resistance of the plant to the attack of gall insects.

The research of Nierenstein (1930) on the enzyme complex of cynipid galls and of Zweigelt (1931) on aphid gall development showed that the plant resists and counteracts the cecidozoan at every step of cecidogenesis. It was actually Zweigelt (1931) who first discarded the teleological approach and asserted that a plant produces a gall *not* primarily for the benefit of the gall-inducing organism, but to isolate the gall insect in space and time by forcing it to extreme specialization. That the cecidozoan seems to benefit is purely incidental and irrelevant and is to be attributed to the narrow anthropocentric approach.

It follows from these conceptual changes that the basic character of a gall is not really its causative organism nor its relationship to the causative factor or even its abnormal structure. The basic problem is to understand the role of the cells in contact with gall inducers and how these cells escape the morphogenetic field of the host plant.

Most plant galls are strictly limited neoplasmic outgrowths; however, there are some types that are subject to a total escape from morphogenetic control. Galls of this type, which include those induced by *Agrobacterium tumefaciens* (Sm & Towns.), can be described as neoplastic outgrowths and cannot be compared to insect-induced galls (Rohfritsch and Shorthouse 1982). As described above, galls result from a continual interaction between two living organisms. No such relationships exist between *A. tumefaciens* and its host plant. In contrast, the tumor-inducing plasmid carried by *A. tumefaciens* is responsible for the growth of crown gall and not the bacteria directly (Braun 1982).

Complexities associated with insect galls extend much further than the relationships between the host plant and the gall inducer. The unusual concentrations of highly nutritious substances within gall tissues, the vulnerability of the cecidozoans, and the presence of cells with near-meristematic condition provide a favorable habitat for the development of microcommunities (see Chapters 14 and 16). The members of these microcommunities belong to astonishing diverse groups including bacteria, fungi, nematodes, mites, spiders, pseudoscorpions, and insects. The associations of bacteria and fungi are often exceptionally complex and there are many unsolved problems of their individual roles. Nearly every mite or insect gall has one or more species of fungi associated with it. Some of these fungi are purely saprophytic or parasitic while others, such as the so-called ambrosia fungi, exist in an intimate symbiotic relationship with the cecidozoa and the gall tissue. Most of the more complex galls contain numerous parasites, predators, parasitoids, inquilines, and successors that appear after the escape of the cecidozoan and the composition and interrelationships of these microcommunities are often characteristic of the species of gall inducer and its gall.

Members of gall communities stand in well-marked hierarchial levels of interrelations not only among themselves, but also with the gall and the gall-inducing species. However, it is important to remember that these complex community interrelationships are based on the primary associations between the gall inducer and cells of the gall. This is the basic foundation on which the vast complex has arisen and is the essential link in the gall community.

As I pointed out previously (Mani 1964), the various branches of cecidology can be conveniently integrated within the science of ecology. This encompassing approach to gall studies, taken by many researchers today, illustrates that cecidology is not merely a multidisciplinary science, but rather a specialized branch of ecology that, perhaps better than any other, embraces all basic ecological principles. Thus approaching gall systems in an ecological manner permits us to relate various studies that may otherwise appear isolated. This shift in emphasis from plant pathology and morphogenesis to the realm of ecology is a recent development, the full significance of which has yet to be widely appreciated.

LITERATURE CITED

Anders, F. 1960–61. Untersuchungen über das cecidogene Prinzip der Reblaus. I–III. Biol. Ztlbl. 79:47–58; 79:679–700; 80:199–253.

Bloch, R. 1965. Abnormal development in plants: A survey. Handbuch der Pflanzenphysiologie, Vol. 15 (2), pp. 145–183. Springer, Berlin.

Böhner, K. 1933. Geschichte der Cecidologie. A. Nemayer, Mittenwald, Bayern.

Braun, A. C. 1982. A history of the crown gall problem. In G. Kahl and J. S. Schell (eds.), The Molecular Biology of Plant Tumors, pp. 155–210. Academic Press, New York.

Buhr, H. 1965. Bestimmungstabellen der Gallen an Pflanzen Mittel- und Nordeuropas, 1–2:1–1572, pl. xxv, fig. 443. Gustav Fischer, Jena.

Cook, M. T. 1902–04. Galls and insects producing them. I–IX. Ohio Naturalist 2(6)(15):263–278; 3(7)(20):419–436; 4(8)(13):125–147.

Cosens, A. 1912. A contribution to the morphology and biology of insect galls. Trans. Can. Inst. 9(3):297–387.

Docters Van Leeuwen-Reijnvaan, J., and W. M. Docters Van Leeuwen. 1926. The Zoocecidia of the Netherlands East Indies. Batavia, The Netherlands

Felt, E. P. 1940. Plant Galls and Gall Makers. Comstock, Ithaca, NY.

Fockeu, H. 1896. Recherches anatomiques sur les Galles. Thèse, Lille.

Hieronymus, G. 1890. Beiträge zur Kenntnis der europaïschen Zoocecidien und Verbreitung derselben. Erg. 68 Jahrb. Schles. Ges. Vater. Kultur. 49:272.

Houard, C. 1908–13. Les Zoocécidies des Plantes d'Europe et du Bassin de la Méditerranée, 3 Vols. Hermann et Fils, Paris.

Houard, C. 1922–23. Les Zoocécidies des Plantes d'Afrique d'Asie et d'Océanie, 2 Vols. Hermann et Cie, Paris.

Houard, C. 1933. Les Zoocécidies des Plantes de l'Amérique du Sud et de l'Amérique Centrale. Hermann et Cie, Paris.

Küster, E. 1911. Die Gallen der Pflanzen. Ein Lehrbuch der Botaniker und Entomologen. S. Hirzel, Leipzig.

Küster, E. 1925. Pathologische Pflanzenanatomie in ihren Grundzügen. Gustav Fischer, Jena.

Küster, E. 1930. Anatomie der Gallen. In Schröders Handbuch der Entomologie, Vol. 1, pp. 1–197. Borntraeger, Berlin.

Lalonde, R. G., and J. D. Shorthouse. 1982. Exit strategy of *Urophora cardui* (Diptera: Tephritidae) from its gall on Canada thistle. Can. Entomol. 114:873–878.

Magnus, W. 1914. Die Entstehung der Pflanzengallen verursacht durch Hymenopteren, pp. 1–160, Jena.

Malpighi, M. 1679. De gallis. In. Anatome Plantarum. II. Johan Martin, London.

Mani, M. S. 1964. Ecology of Plant Galls. Dr. W. Junk. The Hague, The Netherlands.

Maresquelle, H. J., and J. Meyer. 1965. Physiologie et morphogenèse des galles d'origine animale (Zoocécidies Handb. Pflanzenphysiol. 15(2):280–329.

Meyer, J. 1987. Plant Galls and Gall Inducers. Borntraeger, Berlin.

Meyer, J., and H. J. Maresquelle, 1983. Anatomie des Galles. Borntraeger, Berlin.

Molliard, M. 1913. Recherches physiologiques sur les galles. Rev. Gen. Bot. 25:225–252, 285–307, 341–370.

Nalepa, A. 1929. Neuer Katalog der bisher beschriebenen Gallmilben, ihre Gallen und Wirspflanzen. *Marcellia* 25:67–183.

Němec, B. 1913. Pflanzengeschwulste und ihre Beziehung zu den tierischen Geschwulsten. Bot. Ztlbl. 123:409; Ztlbl. Bkt. 58:164.

Nierenstein, M. 1930. Interrelations between gall producers and galls. Nature (London) 125(3149):348–349.

Réaumur, R. A. F. 1737–1742. Des galles des plantes et arbres. Mémoires pour servir à l'histoire des insectes. 3. 12e ed.

Rohfritsch, O. 1971. Développement cécidien et rôle du parasite dans quelques galles d'arthropodes. Marcellia 37:233–239.

Rohfritsch, O., and J. D. Shorthouse. 1982. Insect galls. In G. Kahl and J. S. Schell (ed.), The Molecular Biology of Plant Tumors, pp. 131–152. Academic Press, New York.

Ross, H. 1932. Praktikum der Gallenkunde (Cecidologie). Entstehung, Entwicklung, Bau der durch Tiere und Pflanze hervorgerufenen Gallenbildungen sowie Ökologie der Gallenerreger. Julius Springer, Berlin.

Ross, H., and H. Hedicke. 1927. Die Pflanzengallen Mittel- und Nordeuropas ihre Erreger und Biologie und Bestimmungstabelle. Gustav Fischer, Jena.

Rübsaamen, E. H., and H. Hedicke. 1923–25. Die Zoocecidien, durch Tiere erzeugte Pflanzengallen Deutchlands und ihre Bewohner. Zoologica 77.

Sinnot, E. W. 1962. Plant Morphogenesis. McGraw-Hill, New York.

Steeves, T. A., and I. M. Sussex. 1989. Patterns in Plant Development, 2nd Ed. Cambridge University Press, New York.

Wardlaw, C. W. 1968. Morphogenesis in Plants: A Contemporary Study. Methuen, London.

Weidel, F. 1911. Beiträge zur Entwicklungsgeschichte und vergleichenden Anatomie der Zynipidengallen der Eiche. Flora (NF) 2:279–334.

Zweigelt, F. 1931. Blattlausgallen. Histologische und biologische Studien an Tetraneura- und Schizoneura-gallen. Die Blattlausgallen im Dienste prinzipieller Gallenforschung. Monogr. angew. Ent. 11 (Beihefte zu Z. angew. Ent. 27:1–684.

2

Diversity of Gall-Inducing Insects and Their Galls

F. DREGER-JAUFFRET
AND J. D. SHORTHOUSE

Galls are defined as any deviation in the normal pattern of plant growth produced by a specific reaction to the presence and activity of a foreign organism (animal or plant) (Bloch 1965). The modified plant tissues supply the causative agent with both a suitable shelter and nutritious food.

Insects capable of inducing plant galls are highly specialized plant feeders. In contrast to the majority of phytophagous insects that feed externally and move about their hosts removing vast quantities of plant tissues from leaves, stems, flowers, or roots, gall insects are sedentary and feed only on certain cells within a structure they have "designed" from tissues of the attacked organ. Gall insects feed inside plant tissues; they are most closely related to miners and borers. However, the insects of the mining and boring guilds, like the external feeders, must continuously tunnel to new feeding sites once all tissues in a region are consumed. Gall insects, in contrast, barely move within their galls and instead apply as yet poorly understood stimuli to nearby cells that induce tissues of the host organs into surrounding the feeding sites with masses of enlarged and rapidly dividing cells.

Gall insects do not merely stimulate the formation of amorphous masses of parenchyma cells, but rather they control the growth and differentiation of these cells and direct their morphogenesis into structures foreign to the host plant. Although gall insects become internal feeders as a result of their actions, they do not indiscriminately consume gall cells. Instead, gall insects feed on specialized cells that line their larval chambers. These cells, called nutritive cells by students of gall biology, provide a continuous source of food for the developing inhabitants (Chapter 8). Nutritive cells are of higher nutritional value to insects than unaffected cells of the host organ as they commonly contain high concentrations of starch, sugars, lipids, and protein. According to Bronner (1977) these substances appear to be attracted to the feeding sites from other parts of the plant via vascular bundles within the gall joined to those of the host organ. Gall insects therefore are unique in that they appear to exhibit control over the within-plant movement of assimilates as well as tissue morphogenesis.

Thus a plant gall is an atypical plant growth in which the size, shape, structure, and metabolism are under the control of the phytophagous organisms within. Perhaps the most amazing feature of gall biology is that each of the many hundreds of species within this guild is responsible for galls that are structurally unique to the species of inducer. How this occurs is currently a subject of speculation.

Galls are induced by a wide variety of organisms ranging from viruses and bacteria to advanced insects. Galls induced by viruses, bacteria, or fungi are referred to as "phytocecidia," whereas galls caused by protozoa, nematodes, and arthropods are referred to as "zoocecidia."

Many nematodes are responsible for galls, most of which are on roots, but since these animals are not arthropods, they will not be discussed further here. Acarina, or mites, cause many galls and are present all over the world. Their galls are found mostly on leaves and are very minute structures. Some aspects of their cecidogenesis are discussed in Chapter 9. However, the arthropods and in particular the insects, are by far the most important group of cecidozoans.

GALL INSECTS AND THEIR HOST PLANTS

Gall Insects

Galls are induced by a wide variety of organisms; however, the survey in this chapter is restricted to the insect-induced galls. The purpose is to review the main Orders of gall-inducing insects and to provide an appreciation for the range of gall types in existence. Feeding mechanisms of the insects involved are also described since they are intimately associated with the presence of the gall insects and their activities. Particularly striking or unusual galls in each group are also described.

About 13,000 cecidogenous insects have been recorded, which represents about 2% of the total number of insects (Buhr 1965). The principal gall inducers are found in the following seven Orders:

1. Order Thysanoptera (thrips)
2. Order Hemiptera
    - Family Tingidae (lace bugs)
3. Order Homoptera
    - Superfamily Cicadoidea
        - Family Cercopidae (spittle bugs)
        - Family Cicadellidae (leafhoppers)
    - Superfamily Psylloidea
        - Family Psyllidae (psyllids)
    - Superfamily Aphidoidea
        - Family Aphididae (aphids)
        - Family Eriosomatidae (woolly and gall aphids)
        - Family Adelgidae (pine and spruce aphids)
        - Family Phylloxeridae (phylloxerans)
    - Superfamily Coccoidea
        - Family Coccidae (scale insects)
        - Family Kermidae (gall-like coccids)
        - Family Asterolecaniidae (pit scales)
        - Family Diaspididae (armoured scales)
        - Family Eriococcidae (eriococcid scales)
4. Order Lepidoptera (moths)
5. Order Coleoptera (weevils)
6. Order Diptera
    - Suborder Nematocera
        - Family Cecidomyiidae (gall midges)
    - Suborder Cyclorrhapha
        - Family Tephritidae (fruit flies)
        - Family Chloropidae (chloropid flies)
        - Family Agromyzidae (leaf miners)
        - Family Anthomyzidae (anthomyzid flies)
        - Family Clythiidae (fungi feeders)
7. Order Hymenoptera
    - Suborder Symphyta
        - Family Tenthredinidae (sawflies)
    - Suborder Apocrita
        - Superfamily Chalcidoidea
            - Family Agaonidae (fig wasps)
            - Family Pteromalidae (pteromalids)
            - Eurytomidae (seed chalcids)
        - Superfamily Cynipoidea
            - Family Cynipidae (gall wasps)

Host Plants

The host plants attacked by gall insects belong to nearly all branches of the plant kingdom from mushrooms to angiosperms (Mani 1964; Buhr 1965; Docters Van Leeuwen 1982). However, there are groups such as the algae in which no insect galls have been found. There are also groups such as cryptogams and gymnosperms in which galls are uncommon and groups such as dicotyledons and especially Fagales, Salicales, Asterales, and Fabales in which they are particularly common. Houard (1908) recorded 500 species of gall insects on 10 species of Salicaceae. Mani (1964) reported that about 50% of the European and North American galls are found on Fagaceae. Leguminous plants are most commonly attacked in South America, Africa, and India and more than half the known galls of Australia are found on Myrtaceae. Details of the evolutionary relationships between host plants and families of gall insects are discussed in Chapter 3.

BIOGEOGRAPHY

The influence of geographical localization on predominant types of galls has been described by Felt (1940). He showed that the total number of galls is almost identical in Asia, Central Europe, Southern Europe, as well as in North and South America, but that the ratio of the different types of cecidozoa responsible for these galls, as well as the families of plants attacked, varies. One must nevertheless recognize the universal predominance of Diptera and especially the cecidomyiids, both in Palearctic and Nearctic regions and other regions such as Africa, Australia, and Asia. Some other peculiarities can be pointed out without contradicting the above, such as the wealth of coccid and aphid galls in China, the predominance of coccid galls in Australia and New Zealand, and the wealth of thrips and psyllid galls in Western Asia.

HOST AND ORGAN SPECIFICITY

Most gall-inducing insects are highly host and organ specific, that is, they induce galls on only one species or a closely related group of host species. The cecidomyiid *Taxomyia taxi* (Inch.) for example, lives exclusively on the terminal buds of yew twigs. It also is common for one genus of gall insects to attack only one genus of host plant. Cynipids of the genus *Diplolepis,* for example, are restricted to the genus *Rosa.* Tenthredinids of the genus *Pontania* are found only on willows, and aphids of the genus *Pemphigus* induce galls only on poplars. Also, within an insect genus linked to a given plant genus, different species attack specific organs. For example, *Andricus fecundator* (Hartig) attacks buds, *A. marginalis* Schilde attacks leaves, *A. amenti* Giraud attacks male catkins, and *A. quercuscalicis* (Burgsd.) attacks cupules of various species of *Quercus.*

All plant organs are subject to galling including roots, stems, leaves, flowers, and fruits and as well, the various parts of these organs are preferred by certain gall insects. Leaves are by far host to the most gall insects. According to Mani (1964), about 80 percent are on this organ.

MAIN TYPES OF GALLS

Structurally, galls range in complexity from slight depressions (pit galls) to elaborate growths with numerous adaptations and layers of tissues. Several authors have attempted to classify galls according to morphological criteria and although this kind of classification is rather artificial, it has formed the basis for most developmental research (Maresquelle and Meyer 1965). Certain special terms are so widely used that it is useful to discuss them here.

Küster in his classic 1911 monograph distinguished two large types of galls: organoid and histioid. Organoid galls result from the proliferation of organs or are a modification of organs in which the attacked organs remain recognizable. Examples include witch's brooms and virescences (transformation of floral buds in vegetative buds). Proliferation of roots or buds in abnormal parts of the plant are also organoid effects. Histioid galls, in contrast, are derived from a proliferation of modified cells that leads to a new organ.

In the same monograph, but from a different point of view (and this gives an indication of the complexity of the task), Küster classified galls as either kataplasmas or prosoplasmas. Galls of the kataplasmic type are irregular in size and shape. Their growth is variable and their tissues are scarcely differentiated. Examples include tumor galls or callus-like galls. Among the arthropods, only a few species of Hemiptera and Homoptera provoke this kind of gall.

The prosoplasmic galls are far more complex and their tissues undergo differentiation as the gall matures, leading to specific gall tissues such as nutritive and sclerenchyma cells and, as a result, an entirely new structure is created. All prosoplasmic galls are histioid. Among the arthropods, many cecidomyiids and all cynipids induce prosoplasmic galls. However, all attempts at classification result in overlaps as it is always possible to find transitional forms between the basic types. Some organoid galls, for example, possess some prosoplasmic characteristics and the reverse is also true.

The ultimate form of a gall depends on many factors including the species of plant attacked, the species of gall insect, the attacked organ, and sometimes the stage or even the sex of the gall insect.

In addition to the rather theoretical classification of Küster there is also a more empirical classification based on the mode of gall formation, which in turn is related to the initial position of the gall insect. The following list reviews the basic gall types, each of which is illustrated in Figure 2-1. Most are of the histioid and prosoplasmic type.

Filz galls are characterized by the presence of hairs or hairy outgrowths of the epidermis (Fig. 2-1b), which are sometimes the only sign of cecidogenesis. Most filz galls are induced on the underside of the leaf blade and in this type of gall the inducer is situated externally. Many mite galls are of the filz gall type.

Pit galls (Fig. 2-1c) and *blister galls* (Fig. 2-1d) are simple galls characterized by a slight arching of the leaf blade. The epidermis sometimes splits to form the blister. The simplest of these galls are hardly visible and are called "rudimentary galls."

Pouch galls develop first as a bulging of the leaf blade (Fig. 2-1e and f). They are often induced on young leaves by the very first stimulation of the gall insect. When further stimulation occurs in the same place, the insect induces an intense outward arching of the leaf blade producing a bulge on one side and a corresponding invagination on the other. The arched parts are also swollen, with considerable modification and the bulging and invagination enclose the cecidozoa within the chamber of the gall. As the name implies, pouch galls are typically hollow. There is usually an opening to the outside referred to as an ostiole. Pouch galls are provoked mostly by Acarina, aphids, cecidomyiids, and psyllids.

Roll and fold galls may present different degrees of complexity. Depending on the stage of swelling and the differentiation of the lamina, numerous types of roll and fold galls are produced. In leaf roll galls (Fig. 2-1g), the blade rolls upward from the leaf margin toward the midrib and there is swelling of the rolled-up part of the leaf blade. In fold galls (Fig. 2-1h and i) the unfolding of the leaf rudiment inside the developing bud is suppressed and affected tissues are swollen and lack differentiation. Thysanoptera, aphids, psyllids, and cecidomyiids are mostly responsible for galls of this type.

Covering galls often present a greater degree of differentiation than the previous forms. In this type of gall the inducer, acting externally at first, provokes a strong hypertrophy of the surrounding tissues, which leads to the inducer being quickly enclosed (Fig. 2-1j, k, m, and s). As in the previous type, an opening to the exterior may be retained, but here the ostiole can be minute and often presents special adaptations. Galls induced by cynipids are always completely surrounded by tissues; that is, there is no ostiolar opening to the outside. These galls are always highly specialized and correspond to the prosoplasmic galls of Küster. They are mostly caused by Acarina, aphids, coccids, cecidomyiids, and cynipids.

All of the above galls occur on leaves except for the covering galls, which may also be found on other immature organs.

Mark galls arise mostly on stems, but also can be found on leaves. In this case the egg is deposited inside the tissues (tenthredinids) or the larva bores into the tissues (tephritids and cecidomyiids). The larvae in these galls are completely enclosed within plant tissues from the beginning of gall development (Fig. 2-1n).

Bud and rosette galls range in complexity from simple enlargements of buds, the so-called "big buds," to solid or hollow swellings with spiny or scaly surfaces representing the bud leaves. Most bud galls develop during the spring. They are usually globose, but some are flask shaped (Fig. 2-1q). Some galls are fleshy swellings of all leaf rudiments. Often the gall insect uses the special potential of the vegetative point causing an arrest of the elongation of the internodes and a multiplication of the leaves resulting in rosette galls, which resemble a miniature cabbage or pine cone (Fig. 2-1p). In complex bud galls there is a pronounced swelling and fusion of the parts giving rise to a fleshy or woody mass. In another bud gall formed by adelgids (Fig. 2-1r), the shoot axis immediately below the bud becomes swollen, and the leaves become expanded and swollen basally into pe-

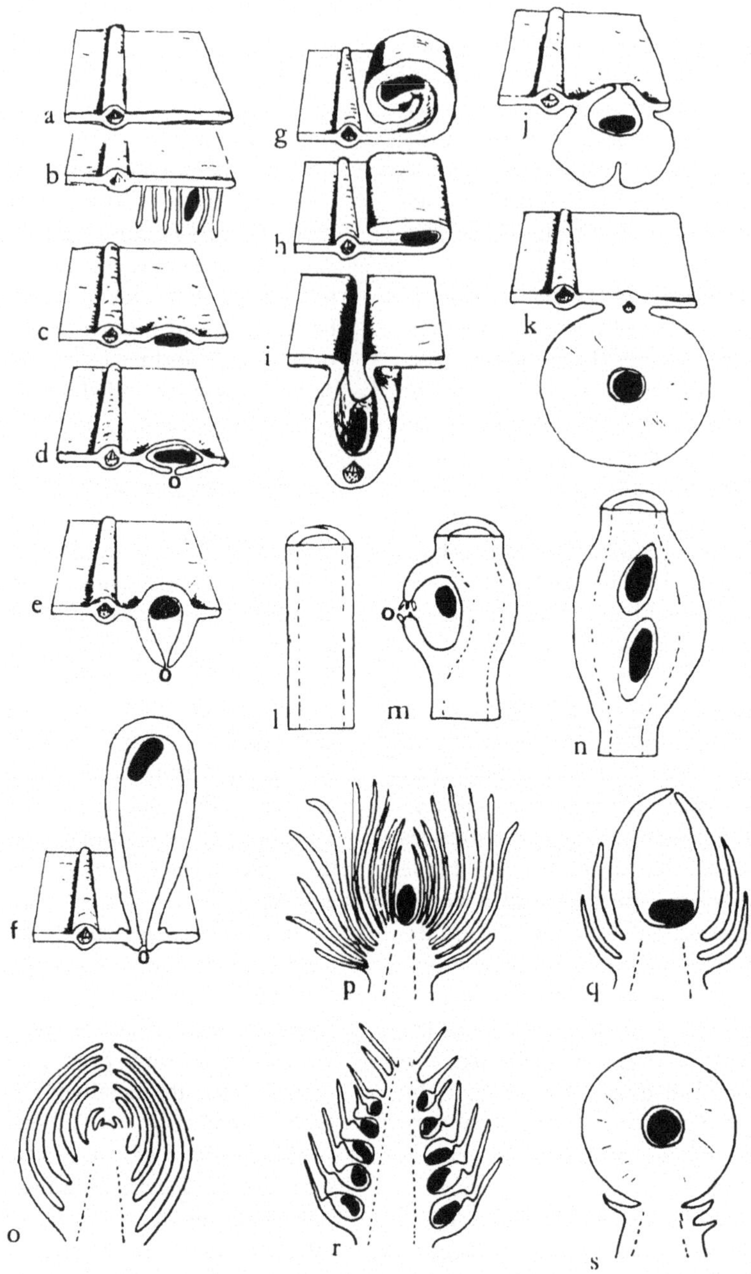

Figure 2-1. Schematic drawings of the main types of leaf, twig, and bud galls. (a) Normal leaf; (b–k) leaf galls: (b) filz gall; (c) pit gall; (d) blister gall; (e, f) pouch galls; (g) roll gall; (h, i) fold galls; (j, k) covering galls; (l) normal twig; (m) covering gall on a twig; (n) multichambered mark gall on a twig; (o) normal bud; (p) rosette gall (multiplication of leaves); (q) bud gall (inhibition of the multiplication of leaves and modification); (r) bud gall (modification of the leaves); (s) covering gall (or mark gall) on a bud. o, Ostiole; the black dot represents the larva.

culiar disc-shaped structures, where the insects live. In some cases buds can even bear covering or mark galls.

BIBLIOGRAPHICAL REFERENCES ON GALLS

Detailed reviews of galls and gall inducers have recently been completed by Meyer and Maresquelle (1983), Ananthakrishnan (1984), and Meyer (1987). These volumes contain invaluable information on the galls of arthropods. Other particularly useful references include the complete collection of *Marcellia* and the book *Ecology of Plant Galls* by Mani (1964). Major catalogues of cecidozoa of the world include

C. Houard (1908–1913) for Europe and the Mediterranean region
C. Houard (1922–1923) for Africa, Asia, and Oceania
C. Houard (1933) for South America
J. Docters Van Leeuwen-Reijnvaan and W. M. Docters Van Leeuwen (1926) for the Dutch East Indies
E. P. Felt (1940) for North America
J. M. Mimeur (1949) for Morocco
M. S. Mani (1948) for India
H. Buhr (1965) for North and Central Europe
W. M. Docters Van Leeuwen (1982) for the Netherlands

SURVEY OF MAIN ORDERS OF GALL INDUCERS

Thysanoptera

Thrips are small insects (0.5–10 mm in length) generally having four narrow, hair-covered wings (Fig. 2-2) that may be folded over the body. The bodies are usually long and slender; however, some species such as *Kladothrips augon saxxos* Moulton develop enlarged abdomens when they are filled with eggs (Fig. 2-3). The mouthparts of thrips are of the punching and sucking type (Lewis 1973). Many species are important pests because of their numbers and wide variety of plants attacked. Approximately 300 species of thrips are associated with galls on various plants (see Chapter 12).

Gall thrips live in colonies composed of numerous individuals of different ages or of different species. Sometimes it is difficult to determine which species is responsible for the initiation of the gall, which species have moved into the gall and contributed to its expansion, or which species are inquilines.

Gall-inducing thrips are particularly abundant in tropical regions, and especially in Asia where they have been studied the most (Karny *et al.* 1913, 1914–1916; Ananthakrishnan 1978). Certain species, such as *Gynaikothrips uzeli* Zimmermann, are rather cosmopolitan. They occur in India, Malaysia, North Africa, and Israel. Others, such as most Australian thrips, are more restricted in their distribution. The first European thrips galls were described by Karny (1911, 1913).

Although most thrips galls develop on leaves, a few are found on flowers and buds. However, thrips galls in India and Australia are often much more complex. *Kladothrips augon saxxos,* for example, induces globose galls on the leaves of *Acacia doratoxylon* Meissn. (Figs. 2-4 and 2-5). However, leaf fold or roll galls, such as those of *Eugynothrips* sp. on *Smilax* sp. (Fig. 2-6) are by far the most common type of gall. In Europe the galls of species such as *Thaeniothrips firmus* Uzel. on *Vicia cracca* L. consist of simple foldings or twistings of host leaves.

Hemiptera and Homoptera

The Orders of Hemiptera and Homoptera include a great number of gall inducers and are remarkable for the diversity of their galls. The following includes only the most important or interesting families with gall inducers that are distributed throughout the world:

Order Hemiptera: Family Tingidae (lace bugs)
Order Homoptera: Family Cercopidae (spittle bugs)
Family Psyllidae (psyllids)
Family Eriosomatidae (wooly and gall-inducing aphids)
Family Adelgidae (pine and spruce aphids)

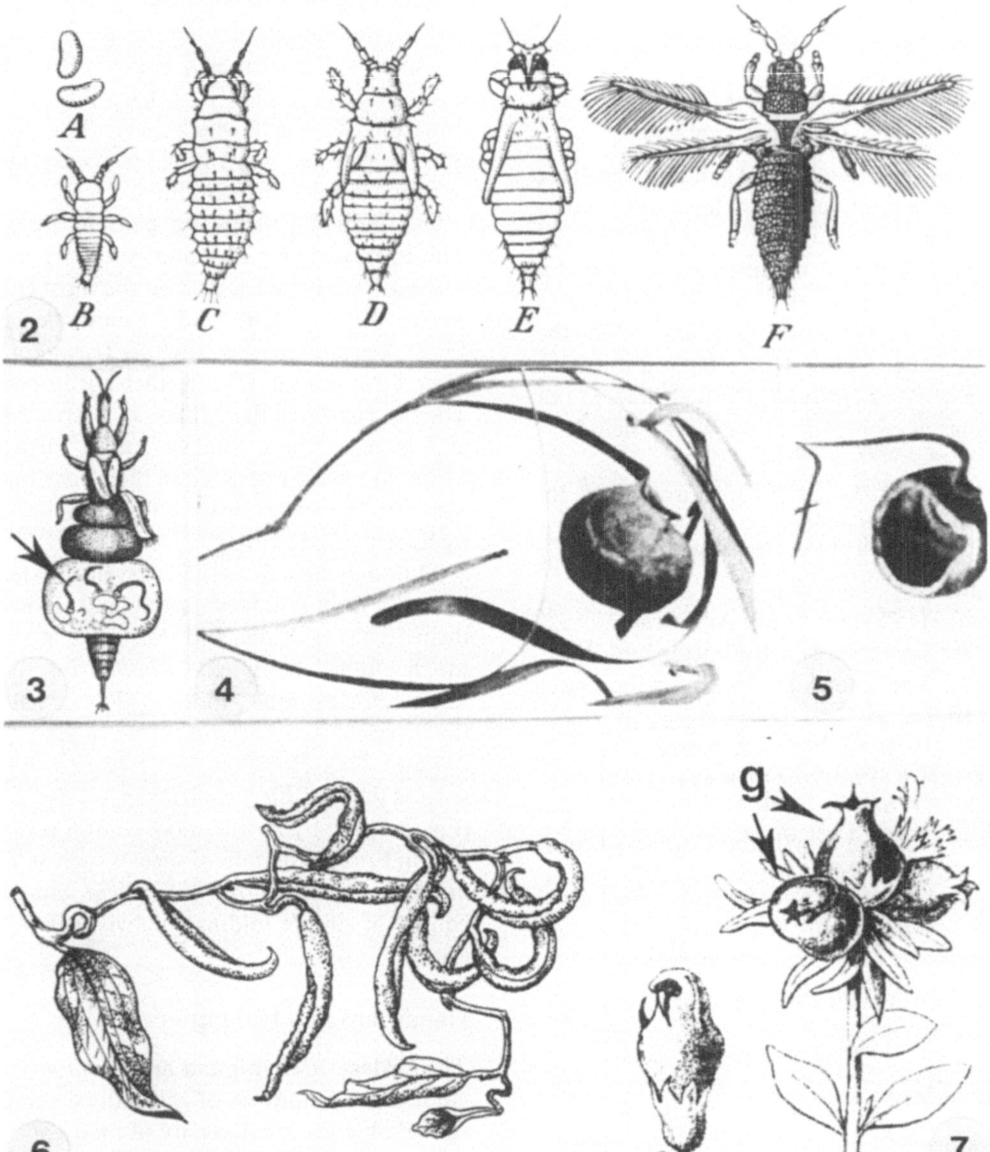

Figures 2-2–2-7. Thysanoptera gall inducers and galls (Figs. 2-2–2-6) and gall of a hemipteran (Fig. 2-7). **Figure 2-2.** Different forms of *Heliothrips haemorrhoidalis.* (A) Egg, (B) young larva, (C) old larva, (D) prepupa, (E) pupa, (F) adult. (From Russell in Ross 1932.) **Figure 2-3.** *Kladothrips augon saxxos* with distended egg-filled abdomen (arrow). (From Lewis 1973.) **Figure 2-4.** Spherical leaf gall of *K. augon saxxos* on *Acacia* sp. **Figure 2-5.** Dissection of a gall of *K. augon saxxos* showing the chamber that may shelter up to 600 thrips. (From Ananthakrishnan 1978.) **Figure 2-6.** Roll galls of *Eugynothrips* sp. on *Smilax* sp. (From Ananthakrishnan 1978.) **Figure 2-7.** Flower gall of *Copium teucrii* on *Teucrium montanum.* a, Detail of giant flower; g, "giant flower." (From Houard 1906.)

Family Phylloxeridae (phylloxerans)
Family Coccidae (scale insects)
Family Kermidae (gall-like coccids)
Family Asterolecaniidae (pit scales)
Family Diaspididae (armoured scales)
Family Eriococcidae (eriococcid scales)

Hemipterans and homopterans have piercing and sucking mouthparts that can be inserted into plant tissues. There are two canals between chitinous stylets that allow salivary injections and the ingestion of sap.

Both hemipterans and homopterans have two to five moults and often have complex life cycles. Some Eriosomatidae and Coccidae have both gall-inducing and non-gall-inducing stages on the same or different hosts (Börner and Schilder in Sorauer 1932). In galls of other species such as the gall aphids, the action of the insect may vary depending on its stage or on the host plant. Such is the case with the adelgid *Adelges strobilobius* Kalt., which lives alternatively on *Picea excelsa* Lmk and *Larix europea* D.C. This insect is cecidogenetic on *Picea*, its primary host, where the stem mother (also called the fundatrix) and its parthenogenetic offspring are successful gall inducers. No galls are induced on *Larix*, its secondary host (Gaumont 1957 and Chapter 7).

Gall Types

The only gall inducers in the Order Hemiptera are found in the family Tingidae. Here the genus *Copium*, which is restricted to the plant genus *Teucrium*, causes mainly floral galls. *Copium teucrii* Host., for example, induces giant flowers on *Teucrium montanum* L. (Fig. 2-7). Development of the flower is inhibited by this tingid and as well provokes cell hyperplasia and hypertrophy.

Homopterans induce a variety of gall types ranging from simple forms to complex structures. *Philaenus spumarius* L. (Fig. 2-8), for example, simply causes growth inhibition and wrinkling (Fig. 2-9) of the host leaves. The larvae, like all other spittle bugs, are protected by a mass of spittle (Fig. 2-9). The European *Philaenus spumarius* is not host specific and lives on monocotyledons, dicotyledons, and even *Equisetum*.

The family Psyllidae contains 350 gall-inducing species (Mani 1964; Hodkinson 1984) with a worldwide distribution. Most are found on the leaves of dicotyledons. These insects cause pit, pouch, and covering galls that shelter a single larva (Hodkinson 1984). Psyllid galls are often complex, multichambered, aggregated galls. European psyllid galls are often simple pit galls such as those induced by *Trioza* sp. or leaf margin roll galls induced by *Trichochermes walkeri* Foerst. on *Rhamnus cathartica* L. (Fig. 2-10). Other species, such as *Livia juncorum* Latr. (Fig. 2-11), induce a witch's broom-like organoid gall (Fig. 2-12) on the inflorescences of *Juncus* sp. (Schmidt and Meyer 1966). An unknown psyllid on *Trichillia* sp. induces a covering gall (Fig. 2-13) from mesophyll cells (Rübsaamen 1911). Some pouch galls are epiphyllous whereas others such as the East Asian gall of *Pauropsylla tuberculata* Crawf. on *Alstonia scholaris* R.Br. (Fig. 2-14) occurs on both sides of the blade. In some cases the pouch galls have a dehiscence mechanism allowing the gall to open at maturity (Docters Van Leeuwen-Reijnvaan and Docters Van Leeuwen 1926; Uichanco 1919).

Most of the homopteran gall inducers are found in the family Eriosomatidae with the main genera being *Hamamelistes, Hormaphis, Byrsocrypta, Eriosoma, Schizoneura, Pemphigus, Forda*, and *Baizongia*. Some Eriosomatidae such as *Anuraphis helichrysi* Kalt. on *Prunus domestica* L. provoke differential growth as they feed, which results in shriveled and curled galls on leaves and young shoots. The feeding activities of some species cause complex galls such as the spirally twisted galls of *Pemphigus spirothecae* Pass. on *Populus* sp. (Figs. 2-15 and 2-16). Other species such as *Melaphis rhois* Fitch on *Rhus* sp. form one of the largest aphid galls (Fig. 2-17) in North America. Also in North America, witch hazel (*Hamamelis virginiana* L.) is host to interesting galls induced by *Hormaphis hamamelidis* (Fitch (Fig. 2-18a and b). This aphid has a nongalling stage on *Betula*.

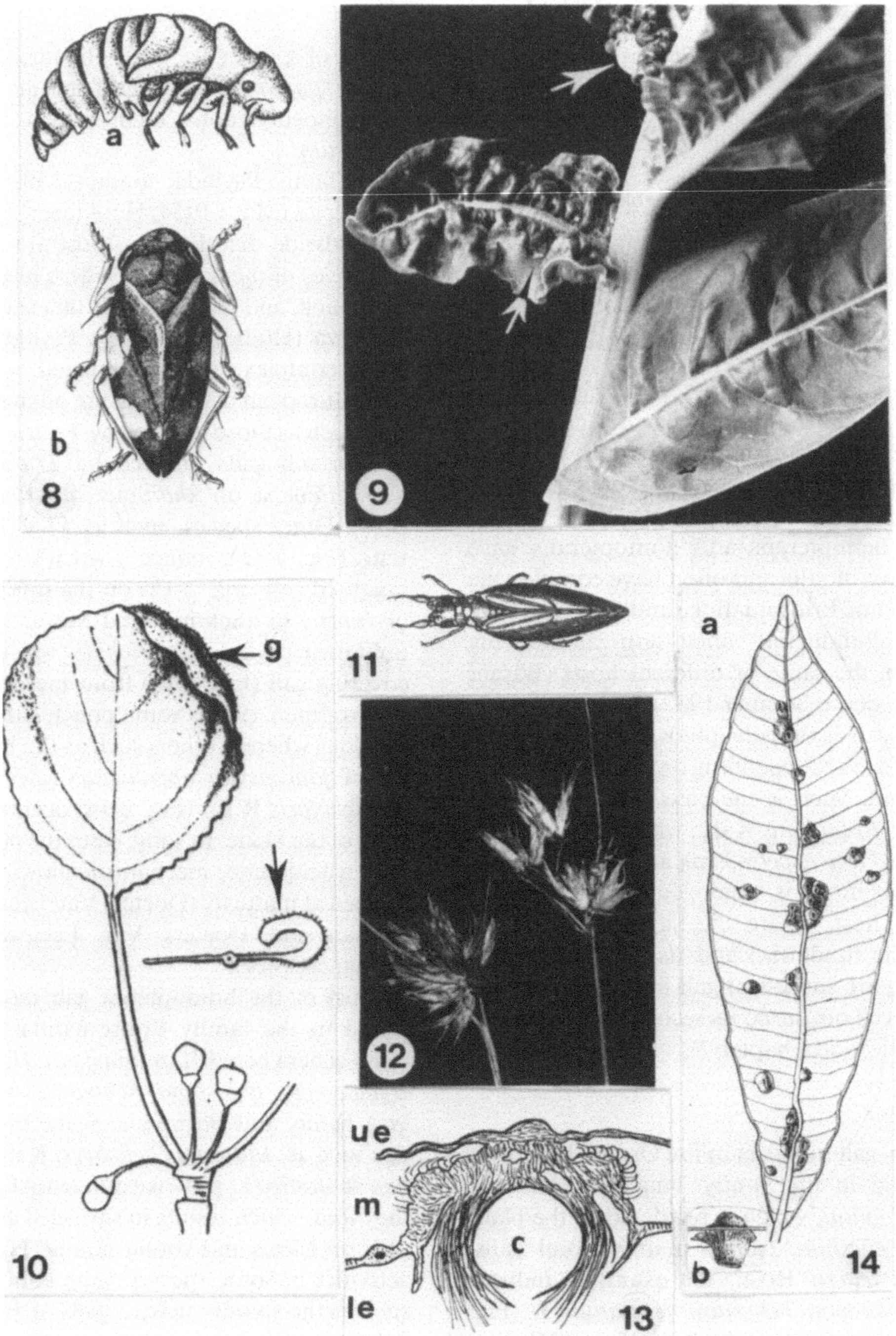

Figures 2-8–2-14. Galls and adults of Cercopidae and Psyllidae. **Figure 2-8.** *Philaenus spumarius.* (a) Larva; (b) adult. (From Ross 1932.) **Figure 2-9.** Wrinkled leaf of *Oenothera* sp. caused by *Philaenus spumarius.* Arrows indicate the spittle mass secreted by the larvae. **Figure 2-10.** Gall of *Thrichochermes walkeri* on *Rhamnus cathartica.* Arrows indicate a schematic section showing the thickening of the margin roll. (From Houard 1909.) **Figure 2-11.** Adult *Livia juncorum.* (From Ross 1932.) **Figure 2-12.** Witch's broom-like organoid gall caused by *Livia juncorum* on inflorescences of *Juncus* sp. (Photograph by E. Westphal.) **Figure 2-13.** Gall of an unknown psyllid on a leaf of *Trichillia* sp. ue, Upper epidermis; le, lower epidermis; m, hypertrophied and transformed mesophyll; c, larval chamber. (From Rubsaamen 1911.) **Figure 2-14.** (a) Truncated cone-shaped pouch gall of *Pauropsylla tuberculata* on leaf of *Alstonia scholaris.* (b) Detail of a gall that is equally developed on both sides of the leaf. (From Mani 1964.)

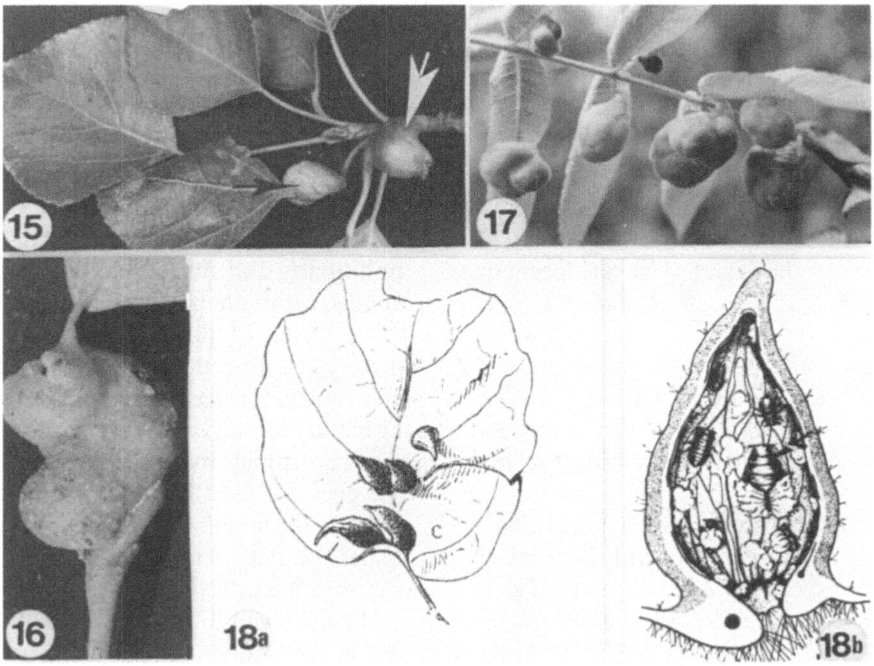

Figures 2-15–2-18. Galls induced by aphids. **Figure 2-15.** Galls induced by two species of *Pemphigus* on *Populus.* Black arrow shows the gall of *P. spirothecae.* The white arrow shows the gall of *P. bursarius.* **Figure 2-16.** Mature gall of *Pemphigus spirothecae.* **Figure 2-17.** Pouch gall of *Melaphis rhois* on *Rhus.* **Figure 2-18.** (a) Pouch gall of *Hormaphis hamamelidis* on leaf of *Hamamelis virginiana.* (From Pergande in Felt 1940.) (b) Dissection of a mature gall of *Hormaphis hamamelidis.* The larger inhabitant is the mother of the first and second instars. (From Lewis and Walton 1958.)

The galls of Adelgidae and Phylloxeridae must be mentioned because of their economic importance and scientific interest. The pineapple galls of *Adelges abietis* L. and *Adelges strobilobius* Kalt. are remarkable covering galls that do damage to the shoots of *Picea excelsa* (Chapter 7).

The galls of Phylloxeridae range in complexity from simple pit galls to pedunculate and dehiscent pouch galls. Most of phylloxerans occur in North America. The best known is *Viteus* (= *Peritymbia*) *vitifolii* Fitch. (formerly *Phylloxera vastatrix* Planch). The remarkable life cycle of this insect includes different forms that attack roots and leaves of American *Vitis riparia* Michx. and the roots of European *Vitis vinifera* L. Feeding on the roots results in swellings whereas covering galls are induced on the leaves. Although the Phylloxeridae are never viviparous (in contrast to other aphids), their proliferation through eight or nine generations results in serious damage to vineyards (Pesson 1951).

Several families of coccids in the superfamily Coccoidea are able to induce galls. Most are found in the families Eriococcidae and Diaspididae (Beardsley 1984); however, a few gall-inducing genera are found in the Asterolecaniidae. Some of these galls have been studied anatomically by Houard (1911) and Goidanich (1960) (see Chapter 7).

Most of the gall-inducing coccids that induce simple galls on some plants are also able to feed on other taxa without noticeably causing distortions (Buhr 1965; Howell and Kosztarab 1972). Gall-inducing coccids display a strong sexual dimorphism (Grassé *et al.* 1961; Jacobs and Renner 1974). The females remain apterous and larval shaped and usually undergo only two moults. Although mobile in the first instar, they later become immobile and undergo a strong morphological regression whereby they lose their legs and antennae. Adult males, in contrast, remain mobile and develop a pair of membranous wings. They also undergo one or two more instars than the females.

Galls of Asterolecaniidae and Diaspididae are most often simple pit galls and sometimes the coccid remains external. The gall depression is sometimes sheltered by a wax or sclerified secretion of the insect. The cecidogenetic action of species such as *Carulaspis visci* (Schrank) Low on *Viscum album* L. is weak, consisting of inhibition under the insect and some growth stimulation around it; however, an African *Asterolecanium* provokes a stronger response of the organoid type resulting in a kind of witch's broom.

The Eriococcidae are by far the most important group of cecidogenetic coccids in terms of the number of species and the diversity of galls induced. As with other coccids, the Eriococcidae are characterized by sexual dimorphism. This dimorphism of the insects is sometimes accompanied by a dimorphism of the galls. This is the case for the Australian genera *Apiomorpha* and *Ascelis* on *Eucalyptus*. Female galls are globular, woody, and shaped like urns. They are sometimes adorned with horns or rib-like structures. Male galls are much smaller and simply tubular. Sometimes male galls arise as satellite galls on the surface of female galls (Beardsley 1984).

A group of Eriococcidae producing neoformed bud galls on leaves of various *Shorea* species in Malaysia was recently discovered by Anthony (1980). These gall insects are thought to belong to the genus *Gallacoccus* described by Beardsley (1971). Beardsley based his observations on the first two specimens discovered by Anthony (1972).

Coleoptera

Considering the size of the Order Coleoptera, very few beetles have evolved the ability to induce galls. Buhr (1965) lists about 135 gall-inducing beetles found in Europe. Most attack dicotyledons and a few are on gymnosperms. No species are known to gall monocotyledons. Most coleopterous gall inducers are in the family Curculionidae, but some interesting galls are caused by Buprestidae. Buprestids of the genus *Agrilus* are found on *Rubus* and *Quercus*. The larvae dig and bore into the shoot axis. A few species of Cerambycidae cause galls. *Saperda populnea* L., for example, is common on *Populus* in Europe and causes a thickening of the young mined shoot axis. Some Scolitidae cause galls on a few species of Labiatae.

The larvae and adults of beetles have chewing mouthparts and do a great deal of structural damage within their galls (Le Pape and Bronner 1987). Layers of nutritive tissues are not found in the galls of beetles; presumably they do not have time to form as a result of the insect's feeding habits. The larvae of several species produce mines or tunnels in different parts of the plant. The eggs are usually thrust into a cavity prepared by the female and then enclosed by callus tissues (Le Pape and Bronner 1987). Pupation occurs in the galls or in the ground.

The main gall-inducing genera of Curculionidae in the temperate region are *Ceuthorrhynchus, Gymnetron, Nanophyes, Apion, Miarus,* and *Lixus*. Among the more interesting galls are those induced by *Gymnetron villosulum* Gyll. on fruits of *Veronica anagallis-aquatica* L. (Fig. 2-19) and an organoid gall induced by *Lixus punctiventris* Bohem. on *Crepis biennis* L. (Fig. 2-20), which tends to cause a virescence and proliferation of flowers. *Nanophyes pallidus* Ol. gives rise to the "jumping gall," which is the fruit of *Tamaris* sp. inhabited by a single larva (Tavares Da Silva 1902). Large galls on the shoot axes and petioles of *Umbilicus rupestois* (Salish.) Danoy (Fig. 2-21) are induced by the weevil *Nanophyes durieui* (Lucas) (Fig. 2-22) and involve hyperplasia and abnormal formation of vascular bundles (Houard 1906).

Lepidoptera

In spite of the small number of galls induced by Lepidoptera, 100 in the entire world according to Mani (1964), many are of great biological interest. Most are induced by microlepidoptera and in Europe most are in the family Tortricidae (Mees 1923). Their feeding habits and chewing mouthparts cause a rapid destruction of tissues lining the larval chamber. As with the galls induced by beetles, differentiation of nutritive tissues generally does not occur. The basic cellular differentiation involves cortical hypertrophy and cambial and medullar hyperplasy (Meyer and Maresquelle 1983). Most dissected lepidopteran galls can be recognized by the large amount of feeding debris left by the larvae.

Like the groups described above, the galls of Lepidoptera show a large diversity of forms. An interesting spherical one induced by *Cecidoses eremita* Curtis on the stems of *Schinus dependens* Ortega (Fig. 2-23) has a plug-filled opening produced by a special differentiation of tissues. To escape from the gall, the adult moth simply pushes the plug out (Karsch 1880; Tavares Da Silva 1915). Another gall on the axis of *Pinus silvestris* L. (Fig. 2-24) induced by *Evetria resinella* L. (Fig. 2-25) develops as a result of lateral thickenings of stem tissues. This gall also is characterized by an accumulation of oozing resin (Ross 1932; Ruppolt 1959). *Coleophora icterella* Toll. induces its gall (Fig. 2-26) on the leaves of *Polygonum convolvulus* L. This moth cuts its gall from the plant and drops to the soil, within the gall, suspended by a silken thread. The larva then wanders about until a suitable substrate is found where the gall is again fastened with silk and then the larva pupates (Goidanich 1951). The North American moth *Gnorimoschema gallaesolidaginis* Riley induces a large stem gall (Figs. 2-27 and 2-28) on *Solidago canadensis* L.

Diptera

The Order Diptera contains more gall-inducing insects than any other Order and all are found in the following two suborders:

Nematocera:	Family Cecidomyiidae (= Itonididae = gall midges)
Cyclorrhapha:	Family Tephritidae (fruit flies)
	Family Chloropidae (chloropid flies)
	Family Anthomyiidae
	Family Clythiidae (fungi feeders).

Dipterous galls are widely distributed throughout the plant world. In contrast to galls induced by other groups of cecidozoans, they are also on monocotyledons and especially Gramineae. They are most often leaf galls, but there also are bud, stem, flower, and even root galls. Dipterous galls show a fantastic diversity of morphology and are almost always characterized by a high level of specialization of their tissues, structures, and opening mechanisms. Cecidomyiids also show complexity in their biological cycles. As in other groups, there may be one or several generations in the same year and successive generations of the same insect may provoke galls on different parts of the same plant (e.g., the first generation of *Rabdophaga heterobia* H. Löw. on willow produces galls on male catkins in the early spring whereas the second summer generation attacks buds and induces development of rosette galls).

Depending on the species, dipterans and especially gall midges may pupate inside their galls, or they may leave them to pupate in the soil. Variations in life cycle strategies are only one indication of the high degree of specialization associated with this group of gall insects (Yukawa 1981, 1987; Ohsako *et al.* 1981; Skuhravá *et al.* 1984; Gagné 1989).

Galls Induced by the Suborder Nematocera

CECIDOMYIIDAE

The family Cecidomyiidae is the only family of gall inducers in the Nematocera. According to Skuhravá *et al.* (1984), 60 percent of the 1200 cecidomyiids described in the panarctic region are gall-inducing insects and 682 and 535 species are found in North America and in the Dutch East Indies respectively (Felt 1940; Docters Van Leeuwen-Reijnvaan and Docters Van Leeuwen 1926). New species are frequently described in North America by Gagné, in India by Grover, and in Japan by Yukawa. It is impossible to give a complete account of the group here and instead only representative and striking examples are given.

The eggs of cecidomyiids are deposited on the surface of plant tissues (Fig. 2-32) and the newly hatched larvae search for the right place to initiate the gall (see Chapter 5). Only the larvae of cecidomyiids are the gall inducers. They have reduced mouthparts and feed by sucking exuding fluids from gall cells. Small feeding punctures have been seen in epidermal cells (Meyer 1957; Rohfritsch 1978, 1987) but extensive cellular damage is rare (Raymond 1927; Meyer 1957). Most cecidomyiid larvae are characterized by a prosternal sclerified spatula (Chapter 13). For further details on the biology of cecidomyiids see Chapters 3, 8, 11, and 13.

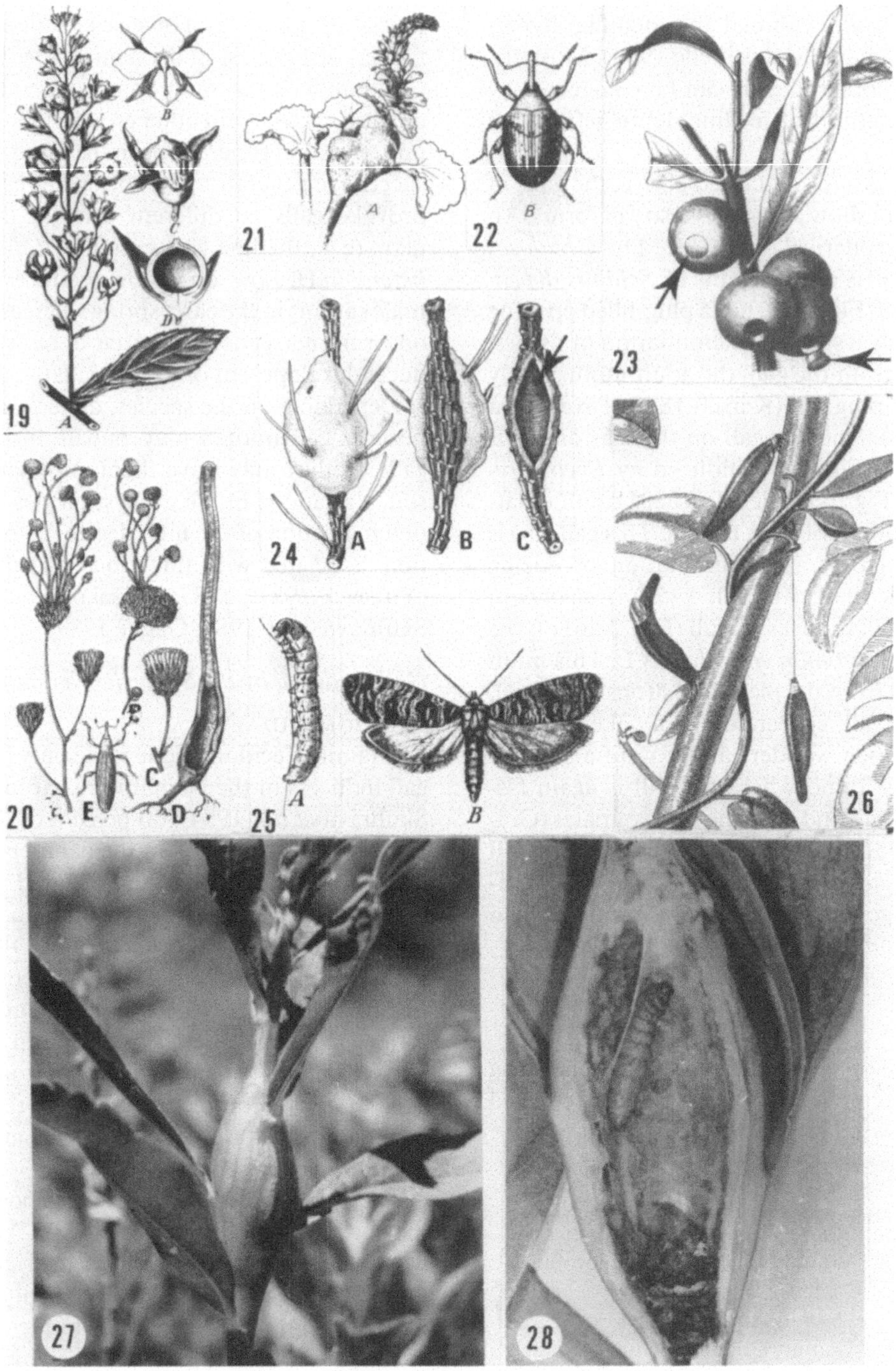

Figures 2-19–2-28. Adults and galls of Coleoptera (Figs. 2-19–2-22) and Lepidoptera (Figures 2-23–2-28). **Figure 2-19.** Fruit gall of *Gymnetron villosulus* on *Veronica aquatica.* (From Ross 1932.) **Figure 2-20.** Galls of *Lixus punctiventris* on *Crepis biennis.* (A) General view showing the organoid effects of the larva (virescence and proliferation of flowers; (B) detail of a proliferating capitulum; (C) detail of a normal flower; (D) larval chamber in the lower part of the stem; (E) adult beetle. (From Ross 1932.) **Figure 2-21.** Stem and petiole gall of *Nanophyes durieui.* (From Mimeur 1949.) **Figure 2-22.** Adult *Nanophyes durieui.* **Figure 2-23.** Spherical gall of *Cecidoses eremita* on *Schinusdependens.* The three galls show the plug (arrow) in place, in the process of being pushed out, and removed. (From Kerner von Marilau 1905.) **Figure 2-24.** Stem gall of *Evetria resinella* on *Pinus sylvestris.*

The galls of cecidomyiids exhibit the widest range in complexity of all gall-inducing groups. This diversity has inspired many workers and as a result the literature on cecidomyiids and their galls is the richest of any gall group. Among the most important works are those of Kieffer (1900, 1913), Rubsaamen and Hedicke (1926), Skuhravá and Skuhravý (1960), Nijveldt (1969), Yukawa (1971), Mamaev (1975), Grover and Kashyap (1984), and Gagné (1989). Over half the cecidomyiid galls are found on leaves and these range from simple swellings (Figs. 2-29, 2-33, 2-45–2-47) to complex pouch galls (Figs. 2-31 and 2-38) and piston galls (Figs. 2-39 and 2-40). Some cecidomyiids induce complex terminal or twig galls (Figs. 2-42–2-44), whereas others induce stem galls (Figs. 2-30, 2-34, and 2-35).

Some cecidomyiids such as *Mayetiola poae* Bosc. (Fig. 2-36) on *Poa nemoralis* L. induce the proliferation of roots on the shoot resulting in a wig-like structure. The rare gall of *Oligotrophus oleariae* on the composite *Olearia paniculata* Druce in New Zealand consists of an epiphyllous bud exhibiting normal phyllotaxy (Philipson 1964).

The simplest midge galls are those such as *Dasineura pustulans* (Rübs). on the leaves of *Filipendula ulmaria* (L.) Maxim (Fig. 2-37). The larva lives in a slight depression of the leaf blade without any protection, but it causes the differentiation of a true nutritive tissue. *Dasineura ulmariae* (Bremi) induces a pouch gall (Fig. 2-38) modifying both sides of the leaf. This gall is a spherical red-green protrusion on the upper surface whereas on the lower surface the protrusion is oblong and covered with hairs and is white-green. The maple-leaf spot gall (Fig. 2-29) induced by *Cecidomyia ocellaris* O.S. in North America is a blister leaf gall of this category. The gall of *Dasineura urticae* Perris (Fig. 2-33), common in the wastelands of Europe, occurs on all young organs of *Urtica dioica* L. In the case of these galls, the larvae are first external and then become surrounded by gall tissues, the growth and differentiation of which is caused by their presence. In other cases such as in the stem galls of *Rabdophaga batatus* Walsh (Fig. 2-30) and *Rabdophaga ramicola* Ruebs. (Figs. 2-34 and 2-35), the larvae work their way into the plant tissues.

Among the most remarkable leaf galls are those of *Contarinia (Lobopteromyia)* on *Acacia ferruginea* D.C. (Figs. 2-39 and 2-40) first described by Mani (1959) and then studied in further detail by Rohfritsch (1971, 1974). This gall results from the transformation and association of two contiguous leaflets. Another peculiar leaf gall is that of *Mikiola fagi* Htg. on *Fagus*. The galls of this species exhibit sexual dimorphism. Galls sheltering males are small and narrow in contrast to the large globose galls sheltering the females (Fig. 2-31) (Coutin and Riom 1967). Mention also must be made of the cecidomyiids such as *Asphondylia sarothamni* H.Lw. (Fig. 2-41) on *Sarothamnus scoparius* Wimmer that do not cause the differentiation of nutritive tissue but instead live in association with fungi (Deuteromycetes), the mycelium of which lines the larval chamber. In these "ambrosia galls" the larvae and fungi live symbiotically. The larvae feed on the mycelium and the gall protects the fungi. The adults also ensure the propagation of the fungi (Richert-Vollert 1964 and Chapters 8 and 13).

Abscission and dehiscence phenomena involve a wide range of mechanisms in leaf galls. The gall of *Didimomyia reaumuriana* F. Lw. on *Tilia* sp. (Figs. 2-45–2-47) provides a good example of the complexity of these mechanisms (Rohfritsch 1967). The "inner gall" in its sclerified sheath is surrounded by an "outer gall." Cellular proliferation occurs in the layers of the outer gall and this is followed by pectic

(A) External view of resinous gall; (B) resinous layer partially removed to show the swelling of the stem; (C) dissection of gall showing the larval chamber (arrow). (From Ross 1932.) **Figure 2-25.** Caterpillar and adult of *Evetria resinella*. (From Ross 1932.) **Figure 2-26.** Floral bud gall of *Coleophora icterella* on *Polygonum convolvulus*. The caterpillar uses its gall as a sheath and can move it when feeding or fix it with a silken thread for metamorphosis. (From Goidanich 1951.) **Figure 2-27.** Stem gall of *Gnorimoschema gallaeosolidaginis* on *Solidago canadensis*. **Figure 2-28.** Dissection of gall of *Gnorimoschema gallaeosolidaginis* showing the caterpillar within its chamber.

Figures 2-29–2-41. Galls and ovipositing adults of Cecidomyiidae. **Figure 2-29.** Blister gall of *Cecidomyia ocellaris* on *Acer saccharum.* **Figure 2-30.** Petiole or twig gall of *Rabdophaga batatus* on *Salix* sp. **Figure 2-31.** Pouch galls of *Mikiola fagi* on leaf of *Fagus silvatica.* Sexual dimporhism of galls is shown in this figure. The smaller gall contains a male, whereas the larger globose gall contains a female. (Photograph by O. Rohfritsch.) **Figure 2-32.** Ovipositing female of *Dasineura urticae* on *Urtica dioica.* The ovipositor (arrow) can be seen between the hind legs. (Photograph by O. Rohfritsch.) **Figure 2-33.** Immature leaf galls of *Dasineura urticae* on *Urtica dioica.* (Photograph by O. Rohfritsch.) **Figure 2-34.** Twig gall of *Rabdophaga ramicola* on *Salix* sp. **Figure 2-35.** Dissection of twig gall of *Rabdophaga ramicola* showing the numerous larval chambers. **Figure 2-36.** Gall of *Mayetiola poae* on *Poa nemoralis.* This wig-like gall consists of a proliferation of neoformed roots around the larval

maceration. Pressure caused by the proliferation provokes the ejection of the inner gall (Fig. 2-46). This mechanical dehiscence occurs normally, but when the weather is dry, the inner gall is released by tearing of the lamina (Garrigues 1951).

Bud galls induced by cecidomyiids differ in their structures from leaf and shoot axis galls described above. Although caused by a limited number of larvae, in contrast to the homopteran bud galls, cecidomyiid bud galls generally involve great disturbances of the morphology and the functioning of the bud. In some cases, there is enlargement of the vegetative tip, which is transformed into a nutritive pad and cessation of shoot axis elongation. The number of leaves may be reduced or largely increased (Dreger-Jauffret 1978). The latter gives the "rosette galls" such as those of *Rabdophaga strobiloides* Walsh. (Figs. 2-42 and 2-43) and *Rabdophaga rosaria* Löw. (Fig. 2-44) on *Salix* sp. In others, as in the case of *Mayetiola rigidae* (O.S.) on *Salix* sp., the transformation is more complete and the bud is no longer recognizable.

Galls Induced by Flies of the Suborder Cyclorrhapha

The larvae of flies in the suborder Cyclorrhapha are characterized by their well-developed mouthparts, which allow them to feed by rasping plant tissues. In contrast to the cecidomyiids, the larvae of these flies leave excreta because they feed on solid foods. Cyclorrhaphous larvae most commonly induce galls on phanerogams. However, one gall is found on Polyporacea, *Agathomyia wankowiczi* Schnabl. on *Ganoderma applanatum* (Pers.)Pat; another one, *Craspedochaeta signata* Brischke, is found on various ferns of temperate forests in Europe.

Most of the galls found on phanerogams are shoot galls such as *Lipara lucens* Mg. on *Phragmites communis* Trin. The larva of this species lives in a large cavity excavated under the apex and tunnels through the pith of the thickened shoot. The actions of the newly hatched larva have been described by Docters Van Leeuwen and Docters Van Leeuwen Reijnvaan (1904) and Ruppolt (1959), before it penetrates the base of the apex.

The Tephritidae have received a great deal of attention because of their economic importance. Gall inducers in this family are closely associated with Compositae and several studies have examined the use of tephritids and their galls in the biological control of weeds (Harris 1980). The adult flies (Fig. 2-48) are among the largest of the gall-inducing insects.

Tephritids cause a large variety of single or multichambered galls consisting of conspicuous overgrowths on stems or young shoots. *Urophora cardui* L., for example, causes a multichambered gall (Figs. 2-49 and 2-50) on different species of *Cirsium* in both Europe and North America (Lalonde and Shorthouse 1984). *Eurosta solidaginis* Fitch. is a common North American species that induces a single-chambered gall (Figs. 2-51 and 2-52) on *Solidago* sp.

Other tephritids produce galls on leaf petioles, midribs, and other veins. Tephritids such as *Urophora affinis* Frld. and *U. quadrifasciata* Meig. induce their galls within flowerheads (Harris 1980). Receptacle galls such as that of *Tephritis dilacerata* Löw (Shorthouse 1980b) on *Sonchus arvensis* L. (Fig. 2-53) are rather rare. There are no known root galls induced by tephritids.

chamber. **Figure 2-37.** Pit gall of *Dasineura pustulans* on a leaf of *Filipendula ulmaria.* This gall consists of a simple depression of the leaf blade where the larva (arrow) develops in a droplet of plant liquid without any protection. (Photograph by O. Rohfritsch.) **Figure 2-38.** Pouch gall of *Dasineura ulmariae* on the underside of a leaf of *Filipendula ulmaria.* (Photograph by O. Rohfritsch.) **Figure 2-39.** Two "piston" galls of *Contarinia* on *Acacia.* (From Rohfritsch 1971.) **Figure 2-40.** Schematic longitudinal section of a mature gall of *Contarinia* sp. showing the association of the two contiguous leaflets. Fs, Upper leaflet; Fi, lower leaflet; P, piston; B, pouch; L, larva. (Drawing by O. Rohfritsch.) **Figure 2-41.** Female *Asphondylia sarothamni* laying eggs on a floral bud of *Sarothamnus scoparius.* (From Richter-Vollert 1964.)

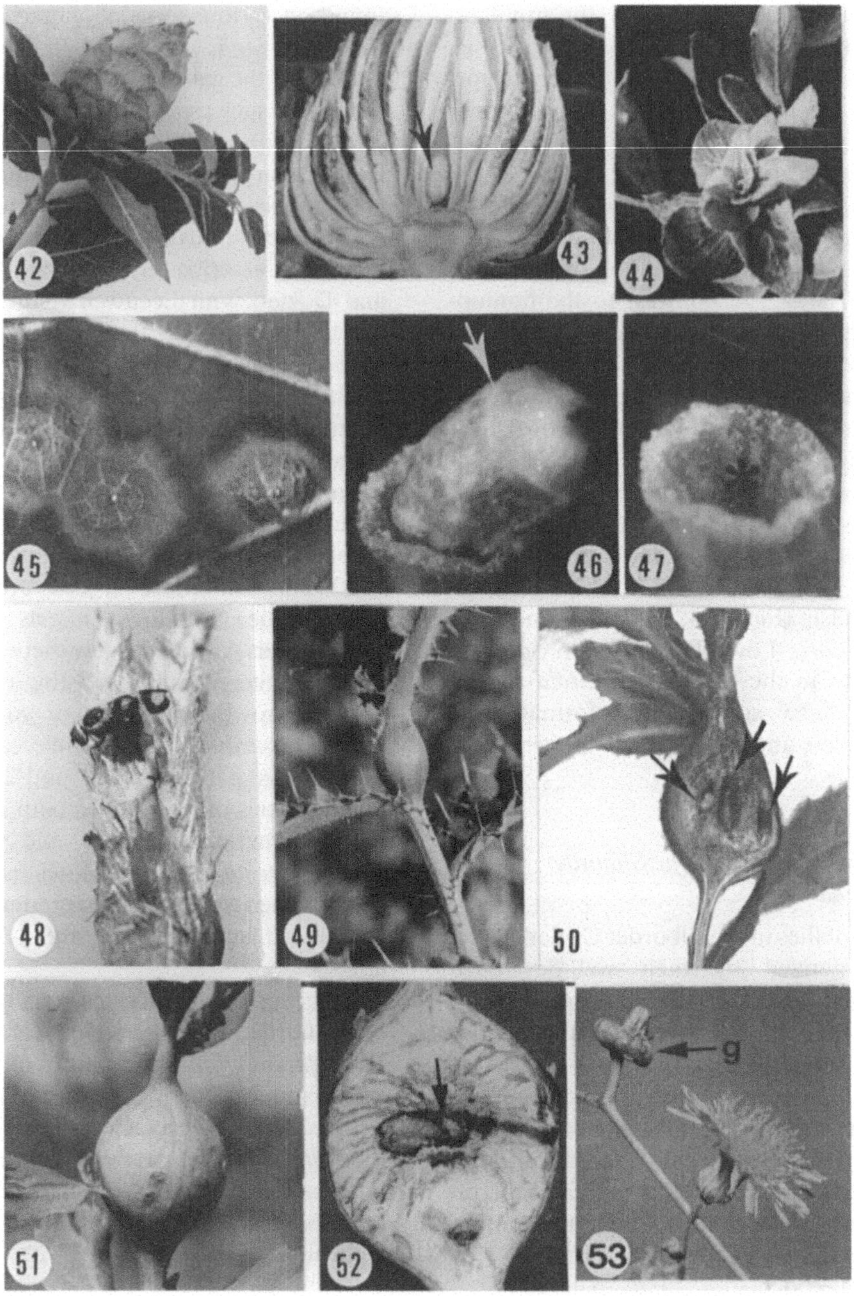

Figures 2-42–2-53. Galls of Cecidomyiidae (Figs. 2-42–2-47) and adults and galls of Tephritidae (Figs. 2-48–2-53). **Figure 2-42.** Bud gall, referred to as the willow cone gall, caused by the larva of *Rabdophaga strobiloides* on *Salix* sp. **Figure 2-43.** Dissection of the willow cone gall showing the thickened leaves and single larva (arrow) in its chamber. **Figure 2-44.** Rosette gall of *Rabdophaga rosaria* on *Salix*. **Figure 2-45.** Immature galls of *Didimomyia reaumuriana* on *Tilia* sp. (Photograph by O. Rohfritsch.) **Figure 2-46.** Mature gall of *Didimomyia reaumuriana* during dehiscence. The arrow indicates the inner gall. (From Rohfritsch 1967.) **Figure 2-47.** Mature gall of *Didimomyia reaumuriana* after the ejection of the inner gall. (From Rohfritsch 1967.) **Figure 2-48.** *Urophora cardui* ovipositing into a shoot of *Cirsium arvense*. **Figure 2-49.** Maturing stem gall of *U. cardui*. **Figure 2-50.** Dissection of a mature gall of *U. cardui* showing larval chambers. **Figure 2-51.** Stem gall of *Eurosta solidaginis* on *Solidago* sp. **Figure 2-52.** Dissection of stem gall of *E. solidaginis* showing the larval chamber, larva (arrow), and the escape route to the outside cut by the larva. **Figure 2-53.** Floral receptacle gall of *Tephritis dilacerata*.

Hymenoptera

The Order Hymenoptera contains the most complex and highly evolved galls and gall inducers in the entire guild. Gall inducers here are found in the two suborders Symphyta and Apocrita:

Symphyta:	Family Tenthredinidae
Apocrita:	Superfamily Chalcidoidea
	Family Pteromalidae
	Family Eurytomidae (seed chalcids)
	Family Agaonidae (fig wasps)
	Superfamily Cynipoidea
	Family Cynipidae

Hymenopterous galls are found throughout the world on the higher Orders of dicotyledons and even on monocotyledons, including Gramineae. Most are highly host and organ specific. Many species of Cynipoidea (the main group of galls inducers in this Order) cause fundamentally different galls on different organs and in alternate generations (Chapters 6 and 16).

Galls Induced by the Suborder Symphyta

FAMILY TENTHREDINIDAE

All of the gall-inducing tenthredinids belong to the subfamily Nematinae. According to Buhr (1965), only 75 species of this group of gall inducers occur in Europe and they were described by Dittrich in 1924. The larvae look like caterpillars of Lepidoptera and are equipped with vigorous chewing mouthparts (Chapters 5 and 6). The females use their saw-like ovipositors to pierce plant tissues to introduce eggs and ovipositional fluids. The presence of the egg induces a "procecidium" (Smith 1970), which is a tumor-like growth. The gall continues to grow under the influence of the larva. The larva may pupate inside the gall or on the ground, depending on the species.

Sawflies of the genus *Pontania* and their galls (Figs. 2-54 and 2-55) are well known in North America and Europe. Most species live on Amentaceae and produce brightly colored, round or reniform galls on leaves, petioles, or stems on *Salix* and *Populus*. A stem gall induced by *Hoplocampoides xylostei* Giraud on *Lonicera xylosteum* L. (Fig. 2-56) occurs in Central Europe.

Galls Induced by the Suborder Apocrita

SUPERFAMILY CHALCIDOIDEA

Most of the galls induced by Chalcidoidea occur on the buds or growing points of the shoot apices of their hosts, but some are found on leaves, roots, and flowers. Galls are characterized by their solid, hard, woody structures with numerous superficially isolated larval chambers in the cortex.

FAMILY PTEROMALIDAE

Nearly all species of this large family [560 genera and more than 2800 species, Narendran (1984)] are parasitic on other insects. In North America only *Hemadas nubilipennis* Ashmead induces a gall. This small wasp (Fig. 2-57) induces a reniform gall (Figs. 2-58 and 2-59) on the vegetative shoots of *Vaccinium angustifolium* Aiton (Shorthouse *et al.* 1986; West and Shorthouse 1989).

FAMILY EURYTOMIDAE

The most common European gall inducers in this family are in the genus *Harmolita* (= *Isthmosoma*) (Hedicke 1924), most of which gall Graminae. Female *Harmolita hyalipennis* Walk. bear a saw-like ovipositor used to introduce eggs into the shoot axis under the apex of different species of Graminae. Growth of the shoot is reduced and affected leaves are broader and shorter than normal.

In Australia and India, the genus *Trichilogaster* provokes hard, woody galls on *Acacia* and *Eucalyptus.* Some are very curious such as that of a *Trichilogaster* species on *Acacia leucophloea* Willd. described by Mani (1964). This large, spherical gall consists of numerous larval chambers arranged within a flattened, cup-like depression. Several of these cup-like depressions are fused to form the spherical gall.

FAMILY AGAONIDAE

The family Agaonidae includes the curious fig wasps of the genus *Blastophaga.* Most of these wasps live in the reproductive structures of different species of *Ficus* and are essential for the maturation of fruits. *Blastophaga psenes* L. has been known since antiquity as the agent responsible for pollination of the flowers of *Ficus carica* L. (Trabut 1901). The scarcely visible galls develop in transformed ovaries of spring

Figures 2-54–2-64. Galls of Tenthredinidae (Figs. 2-54–2-56), adult and galls of Pteromalidae (Figs. 2-57–2-59), and galls of Cynipidae (Figs. 2-60–2-64). Symbols: ♀ ♀ agamic form; ♀ ♂ bisexual form. **Figure 2-54.** Pouch gall of *Pontania pomum* on *Salix* sp. **Figure 2-55.** Dissected gall of *Pontania pomum* occupied by a single larva. Black objects (arrows) are fecal pellets. **Figure 2-56.** Stem gall of *Hoplocampoides xylostei* on *Lonicera* sp. **Figure 2-57.** *Hemadas nubilipennis* ovipositing into a vegetative shoot of *Vaccinium.* (Photograph by R. G. Lalonde.) **Figure 2-58.** Mature gall of *Hemadas nubilipennis.* **Figure 2-59.** Dissected gall of *Hemadas nubilipennis* showing numerous larval chambers. **Figure 2-60.** Foliaceous gall on a leaf of oak induced by *Andricus foliosus* ♀ ♂. This species causes an atypical bud on the leaf blade. (From Felt 1940.) **Figure 2-61.** Galls of two species of European cynipids on the same leaf of oak. The smaller and darker galls (small arrow) are induced by *Cynips disticha*

figs of wild trees. *Blastophaga* adults emerge from the gall and carry pollen to late female flowers of cultivated trees and then pollinate them.

SUPERFAMILY CYNIPOIDEA

The Cynipoidea is a relatively small superfamily of parasitic and phytophagous species. The gall inducers (family Cynipidae) are the best known members of the superfamily, but the majority of species are parasitic. There are approximately 13 natural groups based on biological characteristics but the relationships between these groups have not been determined. Most authors treat each group as a subfamily; however, there is little agreement as to the number and composition of families. Most European workers include both phytophagous forms and parasites in the Cynipidae (see Chapter 3). North American authors limit the Cynipidae to only the phytophagous species and place parasitic taxa in other families.

The most recent work on the world fauna of Cynipoidea is that of Weld (1952). This work gives keys to genera for all families. The only other treatment of the world fauna is the monograph by Dalla Torre and Kieffer (1910). More recently, Eady and Quinlan (1963) and Quinlan (1979) prepared new keys to the families and subfamilies of the Cynipoidea. Most authors recognize six families, with the family Cynipidae containing the gall-inducing and inquiline species (Askew 1984).

FAMILY CYNIPIDAE

The phytophagous Cynipidae either induce galls in various plants or live as inquilines in galls induced by other species. Galls of cynipids were probably the first recognized by early naturalists. Hippocrates, for example, mentioned cynipid galls as early as 460 B.C., often for their medicinal or commercial value. The gall of *Andricus gallaetinctorie* (Oliv.) contains 60 percent tannic acid and was used to make ink and chemicals for tanning hides. Others, such as *Dryocosmus deciduus* Beut. were known to be edible by humans and animals (Felt 1940).

Female cynipids lay their eggs in or on tissues and the eggs themselves induce a "procecidium" and have a lytic action on surrounding cells (Bronner 1985). However, once the eggs hatch, the larvae become the most important cecidogenetic agents. The larvae have strong mouthparts, which allows them to break cell walls and suck juices of the nutritive tissue that is progressively renewed (Bronner 1977 and Chapter 8). For further details on the tissues of cynipid galls see Chapters 5, 6, and 8.

Many species of cynipids are characterized by a complex life cycle known as heterogeny where sexual generations alternate with asexual or agamic generations. The insects of the two generations can attack the same or different organs and produce quite different galls (Chapter 6). In some cases the insects of the two generations have been inadvertently described as different species. An example of this complex life cycle is found in the species *Callirhytis serricornis* Kinsey (= *Dryocosmus* Weld) that causes galls on *Quercus agrifolia* Nee (Lyon 1970). This wasp attacks different organs of the plant, depending on the generation. That is, the first generation attacks flowers and the second attacks leaves. Another example is *Neuroterus quercusbaccum* L. (= *Neuroterus lenticularis* L.) in which both generations produce leaf galls. A berry-like gall is formed in May by the sexual generation and a lenticular gall in October by the parthenogenetic generation (Figs. 2-63 and 2-64).

Cynipids and their galls are uncommon in Asia; however, they are abundant in North America (805 species) (Felt 1940) and North-

♀♀. The larger and lighter galls (large arrow) are induced by *Cynips divisa* ♀♀. **Figure 2-62.** Oak ball galls induced on leaves by *Cynips quercusfolii* ♀♀. **Figure 2-63.** Galls of two species of European cynipids on the same leaf of oak. The small galls with the colored centers (small arrows) are induced by *Neuroterus numismalis* ♀♀. The larger galls (large arrow) are induced by *Neuroterus quercusbacarum* ♀♀ (= *N. lenticularis*). **Figure 2-64.** Drawing from Kieffer and Dalla Torre (1910) showing seven European galls induced by cynipids on the male catkins of oak. (A) *Neuroterus quercusbaccarum* ♀♂; (B) *Neuroterus aprilinus* ♀♂; (C, D) *Andricus foecundatrix* ♀♂; (E) *Andricus quercusramuli* ♀♂; (F) *Andricus quadrilineatus* ♀♂; (G) *Andricus xanthopsis* ♀♂; (H) *Andricus amenti* ♀♂.

ern and Central Europe (170 species) (Buhr 1965). Among the important works on Cynipidae are Dalla Torre and Kieffer (1910), Folliot (1964), and Kinsey (1920 and 1930). Weld (1952) gives keys to all gall formers and inquilines of the United States. Eady and Quinlan (1963) give keys to the European fauna. There are approximately 50 genera and 2000 species throughout the world with 46 genera in North America. Many workers have concentrated on the structure and developmental morphology of these galls (Beijerinck 1882; Molliard 1899; Meyer 1957; Rey 1967, 1969, 1976).

According to Kinsey (1920), three tribes can be distinguished within the Cynipinae by morphological characters: Aulacini, Rhoditini, and Cynipini.

1. Aulacini. Kinsey (1920) included 6 genera in the Aulacini, all of which form rather simple galls on various Compositae, herbaceous plants, and *Rubus* (Rosaceae) (Figs. 2-68–2-70). The genus *Aylax* is associated with the families Labiatae and Papaveraceae, the genus *Aulacidea* on Compositae, the genus *Diastrophus* on *Rubus,* and *Xestophanes* on *Potentilla.*
2. Rhoditini. The Rhoditini is composed of the genus *Diplolepis,* all of which induce their galls on *Rosa* (Shorthouse and Ritchie 1984). All organs of the host plant are attacked by various species of this genus and the galls (Figs. 2-72–2-77) are more complex than those of the Aulacini. European examples include *Diplolepis eglanteriae* Htg. (Fig. 2-74), which causes a simple round leaf gall containing one larva. *Diplolepis rosae* L. causes the common "mossy" rose gall (Fig. 2-73) on leaves, which often develops in large, multilocular masses. *Diplolepis triforma* Shorthouse and Ritchie of North America (Shorthouse and Ritchie 1984) also induces a multilocular gall, but it attacks buds (Figs. 2-71 and 2-72). Other common North American species include *Diplolepis spinosa* (Ashmead) on twigs (Fig. 2-75), *D. polita* (Ashmead) on leaves (Fig. 2-77), and *D. radicum* (O.S.) on roots (Fig.2-76) (Shorthouse 1988).
3. Cynipini. The Cynipini is composed of all the oak gall inducers. These species attack all organs of *Quercus* sp. and form complex galls (Figs. 2-60–2-67). Most oak galls are monothalamous (single chambered galls).

The external form of cynipid galls varies widely. Some are curiously pyriform growths borne on long slender stalks, whereas others are complicated coalesced masses. Some galls, such as those of *Andricus foliosus* Weld (Fig. 2-60) are foliaceous and look like an organoid type. The surfaces of cynipid galls are covered by either hairs, fleshy, or spiny growths, or scaly processes. One of the most striking structural peculiarities of cynipid galls is the presence of concentric zones of differentiated cells around the larval chambers with a zone of sclerenchyma cells.

The majority of cynipid galls arise on leaves (Figs. 2-60–2-63, 2-68, 2-70, and 2-73) and buds (Figs. 2-65–2-67); however, some are found on the stems (Figs. 2-69, 2-72, and 2-75) and even catkins (Fig. 2-64) and roots (Fig. 2-76). Figure 2-64 is from Dalla Torre and Kieffer (1910) and shows seven galls on the male catkins of *Quercus.* It is common to find the galls of two species of cynipids side by side on the same leaf (Figs. 2-61 and 2-63).

Some species of cynipids, such as *Andricus foecundatrix* (unisexual), form rosette-like galls on the vegetative buds of *Quercus* sp. (Fig. 2-65). The modified scaly and feathery leaves shelter an inner, unilocular gall (Fig. 2-66). The massive pink gall of *Biorhiza pallida* (Oliv.) (bisexual) (Fig. 2-67) in contrast harbors a large number of larval chambers; it is polythalamous.

One cannot discuss the galls of cynipids without mentioning that most are inhabited by numerous species of insects besides the gall inducers. These assemblages of insects form microcommunities and have been the subject of many studies (Mani 1964; Shorthouse 1973; see Chapter 16). Cynipid wasps referred to as inquilines are found within the galls of many cynipids. These insects feed both on the gall tissues induced by the gall wasp and often induce tissues and chambers of their own inside the host gall. Although the inquiline never feeds on the gall inducer, the latter is frequently killed by the ovipositing female (Shorthouse 1980a). The tissues of the host gall are frequently modified by the inquiline, sometimes substantially alter-

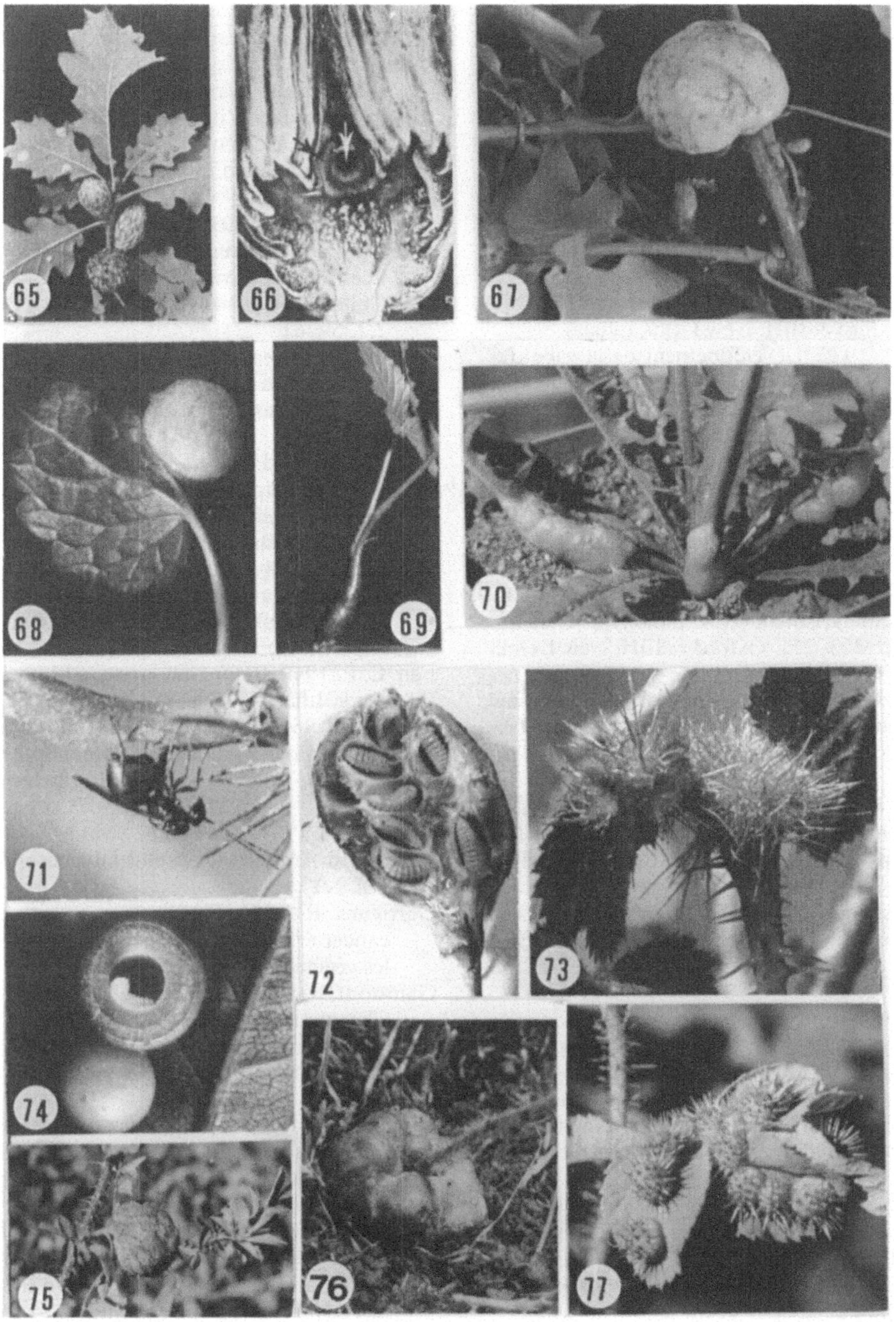

Figures 2-65–2-77. Galls of Cynipidae. **Figure 2-65.** Rosette-like gall on an oak bud induced by *Andricus foecundatrix* ♀♀. **Figure 2-66.** Dissected gall of *Andricus foecundatrix* showing the unilocular larval chamber (white arrow) of the inner gall (black arrow) surrounded by modified leaves of the bud. **Figure 2-67.** Callus-like bud gall of *Biorrhiza pallida* ♀♂. **Figure 2-68.** Leaf gall of *Aylax glechomae* on *Glechoma hederacea.* **Figure 2-69.** Twig gall of *Lasioptera rubi* on *Rubus* sp. **Figure 2-70.** Petiole mark gall of *Phanacis taraxaci* on *Taraxacum officinale.* **Figure 2-71.** *Diplolepis triforma* ovipositing into immature stem tissue of *Rosa acicularis.* **Figure 2-72.** Dissected gall of *Diplolepis triforma* showing larvae in their chambers. **Figure 2-73.** Immature gall of *Diplolepis rosae.* **Figure 2-74.** Dissected unilocular gall of *Diplolepis eglanteriae.* **Figure 2-75.** Stem gall of *Diplolepis spinosa* on *Rosa blanda.* **Figure 2-76.** Root gall of *Diplolepis radicum* on *Rosa acicularis.* **Figure 2-77.** Leaf galls of *Diplolepis polita* on *Rosa acicularis.*

ing the external appearance of the gall. There are 7 genera and about 175 species of inquiline Cynipidae.

LITERATURE CITED

Ananthakrishnan, T. N. 1978. Thrips galls and gall thrips. Zoological Survey of India, Technical Monograph.

Ananthakrishnan, T. N. 1984. Biology of Gall Insects. Oxford & IBH, New Delhi, India.

Anthony, M. 1972. Développement de la galle du *Gallacoccus anthonyae* Beardsley sur la feuille de *Shorea pauciflora* King (Dipterocarpaceae). C. R. Acad. Sci. Paris 275:2263–2266.

Anthony, M. 1980. Etude des bourgeonnements épiphylles déterminés par l'action d'Eriococcides cécidogènes chez les Diptérocarpacees du genre *Shorea.* Thèse de Doctorat d'Etat, Strasbourg.

Askew, R. R. 1984. The biology of gall wasps. In T. N. Ananthakrishnan (ed.), Biology of Gall Insects, pp. 223–271. Oxford & IBH, New Delhi, India.

Beardsley, J. W. 1971. A new genus of gall inhabiting Eriococcidae from Singapore (Homoptera: Coccoidea). Proc. Hawaiian Entomol. Soc. 21:31–39.

Beardsley, J. W. 1984. Gall-forming Coccoidea. In T. N. Ananthakrishnan (ed.), Biology of Gall Insects, pp. 79–106. Oxford & IBH, New Delhi, India.

Beijerinck, M. W. 1882. Beobachtungen über die ersten Entwicklungsphasen einiger Cynipidengallen. Naturk. Verh. Koninkl. Akad. Wetenschap. 22:1–198.

Bloch, R. 1965. Abnormal Development in Plants: A Survey. Handbuch der Pflanzenphysiologie, Bd. 15/2, pp. 145–183, Springer, Berlin.

Börner, C., and F. A. Schilder. 1932. In P. Sorauer (ed.), Handbuch der Pflanzenkrankheiten V. Tierische Schädlinge an Nutzpflanzen, 2er Teil. Paul Parey, Berlin.

Bronner, R. 1977. Contribution a l'étude histochimique des tissus nourriciers des zoocécidies. Marcellia. 40:1–134.

Bronner, R. 1985. Anatomy of the ovipositor and oviposition behaviour of the gall wasp *Diplolepis rosae* (Hymenoptera: Cynipidae). Can. Entomol. 117:849–858.

Buhr, H. 1965. Bestimmungstabellen der Gallen (Zoo- und Phytocecidien) an Pflanzen Mittel- und Nordeuropas. I and II. G. Fischer, Jena.

Coutin, R., and J. Riom. 1967. Dimorphisme des galles provoquées par *Mikiola fagi* Hartig (Dipt. Cecid.) sur *Fagus silvatica* L. Androcécidies et gynocécidies. C. R. Acad. Sci. Paris. 265:975–978.

Dalla Torre, K. W., and J. J. Kieffer. 1910. Cynipidae. In Das Tierreich. R. Friedländer, Berlin.

Dittrich, R. 1924. Die Tenthredinidocecidien, durch Blattwespen verursachte Pflanzengallen und ihre Erzeuger. Zoologica 24(61):585–635.

Docters Van Leeuwen, W. M. 1982. Gallenboek. Revised and edited by A. A. Wiebes-Rijks and G. Houtman. B. V. W. J. Thieme and Cie, Zutphen, The Netherlands.

Docters Van Leeuwen, W. M., and J. Docters Van Leeuwen-Reijnvaan. 1904. Die Entwicklung der Galle von *Lipara lucens.* Rec. Trav. Bot. Neerl. 2:235–261.

Docters Van Leeuwen-Reijnvaan, J., and W. M. Docters Van Leeuwen. 1926. The Zoocecidia of the Netherlands East Indies. Batavia, The Netherlands.

Dreger-Jauffret, F. 1978. Effets du parasitisme sur le méristème apical et le développement des bourgeons de quelques plantes ligneuses et herbacées. Thèse, Strasbourg, France.

Eady, R. D., and J. Quinlan. 1963. Cynipoidea. Key to families and subfamilies and Cynipinae (including galls). Handbook Ident. Insects Gr. Britain VIII(1a).

Felt, E. P. 1940. Plant Galls and Gall Makers. Comstock, Ithaca, NY.

Folliot, R. 1964. Contribution à l'étude de la biologie des Cynipides gallicoles (Hyménoptères, Cynipoidea). Ann. Sci. Nat., Zool., 12e serie. 6:407–564.

Gagné, R. J. 1989. The Plant-Feeding Gall Midges of North America. Cornell University Press, Ithaca, NY.

Garrigues, R. 1951. Recherches sur les cécidies, le cancer et l'action des carbures cancérigènes sur les végétaux. Thèse, Sci. Nat. Lyon.

Gaumont, R. 1957. Etude sur la biologie de quelques Chermesidae. Thèse, Sci. Nat., Paris.

Goidanich, A. 1951. La cecidoforia: fenomeno etologico nuovo. Atti Acad. Asi. Torino 85:312–316.

Goidanich, A. 1960. Specializzazione ecologica e nomenclatura delle *Carulaspis* del *Viscum* e delle Cupressaceae (Hemiptera, Coccoidea, Diaspididae). Boll. Ist. Entomol. Bologna 24:1–38.

Grassé, P. P., R. Poisson, and O. Tuzet. 1961. Zoologie I. Masson, Paris.

Grover, P., and U. Kashyap. 1984. On the study of bio-ecology of certain gall midges. A discussion. Cecidol. Int. 3:73–187.

Harris, P. 1980. Effects of *Urophora affinis* Frfld. and *U. quadrifasciata* (Meig.) (Diptera: Tephritidae) on *Centaurea diffusa* Lam. and *C. maculosa* Lam. (Compositae). Z. ang. Entomol. 90:190–210.

Hedicke, H. 1924. Die Isthmosominocecidien, von Isthmosominen verursachte Pflanzengallen und ihre Erzeuger. Zoologica 24(61):637–674.

Hodkinson, I. D. 1984. The biology and ecology of

the gall-forming Psylloidea (Homoptera). In T. N. Ananthakrishnan (ed.), Biology of Gall Insects, pp. 59–77. Oxford & IBH, New Delhi, India.

Houard, C. 1906. Modifications histologiques produites par des *Copium* dans les fleurs de *Teucrium.* Marcellia 5:83–101.

Houard, C. 1908, 1909, 1913. Les Zoocécidies des plantes d'Europe et du Bassin de la Méditerranée. I, II, and III. Hermann, Paris.

Houard, C. 1911. Action de cécidozoaires externes appartenant au genre *Asterolecanium,* sur les tissus de quelques tiges. Marcellia 10:3–25.

Houard, C. 1922, 1923. Les Zoocécidies des plantes d'Afrique, d'Asie et d'Océanie. I & II. Hermann, Paris.

Houard, C. 1933. Les Zoocécidies des plantes de l'Amerique du Sud et de l'Amerique Centrale. Hermann, Paris.

Howell, J. O., and M. Kosztarab. 1972. Studies on morphology and systematics of scale insects. No. 4. Morphology and systematics of the adult females of the genus *Lecanodiaspis* (Homoptera: Coccoidea; Lecanodiaspididae). Virginia Polytechnic Inst. and State Univ. Res. Div. Bull. 70.

Jacobs, W., and M. Renner. 1974. Taschenlexikon zur Biologie der Insekten. G. Fischer, Stuttgart.

Karny, H. 1911. Ueber Thripsgallen und Gallenthripse. Zentbl. Bakt. Parasit. Kde abt. II, 30:556–572.

Karny, H. 1913. Ueber Gallenbewohnende Thysanopteren. Verh. zool. bot. Ges. Wien. 63.

Karny, H., J. Docters Van Leeuwen-Reijnvaan, and W. M. Docters Van Leeuwen. 1913. Beiträge zur Kenntniss der Gallen von Java. 5. Ueber die Javanischen Thysanopterocecidien und deren Bewohner. Bull. Jard. bot. Buitenzorg. 10(2):1–126.

Karny, H., J. Docters Van Leeuwen-Reijnvaan, and W. M. Docters Van Leeuwen. 1914–1916. Beiträge zur Kenntniss der Gallen aus Java. Zweite Mitteilung ueber javanische Thysanopterocecidien und deren Bewohner. Z. f. wiss. Insektenbiol, Berlin, Vols. X, XI, XII.

Karsch, F. 1880. Neue Zoocecidien und Cecidozoen. Z. ges. Naturwschft. 53:286–310.

Kerner Von Marilaun, A. 1905. Pflanzenleben. Zweite Auflage. 2 Bd. Die Geschichte der Pflanzen. Leipzig und Wien, Germany.

Kieffer, J. J. 1900. Monographie des Cecidomyides d'Europe et d'Algérie. Ann. Soc. Entomol. Fr. 69:181–472.

Kieffer, J. J. 1913. Diptera. Fam. Cecidomyiidae. In P. Wytsman (ed.), Genera Insectorum, Bd. 152. Brussels.

Kinsey, A. C. 1920. Phylogeny of cynipid genera and biological characteristics. Bull. Am. Mus. Nat. Hist. 42:357–402.

Kinsey, A. C. 1930. The gall wasp genus *Cynips.* A study on the origin of species. Indiana Univ. No. 220. (Ent. Ser. No. 7).

Küster, E. 1911. Die Gallen der Pflanzen. Hirzel, Leipzig.

Lalonde, R. G., and J. D. Shorthouse. 1984. Developmental morphology of the gall of *Urophora cardui* (Diptera, Tephritidae) in the stems of Canada thistle *(Cirsium arvense).* Can. J. Bot. 62:1372–1384.

Le Pape, H. and R. Bronner. 1987. The effect of *Ceutorrhynchus napi* on stem tissues of *Brassica napus* var. *oleifera.* In V. Labeyrie, G. Fabres, and D. Lachaise (eds.), Insects–Plants, pp. 207–211. Dr. W. Junk, The Hague.

Lewis, J. F. and L. Walton. 1958. Gall formation on *Hamamelidis virginiana* resulting from material injected by the aphid *Hormaphis hamamelidis.* Trans. Am. Microbiol. Soc. 77:146–200.

Lewis, T. 1973. Thrips, Their Biology, Ecology and Economic Importance. Academic Press, London.

Lyon, R. J. 1970. Heterogony in *Callirhytis serricornis* (Kinsey) (Hymenoptera: Cynipoidea). Proc. Entomol. Soc. Wash. 72:176–178.

Mamaev, B. M. 1975. Evolution of Gall-Forming Insects—Gall Midges. netherby: The British Library, Lending Division. (Russian original: Leningrad).

Mani, M. S. 1948. Cecidozoa and Zoocecidozoa from India. J. R. Asiatic Soc. Bengal. (Sci.). 14:27–195.

Mani, M. S. 1959. Cecidotheca indica. Agra Univ. J. Res. (Sci.). 8:91–220.

Mani, M. S. 1964. Ecology of Plant Galls. W. Junk. The Hague.

Maresquelle, H. J., and J. Meyer. 1965. Physiologie et morphogenèse des galles d'origine animale (Zoocécidies). Handbuch Pflanzenphysiol. 15(2):280–329.

Mees, A. 1923. Die cecidogenen und cecidocolen Lepidopteren, gallenerzeugende und gallenbewohnende Schmetterlinge und ihre Cecidien. Zoologica 24:499–584.

Meyer, J. 1957. Cécidogenèse comparée de quelques galles d'Arthropodes et évolution cytologique des tissus nourriciers. Thèse, Sci. Nat., Strasbourg.

Meyer, J. 1987. Plant Galls and Gall Inducers. Gebrüder Borntraeger, Stuttgart.

Meyer, J., and H. J. Maresquelle. 1983. Anatomie des Galles. Gebrüder Borntraeger, Stuttgart.

Mimeur, J. M. 1949. Contribution à l'étude des Zoocécidies du Maroc. Encyclopedie Ent. A. 24. P. Lechevalier, Paris.

Molliard, M. 1899. Sur la galle de l' *Aulax papaveris* Perris. Rev. Gen. Bot. 11:209–217.

Narendran, T. C. 1984. Chalcids and sawflies associated with plant galls. In T. N. Ananthakrishnan (ed.), Biology of Gall Insects, pp. 273–303. Oxford & IBH, New Delhi, India.

Nijveldt, W. 1969. Gall Midges of Economic Importance. VIII. Crosby Lockwood, London.

Ohsako, S., J. Yukawa and M. Horikiri. 1981. New data on the life history of the *Ligustrum* fruit midge *Asphondylia sphaera* Monzen (Diptera, Cecidomyiidae). Kyushu 27:116–119.

Pesson, P. 1951. Ordre des Homoptères. In P. P. Grassé (ed.), Traité de Zoologie. X. Fasc. 2, pp. 1391–1656. Masson et cie, Paris.

Philipson, W. R. 1964. Morphogenesis as affected by associated organisms. Phytomorphology 14:163–185.

Quinlan, J. 1979. A revisionary classification of the Cynipoidea (Hymenoptera) of the Ethiopian zoogeographical region. Bull. Br. Mus. Nat. Hist. Entomol. 39:85–133.

Raymond, G. 1927. Contribution à l'étude des cécidomyides attaquant les violettes cultivées. Ann. Soc. Linn. 73:87–101.

Rey, L. 1967. Les premiers stades de développement de la galle de *Pontania proxima* Lep. Bull. Soc. Bot. Fr. 114:80–95.

Rey, L. 1969. Les premiers stades du développement de la galle de *Biorhiza pallida* Ol. ♀ ♂. Marcellia 36:119–135.

Rey, L. 1976. Recherches comparées sur trois types d'hymenopterocécidies (Cynipidae et Tenthredinidae). Etude morphologique, histologique, et ultrastructurale du développement et de la différenciation. Essais experimentaux. Thèse, Nantes.

Richter-Vollert, I. 1964. Untersuchungen zur Morphologie und Ökologie von *Asphondylia sarothamni* H. Lw. Zoologica 40:1–54.

Rohfritsch, O. 1967. Ejection de la coque cécidienne d *'Oligotrophus reaumurianus* de la feuille du Tilleul. Marcellia 34:171–182.

Rohfritsch, O. 1971. Etude d'une galle de *Lobopteromyia* sp. sur *Acacia* (*Acacia ferruginea* D. C.). Marcellia 34:171–182.

Rohfritsch, O. 1974. Etude de la galle du *Lobopteromyia ramachandrani* Mani et de son développement sur l *'Acacia ferruginea* D. C. Marcellia 38:67–75.

Rohfritsch, O. 1978. Premières manifestations de l'action parasitaire du *Hartigiola annulipes* Hartig sur la feuille du Hêtre. 103e Congres Soc. Sav. Nancy 3:311–322.

Rohfritsch, O. 1987. Gall development and feeding strategy of *Didymomyia reaumuriana* on *Tilia.* Phytophaga 1:1–20.

Ross, H. 1932. Praktikum der Gallenkunde (Cecidologie). Springer, Berlin.

Rübsaamen, E. H. 1911. Beiträge zur Kenntniss aussereuropäischer Zoocecidien. V. Beitrag: gallen aus Afrika und Asien. Marcellia 10:100–132.

Rübsaamen, E. H., and H. Hedicke. 1926. Die zoocecidien, durch Tiere erzeugte Pflanzengallen Deutschlands und ihre Bewohner. V. Die Cecidomyiden (Gallmücken) und ihre cecidien. Zoologica 29.

Ruppolt, W. 1959. Harzgallen an Kiefern. Orion, Z. Nat. Tech. 8:596–599.

Schmidt, E., and J. Meyer. 1966. Observations sur la structure des galles de *Livia juncorum* Latr. sur *Juncus articulatus* L. Marcellia 33:237–253.

Shorthouse, J. D. 1973. The insect community associated with rose galls of *Diplolepis polita* (Cynipidae, Hymenoptera). Quaest. Entomol. 9:55–98.

Shorthouse, J. D. 1980a. Modifications of galls of *Diplolepis polita* by the inquiline *Periclistus pirata.* Bull. Soc. Bot. Fr. (Actual. Bot.) 127:79–84.

Shorthouse, J. D. 1980b. Modification of the flower heads of *Sonchus arvensis* (family Compositae) by the gall former *Tephritis dilacerata* (order Diptera, family Tephritidae). Can. J. Bot. 58:1534–1540.

Shorthouse, J. D. 1988. Occurrence of two gall wasps of the genus *Diplolepis* (Hymenoptera: Cynipidae) on the domestic shrub rose, *Rosa rugosa* Thunb. (Rosaceae). Can. Entomol. 120:727–737.

Shorthouse, J. D. and A. J. Ritchie. 1984. Description and biology of a new species of *Diplolepis* Fourcroy (Hymenoptera: Cynipidae) including galls on the stems of *Rosa acicularis.* Can. Entomol. 116:1623–1636.

Shorthouse, J. D., A. West, R. W. Landry, and P. D. Thibodeau. 1986. Structural damage by female *Hemadas nubilipennis* (Hymenoptera: Pteromalidae) as a factor in gall induction on lowbush blueberry. Can. Entomol. 118:249–254.

Skuhravá, M., and V. Skuhravý. 1960. Bejlomorky (Gallmücken). Cesk. Akad. Zemědělských Věd. Prague.

Skuhravá, M., V. Skuhravý, and J. W. Brewer. 1984. Biology of gall midges. In T. N. Ananthakrishnan (ed.), Biology of Gall Insects, pp. 169–222. Oxford & IBH, New Delhi, India.

Smith, E. L. 1970. Biosystematics and morphology of Symphyta. II. Biology of gall-making Nematine sawflies in the Californian region. Ann. Entomol. Soc. Am. 63:36–51.

Tavares Da Silva, J. 1902. Descripcao de 6 Coleopterocecidias novas. Revista de Ciencias Naturales.

Tavares Da Silva, J. 1915. Cecidologie Argentine. Broteria Braga Zool. 13:88–128.

Trabut, L. 1901. La caprification en Algérie. Rev. Viticulture. Paris 16:115–138.

Uichanco, L. B. 1919. A biological and systematic study of Philippines plant galls. Philipp. J. Sci. 14:527–554.

Weld, L. H. 1952. Cynipoidea. Privately published. Ann Arbor, MI.

West, A., and J. D. Shorthouse. 1989. Initiation and development of the stem gall induced by *Hemadas nubilipennis* (Hymenoptera: Pteromalidae) on lowbush blueberry, *Vaccinium angustifolium* (Ericaceae). Can. J. Bot. 67:2187–2198.

Yukawa, J. 1971. A revision of the Japanese gall

midges (Diptera: Cecidomyiidae). Mem Fac. Agr. Kagoshima Univ. 8:1–203.
Yukawa, J. 1981. Collecting record of the *Machimus* leaf gallmidge *Daphnephila machilicola* Yukawa (Diptera, Cecidomyiidae) from Formosa. Mem. Kagoshima Univ. Res. Center S. Pac. 2:49–50.
Yukawa, J. 1987. Life history strategies of univoltine gall making Cecidomyiidae (Diptera) in Japan. Phytophaga 1:121–139.

3

Evolution of the Gall-Inducing Guild

J. C. ROSKAM

INTRODUCTION AND DEFINITION OF THE GUILD

Gall-inducing insects are specialized herbivores and are found in seven Orders (see Chapter 2). Because the various groups have different, non-cecidogenic ancestors, they have evolved along parallel lines. Gall inducers are also found among other groups of animals, namely, nematodes and mites. Among plants, many parasitic fungi, some bacteria, green algae, and even angiosperms (Cuscutaceae, Loranthaceae) can induce galls. All organisms capable of inducing galls are considered as belonging to the same guild—a guild being a group of organisms sharing the same trophic type (Root 1973).

In this chapter the evolutionary processes relevant for gall insects and mites are examined. These processes include the origin of the different taxonomic groups that attained the gall-inducing ability, the acquisition of the gall-inducing trophic type, the adaptative radiation of the various taxonomic groups of gall inducers, and, finally, some regressive developments that enabled taxa to abandon the gall-inducing guild. With regard to these evolutionary processes a comparison will be made for the various lines of evolution that allows recognition of some alternative patterns followed along these lines.

The various groups of gall inducers must have come to the gall-inducing guild via different ancestral trophic types (Fig. 3-1). Depending on the substrates available during the different periods of host plant evolution, the trophic types were subject to different evolutionary changes. One can distinguish primitive trophic types that evolved on dead organic substrates in various stages of decay. Detritus feeding (saprophagy, scavenging) is presumed to be the most primitive trophic type in mites, as well as in insects (Smart and Hughes 1972, Zwölfer 1978). Many insects, at least in their immature stages, have retained this type and still feed on decaying plant materials, molds, and bacteria (Southwood 1972).

Feeding on living plant and animal substrates evolved later in geological history and therefore has been considered derived. Within plant feeding a further subdivision can be made in connection to the angiosperms, because these plants are the dominant aspect of our vegetation. Ancient derived feeding modes then evolved during the period of green plant radiation (Devonian) and recently derived modes appeared after the radiation of the angiosperms during the Cretaceous. Spore feeders and later pollen feeders, followed by plant-sap feeders represent anciently derived trophic types, whereas stem borers, foliage feeders, and leaf miners represent recently derived types.

Within the various arthropod families major and minor groups of gall inducers can be distinguished. Most members of a major group (a family or subfamily) are gall inducers and usually they are known from a large diversity of mainly angiospermous host plants. Major groups of gall inducers are supposed to have evolved from an ancestor that already possessed the gall-inducing ability in an early period of angiosperm radiation.

Minor groups of gall inducers evolved among Curculionidae, Tenthredinidae, Cephidae, Gelechiidae, Tephritidae, and Agromyzidae. Gall inducers remained exceptions within these

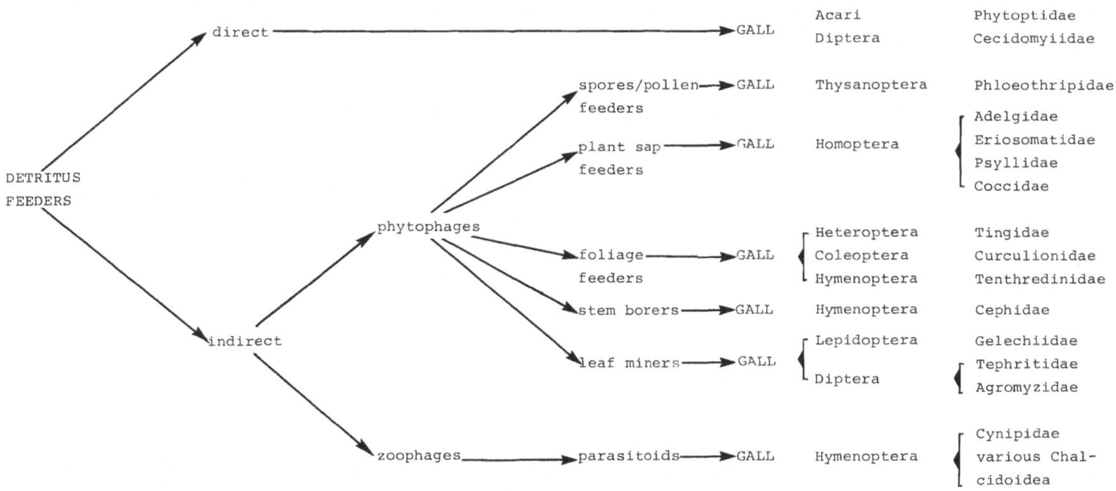

Figure 3-1. Ancestral trophic types of gall-inducing animals.

families where recently derived types of phytophagy are dominant. Host plants spectra are narrow because radiation occurred only on particular plant genera.

Gall thrips and homopteran gall inducers are in a rather intermediate position. Both groups have numerous gall-inducing representatives but they are scattered among non-gall-inducing forms and have no close mutual phylogenetic bonds. Dominating and therefore probably ancestral feeding modes belong to the anciently derived phytophagous categories such as spore, pollen, and juice feeding. Host plant spectra usually are broad, although the various groups are characterized by restrictions. Gall thrips, for example, are especially abundant on tropical host plants. Gall-inducing Homoptera, alternating their host plants during the life cycle, make their galls only on the primary, woody, host plants and not on their secondary, herbaceous hosts. Host plants of Psyllidae and Eriosomatidae mainly belong to the Dicotyledonae whereas Adelgidae are restricted to conifers. Feeding on plant sap offers homopterans extensive opportunities for manipulating host plants and, as a result, gall inducers evolved many times and independently from sap feeders during distant geological periods in which different classes of host plants (conifers or angiosperms) made up the dominant aspects of the vegetation.

Gall wasps (Cynipinae) and fig wasps (Chalcidoidea, Agaonidae) in this respect have an intermediate position. Although they presumably evolved from a derived trophic type (i.e., parasitism of other arthropods) and have a narrow host plant spectrum, they are important groups of gall inducers that include large number of species. As in other major groups of gall inducers, gall wasps and fig wasps both evolved from gall-inducing ancestors. Other chalcidoid gall inducers (i.e., the eurytomid genus *Tetramesa*) belong to groups in which gall induction is an exception. Their status is similar to that of other minor groups of gall insects.

Differences in gall structure also have been used to illustrate evolutionary processes in gall inducers (Cockerell 1890; Wells 1921). Kataplasmic galls have walls composed of undifferentiated parenchyma cells (Küster 1911) and are considered to be primitive; prosoplasmic galls are advanced and have distinctly differentiated layers. The first step in the primitive kataplasmas was inhibition of normal differentiation. This was followed by proliferation of parenchymatous cells and, later in evolutionary time, differentiation of the parenchymatous cells was controlled by the gall inducers and the prosoplasmic galls developed.

The ecological criterion (the trophic type from which the gall inducer originated) and the taxonomical criterion (whether or not all members of a taxon are gall inducers) are both important for the analysis of evolutionary patterns. Such an analysis will be presented in the following sections for gall-inducing Hymenoptera and Diptera that evolved either a primitive trophic type or via a derived one to the gall-inducing guild.

GALL-INDUCING HYMENOPTERA

The Order Hymenoptera may be subdivided into three main groups: the "Phytophaga," "Terebrantes," and "Aculeata." Bionomical characteristics are evident in this old classification proposed by Latreille. The Phytophaga, synonymous today with the "Symphyta," feed on plant tissue. Terebrantes and Aculeata evolved from the Phytophaga via a common ancestor and today are usually united under the name "Apocrita." The larvae of the Terebrantes basically have a parasitoid habit and feed on other arthropods, most of which belong to the holometabolous Orders. The hosts are eventually killed after parasitism. The evolution of the Aculeata is characterized by the advancement of instinct, culminating in the social behavior of ants, bees, and wasps.

Gall insects evolved independently at least three times in the Hymenoptera: among the Phytophaga in the Tenthredinoidea, among the Terebrantes in the Cynipoidea, and in the Chalcidoidea. Gall formers do not occur in the Aculeata. To determine the relationships between hymenopteran gall insects and their noncecidogenic relatives the results of Königsmann (1976, 1977, 1978a,b) are adopted. His conclusions were mainly based upon morphological characters, which were phylogenetically interpreted according to Hennig's (1966) rules.

Tenthredinoidea

Although the Phytophaga are a very old group of insects, saprophagy, the primitive trophic type, is absent. In the Mecopteroidea, including the orders Mecoptera, Trichoptera, Lepidoptera, and Diptera (and, according to Königsmann forming the sister group of the Hymenoptera), detritus feeders are predominant, at least in the primitive forms (Southwood 1972). Hence, the origin of the Hymenoptera was apparently accompanied by a change of the larval feeding habit from saprophagy to plant feeding. The phylogenetic relations of the Phytophaga and Apocrita are given in Figure 3-2. The various feeding habits present in the Phytophaga

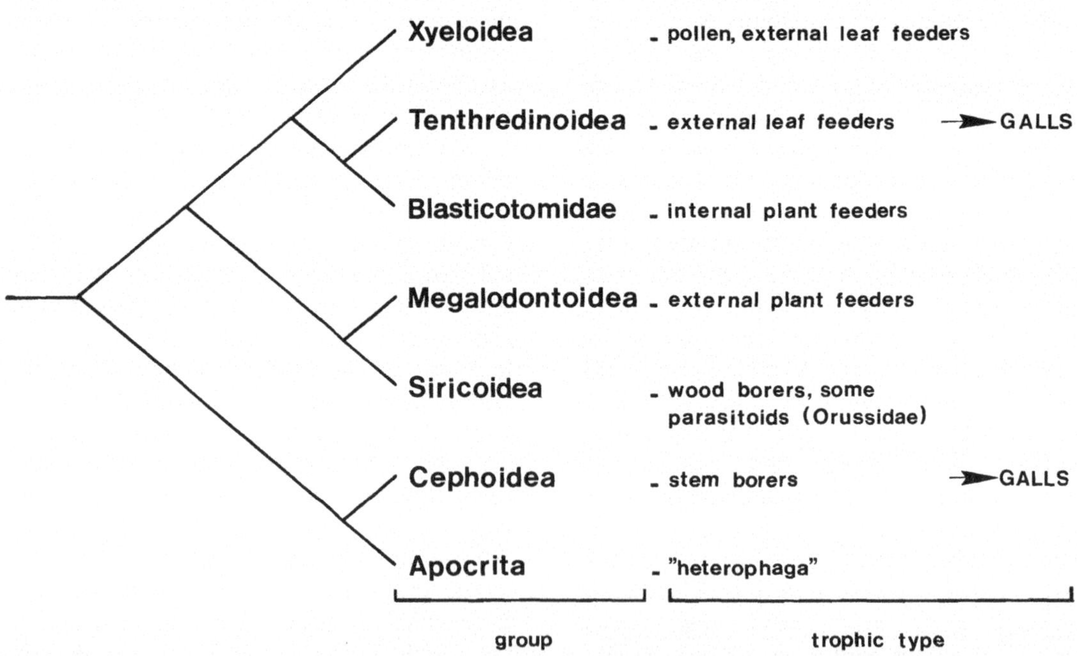

Figure 3-2. Phylogenetic relations between the various groups of Symphyta and Aprocrita. (After Königsmann 1977.)

form a series. Among the Xyeloidae, which are the most primitive, some are pollen feeders on gymnosperms. The phytophagous series, via a number of external plant feeders, lead to groups of internal plant feeders, which in turn gave rise to heterophagous forms.

Gall-inducing Phytophaga are mainly present in the family Tenthredinidae, the sawflies. This very old family originated 200 million years ago, in the Triassic, at which time period ferns and gymnosperms were available as food. At present many species are allied with woody angiosperms (about 70 percent on host plants of the subclasses Magnoliidae, Hamamelidae, and Rosidae), but a relatively large part (24 percent) still occurs on ferns and gymnosperms. Gall-inducing forms presumably were not present before the beginning of the Tertiary, some 75 million years ago (Zwölfer 1978). We must therefore suppose that the group radiated as external plant feeders and secondarily gave rise to gall inducers. This is reflected in the host plant relations of the gall-inducing sawflies, as analyzed from Buhr's (1965) keys (Table 3-1). Not less than 92 percent occurs on woody angiosperms and only 4 percent on ferns, whereas gall inducers are absent from gymnosperms. It is remarkable that sawfly galls occur mainly on Salicaceae. Sawflies of the genera *Pontania* and *Euura* comprise 56 percent of all species and all are found on *Salix* (see Chapter 2).

Sawfly galls are different from most insect galls in that the ovipositing behavior of the female plays an important role in the gall-inducing process (see Chapter 5). In other arthropod groups the gall-inducing process is usually the result of larval activity and associated with feeding behavior. The female sawfly makes a hole in the plant tissue with her ovipositor, and before she lays her egg in the cavity she injects a fluid synthesized in the accessory sex gland (acid gland).

Sawfly larvae also have a different feeding behavior from other larvae in the gall-inducing Hymenoptera. They are eruciformous and defecate during their development. The half-grown larva gnaws a hole in the gall wall and removes the fecal accumulation through this hole (Mani 1964).

The gall-inducing habit of some sawflies forms an interesting specialization, but the significance of gall-inducing forms for the onward evolution of the Order is at least problematic. Königsmann (1978a) considered all Apocrita (Terebrantes and Aculeata) as the sister group of the Phytophaga superfamily Cephoidae (Fig. 3-2). He agreed to some extent with Malyshev (1966), who considered that the Terebrantes evolved from gall-inducing Cephoidea via an archaic inquiline phase. Malyshev theorized that some stem sawflies lost the ability to induce galls and became food parasites of other gall-inducing Symphyta. From food parasites they evolved to parasitoids of the gall inducers themselves. However, there are only a few gall inducers among recent Cephoidea, which occur on modern plants (two *Cephus* species on wheat, rye, and some wild grasses), whereas

Table 3-1. Percentages of Gall Midge, Gall Wasp, and Gall-Making Sawfly Species in Subclasses of Vascular Plants (Ferns and Gymnosperms Together)[a]

Subclasses of vascular plants	Plant species (n = 2796)	Gall midge species (n = 556)	Gall wasp species (n = 139)	Gall-making sawfly species (n = 73)
Ferns and gymnosperms	4.0	1.6	0	4.1
Magnoliidae	0.3	0	0	0
Ranunculidae	3.9	2.3	1.4	2.7
Hamamelidae	1.3	8.3	69.1	13.7
Caryophyllidae	8.2	7.6	0	0
Dilleniidae	12.7	16.7	0.7	58.9
Rosidae	20.3	25.5	11.5	19.2
Asteridae	27.9	29.9	17.3	1.4
Alismatidae	1.7	0	0	0
Liliidae	19.2	9.5	0	0
Arecidae	0.6	0	0	0

[a]Numbers of plant species from Rothmaler (1972) and of gall insect species from Buhr (1965).

gall-inducing forms are abundant in Tenthredinidae, which occur on plants of more ancient groups. Hence, gall inducers are exceptions in the Cephoidea and they are apparently of recent origin. Their relations with an important ancient stock of gall inducers are improbable. However, alternative hypotheses (Ross 1965), which advocate the evolution of parasitoid forms from stem-boring and not gall-inducing cephoids, appear more feasible.

Cynipoidea

Both Cynipoidea and Chalcidoidea belong to the "Terebrantes" and they share a common, parasitoid ancestor (Königsmann 1977, 1978a). Like all other Apocrita, the Cynipoidea and Chalcidoidea have apodous, sedentary larvae that exhibit a peculiar feature of the intestine tract. Whereas the mid and hind gut in Symphyta are only separated during the embryonic stage, in the Apocrita these two intestine parts remain separated until the end of the prepupal stage. The frass in Apocrita is accumulated in the mid-gut during the larval stages and only the prepupa defecates. The retardation in ontogeny of the formation of a junction between mid and hind gut might be the result of a change from plant food to an animal diet. If so, all forms, including phytophagous ones that display this characteristic, evolved from zoophagous forms or more precisely from parasitoid ancestors.

All gall inducers in Cynipoidea belong to the subfamily Cynipinae. With about 800 species, the true gall wasps rank second in number among gall insects, surpassed only by the gall midges (Felt 1940). The closely related families Ibaliidae and Figitidae, as well as the other cynipid subfamilies, viz. Eucoilinae and Charipinae, have a parasitoid habit and, therefore, gall wasps must have evolved from parasitoids. Gall wasps apparently are of recent origin, for even in the Miocene (25 million years ago) the parasitoid Cynipidae dominated (Telenga 1952).

To determine the phylogeny of gall wasp genera, Kinsey (1920) used characters such as wing venation, extension of the second gastral tergite, and complexity of gall structure. The host plant relations of the gall wasp genera are listed in Table 3-2. Primitive character states are mainly encountered in the Cynipinae tribe Aulacini. Some genera of this tribe are oligophagous on Asteraceae (e.g., *Phanacis* and *Aulacidae*). Also, *Aylax*, the only polyphagous genus, belongs to this group and its species are on host plants belonging to Salicaceae, Fabaceae Papaveraceae, Lamiaceae, and Asteraceae. Furthermore, all inquilinous forms (e.g., *Ceroptres*, *Synergus*, and *Periclistus*) are found among the Aulacini (sensu Kinsey 1920, see Table 3-2). All gall inducers of this tribe have relatively simple agglomerate (galls clumped together) or polythalamous (multichambered)

Table 3-2. Classification of Cynipinae[a]

Tribe	Genus	Host plant family
Aulacini—gall makers[b]	*Phanacis* (= *Timaspis*, = *Gilletia*)	Asteraceae
	Aulacidea	Asteraceae
	Antistrophus	Asteraceae
	Aylax	Various
	Cecconia	Valerianaceae
	Isocolus	Dipsacaceae
	Liposthenes	Lamiaceae
	Endocaulonia	Lamiaceae
	Panteliella	Lamiaceae
	Vetusia	Lamiaceae
	Xestophanes	Rosaceae
	Diastrophus	Rosaceae
	Gonaspis	Rosaceae
	Pediaspis[c]	Aceraceae
Aulacini—inquilines[d]	*Saphonecrus*	—
	Synergus	—
	Synophrus	—
	Synophromorpha	—
	Periclistus	—
	Myrtopsen	—
	Ceroptres	—
	Euceroptres	—
Rhoditini	*Diplolepis* (= *Rhodites*)	Rosaceae
Cynipini[e]	*Neuroterus*	Fagaceae
	Callirhytis	Fagaceae
	Andricus	Fagaceae
	Cynips	Fagaceae
	Trigonaspis	Fagaceae
	Biorrhiza	Fagaceae
	Fioriella	Fagaceae
	Chilaspis	Fagaceae
	Disholcaspis	Fagaceae
	Amphibolips	Fagaceae
	Aphelonix	Fagaceae

[a]In the case of gall inducers the host plant family is given.
[b]*Aylax* group according to Weld (1952).
[c]Placed by Kinsey (1920) in Pediaspini and by Weidner (1968) in Aulacini.
[d]Inquiline group according to Weld (1952) and Synerginae in Krombein *et al.* (1979).
[e]For a complete list of Nearctic genera, see Krombein *et al.* (1979).

galls. Weidner (1968) also considered the gall-inducing Aulacini wasps as a primitive group, which presumably evolved from parasitoid cynipid ancestors. The genera *Diastrophus, Xestophanes Gonaspis* (Aulacini), and *Diplolepis* (= *Rhodites*), all of which occur on Rosaceae, belong to an intermediate group with respect to adult morphology and gall construction. All Cynipini exhibit derived character states: they merely make monothalamous galls (single chambered) on Fagaceae (all except for two species on oaks).

Many gall wasps have complex life cycles (see Chapter 16) referred to as heterogeny. They alternate between a normal bisexual generation in the spring and an agamic generation in the autumn. Heterogony is present in Cynipini, e.g., *Neuroterus, Andricus,* and *Cynips,* and in *Pediaspis* (Aulacini). Contrary to aphids, the host species of the bisexual and agamic generations of gall wasps are the same or they are closely related. Both generations of gall wasps induce galls, but in different tissues of their host plants. As in many other Hymenoptera, some Cynipini genera display thelytokous parthenogenesis. In *Diplolepis,* both arrhenotokous parthenogenesis (females develop from fertilized eggs and males from unfertilized eggs) and thelytokous parthenogenesis (only females develop from unfertilized eggs) occur, whereas in the Aulacini mainly arrhenotokous, usually female-biased, forms are present (Folliot 1964). Thelytokous parthenogenesis is supposed to be the derived type of reproduction and it occurs in groups with many other advanced characteristics.

When all evolutionary trends in the gall wasps are combined, the conclusion is that the most primitive conditions are present in the tribe Aulacini. Furthermore, this tribe is characterized by two phenomena, which are absent from the other tribes. First, all inquiline forms belong to the Aulacini. Because all inquilines are allied with other gall wasps, the inquiline trophic type is probably not a transitional stage between cynipid parasites and gall inducers, but an early offshoot of the gall-inducing Cynipinae. Second, the genus *Aylax,* the only polyphagous genus, belongs to the Aulacini. It might be argued that a polyphagous behavior (at the genus level) represents a primitive condition and that as more advanced genera became specialized they became restricted to one family of host plants.

The trends from primitive to advanced tribes in gall wasps seem to be contradictory with respect to the phylogenetic position of their host plants. Malyshev (1966) concluded that the relatively primitive Aulacini are allied with the advanced Asteraceae, while the advanced Cynipini induce galls on species of the relatively primitive Fagaceae. He therefore disagreed with Kinsey (1920), who proposed the evolutionary trend mentioned above. However, the rule that primitive phytophages are predominant on primitive host plants is only true when the radiations of plants and phytophages coincide in geological time. This does not seem to have happened for angiosperms and cynipid gall inducers. The radiation of angiosperms has to be dated in the Cretaceous (135 million years ago) whereas the gall wasps did not become abundant before the Miocene (25 million years ago) at which time Asteraceae also became abundant (Muller 1981). In other words, all suitable host plants were available during the period of gall wasp radiation. Thus bonds between gall wasps and host plants merely were dependent on the dominance of certain plant genera during the geological period in which the gall wasps radiated.

Oaks belong to the genera that produced the main body of the deciduous forests in the Northern hemisphere during the beginning of the Tertiary (Kovalev 1971) and apparently radiation of gall wasps occurred during or following the period of oak abundance. No less than 69 percent of all Palaearctic gall wasp species are allied with oaks, while the remaining species are scattered among the dicotyledonous angiosperms: (from Bühr's 1965 keys). In the Nearctic, 82 percent of the species occur on oaks (Kinsey 1920). The absence of most oak species from the tropics might also be the explanation for the scarcity of cynipid gall inducers in that region (Docters van Leeuwen-Reijnvaan and Docters van Leeuwen 1926).

Chalcidoidea

Chalcidoidea and Cynipoidea presumably are sister groups. All phytophagous forms of the Chalcidoidea, which are found in various families, have parasitoid progenitors. Apparently phytophagy originated independently in the different groups and in different times during the phylogenetic history. Plant feeders among the Chalcidoidea belong to two different guilds, namely seed or ovule eaters and gall inducers. Many phytophagous species belong to three related, rather primitive families: the Torymidae (= Callimomidae), Eurytomidae, and Agaonidae. Incidentally, species belonging to the Perilampidae, Encyrtidae, Eulophidae, and Pteromalidae also have evolved phytophagous habits. Gahan (1922) presented a detailed list of phytophagous Chalcidoidea. No gall insects are found among the Torymidae; all phytophagous forms feed on ovules (e.g., *Megastigmus* on Pinaceae and Rosaceae). Many *Torymus* species are parasitoids of gall insects and some have been shown to develop on the host larva as well as phytophagously in the gall tissue (Grissell 1976).

All Agaonidae, closely related to Torymidae, are true gall insects, developing in the ovaries of special "gall flowers" present in fig syconia. They not only parasitize their host plants but also pollinate them. The mutual adaptations of figs and fig wasps make them one of the finest examples of symbiosis between plants and insects (Janzen 1979; Wiebes 1979). Various gall-inducing forms are present in the Eurytomidae. All species of *Tetramesa* (= *Harmolita, Isthmosoma, Isosoma*) are phytophagous and all live in stems of Poaceae. Some species make conspicuous galls, whereas others hardly cause distortions (Claridge 1961b). There is a large array of feeding habits in *Eurytoma.* Apart from the parasitoids, many species are zoophagous for only a short period and complete their development phytophagously (Claridge 1961a). Some species live in ovules or seeds but only a few are reported to induce galls [*E. longipennis* (Wlk.) on *Ammophila arenaria; E. orchidearum* (Westw.) on various parts of *Cattleya*].

Some gall-inducing Perilampidae are known from Australia. Various *Trichilogaster* species are allied with *Acacia* (Mayr 1905) and *Perilampoides bicolor* Girault induces galls on leaves of *Eucalyptus* (Girault 1913). *Asparagobius braunsi* Mayr is a gall inducer on *Asparagus* in South Africa (Mayr 1905). Gahan (1922) reported some eulophid species from Australia and New Zealand that presumably also induce galls. They belong to the genera *Rhicnopeltella* and *Zagrammosomoides;* all occur on *Eucalyptus.* Finally, one encyrtid species *Mimapis nigra* Brethes (1916) with galls in branches of *Scutia* (Rhamnaceae) and one pteromalid species *Hemadas nubilipennis* (Ashmead) have to be mentioned. The pteromalid gall inducer is very peculiar. The ovipositing female lays her eggs in vegetative shoots of *Vaccinium* species (Ericaceae). After depositing eggs she damages the tip of the shoot by stabbing it with her ovipositor (Shorthouse *et al.* 1986). The place where the larva develops becomes a kidney-shaped gall (West and Shorthouse 1989). As with the cynipid wasp *Biorhiza pallida* (01.), the gall of *H. nubilipennis* is an example in which physical damage by the ovipositor contributes to initiation of the gall. Apparently this ensures that all nutrients become available to the offspring.

The evolution of phytophagous forms among the chalcidoids might be explained as follows. Many parasitoid Torymidae and Eurytomidae first attack immature host larvae and lay eggs on (or in) second-instar larvae (Askew 1975). Contrary to many other parasitoids of young larval stages, they are not able to delay their larval development until the host larva is mature. Depending on the age of the host larva, the food source is either sufficient for complete development or it is not. The larva then has to continue feeding on gall tissue, the only other available food. The ovipositing female then becomes independent of the host larva, resulting in a completely phytophagous habit.

The Chalcidoidea are an old group, the origin of which may be dated in the Upper Jurassic. However, the earliest fossils are from Canadian Cretaceous amber, between 70 and 90 million years old (Yoshimoto 1975). Because recent figs depend entirely on Agaonidae for their pollination, the fig wasps might belong to the oldest known gall insects. Judging from their distribution, they may originate from the Cretaceous. Nikolskaya (1956) considered the gall-inducing Perilampidae and Eulophidae to be ancient forms, based on the tropical area of distribution of the few recent species and the old age of their host plants (Kovalev 1971). Although Eurytomidae and Torymidae are supposed to be ancient families, the gall inducers evolved on rather modern grasses. In these groups the seed-eating habit might be of ancient origin, since the forms belonging to this trophic type attack the Coniferae and Rosaceae, which are ancient plant groups.

GALL-INDUCING DIPTERA

Gall inducers are found in two suborders of the Diptera. The gall midges of the family Cecidomyiidae belong to the Nematocera. There are about 4300 known species of cecidomyiids, of which about the half are gall inducers (Felt 1940, Skuhravá personal communication). Many plant feeders are present in the Brachycera, but contrary to the Nematocera, gall induction here is exceptional. Some gall inducers are present in the Agromyzidae (Buhr 1955); *Lipara* in the family Chloropidae induces galls in stems of *Phragmites*. *Phorbia* of the family Anthomyiidae makes galls on *Senecio* whereas *Eurosta* and *Urophora* in the family Tephritidae induce galls on Asteraceae (see Chapter 2).

The nematoceran complex of the Bibionomorpha consists of 23 families of which the Mycetophilidae and Cecidomyiidae are the largest (Rodendorf 1964). Apart from the gall-inducing Cecidomyiidae, all other species are detritus feeders in their larval stage. They feed on rotting wood, decomposing plant material, decaying fruit bodies of fungi, or live mycelia. Also, many Cecidomyiidae are mycetophagous, viz. all Lestremiinae, and two tribes of the Cecidomyiinae (Table 3-3). The advanced forms belonging to the Cecidomyiinae and Oligotrophini are gall inducers, as are the major part of Lasiopterini and all Asphondyliini. Many Asphondyliini and some Lasiopterini exhibit a peculiar feature that is reminiscent of the original mycetophagous condition: their gall chambers are covered with a layer of fungal hyphae on which they feed. These are ambrosia galls, analogous with ambrosia beetles, which also feed on their own symbiotic fungi. Hence, phytophagy and gall induction are derived trophic types in the Cecidomyiidae, having evolved from primitive mycetophagy.

Appearance of the gall-inducing guild might have occurred either monophyletically or polyphyletically; the few authors that have dealt

Table 3-3. Tribes and Trophic Types of the Cecidomyiidae, According to Mamaev and Krivosheina (1965) and Mamaev (1975)

Subfamily	Tribe	Trophic type
Lestremiinae	Lestremiini	Mycetophagy
	Moehniini	Mycetophagy
	Micromyiini	Mycetophagy
Cecidomyiinae	Heteropezini	Mycetophagy
	Porricondylini	Mycetophagy
	Oligotrophini	Mycetophagy, phytophagy, gall inducing, zoophagy
	Lasiopterini	Phytophagy, gall inducing
	Cecidomyiini	Mycetophagy, phytophagy, gall inducing, zoophagy
	Asphondyliini	Gall inducing

with gall midge phylogeny disagree about this item. Möhn (1961), who treated only the Asphondyliini in detail, advocated a monophyletic route to gall induction. Inquilinous and mycetophagous forms would have been secondarily developed from gall inducers. Panelius (1965), dealing with Porricondylini, had doubts about a monophyletic route for the other Cecidomyiinae tribes. Mamaev (1965, 1975), who treated this problem in detail, concluded that a polyphyletic passage might be possible. His main arguments (in Roskam 1985) are summarized here.

1. The development of a phytophagous (and later a gall-inducing) feeding mode, whether monophyleticly or polyphyleticly, is possible only after obtaining certain prerequisites in the mycetophagous phase. They must allow change to a complex of phytophagous habits, which differ considerably from mycetophagy. First, the gravid female must alight on a strange substrate such as a large plant. She then has to oviposit and finally the hatched larva has to survive on the new food.

In Figure 3-3 the phylogenetic relations between the Cecidomyiinae tribes are presented, as analyzed from Mamaev (1975, Figs. 45 and 49). The adult Cecidomyiinae are characterized by progressive reductions that are adaptations to passive dispersal. This is in contrast to the Lestremiinae, which are primarily adapted to active flight, as are the other Bibionomorpha

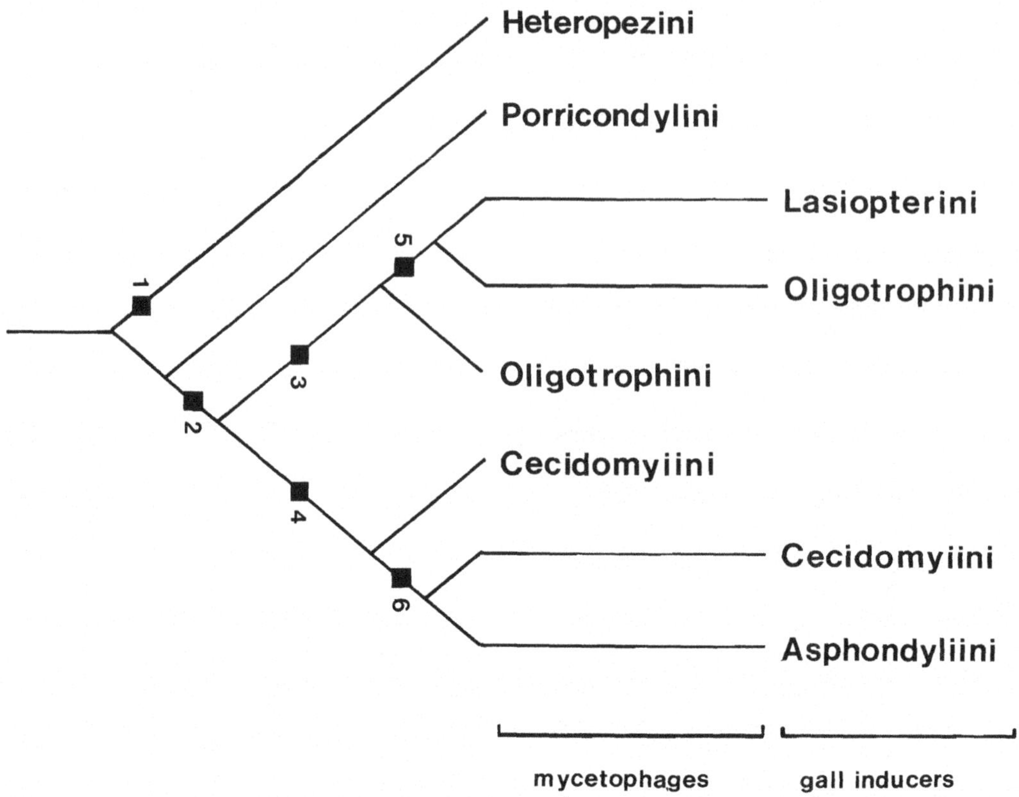

Figure 3-3. Phylogenetic relations between the tribes of Cecidomyiinae. The black squares represent synapomorphous conditions: 1, paedogenesis; 2, simplification of larval intestines connected with extra-intestinal digestion and secondary simplification of the head capsule; 3, antennae with uninodal, stalked flagellomeres; 4, antennae with binodal male flagellomeres and loop-like filaments; 5, appearance of basal outgrowths of basimeres encircling the aedeagus in male genitalia; 6, reduction of the ninth tergite in favor of a lengthened tenth one.

families. Cecidomyiinae have weakly costalized wings, long weak limbs, and fragile antennae. These structures, generally considered unfavorable, enhance the possibilities of an aeroplankton dispersal. The severe loss of individuals due to passive dispersal is compensated by large numbers of eggs per female. With numbers varying between 150 and 250 they are close to the maximum production for many groups of Diptera. Moreover, many forms have their sex ratio shifted in favor of females by an ingenious mechanism (White 1946). The ultimate reduction to lighten the adult body is realized in the Heteropezini, where the number of eggs is diminished (sometimes as low as 2) and the multiplication of individuals is shifted to the larval stage (paedogenesis). All adaptations appear to improve passive dispersal along with increasing the number of individuals.

A number of progressive reductions in the digestive tract are also evident at the larval stage (Fig. 3-4). In the primitive condition, found in the Lestremiinae and the Cecidomyiinae–Porricondylini, the intestines are equipped with blind outgrowths, the mid-gut consists of two sections, there is a well-developed peritrophic membrane, and there are two pairs of Malpighian tubules. In the mycetophagous Cecidomyiini and Oligotrophini, and in the mycetophytophagous Asphondyliini and Lasiopterini, the intestinal tract is more simplified. The cardial sac, peritrophic membrane, and blind outgrowths have disappeared, and there is only one pair of Malpighian tubules.

The reductions of the peritrophic membrane and the cardial sac make extra-intestinal digestion possible. The digestive juices of the mid-gut, to which salivary secretions are added, are

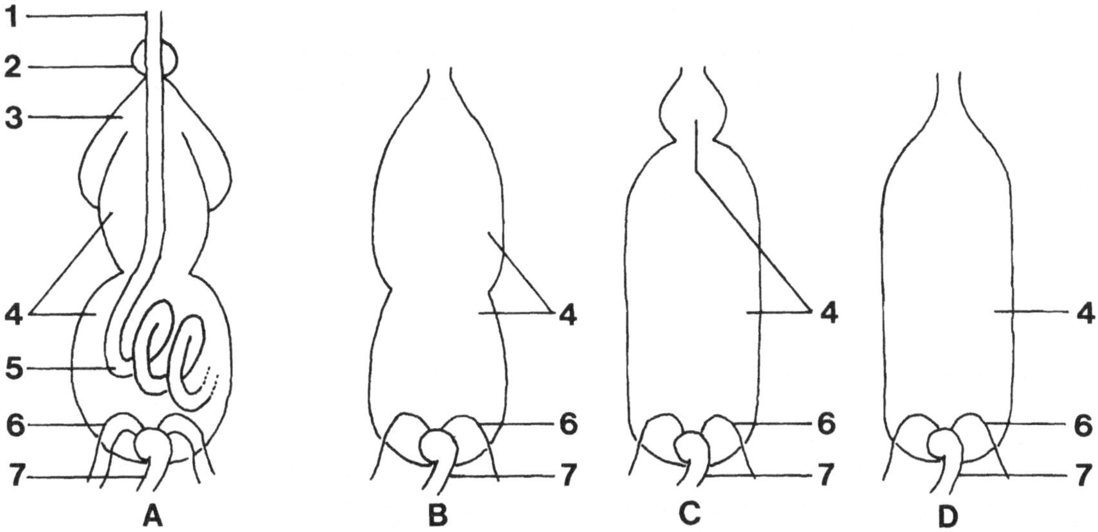

Figure 3-4. Different types of structure of intestines of cecidomyiid larvae: 1, oesophagus; 2, cardial sac; 3, blind appendages of mid-gut; 4, mid-gut; 5, peritrophic membrane; 6, Malphigian vessels; 7, hind-gut; (A) Lestremiinae and Porricondylini; (B) mycetophagous Cecidomyiini and Oligotrophini, phytophagous Asphondyliini and Lasiopterini; (C) phytophagous Oligotrophini; (D) phytophagous and zoophagous Cecidomyiini. (After Mamaev 1975.)

applied by regurgitation onto the food substrate. The food becomes liquified and readily accessible compounds are broken down and easily swallowed. Extra-intestinal digestion is also shown in other groups of insects, although this type of digestion in almost every other group is distinguished by its own characteristics (Wigglesworth 1939). "One may with considerable confidence assert that the intensification of extra-intestinal digestion, which became possible after simplification of the digestive tract, opened the way to the expansion of trophic links; that is, it created conditions for the gall midges to phytophagy and zoophagy" (Mamaev 1975). Hence, progressive reductions gave new possibilities for the biological progress of the group.

2. The radiation of the gall midges coincided with the rise of the angiosperms. Two tissues of the angiosperms (the cambium of broken, damaged stems and flowers) appear to attract special attention as a food substrate for plant-feeding gall midges. Angiosperms are, among other things, characterized by a progressive development of the conducting system, which allows intensive sap streams in their stems. Midges belonging to the Lasiopterini are especially adapted to stem feeding. The transition from mycetophagy to stem feeding is not too difficult to imagine. In the forest litter there is a gradient from almost completely decayed plant material to newly broken branches, stems, or other plant organs. Each phase of degradation has its own saprophages and whole Bibionomorpha families are specialized on particular phases of degradation (Mamaev 1975). *Ledomyia* (Oligotrophini) has an intermediate position between mycetophagy and cambium feeding. The larvae live in colonies in damp rotting wood and the moribund cambial zones of stumps and broken trunks, feeding on fungal mycelium as well as on secreted plant juices. Phytomycetophages also are among the Cecidomyiini. Larvae of the closely related genera *Thomasiana* and *Profeltiella* develop in the damaged, but live cambium of trees and in the decaying cambium of older stumps, respectively.

Many phytophagous gall midges feed on flowers (e.g., 70 percent of the species belonging to *Contarinia,* 53 percent to *Asphondylia,* and 42 percent to *Jaapiella*), whereas some are not able to induce galls. The coincidence of the expansion of gall midges and angiosperms might be due to the development of flowers, which contain readily accessible and highly nutritious substances such as sugars secreted by nectaries. Subsequently, other tissues of high food quality, such as ovaries and fruits, might also be exploited. The transition from mycetophagy to phytophagy might even be facilitated by fungal infections in the distorted flowers. Once the flowers were accepted, other plant organs were exploited as food substrate. Gall induction possibly evolved as a reaction of the host plant to irritation by the larvae, presumably caused by juices excreted during extra-intestinal digestion.

Mamaev's (1975) argument for a polyphyletic origin of gall-inducing cecidomyiids is based on a detailed analysis of larval and adult characters. All forms belonging to the Lestremiinae, as well as those belonging to the Cecidomyiinae, Heteropezini, and Porricondylini, are mycetophagous. Furthermore, in the mycetophagous forms belonging to the Cecidomyiini and Oligotrophini, pleisomorphous character states predominate. Both complexes of gall-inducing Cecidomyiini–Asphondyliini and of gall-inducing Oligotrophini–Lasiopterini possess apomorphous characters of the male genitalia. Gall-inducing and inquiline Oligotrophini and all Lasiopterini share peculiar outgrowths of the gonocoxites (basimeres), which encircle the aedeagus as a sheath. In the male genitalia of gall-inducing Cecidomyiini there is a considerable reduction of the ninth tergite, combined with an intensification and lengthening of the tenth tergite. This apomorphy is shared by all Asphondyliini. Hence, gall-inducing Oligotrophini are closer related to the Lasiopterini than to the primitive mycetophagous Oligotrophini. This means that the Oligotrophini are a paraphyletic group. For similar reasons the Cecidomyiini also are paraphyletic. The gall-inducing Cecidomyiini have a more recent ancestor in common with the Asphondyliini than they have with the mycetophagous Cecidomyiini (Fig. 3-3). A further conclusion is that phytophagy and gall induction independently evolved in Cecidomyiini and Oligotrophini. After the move to phytophagy, a Cecidomyiini form gave rise to the Asphondyliini and an Oligotrophini form to the Lasiopterini.

Although many vascular plants, belonging to all major Orders, bear midge galls, the bulk of the gall-inducing species are found among perennial and woody dicotyledones. Host plants belonging to the relatively primitive Hamamelidae and Rosidae, as well as those belonging to the relatively advanced Asteridae are abundant. Fewer gall midge species are allied with monocotyledones and older conifers and ferns (Table 3-1). These latter groups apparently have been colonized later in geological time.

Wells (1921) introduced evolutionary concepts in gall construction of the various groups of cecidozoa. He pointed out that prosoplasmic galls have arisen from kataplasmic galls. Contrary to gall wasps, where some trends are present, a correlation between the taxonomic position of gall midges and the form of their galls is doubtful. Within the Cecidomyiini and the Oligotrophini both kataplasmas and prosoplasmas are induced. Within certain genera, e.g., *Contarinia* (kataplasma), *Rabdophaga,* and *Rhopalomyia* (prosoplasma), one of the gall types is mostly present. Nevertheless, we have to be careful and much more study is needed to draw evolutionary conclusions concerning gall construction. After all, the gall and its complexity are the result of a coproduction of gall insects and host plants.

There is no correlation either between the state of derivativeness and trophic types such as inquilinism and zoophagy. These trophic types may have evolved from primitive detritus feeding and mycetophagy or from advanced gall inducers. Sometimes different trophic types occur in the same genus and the less frequent types apparently then evolved from the predominant ones. Such a situation is present in the Lasiopterini: three genera, viz. *Exopsyllomyia, Lasioptera,* and *Stephaniola,* together with about 100 Palearctic species, have only six inquilinous species. A similar situation is present in the Oligotrophini. The large genera *Dasineura, Jaapiella,* and *Macrolabis,* together with about 350 Palearctic species, have 17 species with inquilinous habits. *Ledomyia* was mentioned earlier for its intermediate position between mycetophagy and cambium feeding. However, some species of this very diverse and poorly known genus are associated as inquilines and/ or zoophages with gall inducers and scale insects. All five species of *Trotteria* also are inquilinous and/or zoophagous.

The Cecidomyiini are most diverse regarding their feeding modes. Inquilines and gall inducers both occur in the genera *Ametrodiplosis, Anabremia, Contarinia, Diadiplosis, Dicrodiplosis,* and *Parallelodiplosis.* The monotypic genera *Monodiplosis* and *Xenodiplosis* are exclusively inquilinous as is *Tricholaba* (four species). At least 17 genera are exclusively zoophagous, not only on other gall midges (e.g., *Lestodiplosis, Stictobremia*), but also on or in hosts of other arthropod groups, namely mites (e.g., *Acaroletes, Arthrocnodax, Feltiella*), aphids (e.g., *Adelgimyza, Endaphis, Monobremia*), psylloids *(Endopsylla),* and coccoids *(Silvertrina).* Both phytophagous and zoophagous forms are presumably present only in *Kalodiplosis* and *Epidiplosis,* two genera with two species each.

As a rule, we must conclude that inquilinism evolved polyphyletically from gall inducers and that zoophagous forms also evolved polyphyletically. Some genera might be close to mycetophagous forms (e.g., *Acaroletes* to *Mycodiplosis* and *Coquillettomyia*), whereas other genera are presumably closer to gall inducers, or even inquilines.

The oldest fossils of gall midges originated from Lower Cretaceous amber found in Lebanon (Schlee and Dietrich 1970), but these fossils remain undescribed. Gagné (1977) described fossils belonging to Lestremiini and Porriondylini found in Canadian Upper Cretaceous strata (about 73 million years ago). Some of these fossils are more specialized than their modern relatives, which indicates that detritus feeding and mycetophagous cecidomyiids appeared long before the Upper Cretaceous. The transition from mycetophagy to phytophagy presumably was already completed at the end of the Cretaceous (Mamaev 1975). Gagné (1973) reported the earliest fossil Cecidomyiini from Upper Oligocene–Lower Miocene strata. Among these fossils were specimens belonging to *Contarinia,* a genus that contains host-specific gall inducers, along with other forms belonging to predaceous *Lestodiplosis* and mycetophagous *Clinodiplosis.* The latter collections indicate that the cecidomyiid fauna of about 30 million years ago was similar to that found today. The earliest finds of fossil midge galls are apparently of Miocene origin.

Sequoiomyia krauseli Möhn in cones of *Sequoia* was found in brown coal layers, while Marty (1894) reported galls of *Mikiola fagi* Htg. on beech leaves that had a Miocene origin (see Chapter 4).

DISCUSSION

Compared with other herbivorous guilds, gall inducers are of recent origin (Zwölfer 1978). To explain the different patterns of radiation in the various groups of gall inducers, the older feeding modes have been subdivided into primary and derived ones. The primary types are of ancient origin and evolved during the Palaeozoic on primitive green land plants and on fungi. Other trophic types are considerably younger and evolved from the primary types. Many of the derived types are connected with the radiation of the angiosperms in the Upper Cretaceous.

The ability to induce galls has developed in various arthropod Orders in parallel and different patterns are present. For example, the relative abundance (species percentages per category of vascular plants) of gall midges, gall wasps, and sawfly gall makers show some interesting differences (Table 3-1). The relative abundances are rather similar for gall midges and plant species. The gall midges score only very high for the Hamamelidae, as do the other groups of gall inducers (8.3, 69.1, and 13.7 percent of gall midges, gall wasps, and sawflies, respectively, and only 1.3 percent of plant species). The relative abundance of sawflies and gall wasps is not only different from that of the plant species, but also differs among themselves. The gall-inducing sawflies score extremely high for the subclass Dilleniidae, where *Salix* is almost completely responsible for this high abundance. The gall wasps predominate in the small subclass Hamamelidae, with an extremely high occurrence on *Quercus.*

Apparently these three different patterns of host plant relations are the result of combinations of at least three factors, namely the ancestral trophic type of gall midges, gall wasps, and sawfly gall inducers (mycetophagy, zoophagy, and foliage feeding, respectively), the period of radiation in geological time (simultaneous, later or earlier than the radiation of the angiosperms), and, finally, the abundance of some plant taxa, especially at the beginning of the Tertiary. When gall midges shifted from fungi to angiosperms during the Upper Cretaceous, they already possessed prerequisites (extra-intestinal digestion) to induce galls on these plants. Subsequently they followed the radiation of their host plants and in some cases apparent signs of coevolutionary events are present (Roskam 1979; Roskam and van Uffelen 1981; Roskam 1985). The diversity of gall midge species on the geologically older gymnosperms and ferns is low. Moreover, some forms on coniferous hosts possess more derived features than their relatives on angiosperms (Möhn 1960). Apparently, gymnosperms and ferns became colonized later than angiosperms.

There are two groups of gall-inducing cecidomyiids, viz. Cecidomyiini–Asphondyliini and Oligotrophini–Lasiopterini, which independently arrived at the ability to induce galls. There are some indications that these lines are different not only with regard to morphological features but also with regard to host plant adaptation. Large and presumably old genera belonging to the Oligotrophini *(Dasineura, Jaapiella,* and *Macrolabis)* predominate in vegetative tissues, whereas Cecidomyiini *(Contarinia)* and Asphondyliini *(Asphondylia)* prefer flowers and fruits (compilated from Buhr, 1965). This puzzling topic needs further research.

Gall wasps and gall-inducing sawflies have an erratic distribution over the subclasses of their host plants. During the Triassic sawflies radiated as foliage feeders on ferns and gymnosperms and then colonized the angiosperms without changing their feeding mode. They became gall inducers only under particular circumstances. Gall-inducing sawflies are mainly present on *Salix, Betula* and Rosaceae. The radiation of the Cynipidae gall inducers is presumably the most recent one and started after radiation of the angiosperms was completed (Miocene). The primitive forms, Aulacini, are abundant on the advanced, fundamentally herbaceous subclass of Asteridae, whereas the advanced forms (Cynipini) radiated on oaks, which are relatively primitive woody angiosperms. They apparently could radiate in a free space by their ability to live on host plants containing large amounts of tannins, which are repellents for many herbivores. Oaks and willows are examples of taxa that were very abundant in temperate regions of the Northern hemisphere during the beginning of the Tertiary. They produced the main body of deciduous forests that occupied enormous territories. This may have been an important reason for the abundance of gall wasps on oaks and gall-making sawflies on willows. Similar examples are encountered in the Southern hemisphere where, for example, Chalcidoidea gall inducers are abundant on (previously) dominant genera, such as *Eucalyptus, Acacia,* and *Ficus* (Kovalev 1971). Hence, the host plant relations of hymenopterous gall inducers are primarily determined by the dominance of particular host plant species in space and time. The taxonomic diversity seems to be less important; indeed the species richness of gall insects on host plant taxa is determined by many interrelated factors, which seem to work out differently for the various groups of gall insects.

In general, proteins are important nutrients for all phytophages. However, in plants these nutrients are present only in low concentrations and it must be advantageous to the insect to select host plants, or plant tissues, that have relatively high levels of proteins. Apart from protein-rich areas such as seeds, pollen, or special organs in myrmecophilous plants [Beltian bodies (Janzen 1966)], galls provide one of the richest herbaceous substrates with respect to food quality (Llewellyn 1982). Hence, an original defense reaction of the host results in a structure that seems to be extraordinarily beneficial to the phytophage. On the other hand, the phytophage is encapsulated in the galls and this sedentary mode of life restricted the larval morphology to some extremely reductive developments. The gall insect, imprisoned in its "shelter" became a helpless victim for a large diversity of parasitoids and a center of complex insect communities (see Chapter 16). The ecologically low position in the food chain and the vulnerability as highly simplified sedentary organisms (in gall midges this is combined with a passive dispersal) have forced gall insects to produce large numbers of offspring.

Gall induction has not been a blind alley in insect evolution, but instead has given rise to other trophic types. For example, some gall insects have become parasites of other arthropods. If these hosts were gall inducers, either the cecidozoa, or the gall tissues, or both became the target. Inquilinism, which actually is parasitism of the gall tissues, is realized in many cecidomyiid and cynipid genera, but with different results. Either the gall tissue is so excessively present that many commensals can feed without harm to the gall inducer (many gall wasps), or the inquiline, feeding and developing faster than the gall inducer, causes a shortage of nutriment and the gall inducer is eventually killed by starvation (e.g., some inquiline *Dasineura*). Inquilines are sometimes harmful to the gall inducer because they disturb the opening mechanism of the gall and the gall inducer is unable to exit (Mani 1964).

Some gall midges evolved to ecto- and endoparasitoids (e.g., *Endaphis* and *Endopsylla*). Predation and mutualism, two other types of basic interactions among organisms, also have been realized as further derivations of gall-inducing forms (e.g., the gall midges *Lestodiplosis, Aphidoletes,* and *Monobremia*). Mutualistic bonds are encountered in fig insects that offer advantages for the host plant with respect to pollination, but basically remain gall-inducing parasites.

ACKNOWLEDGMENTS

I owe much to the detailed knowledge of and the intensive discussions with Dr. M. Skuhravá and Dr. V. Skukravý of Prague. Furthermore, I am very grateful to Dr. R. J. Gagné of Washington, Dr. A. A. Wiebes-Rijks and Professor J. T. Wiebes of Leiden, Professor B. M. Mamaev of Moscow, Mr. W. C. Nijveldt of Wageningen, and Professor H. Zwölfer of Bayreuth for reading the manuscript and/or providing many valuable comments. Mrs. Stans Kofman abstracted many of the records and prepared the figures.

LITERATURE CITED

Askew, R. R. 1975. The organization of chalcid-dominated parasitoid communities centered upon endophytic hosts. In P. W. Price (ed.), Evolutionary Strategies of Parasitic Insects and Mites, pp. 130–153. Plenum Press, New York.

Brèthes, J. 1916. Hymenoptères parasites de l'Amérique meridionale. An. Mus. Buenos Aires 27:401–430.

Buhr, H. 1955. Minierfliegen als Gallenerzeuger. Nachr.-bl. Bay. Entomol. 4:81–87.

Buhr, H. 1965. Bestimmungstabellen der Gallen (Zoo-und Phytocecidien) an Pflanzen Mittel- und Nordeuropas. I and II. G. Fischer, Jena.

Claridge, M. F. 1961a. An advance towards a natural classification of eurytomid genera (Hym., Chalcidoidea), with particular reference to British Forms. Trans. Soc. Br. Entomol. 14:167–185.

Claridge, M. F. 1961b. A contribution to the biology and taxonomy of some Palearctic species of *Tetramesa* Walker (= *Isosoma* Walk.; = *Harmolita* Motsch.) (Hymenoptera: Eurytomidae), with particular reference to the British fauna. Trans. R. Entomol. Soc. London 113:175–216.

Cockerell, T. D. A. 1890. The evolution of insect-galls. The Entomologist 23:73–76.

Docters van Leeuwen-Reijvaan, J., and W. M. Docters van Leeuwen. 1926. The Zoocecidia of the Netherlands East Indies. Batavia, The Netherlands.

Felt, E. P. 1940. Plant Galls and Gall Makers. Comstock, New York.

Folliot, R. 1964. Contribution a l'Etude de la Biologie des Cynipides gallicoles (Hymenoptères, Cynipoidea). Ann. Sci. Nat. Zool. 12e ser. 6:407–564.

Gagné, R. J. 1973. Cecidomyiidae from Mexican Tertiary amber. Proc. Entomol. Soc. Wash. 75:169–171.

Gagné, R. G. 1977. Cecidomyiidae (Diptera) from Canadian amber. Proc. Entomol. Soc. Wash. 79:57–62.

Gahan, A. B. 1922. A List of phytophagous Chalcidoidea with description of two new species. Entomol. Soc. Wash. 24:33–58.

Girault, A. A. 1913. Australian Hymenoptera V. The family Perilampidae with the descriptions of one new genus and four species. Mem. Queensl. Mus. 2:297–302.

Grissell, E. E. 1976. A revision of Western Nearctic species of *Torymus* Dalman. University of California Publications in Entomology. Univ. of California Press, Berkeley.

Hennig, W. 1966. Phylogenetic systematics. Univ. of Illinois Press, Urbana.

Janzen, D. H. 1966. Coevolution of mutualism between ants and acacias in Central America. Evolution 20:249–275.

Janzen, D. H. 1979. How to be a fig. Annu. Rev. Ecol. Syst. 10:13–51.

Kinsey, A. C. 1920. Phylogeny of cynipid genera and biological characteristics. Bull. Am. Mus. Nat. Hist. 42:357–402.

Königsmann, E. 1976. Das phylogenetische System der Hymenoptera 1: Einführung, Grundplanmerkmale, Schwestergruppe und Fossilfunde. Dtsch. Entomol. Z. N.F. 23:153–179.

Königsmann, E. 1977. Das phylogenetische System der Hymenoptera 2: Symphyta. Dtsch. Entomol. Z. N.F. 24:1–40.

Königsmann, E. 1978a. Das phylogenetische System der Hymenoptera 3: Terebrantes (Unterordnung Apocrita). Dtsch. Entomol. Z. N.F. 25:1–55.

Königsmann, E. 1978b. Das phylogenetische System der Hymenoptera 4: Aculeata (Unterordnung Apocrita). Dtsch. Entomol. Z. N.F. 25:365–435.

Kovalev, O. V. 1971. Host dominance as an evolution of specific parasites. Proc. XIII Int. Congr. Entomol. Moscow 1:260–261.

Krombein, K. V., P. D. Hurd, Jr., D. R. Smith, and B. D. Burks (eds.) 1979. Catalog of Hymenoptera in America North of Mexico. Vol. 1: Symphyta and Apocrita (Parasitica). Smithsonian Institution Press. Washington, D.C.

Küster, E. 1911. Die Gallen der Pflanzen. Hirzel, Leipzig.

Llewellyn, M. 1982. The energy economy of fluid-feeding herbivorous insects. In Insect-Plant Relationships. Proc. 5th Int. Symp. Insect-Plant Relationships, pp. 243–251, Wageningen. Pudoc, Wageningen.

Malyshev, S. I. 1966. Genesis of the Hymenoptera and the Phases of Their Evolution. Methuen, London.

Mamaev, B. M. 1965. Ecological aspects of the evolution of gall midges (Diptera, Itonididae-Cecidomyiidae). Zool. Zh. 43:1161–1172. (In Russian with English summary.)

Mamaev, B. M. 1975. Evolution of gall forming insects—gall midges. Wetherby: The British Library, Lending Division. (Russian original: Leningrad.)

Mamaev, B. M., and N. P. Krivosheina. 1965. Litsinki Galits (Diptera, Cecidomyiidae). Moscow: Izdatel'stvo Nauka. (In Russian.)

Mani, M. S. 1964. The Ecology of Plant Galls. Monographiae Biologicae. W. Junk, The Hague.

Marty, P. 1894. De l'ancienneté de la *Cecidomyia fagi*. Feuille Jeunes Nat. Mulhouse.

Mayr, G. 1905. Hymenopterologische Miscellen IV. Verh. Zool.-bot. Ges. Wien 55:529–575.

Möhn, E. 1960. Eine neue Gallmücke aus niederrheinischen Braunkohle, *Sequoiomyia kraeusel* n.g., n.sp. (Diptera, Itonididae). Senck. Leth. 41:513–522.

Möhn, E. 1961. Gallmücken (Diptera, Itonididae) aus El Salvador 4: Zur Phylogenie der Asphondyliidi der neotropischen und holarctischen Region. Senck. Biol. 42:131–220.

Muller, J. 1981. Fossil pollen records of extant angiosperms. Bot. Rev. 47:1–142.

Nijveldt, W. C. and J. A. Hulshoff. 1968. Observations on the saddle gallmidge in the Netherlands. Mededeling Inst. Plantenziek-tekundig Ond. Wageningen 459:1–77. (In Dutch with English summary.)

Nikolskaya, M. N. 1956. Seed-eating chalcids of the fauna of the USSR and the role of phytophagy in the evolution of the group (Hymenoptera, Chalcidoidea). Entomol. Obozr. 35:570–581. (In Russian with English summary.)

Panelius, S. 1965. A revision of the European gall midges of the subfamily Porricondylinae (Diptera, Itonididae). Acta Zool. Fenn. 113:1–157.

Rodendorf, B. B. 1964. Historical development of dipteran insects. Trudý paleont. Inst. Akad. Nauk SSSR. Moscow 100:1–312 (In Russian.)

Root, R. B. 1973. Organization of plant-arthropod association in simple and diverse habitats: The fauna of collards *(Brassica oleracea)*. Ecol. Monog. 43:93–124.

Roskam, J. C. 1979. Biosystematics of insects living in female birch catkins. II: Inquiline and predaceous gall midges belonging to various genera. Neth. J. Zool. 29:283–351.

Roskam, J. C. 1985. Evolutionary patterns in gall midge-host plant associations (Diptera, Cecidomyiidae). Tyoschv. Entomol. 128:193–213.

Roskam, J. C. and G. A. van Uffelen. 1981. Biosystematics of insects living in female birch catkins. III: Plant-insect relation between white birches, *Betula* L., Section Excelsae (Koch) and gall midges of the genus *Semudobia* Kieffer (Diptera, Cecidomyiidae). Neth. J. Zool. 31:533–553.

Ross, H. H. 1965. A Textbook of Entomology, 3rd. ed. Wiley, New York.

Rothmaler, W. 1972. Exkursionsflora fur die Gebiete der DDR und der BDR (Gefäszpflanzen). Volk and Wissen Verlag, Berlin.

Schlee, D. and M. G. Dietrich. 1970. Insektenführender Bernstein aus der Unterkreide des Libanon. Neues Jahrb. Geol. Palaontol. Monatsch. 40–50.

Shorthouse, J. D. 1982. Resource exploitation by gall wasps of the genus *Diplolepis*. In Insect-Plant Relationships. Proc. 5th Int. Symp. Insect-Plant Relationships, pp. 193–198, Wageningen. Pudoc, Wageningen.

Shorthouse, J. D., A. West, R. W. Landry, and P. D. Thibodeau. 1986. Structural damage by female *Hemadas nubilipennis* (Hymenoptera: Pteromalidae) as a factor in gall induction on lowbush blueberry. Can. Entomol. 118:249–254.

Smart, J., and N. F. Hugues. 1972. The insect and the plant: Progressive palaeoecological integration. In van Emden (ed.), Insect/Plant Relationships, pp. 143–155. Blackwell, Oxford.

Southwood, T. R. E. 1972. The insect/plant relationship—an evolutionary perspective. In Van Emden (ed.), Insect/Plant Relationships, pp. 3–30. Blackwell, Oxford.

Telenga, N. A. 1952. Origin and evolution of parasitism in Hymenoptera Parasitica and development of their fauna in the USSR. Jerusalem: Israel Program for Scientific Translations. (Russian original: Kiev.)

Weidner, H. 1968. Zur Kenntnis der Gallwespentribus Aulacini (Hymenoptera, Cynipidae). Entomol. Zeitschr. 78:105–120.

Weld, L. H. 1952. Cynipoidea. Privately published. Ann Arbor, MI.

Wells, B. W. 1921. Evolution of zoocecidia. Bot. Gaz. 71:358–377.

West, A. and J. D. Shorthouse. 1989. Initiation and development of the stem gall induced by *Hemadas nubilipennis* (Hymenoptera: Pteromalidae) on lowbush blueberry, *Vaccinium angustifolium* (Ericaceae). Can. J. Bot. 67:2187–2198.

White, M. J. D. 1946. The cytology of the Cecidomyiidae (Diptera) II: The chromosome cycle and anomalous spermatogenesis of *Miastor.* J. Morphol. 79:323–369.

Wiebes, J. T. 1979. Co-evolution of figs and their insect pollinators. Annu. Rev. Ecol. Syst. 10:1–12.

Wigglesworth, V. B. 1939. The Principles of Insect Physiology. Methuen, London.

Yoshimoto, C. M. 1975. Cretaceous chalcidoid fossils from Canadian amber. Can. Entomol. 107:499–528.

Zwölfer, H. 1978. Mechanismen und Ergebnisse der Co-evolution von phytophagen und entomophagen Insekten und höheren Pflanzen. Sonderbd. naturwiss. Ver. Hamburg 2:7–50.

4
Fossil Galls

H. G. LAREW

Present-day galls are physical records of plant interactions with other organisms. Intimacy of the interactions is indicated by gall morphology and by host and organ specificity exhibited by the gall inducer. Presumably, then, a *fossil* gall can indicate a great deal about the history of both cecidogenetic disruption and stability of interactions.

However, the full potential of fossil gall material has yet to be realized for several reasons. First is the fact that almost all known fossil galls are two-dimensional impressions in which neither gall inducer nor gall structure is preserved in detail. The possibility exists, however, that information on the chemical nature of impression galls may be obtainable; Giannasi and Niklas (1981) successfully extracted taxonomically useful flavonoids from fossil leaf impressions. Second, the identity of the fossil host plant may be unclear and this decreases the usefulness of specimens. Descriptions of specimens or specimens themselves are often not of the detail and clarity needed to distinguish a gall from an artifact of preservation. Lastly, there is a scarcity of described specimens.

Even with these limitations, fossil galls that have been described suggest a great deal about the gall-inducing guild. In Chapter 3, Roskam uses evidence primarily from the extant fauna to discuss the evolution of gall-inducing insects. The purpose of this chapter is to review literature and observations made on fossil galls and to assess the significance of these paleobotanical specimens.

HIGHLIGHTS FROM THE FOSSIL GALL RECORD

Published accounts of fossil galls are arranged in Table 4-1 by age of specimens. Host plant and gall inducer are listed if either is known or suggested by the describer. This updates earlier listings (Buhr 1964/1965; Küster 1910; Larew 1986; Trotter 1903).

Several general and specific observations about the fossil gall record may be made from a study of Table 4-1.

The Oldest Gall

The oldest gall known was described from sections of extinct *Stigmaria* [the roots of the arborescent *Lepidodendron* (Order: Lepidodendrales)] taken from the Halifax Hard Beds in England (Weiss 1904). The specimens are of Upper Paleozoic–Triassic (200–300 million years ago). Root tissue disruption, as seen in sectioned material, resembles that caused by the extant fungus, *Urophlyctis* (Order: Chytridiales). Weiss proposed the name *Urophlyctites stigmariae* (Order: Chytridiales) for the fossil gall inducer. Interestingly, no fungal structures were observed in the root sections. The evidence nonetheless suggests that a *Urophlyctis*-like fungus survived the extinction of *Stigmaria* to flourish today on the roots of *Medicago* (Family: Leguminosae).

The Oldest Arthopod Galls

What may be the oldest known arthropod-induced galls were discussed by Potonié (1893). He suggested that there was a similarity between extant mite, insect, and fungal galls, and the "runzel" (wrinkled) galls found on the fossil seed fern, *Odontopteris osmundaeformis* (Schloth.) (Order: Pteridospermae) from the Permian (280 million years ago) Rothliegenden formation of Thuringen, East Germany. Figures of the "runzel" galls accompanying Potonié's article depict unusual pinnule deformations that resemble concentric mines. Thus, these important specimens should be reexamined before being considered definitive and photographs of them should be published.

The Oldest Insect Galls

The oldest suspected insect galls were described by Grauvogel-Stamm (1978) from the early Triassic (225 million years ago) in France on cone stalks of *Aethophyllum stipulare* Brogniart (Order Coniferales) Fig. 4-1, 4-1a, 4-1a', 4-1b). Stalk swellings in these specimens resemble galls induced by present day beetles, sawflies, midges, flies, bacteria, or fungi. Thus, identity of the inducer is not clear. The specimens, however, highlight the fact that structurally simple galls, probably induced by insects, were present in the Triassic.

Hickey and Doyle (1977) described what they believed to be the oldest insect galls (Upper Cretaceous of Maryland, U.S.A.; 115 million years ago). In size and imprint structure the galls resemble relatively complex cynipid spangle galls on extant oaks [e.g., small, circular impressions with a hard central capsule (Figs. 4-2 and 4-3)], but were found on the palmately lobed leaves of *Sassafras potomacensis* Berry. Extant *Sassafras* (Family: Lauraceae) bears no such galls. The host plant may have in fact been a primitive member of the Hamamelididae or Rosidae (L. J. Hickey, personal communication). This is a rich collection; a total of 49 gall impressions are available in the Hickey and Doyle material at the U.S. National Museum (Larew, unpublished data). Uncertainty about the host plant makes it difficult to comment on longevity of the galling relationship. The specimens, however, are significant in their suggestion that structurally sophisticated galls were present at a relatively early date.

Möhn (1960) mentioned sawfly-like galls on willow *(Salix)* leaves from the Upper Cretaceous in West Germany, but did not figure or describe them.

The Most Valuable Specimens

In many respects the most valuable fossil galls are those found in the Tertiary ["Miocene(?)": 25 million years ago] lignites of the Duren mines, Rheinland, West Germany (Möhn 1960). These are the only preserved galls known to contain the gall inducer. Möhn found larval and pupal cecidomyiids in the cone seeds of the fossil plant, *Sequoia langsdorfii* (Brogn.) Heer (Family: Taxodiaceae), and erected the new cecidomyiid genus and species, *Sequoiomyia krauseli* (Möhn) for his fossil midges. As far as is known, extant *Sequoia* are not hosts to seed-galling midges. The closely related tree, *Taxodium distichum* Rich. (Family: Taxodiaceae), however, does host seed midges. Because of morphological similarities to fossil midges, Gagné (1968) named the *Taxodium*-attacking midge *Sequoiomyia taxodii* (Felt). Younger (Pleistocene) midge-induced seed galls on *Taxodium imbricatum* were described by Berry (1924). Gagné (1968) suspected that these galls were also caused by the extant *S. taxodii*. Thus the cone galls provide one of the richest collections of fossil galls and allow for a partial reconstruction of events; *S. krauseli* apparently did not survive the reduction in *Sequoia*'s range. It was survived, however, by a sister species on the related host plant, *Taxodium*.

The Largest Collection

Straus's (1977) record of 34 impressions of galled leaves from the Pliocene Willershausen formation in West Germany is remarkable for size and richness. The collection includes 2 fungal, 12 eriophyoid, 1 eriosomatid, 1 aphid, 1 adelgid, 6 cecidomyiid, and 3 cynipid galls. In 13 cases, Straus confidently ascribes gall inducers to extant genera or species; apparently, genera of modern day gall-inducing fungi, eriophyoids, eriosomatids, and cynipids were present 13 million years ago on host plant genera and organs on which they occur today. The richness of the eriophyoid fauna stands out in the Willershausen; perhaps the mites were the most prevalent gall inducers in the German Pliocene (but see eriophyoid galls below). Confirmation of Straus's identifications by other workers is, as yet, unavailable.

Two Noteworthy Specimens

A few described specimens deserve individual comment. For example, a clear imprint of a gall at the petiole–leaf junction on *Populus latior* A.

Table 4-1. Published Accounts of Fossil Galls

Host plant	Host organ	Gall inducer	Age	Place	Source
Pre-tertiary					
Stigmaria	Root	Fungus: *Urophlyctites*	Upper Paleozoic to Triassic	Halifax, England	Weiss, 1904
Odontopteris	Leaf	Arthropod?	Lower Permian	Thuringen, East Germany	Potonié, 1893
Aethophyllum	Stem	Arthropod?	Triassic	Alsace, France	Grauvogel-Stamm, 1978
Acer, Credneria, etc.	Leaves	Mites	Jurassic and Cretaceous	?	Rübsaamen, 1929[a]
"*Sassafras*" *potomacensis*	Leaves	Insect?	Upper Cretaceous	Stump Neck, Maryland, USA	Hickey and Doyle, 1977
?	?	"Galla quercina"	Upper Cretaceous	Ellsworth County, Kansas, USA	Lesquereux, 1892; Berry, 1923
Tertiary					
Fagus	Leaf	*Mikiola fagi*	Tertiary	?	Rübsaamen, 1929[a]
Populus	Leaf	*Cecidomyia*	Tertiary	Oeningen, Switzerland	Heer, 1855 (in Trotter, 1903; Madler, 1936)
Sarothamnus	Leaf	*Asphondylia mayeri*	Tertiary	?	Rübsaamen, 1929[a]
Cedrela puryearensis	Petiole	Cynipid or hemipteran	Lower Eocene	Puryear, Tennessee, USA	Berry, 1916
Rhamnus	Leaf	Cecidomyiid or aphid	Lower Eocene	Puryear, Tennessee, USA	Berry, 1916
Cupanites formosus	Leaf	Stylet-bearer	Lower Eocene	Puryear, Tennessee, USA	Brooks, 1955
Nectandra pseudocoriacea	Leaf	Arthropod	Lower Eocene	Puryear, Tennessee, USA	Brooks, 1955
Quercus pseudolyrata	Leaves	?	Oligocene	Eagle Creek, Oregon, USA	Chaney, 1920[a]
Juglans acuminata(?)	Leaf	*Cecidomyia? dubia*	Upper Oligocene	?	Heyden, 1860
Daphnogene bilinica	Leaf	*Eriophyes daphnogene*	Upper Oligocene	Baromallas, Hungary	Ambrus and Hably, 1979
?	?	Cynipid	Oligocene–Miocene	Florissant, Colorado, USA	Scudder, 1886
Myrica drymeja	Leaf	*Cecidomyia* (?) *pontaniiformis*	Oligocene–Miocene	Florissant, Colorado, USA	Cockerell, 1908
Myrica obscura	Leaf	*Andricus myricae*	Oligocene–Miocene	Florissant, Colorado, USA	Brues, 1910
Quercus cognatus	Leaf	?	Lower Miocene	Douglas County, Washington, USA	Hoffman, 1932
Sequoia langsdorfii	Cones	*Sequoiomyia krauseli*	Miocene?	Duren, West Germany	Möhn, 1960
Many genera	Leaves	Eriophyoids?	Mostly Miocene	Europe	Meschinelli, 1892[b]
?	Leaves	*Pemphigus*	Upper Miocene	Oeningen, Switzerland	Hoernes, 1886[a]
Populus	Leaf	*Pemphigus*	Upper Miocene	Oeningen, Switzerland	Mädler, 1936
Acer osmonti	Leaf	*Cecidomyia ocellaris*	Upper Miocene	Bridge Creek, Oregon, USA	Cockerell, 1927
Alnus carpinoides	Leaf	*Eriophyes laevis*	Upper Miocene	Bridge Creek, Oregon, USA	Cockerell, 1927
Fagus pliocenica	Leaf	*Cecidomyia fagi*	Pliocene	Pas-de-la-Mougudo, France	Marty, 1894
Quercus	Leaf	*Neuroterus*	Pliocene	Laaer Berges, Austria	Berger, 1949
Parrotia	Leaf	*Taphrina?*	Pliocene	Willershausen, W. Germany	Straus, 1977
Crataegus	Leaf	*Gymnosporangium*	Pliocene	Willershausen, W. Germany	Straus, 1977
Tilia	Leaf	*Eriophyes lateannulatus?*	Pliocene	Willershausen, W. Germany	Straus, 1977
Tilia	Leaf	*E. exilis* or *Thyphlodromus*[c]	Pliocene	Willershausen, W. Germany	Straus, 1977
Amelanchier or *Sorbus*	Leaf	*Aceria aroniae*	Pliocene	Willershausen, W. Germany	Straus, 1977
Juglans	Leaf	*Aceria erinea*	Pliocene	Willershausen, W. Germany	Straus, 1977

Carya	Leaf	*A.e.f. caryae-pterocaryae*	Pliocene	Willershausen, W. Germany	Straus, 1977
Alnus	Leaf	*A. inangulis*	Pliocene	Willershausen, W. Germany	Straus, 1977
Betula	Leaf	*A. leionota*	Pliocene	Willershausen, W. Germany	Straus, 1977
Fagus	Leaf	*A. nervisequa faginea*	Pliocene	Willershausen, W. Germany	Straus, 1977
Acer	Leaf	*A. macrochela?*	Pliocene	Willershausen, W. Germany	Straus, 1977
Fagus	Leaf	*A. nervisequa nervisequa*	Pliocene	Willershausen, W. Germany	Straus, 1977
Fagus	Leaf	*A. stenaspis stenaspis*	Pliocene	Willershausen, W. Germany	Straus, 1977
Carpinus	Leaf	*A. tenella*	Pliocene	Willershausen, W. Germany	Straus, 1977
Ulmus	Leaf	*Schizoneura ulmi*	Pliocene	Willershausen, W. Germany	Straus, 1977
Prunus	Leaf	*Myzus?*	Pliocene	Willershausen, W. Germany	Straus, 1977
Conifer	Needle	*Dreyfusia*	Pliocene	Willershausen, W. Germany	Straus, 1977
Carpinus	Leaf	*Dasyneura ruebsaameni?*	Pliocene	Willershausen, W. Germany	Straus, 1977
Fraxinus	Leaf	*D. fraxini?*	Pliocene	Willershausen, W. Germany	Straus, 1977
Quercus	Leaf	*Phegomyia fagiocola?*	Pliocene	Willershausen, W. Germany	Straus, 1977
Betula	Leaf	*Contarinia carpini?*	Pliocene	Willershausen, W. Germany	Straus, 1977
Tilia	Leaf	*Didymomyia reaumuriana?*	Pliocene	Willershausen, W. Germany	Straus, 1977
?	Leaf	*Mikiola fagi?*	Pliocene	Willershausen, W. Germany	Straus, 1977
Quercus praeerucifolia	Leaf	*Neuroterus albipes*	Pliocene	Willershausen, W. Germany	Straus, 1977
Q. praeerucifolia	Leaf	*Andricus curvator?*	Pliocene	Willershausen, W. Germany	Straus, 1977
Q. praeerucifolia	Leaf	*Andricus quercus-radicis*	Pliocene	Willershausen, W. Germany	Straus, 1977
Q. praeerucifolia	Leaf	*Andricus* sp.?	Pliocene	Willershausen, W. Germany	Straus, 1977
Quaternary					
Cupressus	Leaves	Chalcididae	Pleistocene	Santa Cruz Island, California, USA	Chaney and Mason, 1934[d]
Taxodium imbricatum	Seed	*Sequoiomyia taxodii*	Pleistocene	Washington, D.C., USA	Berry, 1924; Gagné, 1968
Moniliform Taxodium	?	Cecidomyiids	Pleistocene	Maryland, USA	Berry, 1923
Quercus agrifolia	Acorn	*Callirhytis*	Pleisto–Holocene	Los Angeles, California, USA	Larew, 1987b
Quercus	Leaf	*Adlevia kollari*	Holocene	Herculanum, Italy	Larew, 1987a
Quercus	?	*Cynips*	Holocene	Southern England	Cunliffe, 1976
Quercus	?	*Biorhiza*	Holocene	Chew, England	Rahtz and Greenfield, 1977
Quercus	?	*Neuroterus*	Holocene	Abindon, England	Robinson, 1980
Quercus	?	*Andricus*	Holocene	Abindon, England	Robinson, 1980

[a]Casual mention of fossil galls.
[b]A checklist of several specimens previously described by others.
[c]May be an acarodomatium.
[d]According to the authors, "similar galls have been found in the Pleistocene deposits at Carpinteria and Millerton, California."

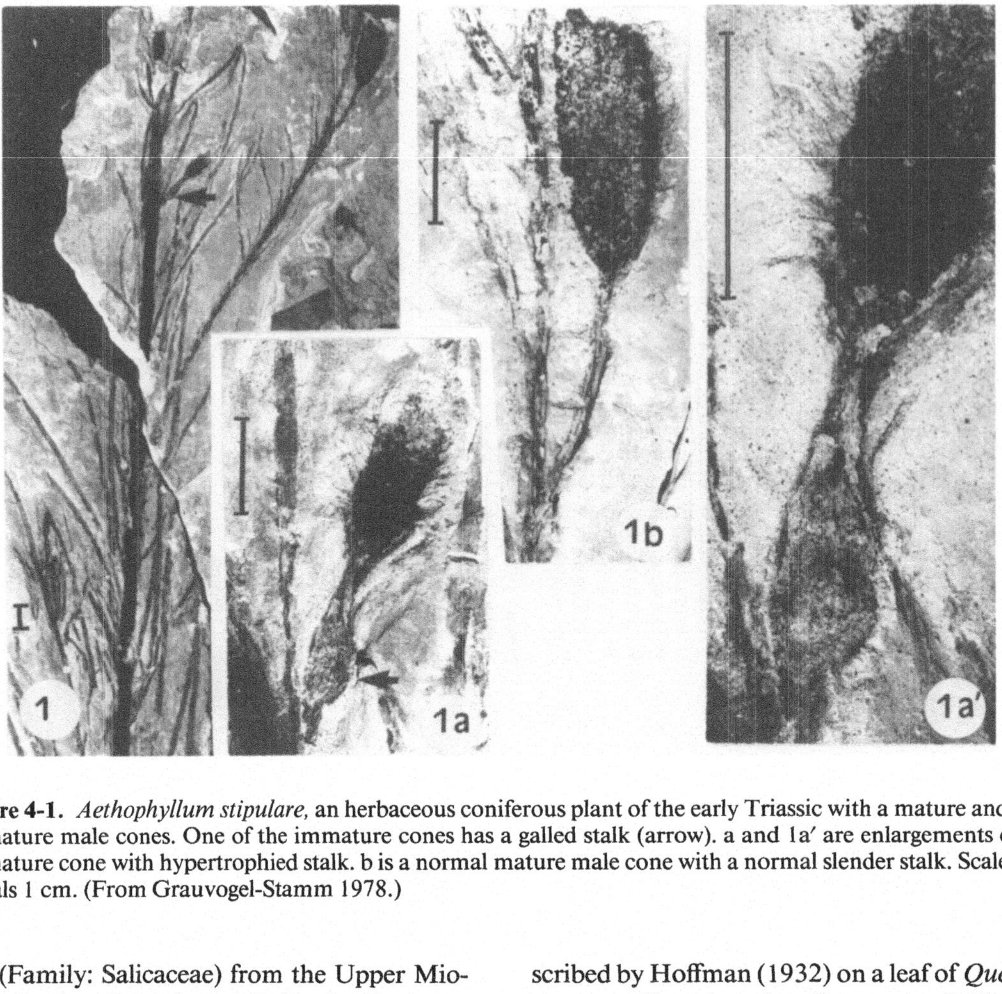

Figure 4-1. *Aethophyllum stipulare,* an herbaceous coniferous plant of the early Triassic with a mature and two immature male cones. One of the immature cones has a galled stalk (arrow). a and 1a′ are enlargements of an immature cone with hypertrophied stalk. b is a normal mature male cone with a normal slender stalk. Scale line equals 1 cm. (From Grauvogel-Stamm 1978.)

Br. (Family: Salicaceae) from the Upper Miocene lignite of Oeningen, Switzerland was ascribed to a *Pemphigus* aphid (Mädler 1936). Strikingly similar galls are induced today by gracillariid moths in the genus *Ectodoemia.* The fossil indicates that in the Miocene, galls similar to extant ones were found at a rather specific spot on the plant. The specimen also serves to emphasize the possible importance of any irregularities in fossils of *Populus* petioles. Because this is a common location for many extant galls, fossil specimens showing unusual (e.g., curled, thickened) petioles, such as described in Smith (1932), should be closely examined.

Berger (1949) pointed out the similarities between circular depressions on imprints of an oak leaf from the Pliocene of Laaer Berges, Austria and the spangle gall induced by the extant cynipid *Neuroterus nummismalis* Oliv. Similar examples of possible spangle galls were described by Hoffman (1932) on a leaf of *Quercus cognatus* Knowlton (= *Q. payettensis* Knowlton; LaMotte 1952) from lower Miocene shales in Washington State, U.S.A. In this last specimen there are 25 galls on the leaf, a pattern of abundance similar to that of spangle galls on extant oak leaves.

Eriophyoid Galls

Considering that it is sometimes difficult to distinguish between present day mite- and fungal-induced galls (blisters, erinea, etc.), it is not surprising that uncertainty exists about the causative agent of small galls on fossil leaves. For example, in order to be correctly assigned, the many "entomocecidia" presumably caused by *Phyllerites* [listed under "entomocecidia" by Meschinelli (1892) but as fungal by Tiffney and Barghoorn (1974)] should be reexamined. Small depressions in fossil leaf imprints often

Figures 4-2–4-3. Figure 4-2. Gall impressions on *Sassafras potomacensis* leaves from the collection of Hickey and Doyle (1977) in the U.S. National Museum, Washington, D.C. Paleobotanical specimen 7316S. Several galls (arrows) are visible. Scale line equals 5 mm. **Figure 4-3.** Gall impression on *Sassafras potomacensis* from the collection of Hickey and Doyle (1977) in the U.S. National Museum. Paleobotanical collection specimen 7316D. Impressions of outer rim (lower arrow) and central circular capsule (upper arrow) are visible. Scale line equals 1 mm.

have been considered eriophyoid galls (Ambrus and Hably 1979; Cockerell 1908, 1927; Straus 1977). The burden is on the describer to clearly indicate why such depressions should not be considered fungal galls or thrips galls or artifacts.

Temperate and Subtropical Fossil Galls

Several temperate plant genera (e.g., *Acer, Quercus, Alnus,* and *Salix*) that host fossil leaf galls have an extant gall fauna (Buhr 1964/1965; Felt 1965; Houard 1922/1923, 1933, 1940). Subtropical plant genera with fossil galls [e.g., *Cedrela* (Family: Meliaceae) and *Myrica* (Family: Myricaceae)] do not have extensive extant gall faunas. Furthermore, the fossil galls on these plant genera do not resemble extant galls on these genera (Houard 1922/1923, 1933). This suggests that since mid-Cretaceous, gall faunas have been more stable on plant genera of temperate climates than on those of tropical climates. A possible exception to this pattern was described by Brooks (1955). He observed 11 leaf galls with short stalks (constrictions) on *Nectandra pseudocoriacea* Berry (Family: Lauraceae), a subtropical plant. Brooks compared these to galls caused by mites, psyllids, gall midges, and gall wasps. Houard (1933) listed several stalked midge galls on leaves of *Nectandra* from Central and South America. If the fossil galls were induced by gall midges, then we have one example of galls of similar type occurring in the sub-tropics of both past and present.

Quaternary Galls

There is an unexplained scarcity of Quaternary (recently preserved) specimens. The fossil gall record is richest in the Tertiary. Undoubtedly as archeological sites continue to be excavated, more fossil galls, such as those found in a fourth-century Roman well in Britain by Robinson (1980), will be discovered. Oak galls preserved in the eruption of Mt. Vesuvius highlighted the significance of galls in ancient commerce (Larew 1987a).

Unrepresented Gall Inducers and Galled Plant Parts

Galls of some extant gall-inducing Orders of insects are not represented in the fossils. There are, for example, no known fossil galls induced by Coleoptera, Lepidoptera, or Thysanoptera. There are no fossil galls that were clearly induced by eriococcids, tenthredinids (see Trotter 1903 "Imenotterocecidii"), nematodes, or adelgids. Furthermore, fossil galls of vegetative stems and of flowers have not yet been found. Several possible explanations as to why some gall-inducing taxa and plant parts are unrepresented in fossils can be suggested. For example, these types of galls may have been rare, as some are today, and thus their chances of being preserved were slight. Perhaps fossil galls that have been ascribed to the mites or cecidomyiids were in fact induced by members of other taxa. In some cases the lack of representation most likely reflects the relatively small paleobotanical collections assembled from parts of the world where certain gall-inducing taxa are presently prevalent. For example, a search through plant fossils from Australia, India, China, or Africa should be profitable for the fossil gall hunter. The present day gall fauna in Australia and India is rich (Gullan 1984; Ananthakrishnan 1978), but no specimens of fossil galls are presently known from these regions.

"Unnoticed" Fossil Galls

Undoubtedly more fossil galls will be described from new field collections. However, there probably are many specimens in existing paleobotanical collections that have not been recognized as galls. For example, while studying the fossil seeds of the Rancho La Brea Tar Pits, Templeton (1964) extracted a specimen from the skull of a sabre-toothed lion that she described as a fossil acorn of *Quercus agrifolia* Nee "showing insect infestation." This acorn (Fig. 4-4) was estimated by Templeton to be between 13,000 and 15,000 years old (Late Pleistocene–Early Holocene). I found another damaged acorn (Fig. 4-5) in the Templeton collection at the G. C. Page Museum (Los Angeles, California, U.S.A.) that shows similar damage (Larew 1987b). There is a strong similarity between these damaged fossil acorns and acorns galled by the extant cynipid *Callirhytis milleri* Weld. The damage by *C. milleri* (Fig. 4-6) consists of a "compact stony-hard mass containing four to a score or more confluent cells . . . more or less filling the interior of the acorn, which is frequently reduced in size" (Weld 1922). It is this woody mass that has been preserved in the La Brea specimens.

Chaney (1920) reported that "leaf galls are present on a number of specimens" of *Quercus pseudo-lyrata* Lesquereux taken from the Eagle Creek formation (Oligocene) of Oregon, U.S.A. Because he collected 1600 leaves of this species his "a number of" becomes significant; however, these galls remain unstudied. Undoubtedly there are many other unnoticed galls in museum collections. The number of useful fossil galls will surely increase as paleobiologists take an interest in them and learn to search for them.

CONCLUSIONS

If we assume that gall inducers presently attack the same plant genera as they did in times past, then it is clear that fossil material of some plant genera needs more careful examination for fossil galls. Of the 10 extant plant genera *(Acacia, Acer, Eucalyptus, Eugenia, Ficus, Populus, Rhus, Salix, Quercus,* and *Vitis)* galled by at least 10 species of insects (Houard 1922/1923, 1933, 1940) 2, namely *Acacia* (Family: Leguminosae) and *Eugenia* (Family: Myrtaceae), are poorly represented as fossils of any kind in North America (Knowlton 1898). Thus the likelihood of finding fossil galls on these genera from this region is slight. Four genera, *Acer* (Family: Aceraceae), *Populus* (Family: Salicaceae), *Salix* (Family: Salicaceae), and *Quercus* (Family: Fagaceae), are represented by abundant North American fossil remains (Knowlton 1898) and have provided several fossil galls. We should continue to search fossil material of these genera for additional galls. Four genera, *Eucalyptus* (Family: Myrtaceae), *Ficus* (Family: Moraceae), *Rhus* (Family: Anacaridaceae), and *Vitis* (Family: Vitaceae) are represented by numerous Cretaceous and Tertiary fossil specimens from North America, but have provided few if any fossil galls (Meschinelli's *Phyllerites*

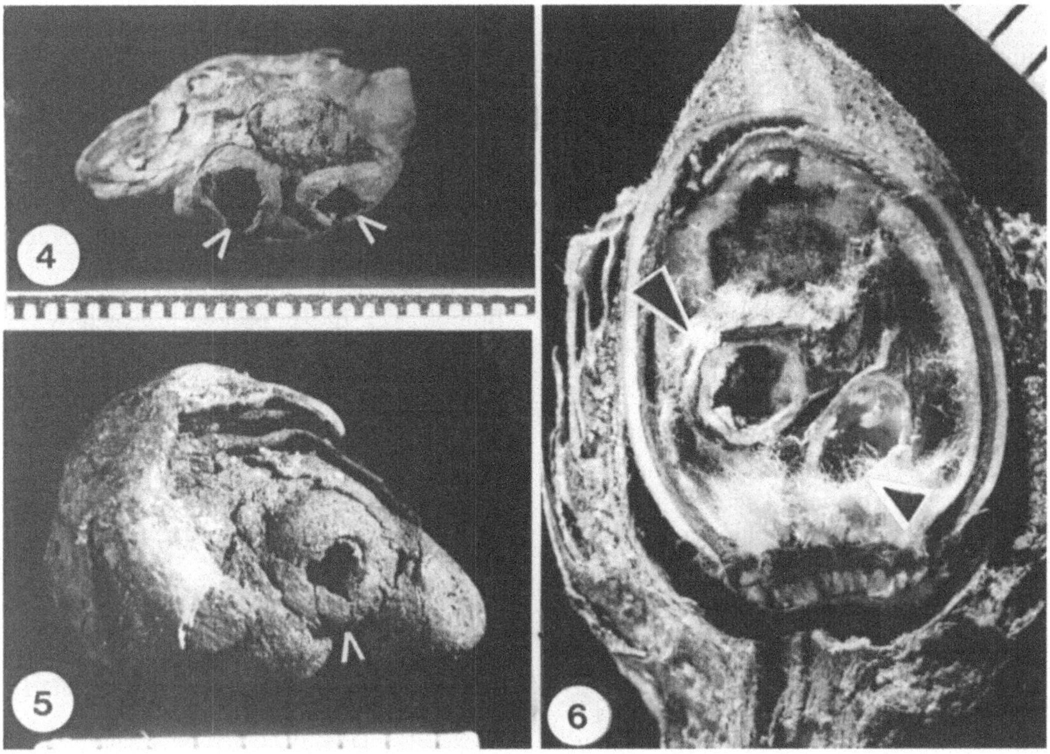

Figures 4-4–4-6. Figure 4-4. A fossilized galled acorn of *Quercus agrifolia* from the La Brea Tar Pits. Pleistotype Number PB 1481, G. C. Page Museum, Los Angeles, California. Note the two opened larval chambers (arrows). Scale line in millimeters. (Photograph reprinted with permission of the Entomological Society of Washington.) **Figure 4-5.** A fossilized galled acorn of *Quercus* sp. from the La Brea Tar Pits. Specimen Number 1418B, Pit 3 (2001/495), G. C. Page Museum, Los Angeles, California. Note opened larval chamber (arrow). Scale in millimeters. **Figure 4-6.** Acorn of *Quercus wislizenii* A. DC. galled by *Callirhytis milleri* Weld. Note exposed larval chambers (arrows). Specimen from the California Academy of Sciences, Golden Gate Park, San Francisco, California. Scale line in millimeters.

excepted; see Trotter, 1903). Fossil material of these four genera should be carefully searched.

The fossil gall record is sketchy and conclusions drawn from it should be considered tentative. However, the record does suggest that as early as the Permian, cecidogenetic relationships existed that resemble those that occur today. The gall-inducing habit in insects, although possibly seen as early as the Triassic, apparently developed to some richness in mid-Tertiary with most examples coming from the Pliocene. From the Pliocene indications are that relationships have been stable.

Fossil evidence of lepidopterous leafminers of *Quercus* demonstrate a similar persistence of an insect–plant relationship from the middle Miocene (Opler 1973). Crane and Jarzembowski (1980) summarized available information of fossil leafmines, and state that "leafmining was well established by the early Caenozoic." Typical damage of cambium-mining insects (Diptera: Agromyzidae: *Phytobia*) has been studied in fossil woods (Süss and Muller-Stoll 1975; Süss 1979). The evidence suggests that these fly larvae caused recognizable damage in the Tertiary, but on plants that are not attacked today. Thus, in general, damage caused by endophytophagous insects is recognized from as early as the Tertiary. Nonetheless, additional specimens of fossils in which the damage and the damager are preserved are needed in order to definitively outline evolutionary patterns.

Analysis of fossilized arthropods belonging to gall-inducing taxa may further elucidate evolutionary trends (Kinsey 1919, 1938; Brooks

1955; Keilbach 1982). However, one must keep in mind that by themselves, preserved insects do not definitely indicate behavior; extant taxa that contain gall inducers also contain non-gall-inducing members (e.g., free-living parasitic or inquiline members). Fossilized gall specimens with gall inducers preserved within the gall provide the best associational evidence.

Weaknesses in the fossil gall record provide opportunities for paleoentomologists and paleobotanists. There is both a need and a potential for strengthening the record through additional specimens collected from all parts of the world. It seems unlikely that all fossil gall specimens will ever be assembled into a single collection; however, a compiled photographic record of known specimens would greatly assist our efforts to assess and comparatively study fossil galls.

ACKNOWLEDGMENTS

Dr. Arthur Boucot, Department of Geology, Oregon State University, Corvallis, Oregon, freely shared information and citations on fossil galls with me. Dr. Harry Phinney, Department of Botany and Plant Pathology, Oregon State University, Corvallis, Oregon, brought B. C. Templeton's Dissertation to my attention. The G. C. Page Museum, Los Angeles, California, and the California Academy of Sciences, Golden Gate Park, San Francisco, California, provided specimens for study. Dr. Odette Rohfritsch provided information on French specimens. Scott Wing assisted with U.S. National Museum specimens.

LITERATURE CITED

Ambrus, B., and L. Hably. 1979. *Eriophyes daphnogene* sp. n., a fossil from the Upper Oligocene in Hungary. Ann. Hist. Nat. Mus. Nat. Hung. 71:55–56.

Ananthakrishnan, T. N. 1978. Thrips galls and gall thrips. Tech. Monogr. No. 1, Zool. Survey of India, Calcutta.

Berger, W. 1949. Lebensspuren schmarotzender Insekten an jüngster Jahren Laubblättern. Sitz. Ber. öster. Akad. Wiss., Vienna, Abt. I, 158:789–792.

Berry, E. W. 1916. The Lower Eocene floras of Southeastern North America. U.S. Geol. Surv. Prof. Paper 91.

Berry, E. W. 1923. Pathological conditions among fossil plants. In R. L. Moodie. (ed.), Paleopathology: An Introduction to the Study of Ancient Evidence of Disease, pp. 99–115. Univ. Illinois Press, Urbana.

Berry, E. W. 1924. Organic remains, other than diatoms from the excavation. J. Wash. Acad. Sci. 14:12–25.

Brooks, H. K. 1955. Healed wounds and galls on fossil leaves from the Wilcox deposits (Eocene) of Western Tennessee. Psyche 62:1–9.

Brues, C. T. 1910. The parasitic hymenoptera of the Tertiary of Florissant, Colorado. Bull. Mus. Comp. Zool. 54:1–125.

Buhr, H. 1964/1965. Bestimmungstabellen der Gallen (Zoo- and Phytocecidien) an Pflanzen Mittel- and Nordeuropas. I-II. G. Fischer, Jena.

Chaney, R. W. 1920. The flora of the Eagle Creek Formation. Univ. Chicago Walker Museum Contrib. 2:5–181.

Chaney, R. W., and H. L. Mason. 1934. A Pleistocene flora from Santa Cruz Island, California. Carn. Inst. Wash. 415 (Contrib. to Paleontol.) I:1–24.

Cockerell, T. D. A. 1908. Fossil insects from Florissant, Colorado. Bull. Am. Mus. Nat. Hist. 24:59–69.

Cockerell, T. D. A. 1927. In R. W. Chaney (ed.), Geology and Palaeontology of the Crooked River Basin with Special Reference to the Bridge Creek flora. Carn. Inst. Wash. 346:45–138.

Crane, P. R., and E. A. Jarzembowski. 1980. Insect leaf mines from the Paleocene of southern England. J. Nat. Hist. 14:629–636.

Cunliffe, B. W. 1976. Iron Age sites in central southern England. CBA Research Report 16, London.

Felt, E. P. 1965. Plant Galls and Gall Makers. Hafner Press, New York.

Gagné, R. J. 1968. Revision of the gall midges of bald cypress (Diptera: Cecidomyiidae). Entomol. News 79:269–274.

Giannasi, D. E., and K. J. Niklas. 1981. Comparative paleobiochemistry of some fossil and extant Fagaceae. Am. J. Bot. 68:762–770.

Grauvogel-Stamm, L. 1978. La flore du grès à Voltzia (Bundsandstein supérieur) des Vosges du Nord (France). Morphologie, Anatomie, Interprétations phylogénique et paléogéographique. Th. Dr. Sc. Un. Louis Pasteur, Strasbourg, France.

Gullan, P. J. 1984. A revision of the gall-forming coccoid genus *Apiomorpha* Rübsaamen (Homoptera: Erioccocidae: Apiomorphinae). Aust. J. Zool., Suppl. Ser. 97, 203p.

Heer, O. 1855. Flora tertiaria Helvetiae (Die tertiäre Flora der Schweiz). Wurster and Co., Winterthur.

Heyden, C. von (Heiden). 1860. Achter Bericht. d. Oberhessisch. Gesellsch f. Natur-und Heilkunde. 8:63.

Hickey, L. J., and J. A. Doyle. 1977. Early Cretaceous fossil evidence for angiosperm evolution. Bot. Rev. 43:3–104.

Hoernes, R. 1886. Manuel de Paléontologie. Trans. by L. Dollo. Paris.

Hoffman, A. 1932. Miocene insect gall impressions. Bot. Gaz. 93:341–342.

Houard, C. 1922/1923. Les zoocécidies des plantes d'Afrique, d'Asie et d'Océanie. Librairie Sci. Jules Hermann, Paris. Vols. I–II.

Houard, C. 1933. Les zoocécidies des plantes de l'Amerique du Sud et de l'Amerique Centrale. Librairie Sci. Jules Hermann, Paris.

Houard, C. 1940. Les zoocécidies des plantes de l'Amerique du Nord: Galles des Chênes. Librairie Sci. Jules Hermann, Paris.

Keilbach, R. 1982. Bibliographie und Liste der Arten tierischer Einschlüsse in fossilen Harzen, ihrer Aufbewahrungsorte. St. Entom. Z. N.F. 29, Heft 1–3:129–286.

Kinsey, A. C. 1919. Fossil cynipidae. Psyche 26:44–49.

Kinsey, A. C. 1938. Cynipidae from Canadian amber. Univ. Toronto Stud. Geol. Ser. 40:21–27.

Knowlton, F. H. 1898. A catalogue of the Tertiary and Cretaceous plants of North America. U.S. Geol. Surv. Bull. 152:1–247.

Küster, E. 1910. Die Zoocecidien. In E. H. Rübsaamen (ed.), Die Zoocecidien, durch Tiere erzeugte Pflanzengallen Deutschlands und ihre Bewohner. Zooligica 61:159–160.

LaMotte, R. S. 1952. Catalogue of the Cenozoic plants of North America through 1950. Geol. Soc. Am. Mem. 51:1–381.

Larew, H. G. 1986. The fossil gall record: A brief summary. Proc. Entomol. Soc. Wash. 88:385–388.

Larew, H. G. 1987a. Oak galls preserved by the eruption of Mount Vesuvius in A. D. 79, and their probable use. J. Econ. Bot. 41:33–40.

Larew, H. G. 1987b. Two cynipid wasp acorn galls preserved in the La Brea tar pits (Early Holocene). Proc. Entomol. Soc. Wash. 89:831–833.

Lesquereux, L. 1892. The flora of the Dakota Group. Monogr. U.S. Geol. Surv. 17:19–400.

Mädler, K. 1936. Eine Blattgalle an einem vorweltlichen Pappel-Blatt. Natur Volk. 66:271–274.

Marty, P. 1894. De l'ancienneté de la "Cecidomyia fagi." Feuilles Jeune Nat. 24:173.

Meschinelli, L. 1892. Fungi fossiles. In P. Saccardo (ed.), *Sylloge Fungorum,* 10, pp. 805–808. Friedlander & Sohn, Berlin.

Möhn, E. 1960. Eine neue Gallmucke aus der niederrheinischen Braunkohle, *Sequoiomyia krauseli* n. g., n. sp. (Diptera, Itonididae). Senck. leth. 41:513–522.

Opler, P. A. 1973. Fossil lepidopterous leaf mines demonstrate the age of some insect-plant relationships. Science 179:1321–1323.

Potonié, H. 1893. Ueber das Rothliegenden des Thüringer Waldes. Teil II: Die Flora des Rothliegenden von Thüringen. Preuss. Geol. Landesanst. Abh. 9.

Rahtz, P. A., and E. Greenfield. 1977. Excavations at Chew Valley Lake. Department of the environment Archaelogical Report 8, London: H.M.S.O.

Robinson, M. A. 1980. Archaeological finds of wasp galls. J. Archaeol. Sci. 7:93–95.

Rübsaamen, E. H. 1929. Gallbidungen. In C. Schröder's Handbuch der Entomologie. G. Fischer, Jena. II:219–248.

Scudder, S. H. 1886. Systematic review of our present knowledge of fossil insects. U.S. Geol. Surv. Bull. 31:1–131.

Smith, H. 1932. The fossil flora of Rockville, Oregon. Masters of Science Thesis, Univ. of Oregon, Eugene, Oregon.

Straus, A. 1977. Gallen, Minen und andere Frassspuren im Pliokän von Willershausen am Harz. Verh. Bot ver. Prov. Brandenburg 113:43–80.

Süss, H. 1979. Durch *Protophytobia cupressorum* gen. nov., sp. nov. (Agromyzidae, Diptera) verursachte Markflecken in einem Holz von *Juniperoxylon* aus dem Tertiär von Sud-Limburg (Niederlande) und der Nachweis von Markflecken in einer rezenten *Callitriche*-Art. Feddes Rep. 90:165–172.

Süss, H., and W. R. Müller-Stoll. 1975. Durch *Paleophytobia platana* n. g., n. sp. (Agromyzidae, Diptera) verursachte Markflecken im Holz fossiler Pflanzen aus dem ungarischen Miozän. Wiss. Zeit. Humboldt-Univ., Berlin, Math.-Nat. R. 24:515–519.

Templeton, B. C. 1964. The fruits and seeds of the Rancho La Brea Pleistocene deposits. Ph.D. Thesis, Oregon State University, Corvallis, Oregon.

Tiffney, B. H., and E. S. Barghoorn. 1974. The fossil record of the fungi. Occas. Papers Farlow Herb. Crypt. Bot. 7:1–42.

Trotter, A. 1903. Studi cecidologici: Le galle ed i cecidozoi fossili. Riv. Ital. Paleontol. 9:12–22.

Weiss, F. E. 1904. A probable parasite of Stigmarian rootlets. New Phytol. 3:63–68.

Weld, L. H. 1922. Notes on American gall flies of the family Cynipidae producing galls on acorns, with descriptions of new species. Proc. U.S. Nat. Mus., No. 2440 61:1–32.

5

Patterns in Gall Development

O. ROHFRITSCH

One of the most striking aspects of gall biology is the rich diversity of forms, colors, sizes, surface projections, and internal structures associated with the galls themselves. Some galls resemble fruits or seeds, others are membranous bags or woody swellings; however, the most basic structural attribute is that each gall inducer solicits galls that are characteristic to the species of inducer. That is, all the galls induced by the individuals of one species are identical and yet they may differ markedly from the galls of closely related species (see Chapter 2).

There is a vast literature explaining how gall inducers structurally alter their host organs (i.e., Beijerinck 1882, 1888; Küster 1911; Magnus 1914; Mani 1964; Maresquelle and Meyer 1965; Rohfritsch 1971b; Meyer and Maresquelle 1983; Meyer 1987); however, even a cursory examination of this literature reveals that related groups of insects have a similar influence on plant tissues. There are numerous patterns or models of gall development but three, one induced by midges, the other by sawflies, and the third by cynipid wasps, are the most basic and common and are referred to as the cecidomyiid, sawfly, and cynipid models in this chapter. The purpose of this chapter is to review these three patterns of gall development and in doing so illustrate how galls are specific expressions of the insect's biology. A review of the basic events associated with the development of galls also is presented.

CLASSIFICATION OF GALLS BASED ON STRUCTURE

Plant galls can always be distinguished from other growth and developmental anomalies by the existence of specific nutritional relationships between the gall inducer and the host plant (Küster 1911). These relationships are usually established by the presence of specialized cells called nutritive cells (see Chapter 8), which signify an active reaction by the plant to the insect's presence. It is best to avoid referring to galls as pathological growths since this terminology implies the involvement of disease. Terms such as atypical or abnormal, which refer to deviations from the norm, are far more appropriate for characterizing gall growth.

According to Küster (1911), all galls can be classified as either organoid or histioid, with the latter group being further divided into kataplasmas and prosoplasmas. Organoid galls are abnormal plant growths that structurally differ only slightly from the normal growth pattern. Instead, they differ in the degree to which developmental potentialities of the cells manifest themselves within certain reactive zones. Thus intumescences, wound callus, teratomas, and in vitro regeneration are of this type of abnormality. Insect feeding or the growth of fungi can induce organoid galls.

Histioid galls, in contrast, exhibit a great variety of abnormal growth phenomena that have qualitative characteristics; that is, in galls of this type, the basic growth patterns of the host organ are altered. These growths are either a result of etiological factors that may have an internal (genetic or epigenetic) base (i.e., genetic tumors, *Agrobacterium*-induced tumors), or they originate externally and are chemical or parasitic in nature.

Kataplasmic galls are histioid growths that are less organized and differentiated than the host organ and consist mainly of several layers of parenchymatous cells. Galls induced by Hemiptera are often of this type. Prosoplasmic galls, in contrast, are highly and harmoniously organized growths with the extent of tissue differentiation dependent on the species of inducing organism. The best examples of prosoplasmic galls are those induced by cecidomyiids and cynipids. Thus prosoplasmic galls are a result of a selective action by feeding insects applied to the pattern-producing potentialities of the host (Magnus 1914; Boysen-Jensen 1948; Bloch 1965; Rohfritsch 1971b). This process appears to be a result of a defensive reaction of the host plant and a specific adaptation to the requirements of the insect. However, this adaptation is apparent only because it is the insect that has had to adapt its life-style and development to the type of plant reaction it induces (Zweigelt 1931). Galls of this type have never been produced experimentally. All the galls considered in this chapter are clearly of the prosoplasmic type.

GENERAL PATTERNS OF GALL DEVELOPMENT

Seasonal Appearance of Galls

The emergence of most zoocecidia is synchronized with the active growth period of the host plant. Galls on trees and shrubs, for example, appear in the spring or at a time when the leaves flush and when food quantity and quality are highest. A second group of galls is initiated later in the season when a second flush of leaves appears or when the fruits and cones appear. As with the spring leaf gall inducers, substantial amounts of food are available in such tissues (Hill 1982; NcNeil and Prestige 1982).

Some species initiate galls at two different periods of the season. The bisexual generation of cynipids on *Quercus*, for example, develops in the early spring whereas the unisexual generation develops at the time of the second flush. Some galls, such as those of *Hartigiola annulipes* (Hartig) and *Adelges abietis* (L.), are initiated in the spring but do not develop until late summer. Several sequential generations of cecidomyiids may be found on herbacious plants and grasses that continue growing throughout the season (Skuhravá and Skuhravý 1973; Birch 1974).

Gall inducers take large amounts of sugars and proteins from the plant sap and the highest concentrations are found near the meristematic tissues. However, the general view that only meristematic tissues react to gall insects is now known to be erroneous. The young stem mother of *Adelges*, for example, can develop a typical nutritive tissue by modifying the cortical cells of a 1-year-old stem (see Chapter 7). Cecidomyiids attack young organs and tissues, whereas some cynipids are adapted to attacking organs with differentiated tissues (see Chapter 6).

Basic Stages in Gall Development

The four basic stages in gall development are initiation, growth and differentiation, maturation, and dehiscence. Gall initiation is characterized by a succession of insect–plant, insect–organ, or insect–tissue recognition systems and very specific behavior of the gall insect (Rohfritsch 1975b, 1980, 1988b, 1990; Skuhravý 1975; Tscharntke 1988). Also of importance in initiation is the isolation and withdrawal of a group of cells from the normal growth patterns of the host organ. These cells are physiologically modified and serve as the source of nearly all the cells that form the gall (Rohfritsch 1971b). Initiation is either associated with oviposition by the adult (i.e., sawflies, cynipids, and some beetles) or with the activity of first-instar larvae (i.e., cecidomyiids, aphids, and coccids). In some cases initiation occurs as a result of the imago feeding before or during oviposition (i.e., mites, some psyllids, and thrips); however, in all cases it is the immature stage that benefits from the events in gall initiation. Initiation is the most important stage in gall development and delicate events during oviposition or feeding by first-instar larvae introduce important changes in the attacked plant tissues.

Gall growth and development are the stage when the biomass of the gall is vastly increased by cellular hyperplasy (cell division) and hypertrophy (cell enlargement). The rate of this increase is generally dependent on the feeding activities of the insect (some *Pontania* galls are the exception). Depending on the type of mouthparts, gall insects either feed on solids or suck liquids from the plant cells. The insects pour salivary fluids on the plant cells either to modify the cell walls (aphids/cecidomyiids) or to liquify the cell contents (cynipids) (see Chapter 8), and the cells react to this stimulation by cell dedifferentiation and further away by cell proliferation. A nutritive tissue is soon present along the inner surface of the larval chamber and vascular tissues arise among the mass of growing cells and join the normal vascular tissue of the attached organ. Thus larval feeding activity is mainly responsible for modeling the shape of the larval chamber (see below).

The maturation stage occurs while the insect is in the last larval instar and a period in its life when it consumes the largest amount of food. This is the main trophic phase of the insect as it actively feeds on the masses of rich nutritive tissues that have accumulated around it. A sclerenchyma sheath often develops around the layers of nutritive tissue near the vascular tissue. The presence of this sheath separates the gall into two regions referred to as the inner and the outer gall. The inner region is directly influenced by the cecidozoa, whereas the outer gall, or gall cortex, is more under the influence of the host plant. The outer gall is often involved in gall opening and/or dehiscence from the host plant (Rohfritsch 1967, 1971b).

Gall dehiscence or gall opening occurs toward the end of the maturation phase and is a period of major physiological and chemical change in the gall tissues. For example, the flow of sap and water to the gall stops and the gall no longer is a nutrient sink (Chapter 11). In some galls an abscission layer may appear within gall tissues separating the inner from the outer gall. The inner gall may fall to the ground while the outer gall remains on the host plant (Rohfritsch 1967). All of these phenomena are strictly correlated with the development of the gall inhabitants and facilitate their escape. In some galls the larvae exit when fully grown and pupate on the plant or ground. However, in other cases the larva pupates in the gall on the host plant (Yukawa 1984, 1987).

Although it is known that ripening of gall tissues allows the escape of some insects, the factors that trigger ripening are not known. It has been demonstrated that the last instar of *Adelges abietis* induces gall maturation and opening of the gall chamber. If the nymphs are killed before the last instar, the gall does not open. However, if several individuals of the last instar are mixed with a large number of individuals of younger instars, the gall will open (Rohfritsch 1971b). The galls of most mites, thrips, and aphids provide food and shelter to several successive generations and in these cases the ripening processes occur when the last generation matures. Many midges, all cynipids and chalcids, and some moths overwinter and pupate in their galls and escape through exit holes. The larvae of tephritids, Lepidoptera, and sawflies chew the exit routes with their mouthparts. Some cecidomyiids prepare an exit route for the adult with their spatula (Loux 1967, Skuhravá and Skuhravý 1981).

Since a gall is considered the result of a specialized reaction by plants to internal and environmental changes caused by the gall inducer, cecidogenesis embraces a complex of interactions between plant cells and the gall-inducing organism (Magnus 1914). Cecidogenesis does not involve processes unknown in normal ontogenesis; however, the various cytogenetic and morphogenetic events often overlap, or are combined, and so are often difficult to analyze (Küster 1911; Bloch 1965; Mani 1964).

Cecidogenic processes are not the same throughout the gall-inducing guild. Indeed several groups have evolved distinct mechanisms whereby they initiate and control the growth of their galls. The cecidomyiid type, which can be extended to include the mites, thrips, and homopterans, is considered the most basic or general type. The tenthredinid type also includes the moths, beetles, and dipterans other than cecidomyiids. The cynipid type is formed only by hymenopterans.

The easiest way to study the modifications that an insect larva induces on host tissues, and to separate and analyze the different events in gall initiation is to kill or remove the gall inducer by mechanical or chemical means at different stages of gall development (Rohfritsch 1971b; Westphal 1977). In this way the host's response to "messages" delivered by the larva can be determined at each step in cecidogenesis. In addition to this main experimental technique other approaches have been used. For example, heterospecific adelgid galls have been artificially created by experimentally transferring the offspring of one species into the galls induced by a different species (Rohfritsch 1971b). Much information on the determinism of gall tissues and on the stability of the insect's message has been obtained by these experiments. Shorthouse (1975) was able to determine the influence of *Diplolepis* larvae on gall morphogenesis by examining galls at various stages of development in which the larvae had been killed by parasites. Lastly, in vitro cultures of galls or gall fractions, with or without the gall inducer, illustrate the essential organizing role of the insect in gall growth and morphogenesis (Lutz and Meyer 1965; Rohfritsch 1971a).

THE CECIDOMYIID MODEL

Covering Galls

Larvae of most cecidomyiids initially feed externally on epidermal cells and become covered by tissues growing around the feeding site. Covering galls are used as a model here because they are the most common.

Cecidomyiids emerge from their galls or the soil usually from late April to early May in western Europe (Skuhravá and Skuhravý 1973; Birch 1974; Rohfritsch 1980). When mated, females locate a suitable organ and the eggs are laid in a locality and manner characteristic to the species (i.e., between bracts, beneath or between hairs of immature leaves).

Some species deposit their eggs directly on the most appropriate sites [i.e., *Geocrypta galii* (H. Lw.), *Dasineura affinis* (Kieffer) and many *Contarinia* species], whereas in other species the larva must locate the feeding site. During this dispersal stage, the larva is agile and resistant to desiccation. In some species, such as *Hartigiola annulipes,* the larvae are solitary, whereas in others, such as *Giraudiella inclusa* (Von Frauenfeld) and *Lasioptera arundinis* (Schiner), the larvae follow each other to a group feeding site.

Once the larva locates the feeding site it stops and begins a wiggling action on a small area of cells that will eventually provide tissues of the future gall. This wiggling activity immediately activates the cecidogenic processes within the plant cells in contact with the larva. After a few hours (6–8 for *H. annulipes*), the larva ceases all movement, and its body enlarges, flattens, and becomes transparent. Once the body is tightly adhered to the plant tissues, it begins to feed. The only detectable movement at this early stage of initiation is a sucking motion of the pharynx. If the larva is artificially moved at this stage, it cannot repeat the prefeeding behavior and is unable to initiate a new gall (Rohfritsch 1980). This first stage of gall initiation is very critical. In some cases the larva is very aggressive and kills plant cells (Solinas and Bucci 1982; Hatchett and Kreitner 1988; Rohfristch 1988b, 1990). The larva must quickly escape the adverse effects of rain, wind, dryness, and predators and locate responsive host tissues. Tissues of the wrong host or the wrong organs of the host plant do not respond to the larva and it dies.

Initiation begins with both physiological and cytological changes in the cells around the larva. If a very young leaf is attacked, the whole leaf lamina becomes modified whereas in other cases reaction by the host is more restricted. However, in both cases there is an homogenization of the tissues, cell types appear identical, the chloroplasts of the mesophyll cells lose their high contrast, the intercellular spaces disappear, and the vacuoles become smaller. The enlarged centrally located nucleus has an extended chromatin and a large nucleolus. The cells also are enriched in various organelles, especially ribosomes. In many cases it takes less than 24 hours for the larva to induce these changes. This early stage in cell modification and gall development is referred to as "metaplasy" (Meyer 1957). From this early stage in gall growth, a number of cells around the larva become "disconnected" from the normal growth pattern of the host organ. Normal growth polarity remains disturbed and there may be a little cell hypertrophy around the site (Fig. 5–3A).

By removing the gall inducer at this early stage, it has been shown that cell metaplasy is a very unstable phenomenon. The cytological characteristics described above are lost within 24 hours. Phenolic substances, a thick cuticula, and sometimes lignified cell walls characterize

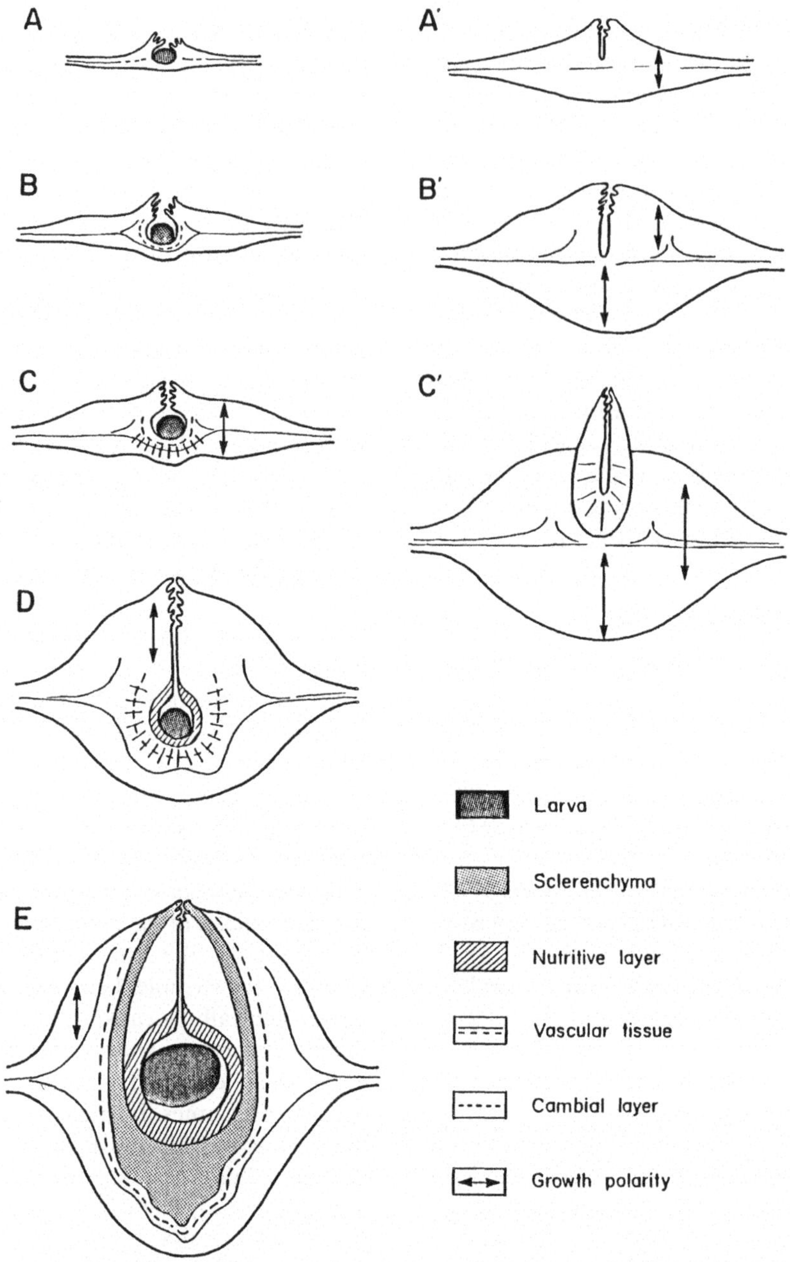

Figure 5-1. Stages in the developmental morphology of galls induced by the cecidomyiid *Didymomyia reaumuriana* on the leaves of *Tilia* sp. (Redrawn from Rohfritsch 1971b.) (A–E) Normal gall development; (A′–C′) galls that have developed without a larva. (A) Gall at 2 days. The larva, which is intimately in contact with the leaf epidermis, produces a dedifferentiation of cell structures and growth inhibition. New tissues are formed by proliferation of epidermis and mesophyll cells around the larva. (A′) Gall without a larva showing that from the second day, the leaf tissue influenced by the larva is partially disconnected from the normal growth pattern. Cell hypertrophy perpendicular to the normal leaf is very important, even beneath the larval chamber. (B) Gall at 5 days. The larva is nearly enveloped by the growing wall of cells and by hairs that develop from epidermal cells at the top of the wall. All the cells of the leaf that will be incorporated in gall morphogenesis are now modified in their growth polarity and in their content. (B′) Gall without a larva at the fifth day. The gall is composed of relatively homogeneous parenchyma. (C) Gall at 1 week. The larva, still intimately in contact with the modified dedifferentiated epidermal cells of the leaf, feeds on cells at the base of the wall. Growth, by cell division, appears between the larva and the vascular tissue of the leaf. (C′) Gall with larva killed at the stage shown in C. The gall

the abandoned metaplasied cells in woody plants, whereas in herbaceous plants the cells return to a nearly normal appearance (Rohfritsch 1971b).

The second step in cecidogenesis, which occurs very rapidly (within the second or third day of larval feeding), is characterized by cellular hyperplasia and results in a wall of cells around the feeding insect. The new growth polarity is oriented perpendicular to the surface of the attacked organ (Figs. 5–1A, 5–2B, and 5–3B). By removing the larva at this stage, one can ascertain that all cells of the host plant participating in cecidogenesis are now withdrawn from the growth pattern of the normal organ. They remain in the gall pattern and do not return to a normal type of cell growth. The cells of the insect-free gall then follow a growth pattern oriented perpendicular to the normal organ surface (Figs 5–1A′ and 5–2B′). All cell divisions cease and any further growth is a result of cell hypertrophy. The epidermal cells are usually transformed into trichomes; however, some become lignified.

During the third step in cecidogenesis, growth of gall tissues is amplified and becomes redirected. In most cases gall tissues curve around the gall inducer (Figs. 5–1B, 5–2C, and 5–3C). This reorientation of growth polarity is strictly related to the location of the larval feeding sites. If the larva always feeds on the same cells located at the top of the larval chamber, the growth polarity of the walls will remain perpendicular to the attacked organ. However, in most cases larval feeding occurs on new cells at the base of the enveloping walls. This results in a large number of stimulated cells and a greater extension of the nutritive tissue. The enveloping tissues then curve obliquely around the larva. Hairs often arise from modified epidermal cells of the distal area and vascular bundles begin to differentiate within the new tissues (Fig. 5–3C).

Gall tissues become further differentiated during the fourth step in cecidogenesis. In some galls, particularly those on herbaceous plants, gall morphogenesis is completed after 5 days of larval feeding and any further gall development is essentially extension growth (cell divisions and cell hypertrophy) and tissue differentiation. However, in most others growth and differentiation occur several weeks after initiation (Figs. 5–1C, D and 5–2E, F). The larva becomes completely enclosed in the narrow gall chamber. Only a thin, hair-covered ostiolar opening remains, which allows the entry of air. In some galls this channel disappears, obliterated by growing masses of parenchymatous cells. Vascular cambium appears as soon as the cell walls begin to envelop the larva (3 days after the beginning of cell growth) and the vascular bundles join those of the host organ (Fig. 5–3C, D). Layers of cytoplasmically dense cells that appear between the vascular tissue and the larval chamber constitute the nutritive cells.

Cells destined to become the nutritive layer in galls of cecidomyiids pass through two stages during their transformation. During metaplasy the cells become rejuvenated and during this early period it is possible for the cells to revert back to near normal. Once the cells are transformed to nutritive tissue cells, a reversion is no longer possible. The cytological and histochemical aspects of nutritive tissue are described in Chapter 8; however, further details on the types found in galls of cecidomyiids are described here.

The first type of nutritive tissue is composed of only a few cell layers (3–5) and is in close contact with vascular tissue. Sometimes this nutritive tissue can be very localized and is produced only by transformation of leaf tissue. In other

expands by hypertrophy such that the "outer gall," which is growing, pushes the "inner gall" out of the leaf. (D) Gall at 2 weeks. The larva and larval chamber are very small and deeply buried in the mass of growing tissues. Gall growth is partially due to a cambial tissue that has appeared between the larval chamber and the vascular tissue of the gall. (E) Mature gall at 4 weeks. The first cambial layer gives rise to the nutritive tissue and the starch parenchyma toward the larva and a sclerenchyma sheath in the opposite direction. As the gall matures, a new cambial layer appears between the sclerenchyma sheath and the vascular parenchyma of the gall. This layer is responsible for abscission of the inner gall. Both the larva and larval chamber enlarge as the larva feeds on cells of the nutritive tissue. These cells are progressively emptied while the cells more inward become stimulated to nutritive tissue cells.

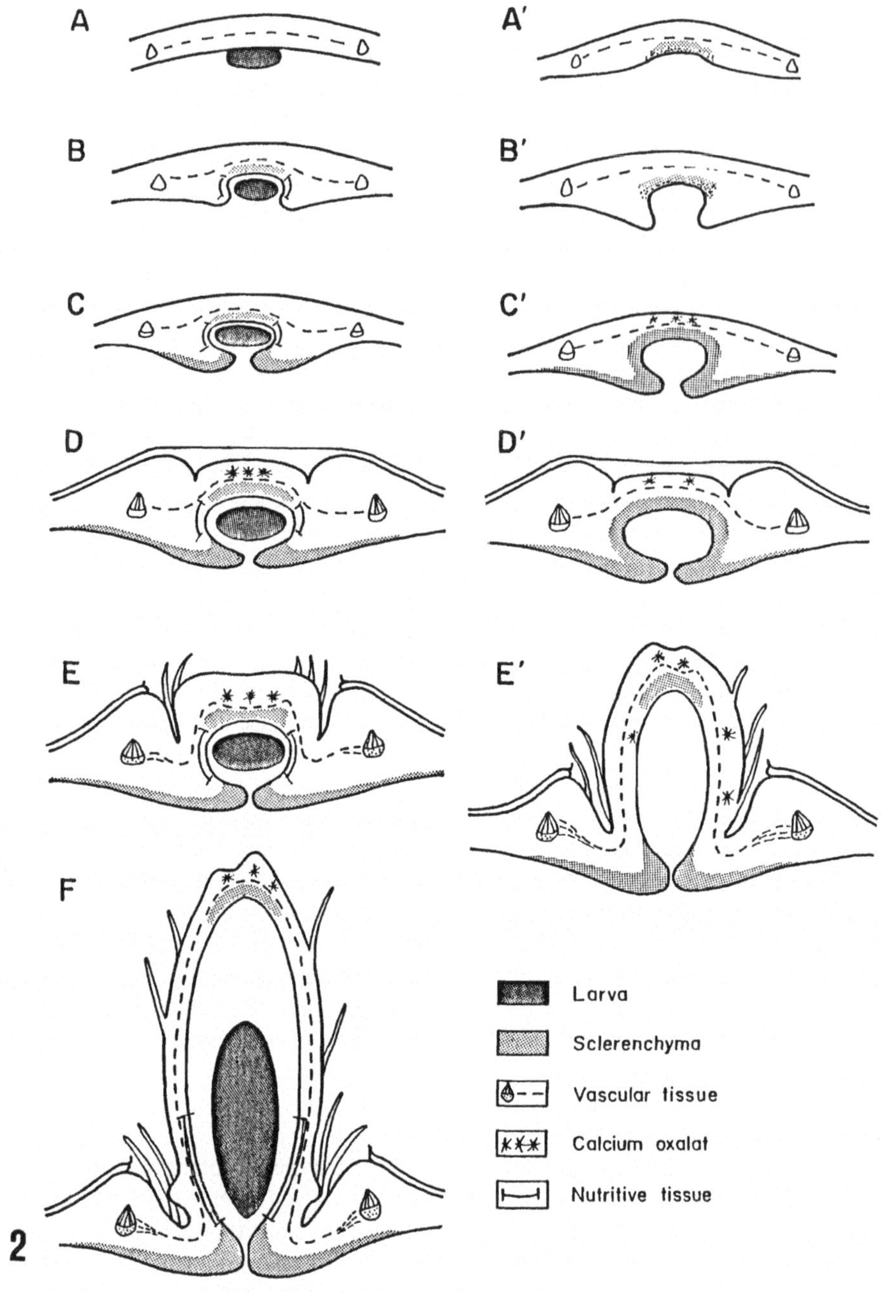

Figure 5-2. Stages in the developmental morphology of galls induced by the cecidomyiid *Hartigiola annulipes* on the leaves of *Fagus silvatica.* (Redrawn from Rohfritsch 1971b.) (A–F) Normal gall development; (A′–E′) galls that have developed without a larva. (A) Gall at 24 hours. Only modifications of cell organelles occur in layers of cells beneath the larva forming the "cell metaplasia" or cell dedifferentiation. (A′) Gall at 24 hours with larva removed. The normal growth polarity of nearby cells has been disturbed and the modified epidermal cells are lignified and contain tannins. (B) Gall at 3 days. A wall of cells is forming around the larva that is feeding on the new cells. The earlier stimulated cells at the top of the larval chamber become sclerified. (B′) Gall at 3 days with the larva removed. Cells at the top of the chamber are sclerified. (C) Gall at 1 week. The larva is now enclosed by the growing wall of cells. Sclerification of cell walls has occurred at the top and bottom of the larval chamber; only a narrow band of nutritive tissue cells remains free of lignin. (C′) Gall with larva removed at 1 week. All the nutritive tissue cells become sclerified. (D) Gall at 2 months. The gall and larva are very small. Tissue proliferation begins around the vascular bundles. The epidermis becomes disconnected from the palisade layer, which begins

galls it is more extensive and may cover most of the inner surface of the larval chamber (Fig. 5–3D). This type of nutritive tissue usually contains starch in the layers near the vascular tissue. The key characteristics of this nutritive tissue includes a prominent nucleus with a large nucleolus, the presence of a large number of mitochondria, abundant endoplasmic reticulum localized along the cell walls, and frequent communications between cell layers and vascular tissues through numerous plasmodesmata. The cell wall is irregularly thickened and sometimes there is an enlargement of plasmodesmata and very large cytoplasmic channels between cell layers. The cells always remain alive and active during larval development (Rohfritsch 1987a, 1988a).

The second type of nutritive tissue is composed of numerous layers of cells and may be formed by a cambium or by active cell proliferation of the attacked organ (Fig. 5–1D). This nutritive tissue constitutes a mass of cells that develops early in cecidogenesis when the larva and growing gall tissue constitute a powerful sink for nutrients. The gall tissues draw the solutes and reserves from neighboring cells (Kirst and Rapp 1974). The nutritive tissue formed during the first two larval stages will be consumed by the last larval instar, which needs a large amount of food. During this last stage, the larva will draw on solutes, reserves, and even the protoplasm of distant cells. More and more cell layers become empty and the area of stimulated cells progresses toward the periphery of the nutritive tissue. The cells are characterized by a relatively large nucleus with a very small nucleolus and by a cytoplasm poor in organelles. The cell wall is regular and thin and has few randomly distributed plasmodesmata. To reach the larva, solutes now have to cross several layers of depleted cells. This type of nutritive tissue is not unique to plants since the physiological and cytological characteristics are similar to endosperm cells of wheat embryos (Smart and O'Brien 1983).

If we consider these two main types of nutritive tissue from a functional standpoint, we see that the first functions as a short distance transport tissue between the vascular tissue of the gall and the insect. The second type acts first as an attractor (an active converter of simple molecules into immobile molecules) and later as a provider of food by hydrolyzing its reserves and cytoplasm (Rohfritsch 1987a, b).

The vascular tissues of cecidomyiid galls are usually similar to those of normal tissues (Meyer 1969); however, the phloem is more developed than the xylem. The tracheids show a helical pattern of lignification in their cell walls that allows cell elongation during the final growth of the gall. The vascular bundles of cecidomyiid galls generally envelop the nutritive tissue (Fig. 5–3D). Calcium oxalate is often present in the proximity of the vascular tissue (Fig. 5–2D, E).

A layer of sclerenchyma cells occurs in many cecidomyiid galls (Fig. 5–1E). If the sclerenchyma occurs between the vascular tissue and the nutritive layer, the cells remain active and may have characteristics similar to those of nutritive cells. The sclerenchyma cells have numerous connections with surrounding cell layers. The wall openings, the distribution of plasmodesmata, and the cell shape allow the movement of solutes from vascular tissue toward cells of the nutritive layer.

The layer of cells beyond the vascular tissue remains composed of parenchyma cells and is referred to as cortical parenchyma. Sometimes the cells are lignified or they may contain phenolic substances or tannins. The cortical tissue often contains chloroplasts. Epidermal cells may be covered with a thick cuticle, lignified, or covered with hairs.

Thus all events in the cecidogenesis of cecidomyiid covering galls require the activity of the larva. Cecidogenesis consists of a series of processes following or overlapping each other with each process induced by the insect. At no time during early cecidogenesis is there disor-

to proliferate and form hairs. (D′) Gall with larva removed at 2 months. The only change which occurs is a sclerification of the entire nutritive tissue. (E) Gall at 4 months. A pouch is now growing from the center of the gall and the larval chamber is enlarged. (E′) Gall with the larva removed at 4 months. Only a small pouch is formed compared to F. (F) Mature gall at 5 months. The nutritive tissue is now of great importance to the larva and is localized at the base of the pouch.

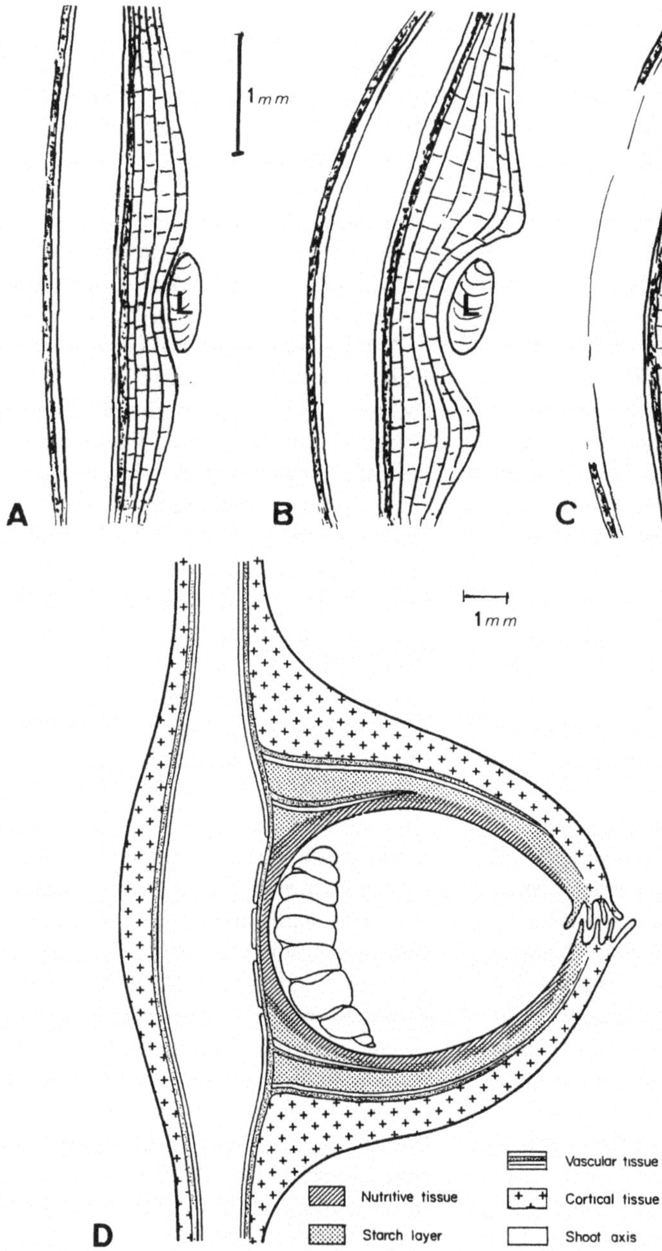

Figure 5-3. Stages in the developmental morphology of galls induced by the cecidomyiid *Geocrypta galii* on the stems of *Gallium mollugo.* A to C are early stages in gall development; only the attacked side is represented along the stem axis. (A) Gall at 1 day. Normal growth and cellular differentiation are inhibited beneath the larva. The cells become dedifferentiated and their growth is strongly inhibited whereas cell hypertrophy can be observed around the larva. (B) Gall at 2 days. The stem bends because of unilateral growth inhibition. The curvature is most important at 2 and 3 days. New tissues are formed around the larva and the growth polarity of the new growth is oriented perpendicular to the growth polarity of the normal stem. The larva is feeding on dedifferentiated epidermal cells of the stem which become nutritive tissue cells. (C) Gall at 3 days. The larva feeds on cells of the developing wall. The growth polarity of the new tissue becomes more oblique and the young larva is enveloped. Vascular tissue appears in the growing cell mass and joins that of the stem. Hairs develop on the cell epidermis around the ostiolar opening. (D) The larval chamber of the mature gall is large and opens to the outside through the enlarged ostiole. The entrance of the gall is protected by numerous hairs. The phloem of the stem behind the 4–5 layers of nutritive tissue cells is well developed whereas the xylem is of less importance than it is in normal stems.

ganization of the proliferating cells. Once the cells are removed from the normal control of the host organ, they are engaged in a new system controlled by the insect (Rohfritsch and Shorthouse 1982). By isolating the "gall cells" early in gall development from normal tissue, the larva has a strong inhibiting action on normal growth and tissue differentiation. Similar growth inhibition is observed in hypocotyl growth of young *Bidens pilosus* L. plants some distance away from cotyledons that have been wounded (Desbiez *et al.* 1983). This effect can be dramatic on very young organs whereas on mature organs it is of less importance. Early attacks result in bending of axes, prevention of unfolding, or unrolling of young leaves, or unopened buds or flowers. If a very young organ or bud is attacked, cecidogenesis has an organoid effect.

This first effect on normal growth is essentially inhibitory and is followed by active growth of gall tissues. The most active growth (cell division) occurs close to the highly stimulated cells on which the larvae feed and inhibit growing. Cell growth is continuously controlled, including control of growth polarity, growth localization, and intensity. The larva exercises such control through the cells on which it feeds and which generally are highly stimulated to proteosynthesis and to synthesis of RNA. Other plant responses such as the formation of ostiolar and epidermal hairs, phenolic and lignin synthesis, are also under the control of the insect. Thus cecidomyiids control all stages of gall morphogenesis via the "metaplasied cells" and later through the "nutritive tissue cells," both of which are in contact with the insect's mouthparts and saliva and which receive different types of mechanical and chemical stimuli. The cecidozoa keeps these cells permanently under stress without killing them. The differentiating nutritive tissue cells will then become specialized for very active transport or as storage tissues.

Variations of the Cecidomyiid Model

Pouch Galls

Pouch galls are formed by many species of cecidomyiids and are the most common variant of the basic cecidomyiid model. Pouch galls result from tissue extensions at minute areas of immature leaves. Initial activity of the larvae is localized at the top of the pouch (Doctors van Leeuwen 1956; Meyer 1957; Rohfritsch 1971b). In some mite and thrips galls the pouch is initiated by a single puncture. Gall growth in restricted areas around the larvae produces an evagination or invagination of the leaf blade from differential surface growths on both sides of the lamina. The outer side (not in contact with the insect) grows more extensively by cellular hypertrophy than the inner side. Although cell hypertrophy and hyperplasia may occur, the wall of the pouch remains thin. Sometimes, as in the galls of *Rondaniola bursaria* (Bremi) and *Phytoptus padi* Nal., they have the same number of cell layers as the host leaf (Meyer and Maresquelle 1983). However, the pouch galls of other species, such as *Mikiola fagi* (Hartig), have thick walls with new cell layers and differentiated gall tissues.

The tissues of pouch galls are generally less differentiated than those of covering galls. In some galls, such as those of *Rondaniola bursaria* on *Glechoma hederacea,* the tissues of the host leaf can be recognized within the walls of the gall. The nutritive tissue of pouch galls is usually less differentiated. There is no starch sheath or sclerenchyma layer; however, the cells of the outer part of the pouch may have lignified walls with lignification beginning at the top of the pouch and extending downward. Other pouch galls, such as those of *Mikiola fagi,* have layers of more or less differentiated nutritive tissues, vascular tissue, and a lignified cortex. Pouch galls often have relatively large ostioles. The ostiolar epidermis is usually covered with hairs, but sometimes the epidermis and other tissues around the ostiole are lignified. Pouch galls usually fall from host leaves and an abscission mechanism is involved.

Mark Galls

Mark galls are induced by only a few cecidomyiids. In this type of gall, the eggs are laid on the outside surface, but then the young mine into the host organ without being enveloped by growing gall tissues. The mining larvae do not induce cecidogenesis; this occurs when the insect becomes sedentary. Tissue proliferation then occurs around the young larva resulting in an oblong form. Several larvae are usually found together and as a result most galls are composed of many larval chambers. This type of gall is found on stems, shoots, and petioles and they are induced by members of the genera *Rabdophaga* and *Lasioptera*. *Lasioptera* larvae often are associated with fungi and induce "ambrosia" galls (see Chapters 8 and 13).

Extension of the Cecidomyiid Model to Other Insect Groups

The covering type of cecidomyiid galls also serves as a general model for galls of mites, homopterans, and thrips; however, there are a few differences related to peculiarities of these insect groups. Whereas the larvae of cecidomyiids select the sites of gall initiation, it is the adults or immature fundatrices of these other groups that induce the galls and their offspring find themselves in galls before they begin to feed. Furthermore, the larvae of cecidomyiids "puncture" cells without provoking cell necrosis, taking their food by sucking on cells of the nutritive tissue whereas in these other groups, the homopterans excepted, feeding by the cecidozoa causes cellular necrosis.

Galls of Thrips

Thrips galls usually develop on leaves and in all cases cell wounding and necrosis takes place. Thrips wound cells and feed on perivascular parenchyma, which is induced to become nutritive tissue (Raman and Ananthakrishnan 1983). Before reaching this more centrally located tissue, they destroy the epidermis and underlying mesophyll cells (Lewis 1973). Thrips galls are not the result of a single individual; instead they are induced by the collective feeding behavior of several adults and a large number of larvae of several subsequent generations (see Chapter 12).

Thrips galls are the most open of all gall systems. The larval chamber remains exposed throughout gall development and there are usually no ostiolar trichomes. Mature thrips freely leave the gall and other species of thrips may enter and live as inquilines.

Each of the feeding sites of thrips constitutes a "micro-gall" with a centrally located wounded zone where surrounding cells are metaplasied, and somewhat inhibited in their growth. There is cell hyperplasia and hypertrophy some distance from the attacked area. The epidermal cells are only slightly or not metaplasied. They may divide but they do not produce hairs as outgrowths.

In general the form of thrips galls is related to the age of the attacked leaf and whether or not it has yet to unroll or unfold, the number of thrips present to initiate the gall, the localization of the feeding sites along the midrib or leaf margins, the population density, and the number of different species living together in the same gall (Ananthakrishnan 1980).

Galls of Homopterans

Homopterans induce a few covering galls, but most are pouch galls. Some homopterans prevent leaves from unfolding whereas others cause folding or rolling. The homopteran fundatrix is usually mobile in its first larval stage but becomes immobile by the last nymphal stage. At the beginning of cecidogenesis, the fundatrix shows a very specific behavior, which varies between species and is responsible for gall morphogenesis (see Chapter 7).

Homopterans always secrete a salivary sheath during the process of probing and feeding and this sheath remains in the tissues marking the pathway followed by the stylets. Cell metaplasia and nutritive tissue cells may occur around the feeding sites. No cells are wounded and there is no cell necrosis. The stylets always travel intercellularly. A well-characterized nutritive tissue occurs in the galls of Adelgidae. The galls of Eriococcidae show a less characteristic nutritive tissue, but one similar to that in galls of *Viteus vitifolii* Fitch (Meyer 1957). In these cases all cells remain very small. They may contain phenolic substances and a kind of transfer tissue is differentiated around these cells.

Nutritive tissue is not differentiated in the galls of many aphids and psyllids. Gall cells here remain parenchymatous, vacuolate, and have little cytoplasm. In these galls the insects take their food in or near the vascular tissue of the attacked organ or of the gall.

Trichomes are usually present in the ostiolar area of homopteran galls except for those of psyllids and coccids. Some pouch galls become bags and sometimes can be very large but they always remain fleshy. Only a few species of homopterans induce sclerenchyma cells in their galls. The largest galls known are caused by a homopteran on *Styrax sumatranus* J.J.S. (Docters Van Leeuwen Reijnvaan and Docters Van Leeuwen 1926). Some may reach 200 mm in length. Other large bag-type galls are induced by *Schlechtendalia chinensis* Bell. on *Rhus semialata* D.C. (Houard 1922).

Galls of Mites

Mite galls are the most minute galls known. They are usually covered with hairy outgrowths. Hairs are not only localized in the ostiolar region, but also can be found along the inside of the gall all over the surface. Sometimes they constitute parts of the nutritive tissue, as is the case with excrescences.

Mites puncture epidermal cells as they feed and often cause cell necrosis (Bronner *et al.* 1989). Callose is synthesized by the punctured cells around the feeding site. Metaplasia also occurs (see Chapter 9). Pouch galls are the most common type caused by mites, but fold, roll, and covering galls also occur. Mites also induce organoid galls by attacking buds (Westphal 1977).

A single mite can induce several galls on young leaves, but later on completely formed galls will develop only under the influence of successive generations all feeding at the same locality. The nutritive tissue of mite galls is parenchymatous or "activated" depending on the developmental stage of the gall and the number of inhabitants. These factors also determine the final form and shape of the gall.

Concluding Comments

All the gall inducers following the cecidomyiid model induce their galls by feeding activities. Generally these galls are not initiated by oviposition or by the eggs. Wounding of the cells is discrete and usually neither necrosis nor cell death is evident at the beginning of cecidogenesis or during proliferation of gall tissues. However, the gall inducers all have a physically injurious action on the plant tissues as a result of contact with the mouthparts. The saliva that bathes the attacked cells is mixed with regurgitated plant sap.

All the gall inducers discussed so far need predigested food. This predigestion occurs in the host cells under the influence of hydrolases coming from the host cell and from the larva (Bronner 1976). The larvae may also stimulate the gall in such a way that specialized cells play the role of a kind of transport tissue that carries solutes from the phloem to the cecidozoa. However, many psyllids and aphids take plant sap directly from the vascular tissues.

THE SAWFLY MODEL

Sawflies of the family Tenthredinidae constitute a large group of insects, most of which are external leaf feeders or leaf miners. Only a few of the approximately 4000 species are gall inducers and nearly all of these are restricted to *Salix* and *Populus.* Gall-inducing sawflies are found in the genera *Phyllocolpa, Pontania,* and *Euura.*

There are three levels of gall complexity and host plant specificity in the tenthredinid gall inducers (Smith 1970). The *Phyllocolpa* on *Salix* and *Populus* are not restricted to any particular species; they form simple fold galls that always remain open. The *Pontania* form a wide variety of closed leaf galls on *Salix,* but they are not host specific. The *Euura* incite closed bud, petiole, or stem galls on *Salix* and of these the stem gall inducers are the most host specific.

Galls induced by sawflies are relatively unstructured (Fig. 5–4) compared to the galls of other insects; most are little more than simple wound callus. There is no cecidogenetic differentiation and often the cell layers keep the characteristics of the host organ on which they form (see Chapter 6).

Sawfly larvae have large, strong mandibles and consume plant tissues by masticating cells, just as do the external, non-gall feeders. Indeed, gall-inducing sawflies live and behave like leaf and stem miners. They do not need predigested food and nutritive cells are not differentiated. Feces are eliminated in the larval chamber. During the third larval instar, the larva bites an exit hole in the gall wall (Fig. 5–4E) and pushes out some of its feces. In the *Pontania* group, the larvae pupate in the soil whereas *Euura* larvae pupate in the gall (Jacobs and Renner 1974, and Chapter 14). Sawfly galls are relatively large and they do not have hairy outgrowths. There is only one larva per gall, but galls sometimes coalesce.

All sawfly galls are initiated by a reaction to the wounding that occurs at oviposition. The act of oviposition is accomplished by a rapid vibration of the gonapophyses and by a sawing motion (hence the name sawflies). The disruption of plant cells is very slight and the female cleans the wounded area before placing an egg into the wound. The egg is bathed in a colleterial fluid secreted as the egg is placed parallel to the split. The oviposition fluid allows the female to seal the wound (Fig. 5–4A).

The plant reacts to the ovipositing action and fluid by slight cell "dedifferentiation," cell hypertrophy, or an outgrowth called a "procecidium," but a complete gall rarely develops. Many species leave the egg pouch and become leaf miners or external leaf feeders. The larvae of *Pontania* remain inside the procecidial growth and feed on the callus cells. In some species, such as *P. proxima* Lep., the young larva has very little influence on gall growth since even without the egg or the larva, the gall almost develops normally (Fig. 5–4A′) (see Chapter 6). In other *Pontania* galls, such as those caused by *P. salicis* (Christ.) and *P. vesicator,* (Bremi-Wolf) the procecidium will develop to a mature gall only under the influence of the feeding larva (Magnus 1914).

Many gall workers have studied the oviposition fluid of *Pontania* females, following the 1888 publication of Biejerinck who suggested that the fluid was responsible for the gall. Küster (1911) agreed with Beijerinck and called the material a "cecidogenic substance." Magnus (1914) repeated Beijerinck's experiments but never succeeded in inducing a leaf gall with secretions from the colleterial gland. However, he studied a large number of *Pontania* galls and was able to explain their sophisticated ovipositional behavior. He found that wounding is very precise and species specific and concluded that it was all but impossible to mimic. He suggested that the variety of growth reactions among the *Pontania* species could be explained by the differences in the type and location of the wound. He also suggested that the time of eclosion was important. In some species such as *P. proxima,* embryogenesis takes about 15 days, whereas in *P. salicis* it takes only 2 days. Magnus did not totally exclude the influence of the ovipositional fluid on procecidial growth, but he denied that it was of major importance in inducing "specific gall morphogenesis."

Hovanitz (1959) and McCalla *et al.* (1962) ignored Magnus and followed Beijerinck's hypothesis and analyzed the secretion of the colleterial glands. They did not take into account the polysaccharide fraction of the secretion, but instead analyzed only the protein fraction. From this they isolated uric acid, adenine derivatives, glutamic acids, and uridine. These substances were then injected into young galls from which the larvae had been removed and they found that the isolated components, under nonsterile conditions, promoted gall growth. They never tried to induce a gall from normal tissues. Thus McCalla *et al.* (1962) did not demonstrate the role of oviposition fluid in gall induction.

It seems clear that Magnus was correct in suggesting that gall growth and development in sawfly galls consist of two distinct stages. The first, or gall induction stage, results from the ovipositional behavior and is dependent only on the ovipositing female. The "procecidial" type of gall is formed (Thomas 1893). The second stage, which leads to a mature gall, is under the influence of the feeding larva. However, in a few cases, such as with *P. proxima,* complete gall development can occur following procecidial growth (Fig. 5–4A′). Here embryogenesis is slow and the larva hatches only after the gall is almost completely developed (Fig. 5–4D). In these cases the eggs are laid on young organs that grow fast and react to the wounding that is localized in the vascular cambium of the leaf. The wounding causes a wound reaction among the surrounding cells. The wounded area escapes normal growth polarity and becomes a sink (Rosenstock and Kahl 1978). When leaf tissues are wounded, the usual events in sugar metabolism and transport are reversed. Soluble sugars become abundant and if starch is present it is hydrolyzed. Cell organelles become modified and dedifferentiated (Barckhausen 1978).

Magnus described a very short period of dedifferentiation that starts before gall initiation. Nearby cells increase in size as do their nucleus and nucleoli before they start dividing. All the layers of cells in the attacked leaf react in a sim-

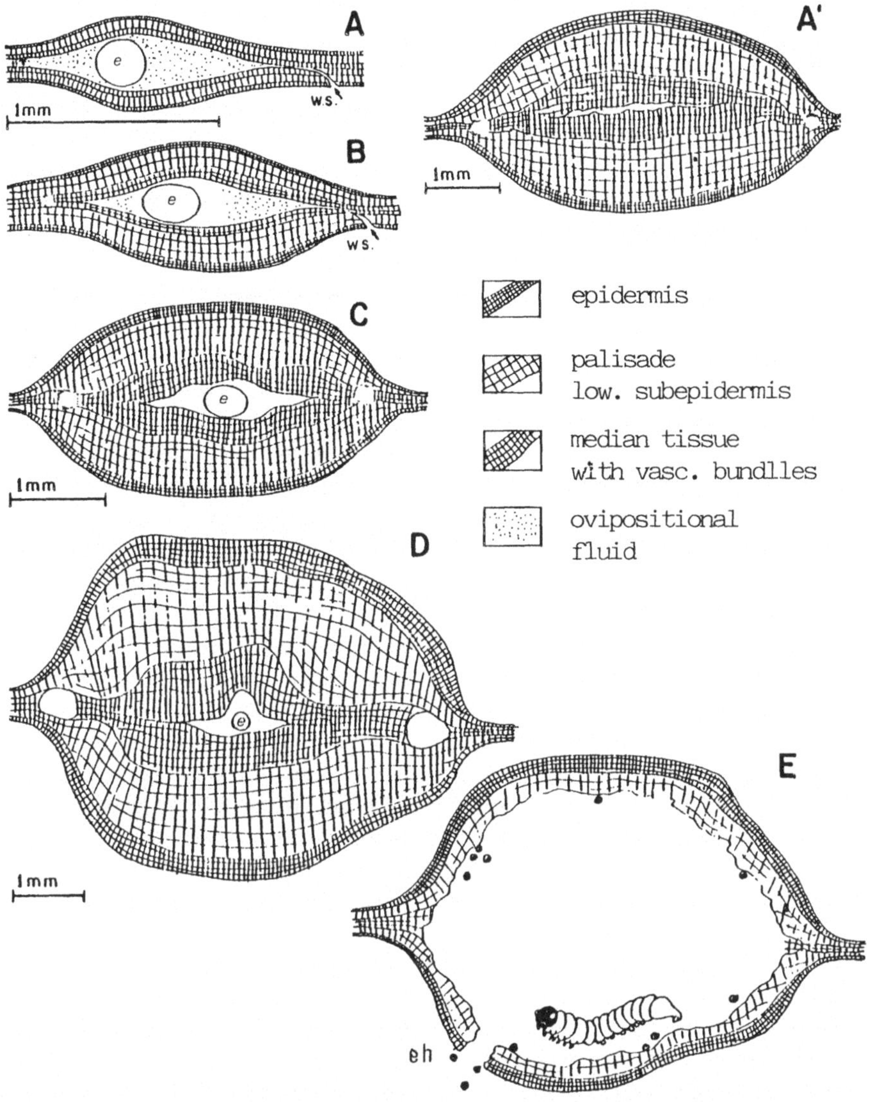

Figure 5-4. Stages in the developmental morphology of galls induced by *Pontania proxima* on the leaves of *Salix*. (A, B, C redrawn from Magnus 1914 and Rey 1976.) (A–E) Normal gall development; (A′) gall that has developed without an egg. e, Egg; ws, wounding scar; eh, exit hole. (Redrawn from Magnus 1914.) (A) Gall a few hours after oviposition. The egg, surrounded by ovipositional fluid, is located in a cavity in the central layer of leaf mesophyll cells. The cavity is prepared by wounding with the ovipositor. The cavity is parallel to the leaf blade and only one layer of cells is destroyed. (B) Gall 3 days after oviposition. The ovipositional fluid has nearly disappeared and the gall appears white with a red halo. All layers of cells, especially the central mesophyll and vascular tissue, show active cell division. (C) Gall 9 days after oviposition. The gall is now half grown, greenish-red, and the central mesophyll cells are small and contain a large amount of tannin and chlorophyll. (D) Gall 2 weeks after oviposition. The gall has now reached its maximum size. Even though the egg is ready to hatch, the egg chamber is still very small. (E) Gall 4 weeks after oviposition. The gall is now mature with a near full grown larva. The larva has consumed the masses of tissues that had formed around the egg. Feces remain in the larval cavity; however, during the third instar, the larva bites a hole in the cavity wall to allow the elimination of some feces and for a later larval exit. The mature larva drops to the soil to pupate. (A′) A small gall develops around the site of wounding by the ovipositor even when the egg is absent.

ilar manner, but those of the central mesophyll and vascular tissues react the most (see Chapter 6). The layers of cells multiply but retain their characteristics (Figs 5–4B, C, and D). Horizontal growth of the gall is influenced by normal leaf growth and lamina extension. At first the young gall is yellow marked by a red ring, but after 3–4 days it becomes green.

Sawfly galls grow essentially by cell divisions. The cells remain small in the central part of the gall where chlorophyll and tannins are abundant and the vascular strands more or less differentiated. Cell hypertrophy occurs at the gall periphery. The epidermis of the host organ is not very modified; the cells may divide but they remain recognizable. There are no epidermal proliferations such as hairs or trichomes.

The central region of the gall, along the oviposition wound, is composed of small densely green cells that appear like "green islands." The same type of tissue is described by Engelbrecht (1970) around wounded tissues, leaf mines, and plant galls. High levels of cytokinines have been found in the tissues of "green islands." The cytokinines are thought to be responsible for the active cell proliferation and for the presence of high levels of chlorophyll. Sziráki *et al.* (1980) demonstrated that the levels of cytokinines increase in stressed tobacco leaves together with a systemic acquired resistance to the mosaic virus.

The gall tissues fed on by the larvae of *P. proxima* remain totally non-responsive, having acquired a type of resistance to the insect-induced change. The larvae have no cecidogenic action on the procecidial tissues. There is no cell dedifferentiation and no cell proliferation, and starch does not accumulate around the larva. The larvae of *P. vesicator,* in contrast, have a strong influence on the tissues of the procecidia. According to Magnus, the gall of *P. vesicator* is induced in a manner similar to *P. proxima.* The wound is restricted to the upper layer of the mesophyll cells and not in the central layer. The first developmental stage of the young gall is of the procecidial type, but after the fifth day, the gall becomes circular and protrudes on the basal side of the leaf. The larva hatches 3–4 days after oviposition and begins to feed on the procecidial callus cells surrounding the larval chamber. New cells are formed behind those eaten as a type of wound callus continuously regenerates new cells. Gall cells away from the regeneration zone are large and round. They have a larger nucleus and nucleolus than the dividing cells of the regenerative area. Starch appears in this gall tissue. Chlorophyll is absent from the inside layer and the tissue remains white. By killing the very young larva, Magnus observed the formation of a chlorophyll-rich tissue composed of numerous cell layers that looked like a normal *P. proxima* gall, demonstrating that the young larva of *P. vesicator* controls tissue growth and differentiation. Thus the inner tissue remains in a more dedifferentiated state, free of tannins and poor in chloroplasts, and continuously responsive to larval feeding activity. By continuous wounding, the larva keeps the cells stimulated and thus *P. vesicator* has real cecidogenetic action on the procecidial tissues. The gall tissues become distinct from normal layers of tissue and a kind of nutritive tissue appears. *P. vesicator* therefore induces a true gall, which develops from a procecidial growth.

Thus cecidogenesis in the sawflies results from a continuous wounding of masses of growing young cells kept in a near undifferentiated stage. As suggested by Magnus, the ovipositional fluid introduced into the wound with the egg has a protective role. The mucilaginous components of the fluid may also help the egg to conserve water. However, the ovipositional fluid cannot be considered the sole "cecidogenic substance," but rather is a component in the cecidogenic interaction. Other factors such as responsiveness of the host plant, the wound, the location of the wound, and the action of the chitinous ovipositor are as important.

Extension of the Tenthredinid Model to Other Insect Groups

Most of the galls induced by beetles and moths, along with those of cyclorrhaphous flies, develop in a manner similar to those of sawflies.

Galls of Beetles

Most of the coleopterous gall inducers belong to the family Curculionidae (see Chapter 2). In this group the eggs are usually laid within stem tissues or on flowers. In all cases the female wounds the tissues with her mouth parts to prepare a hole and then the egg is introduced with ovipositional fluid. Wounding by the ovipositing female and the influence of the eggs are known to be responsible for the host reaction to *Ceutorrhynchus napi* Gyll. on *Brassica napus* L. Here the specific location of the wound and the egg are responsible for gall induction (Deubert 1955; Le Pape and Bronner 1987). When the larva hatches, it leaves the procecidial outgrowth and bores into the pith parenchyma without further gall development. *C. chalybaeus* Germ. feeds on modified pith tissues of *Thlaspi arvense* L., which are similar to the tissues of the sawfly galls of *Pontania proxima.* Another species, *Nanophyes telephii* Bedel., feeds within stems of *Sedum telephium* L. where the cecidogenic growth modifies the vascular ring. Cells fed upon by the larva are somewhat similar to nutritive cells (large nucleus and starch granules); however, no typical gall tissues differentiate (Houard 1903). The larvae of *Rhinocyllus conicus* Froelich feed on receptacle tissues within the flowerheads of *Carduus nutans* L. and induce vast amounts of callus (Shorthouse and Lalonde 1984), but no typical gall tissues appear.

Like the sawflies, beetle larvae have strong mandibles and consume all cells around them. They apparently do not need specialized nutritive cells. The feces remain within the gall. These galls are simply masses of growing cells that allow the insect to obtain the necessary food to develop in specific microhabitats.

Galls of Moths

Gall-inducing moths are more common in tropical regions than they are in Europe and North America, where they have radiated on some genera of plants such as the *Glochidion* of the family Euphorbiaceae (Docters Van Leeuwen-Reijnvaan and Docters Van Leeuwen 1926). Galls of moths are usually fusiform stem galls and are relatively large; however, those that occur on the leaves of *Glochidion* are reniform.

Some lepidopterous galls such as *Heliozela stanneella* F.V.R. on *Quercus suber,* L. are induced by ovipositional behavior, wounding, and ovipositional fluid, which results in a procecidial growth (Prota 1963). The larvae feed on the procecidial growth early in their development but then finish their larval life as a miner. However, the most common case is for the larvae to induce simple galls in stem tissues where they feed as miners. The young larvae of *Gnorimoschema gallaesolidaginis* Riley on *Solidago altissima* L. mine close to the vascular cambium and induce a wound cambiun reaction (Beck 1954). The wall of the mature gall becomes woody.

No tissues typical of most galls develop around the feeding moth larvae. The larvae are always mobile and feed on the callus growths found from the interior to the exterior of the gall. Usually there is only one larva per chamber.

Galls of Cyclorrhaphous Flies

There are few gall-inducing flies outside the family Cecidomyiidae and they are widely dispersed in six families, the most important of which are Tephritidae, Chloropidae, and Agromyzidae. As with the beetles and moths, the larvae of cyclorrhaphous flies have a primitive mode of feeding; they feed by slashing host tissues and ingesting the liquid cell contents. They cause severe damage to the tissues and leave frass in the larval chamber.

Galls of tephritids are usually induced on flowerheads (Freidberg 1984; Arnold-Rinehart 1989); however, two of the best known galls, *Urophora cardui* (L.) on *Cirsium arvense* (L.) Scop. and *Eurosta solidaginis* (Fitch) on *Solidago canadensis* L., are formed on stems. The eggs of most cyclorrhaphous flies are laid externally on immature organs or between unfolded, immature leaves. However, the eggs of *Agromyza erythrinae* de Meijere are laid into the tissues of young leaf petioles (Docters van Leeuwen 1909) and then the larvae mine the cambial tissues. The larvae of *Eurosta solidaginis* and *Urophora cardui* (Lalonde and Shorthouse 1984) first behave as miners, consuming young cells as they tunnel. The larvae stop mining when they locate the cambium near the vascular bundles. However, the new tissues induced are not simple parenchyma as in most galls of moths and beetles. Instead a specialized nutritive tissue is induced as, for example, in the galls of *Agromyza erythrinae* (Docters Van Leeuwen 1909) and in most *Urophora*-induced galls (Lalonde and Shorthouse 1984; Arnold-Rinehart 1989). Some of the cells close to the larvae in maturing galls, referred to as secondary nutritive cells by Lalonde and Shorthouse (1984), are similar to the highly activated cells of the nutritive zone in cecidomyiid galls. However, these secondary nutritive cells are usually not arranged in layers and instead appear in patches formed from the proliferation of vascular cambium. A zone of sclerenchyma cells forms beyond the secondary nutritive tissues and incorporates the entire gall parenchyma, except for the exit channel (Lalonde and Shorthouse 1982).

It appears that the galls of cyclorrhapha represent a transition group between the tenthredinids and cecidomyiids. As with the gall midges, a nutritive tissue is formed and sclerenchyma appears around the nutritive tissue without chloroplasts. The larvae not only have a growth-promoting action, but they also can organize growing tissues. However, in contrast to the cecidomyiids, the nutritive cells of other dipterous galls are not completely "dedifferentiated," that is, they always retain some characteristics of their tissue of origin. Even though patches of nutritive cells are differentiated in these galls, the parenchyma cells beyond are not influenced by the larvae. Thus the nutritive cells in these galls generally have an heterogenous appearance.

Concluding Comments

Most of the gall inducers following the sawfly model belong to families that are plant miners and it can be assumed that the gall-inducing ability evolved from insects of the leaf and stem-mining guild. That is why in all cases gall development begins with wound or callus-like overgrowths induced by oviposition or by tunneling of the young larva. Complexity in gall structure and nutritive tissue differentiation may appear in some groups, but later in gall development.

THE CYNIPID MODEL

Galls induced by cynipids are perhaps the best known of prosoplasmic galls in both Europe and North America. Most have only one larval chamber, are spherical, and have an axial symmetry. The most prominent structural characteristic of cynipid galls is the presence of concentric layers of differentiated cells around the larval chamber (see Chapter 8).

The presence of highly nutritious plant tissues in cynipid galls attracts numerous phytophagous insects and these in turn attract entomophagous species. Thus cynipid galls commonly house a small microcommunity (see Chapter 16). Some of the phytophagous species that inhabit these galls have a cecidogenic action and modify the form and structure of their host galls (Shorthouse 1980).

Another curious feature of cynipids is the phenomenon of alternation of generations on the same host in which a bisexual generation alternates with a unisexual, parthenogenetic generation. Both the insects and galls of these generations are so different (see Chapter 6) that many workers have mistakenly identified them as different species.

Gall initiation by cynipids starts with the act of oviposition and each species lays it eggs in a specific manner and location. The host organ must also be in a particular stage of development at oviposition (Magnus 1914). The highly specific shapes, structures, and epidermal outgrowths are influenced by these early processes (Folliot 1964).

Many galls of the spring generation on *Quercus* appear and develop quickly with the spring flush of leaves since their eggs are laid either during the winter or in the very early spring. Many of these galls occur on flower buds or vegetative buds; few appear on leaves. Bud galls often mature even before the first leaves of spring mature. The eggs of this generation are tightly attached to the plant with an ovipositional fluid secreted by the colleterial gland (Magnus 1914; Meyer 1957). Some cells along the ovipositional channel can be damaged; however, the cells immediately around the egg are not destroyed. In contrast, the eggs of *Biorrhiza pallida* (Olivier) are laid in a cluster between two parts of the sectioned buds of *Quercus* (see Chapter 6). A glue is introduced with the eggs but the glue fixes the egg stalks to the dead part of the bud.

The ovipositional behavior of several *Diplo-*

lepis species on *Rosa* has been studied (Magnus 1914, Meyer 1957, Bronner 1985) and in this group the eggs are introduced into the bud between the young folded leaves. The eggs are not only attached with the glue but also are "pinned" by the ovipositor. In the cases of *D. rosae* (L.), *D. mayri* (Sch.), and *D. spinosissimae* (Giraud) only one epidermal cell is killed (Fig. 5–5A).

The summer gall generations of cynipids develop from July to October. They often appear on leaves or shoots; some are on buds and a few are on roots. The development of this summer generation is slower (Fig. 5–6) than the spring generation. Eggs are introduced into the plant tissue in a very specific manner and are usually laid near a vascular cambium, the cells of which are able to divide even if the leaf, shoot, or roots are mature.

The initiation of cynipid galls is thought to involve several factors including wounding of the tissues with the ovipositor, application of ovipositional fluids, activity of the egg, and activity of the freshly hatched larvae. The eggs are known to exert a proteolytic, cellulolytic and pectinolytic activity in vitro (Bronner 1976; Rey 1976). During embryological development the eggs also exert a polarizing action on the plant tissue. This action is the most active at the distal (opposite to the egg stalk) pole of the egg (known as the posterior larval pole) where the egg is attached to the plant tissue and where the ovipositional fluid is most abundant.

The egg shell ruptures at the edge of the area that is in contact with the plant tissue (Fig. 5–5D) and the young larva turns in the open shell so that it penetrates the plant tissues head first (Magnus 1914; Rey 1976). The influence of externally located eggs on plant tissues has been studied for galls of *Diplolepis* (Magnus 1914), *Neuroterus quercus baccarum* (L.) (bisexual generation) (Meyer 1957), and *Liposthenes glechome* (L.) (Taylor 1949).

When the eggs of *Neuroterus quercusbaccarum* (unisexual generation) are introduced into the plant tissue near a vascular cambium of a leaf (Hough 1953), there is an asymmetric reaction of host tissues between the distal pole of the egg and vascular tissues (Fig. 5–6A). The unisexual generation of *Biorrhiza pallida* induces lignification in root cells contacting the egg. Wound callus appears at the stalk end of the egg and along the oviposition channel. Rey (1976) observed that many eggs die and become totally encircled by the lignified sheath.

The ovipositional fluid of cynipids apparently is used to fix the eggs to the plant tissues. The fluid also allows exchanges of water and substances between the egg and the plant tissue during embryogenesis and gall initiation. Together with the egg chorion, it is the first interface between the insect and the host. However, the egg itself is the most important component in the initiation of cynipid galls.

If the tissue has not been wounded around the egg during oviposition, cell dedifferentiation or metaplasy occurs. The cells become reactivated to proteosynthesis and RNA synthesis. They develop an enlarged nucleus and nucleolus, a fragmented vacuole, the cytoplasm becomes enriched in ribosomes, and if tannins are present they also become fragmented. This first stage appears within 18 hours of oviposition. Cell hypertrophy occurs, but only two cell layers are involved in the case of *Diplolepis rosae* on *Rosa canina* L. (Fig. 5–5B). Cell hyperplasia begins a day later and cell growth beneath the egg gives rise to a pad of tissue supporting the egg (Fig. 5–5C). Cells beneath the egg then begin to lyse. The nucleus and nucleolus become less apparent, vacuoles appear, the cell contents break down, and finally the cell walls collapse. However, the cells around this lytic area grow in such a way that they obliterate the wound (Fig. 5–5C, D). A few cells that are highly stimulated to proteosynthesis (large nucleus and nucleolus and rich in cell organelles and ribosomes) appear beneath the wound. These highly stimulated cells do not appear equally around the cavity; most are beneath the contact zone with the egg shell. Thus two complementary mechanisms—cell lysis and regeneration—are responsible for the formation of the larval cavity within a newly formed tissue called the "lysenchym" by Magnus (1914) and the "plastem" by Beijerinck (1882). Once the egg hatches the larva moves toward this cavity, which is now surrounded by young growing tissues (Fig. 5–5D, E).

The egg chorion remains at the gall surface

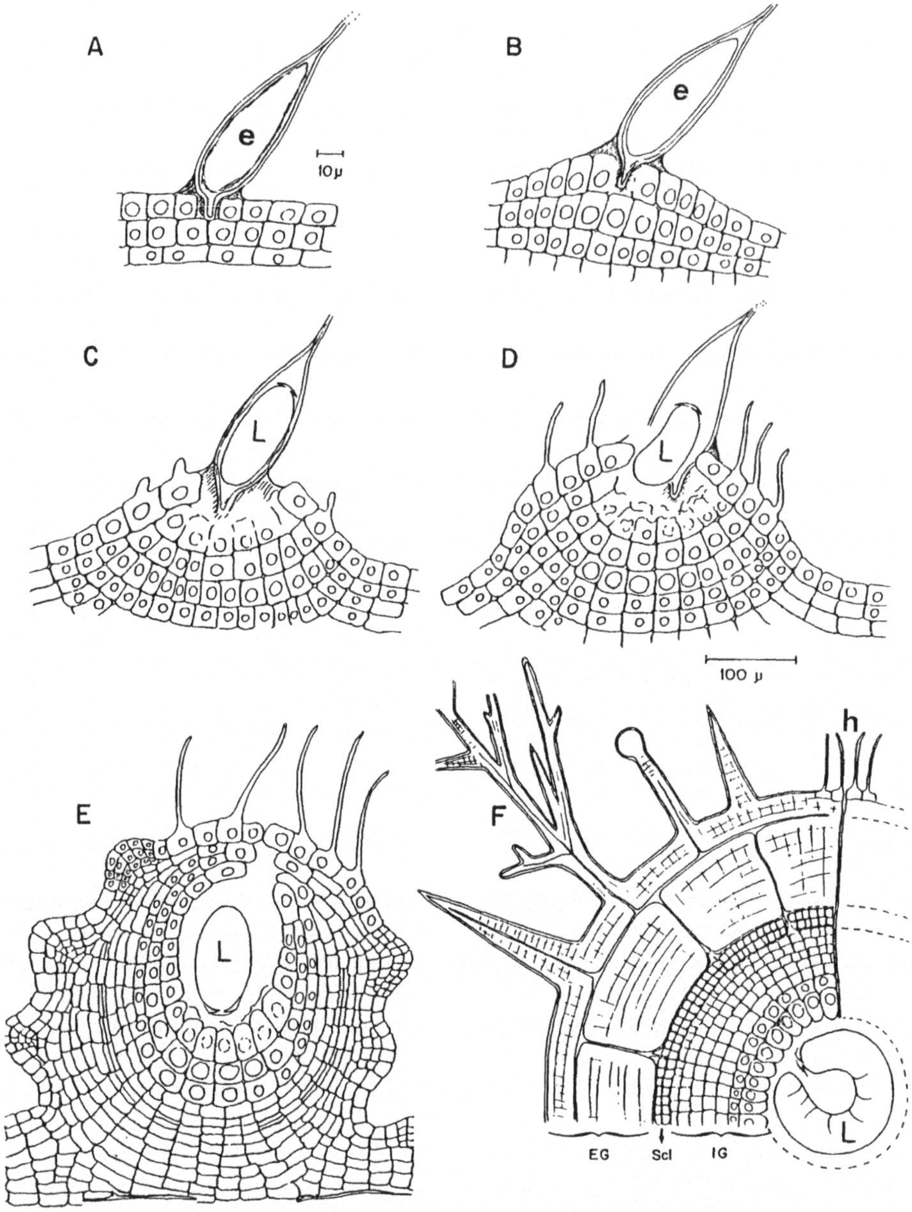

Figure 5-5. Stages in the developmental morphology of galls induced by the cynipid *Diplolepis rosae* on the leaves of *Rosa canina.* (A) Freshly laid egg pinned to the epidermis of a young leaf. The fluid produced by the colleterial gland glues the egg chorion to the plant. The egg has a long stalk at the anterior end. (B) One day after oviposition. The leaf epidermis and subepidermis become dedifferentiated and cell hypertrophy begins. (C) Three days after oviposition and before the egg hatches, the egg induces cell lysis on the pad of young growing cells beneath the egg. A lytic cavity begins to form. The head of the larva is turned toward the egg stalk before hatching. (D) The egg chorion ruptures and as the larva hatches, it moves into the small cavity beneath. The larva turns upside down just as the egg hatches and enters the cavity head first. Active cell growth and cell lysis beneath the larva produces the young gall. Epidermal hairs appear on the gall near the shell. (E) Immature gall 12 days after oviposition. The larva is deeper into the neoformed gall tissue. Cell growth behind the larva closes the larval chamber. Nutritive tissue cells appear in front of the head and a cambial layer is formed in the center of the mass of growing cells. The gall epidermis gives rise to the hairy outgrowths. Unicellular hairs indicate the entrance channel of the larva. (F) Portion of a mature gall. The outer gall is composed of large parenchyma cells and contains vascular tissue. The outer gall is separated from the inner gall by a layer of sclerenchyma. The inner gall is composed of a starch parenchyma and the layers of nutritive tissue. Each species of *Diplolepis* induces different types of superficial outgrowths as indicated on this figure. e, Egg; EG, outer gall; h, hairs; IG, inner gall; L, larva; Scl, sclerenchyma.

and a few trichomes appear near it. If the embryo dies before hatching, the egg will remain fixed in the leaf tissue surrounded by a small partially elevated pad. This pad is often covered with hairs and is red due to anthocyanins.

Once the larva is enclosed in the larval chamber, all the cells that border the chamber are equally stimulated. This may result from the larva continuously turning in the small chamber. The larva grows slowly during the period of most rapid gall growth (Roth 1949). All cells that border the larval cavity and that are stimulated to proteosynthesis and appear to have a very high turnover before becoming lysed are referred to as nutritive cells. The cells closest to the larva are the largest and the most highly stimulated (Figs. 5–5E, F). This stimulation is a result of continual wounding with the chitinous mandible, the presence of salivary enzymes such as amylases and proteases (Bronner 1976), and various plant cell substances that are liberated into the larval cavity. Hydrolases together with amino acids, soluble sugars, different metabolites, and cell wall fragments are liberated from partially lysed cells. All these substances constitute a "cocktail" that the larva activates, manipulates, and on which it feeds. Behind the nutritive tissue cells there are actively dividing cells generating radially orientated lines of parenchymatous cells (Figs 5–5E, F). The epidermis gives rise to different types of surface projections such as trichomes, spines, and glandular hairs. Vascular cambium appears in this mass of growing tissue. There is a peripheral vascular cambium in the outer part of the young gall and it is generally more abundant in the region where the gall is attached to the plant.

When the eggs of some cynipids are deposited into plant tissues, the larva moves to a new site after leaving the egg since the oviposition site often becomes partly surrounded by lignified cells. The larva moves toward the young newly formed tissue, which generally develops only at one pole of the egg. The larvae of some species exerts an interesting morphogenetic action during this early stage. The larvae of *N. quercusbaccarum* (unisexual generation) shapes the peculiar spangle gall on *Quercus* (Hough 1953) as it moves into the young growing cell mass (Fig. 5–6). Inhibition of attacked organs is unusual during morphogenesis of cynipid galls, except in the case of bud galls.

If the larva is killed during early periods the entire gall tissue will continue to grow through general cell hypertrophy but mitosis ceases. The gall tissue gives rise to a uniform parenchyma on which the earlier formed epidermic projections and vascular tissues can be recognized. The cavity around the dead larva is often totally obliterated by growing parenchyma cells (Magnus 1914; Rohfritsch 1975a; Bronner 1976). The external appearance of such a "dead gall" is normal although it may be smaller. Such galls appear kataplasmic, that is, they do not show the characteristic layers of cells found in cynipid galls.

The nutritive tissue is not well structured during the early stages of gall development. Patches of nutritive cells are usually present around the larval chamber or they may be abundant in one region (Figs. 5–5E and 5–6F). However, as larval feeding increases, the tissues become more homogeneous and often cover the entire larval chamber. The layer also becomes thicker. At this stage cell divisions stop at the gall periphery. In some galls such as that of *Liposthenes glechomae,* a new cambial zone appears 5–6 layers behind the nutritive tissue cells. Radial vascular tissue is directed toward the cambial area and toward the larval chamber from the peripheral vascular tissues. The cambial tissue gives rise to parenchyma cells, which contain starch at first, but as soon as mitosis stops and tissue differentiation occurs, the periphery of these cells lignifies. The inner part, referred to as the inner gall, remains parenchymatous and continues to give rise to nutritive tissue. The outer gall is composed of peripheral vascular tissue, cortical parenchyma, and an epidermis with hairy or glandular outgrowths. The outer gall or cortex differentiates first, while the cells of the inner gall are still actively dividing. The cortex contains the peripheral vascular tissue of the gall, the connection with the vascular tissue of the host organ, and the vascular connection with the inner gall. Often the cortex is a water and air reserve. A sheath of sclerenchyma cells sometimes differentiates in the proximity of the vascular tissue and as a result the gall becomes a series of concentric layers (Fig. 5–5F). Gall cortex and external gall

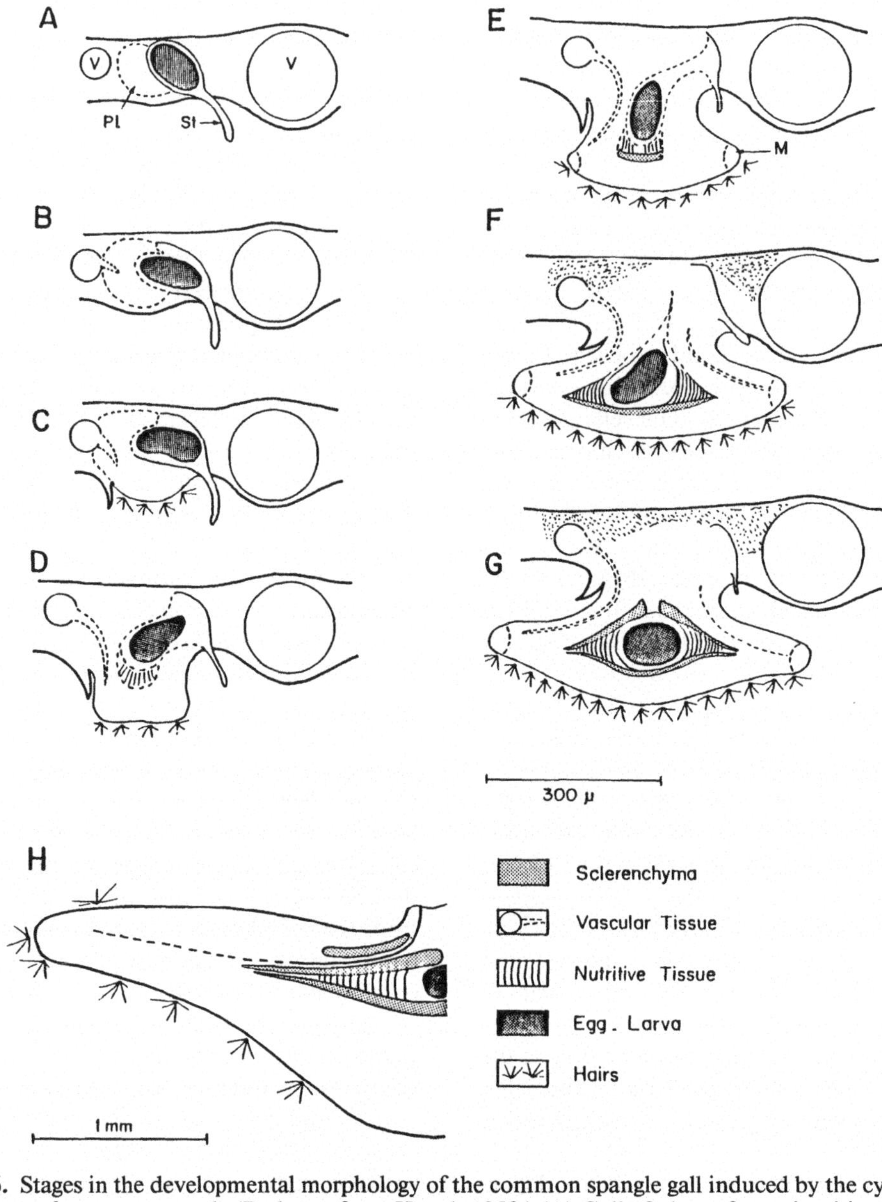

Figure 5-6. Stages in the developmental morphology of the common spangle gall induced by the cynipid *Neuroterus quercusbaccarum* on oak. (Redrawn from Hough 1953.) (A) Gall 12 days after oviposition. A zone of callus tissues or plastem appears between the egg and a small vein near the distal end of the egg. (B) Gall 21 days after oviposition. The larva hatches and penetrates head first into the lytic cavity formed at the distal end of the egg. A vascular cambium develops in the gall plastem from the vascular tissue of a small vein. (C) Gall 25–29 days after oviposition illustrating the rupture of the leaf epidermis and appearance of hairs. (D) Gall 1 month after oviposition showing developing vascular tissue and new orientation of the larva. (E) Gall 5 weeks after oviposition showing differentiation of the marginal meristem and gall sclerenchyma in front of the head of the larva and beneath the first nutritive tissue. (F) Gall 6 weeks after oviposition showing differentiation of the sclerified platform (shaded area in the leaf) and further extension of nutritive and sclerenchyma tissues. (G) Gall 7 weeks after oviposition showing sclerification of the second plate and the final location of nutritive tissue. (H) Mature gall in the fall when it drops to the ground. M, Meristem; Pl, plastem; St, egg stalk; V, vein.

structures (spines, glandular hairs, etc.) are thought to have evolved as protection from parasitic Hymenoptera (Askew 1984).

Rohfritsch (1971a) was able to show with a series of experiments that there is competition between the gall cortex and the layers of the inner gall. The gall cortex is dependent on the host plant and it may be large or small depending on the level and quality of nutrition it receives from the host (Bronner 1983). However, the size of the cortex has no influence on the larval size. The cortex differentiates the abscission layer and it often contains tannins or lignins. The food and water available to the gall are directed toward the inner gall, which acts as a sink (Jankiewicz *et al.* 1969). If the larva is killed the sink effect ceases and all nutrition is used by the cortex. Thus postmortem gall growth is essentially growth of the cortical tissue. Such half-galls, with or without larvae, have been cultured on an artificial medium. If a larva is not present, only the cortex of the gall will develop (Rohfritsch 1971a). When the larva is killed, the tissue of the outer part of the gall realizes its total growth potential, which usually is partly inhibited by the growth and differentiation of the inner gall.

Thick-walled sclerenchyma cells make up the sclerenchyma sheath, which is characteristic of many cynipid galls. In galls without sclerenchyma, a layer of cells with tannins is usually present. The sclerenchyma sheath may be circular or oval. Sometimes it is incomplete along the vascular bundles. In lenticular galls (Fig. 5–6H), the sclerenchyma consists of two large plates of cells, one above and one below the larval chamber.

The sclerenchyma layer is synthesized only if the larva is actively feeding. If the larva is killed just before differentiation, or if the in vitro grown gall receives a weak sugar supply, the sclerenchyma does not differentiate.

The cells of the sclerenchyma layer usually connect through numerous plasmodesmata (Bronner 1976). Often the sclerenchyma acts like a transfer tissue for water and solutes (Fourcroy and Braun 1967). In some galls, these sclerenchyma cells are small, tangentially elongated with heavily lignified walls and in these cases they do not appear very active.

In most maturing cynipid galls the larval chamber is bordered by recently emptied cells and remnants of others while the adjacent cells are either in various stages of dissolution or are highly active (Bronner 1976; Rohfritsch 1974, 1978). These cells are characterized by an increase in size, enrichment in cytoplasm, large nuclei, and prominent nucleoli (see Chapter 8). The enlarged nucleus and nucleolus are most striking in the cells nearest the region of contact with the larva. The thickness of the nutritive zone is greatest at the time of lignification of the hard layer and then it is subject to increasing "dissolution" as the larva makes its demands for nourishment. By the time the insect completes the pupal stage, there is nothing left of the nutritive zone but a lining of a few partially eaten cells and remnants of cell walls bordering the sclerenchyma sheath. Thus the early nutritive tissue functions as a food attractor that allows tissue growth and differentiation. During the last larval stage, the nutritive tissue functions as a provider of food.

The galls of cynipids exhibit the most perfect morphogenesis of all galls. They also have the most original nutritive tissue that shows the same characteristics in all the galls studied to date.

Extension of the Cynipid Model to Other Insect Groups

Some of the wasps in the superfamily Chalcidoidea, most of which are entomophagous parasites, have the ability to induce galls. Little is known about the galls of chalcids, the most important works being those of Magnus (1914), Docters van Leeuwen (1910), and Docters van Leeuwen-Reijnvaan and Docters van Leeuwen (1926). Most studies of chalcidoid galls have been made on the genus *Harmolita* (Eurytomidae); however, little is known about the galls of these species. It is known that as in galls of cynipids, a nutritive tissue differentiates around the young larva (Van Staden *et al.* 1977; West and Shorthouse 1989). The nutritive tissue originates from a plastem; however, the histology of these nutritive cells has been poorly studied. As in the cynipid galls, the larva consumes the entire nutritive layer and a sclerenchyma sheath forms behind the nutritive tissue and it is in close contact with the vascular bundles of the gall. However, in the galls of most species the vascular bundles run parallel to the larval chamber; that is, they do not extend toward the larval chambers as in cynipid galls.

Chalcid females deposit their eggs deep within plant tissues with a long and sharp ovipositor (Magnus 1914). The eggs are usually deposited in a row with each oviposition wound and egg being separated by a few millimeters. The eggs, together with the wound made at oviposition, stimulates cell growth. A large plastem is found, but the chalcid egg does not have the lytic and polarizing action of cynipid galls (West and Shorthouse 1989). Swelling of the attacked organ appears opposite to the oviposition scar. This asymmetrical swelling bends the tip of the shoot horizontally or allows the gall tissue to protrude out of the shoot or of the leaf vein.

Concluding Comments

Galls of the cynipid type are initiated under the influence of the ovipositing female who influences ovipositional sites, degree of wounding, amount of ovipositional fluid deposited, and the egg chorion. However, it is the larval stage of cynipids that provides the main influence on gall growth and tissue differentiation. All of these influences have been clearly demonstrated by Folliot (1964) who obtained interspecific galls of the unisexual hybrid *Andricus kollari* (Htg.) (male) and *Andricus lignicola* Htg. (female). The maternal influence on gall morphology is of great importance in the galls induced by chalcids.

DISCUSSION

It is well known that plant tissues react to a wide variety of externally applied chemical, physical, and biological stimuli. In most cases the plant either reacts by atypical growth, the synthesis of stress metabolites, or cell autolysis. Insects are one of the many biological factors able to induce plant growths and they have been able to build their ecological niches by exploiting this plant reaction. Gall tissues become physiological sinks, as do wounded plant tissues, drawing photoassimilates from other parts of the plant. The sinks occur as a result of insect wounding and feeding, cell proliferation, and synthesis of starch and stress metabolites. The accumulating tissues provide the insect with a shelter and food.

The types of gall-inducing factors, whether they be physical or chemical, appear to be characteristic to certain gall insects. Plant wounding and ovipositional fluids are important growth-promoting factors in the sawflies, whereas in the cynipids the eggs and young larvae possess strong lytic action. In cecidomyiids, light mechanical wounding and salivary enzymes seem to be the most important factors.

Once the gall insect induces a growth reaction on the plant tissues it has to control the growth. Of interest, the synthesis of stress metabolites such as tannins and lignins also appears. Both types of mechanisms isolate the insect from the vascular tissue of the host plant. In the cecidomyiid model, growth between the nutritive layer and the vascular tissue is strongly inhibited. In the cynipid model, cellular lysis allows the insect to overcome growth reactions and in the sawflies the larva consumes all proliferating cells. The systhesis of tannins and lignins are usually inhibited by nutritive tissue, but they appear within the gall more distal from the larval chambers.

Several authors have suggested that there is a correlation between the mode of feeding of gall insects and the complexity of their galls (Wells, 1921; Maresquelle and Meyer 1965). Insects with primitive mandibulate mouthparts such as sawflies, moths, and beetles tend to induce simple kataplasmic galls whereas the more advanced prosoplasmas are induced by cecidomyiids and cynipids, both of which employ a suctorial mode of feeding. Most homopteran galls also are very simple kataplasmas.

The occurrence of extra-intestinal digestion also has to be considered in gall induction. Some insects such as the homopterans often take their food directly from the sap of vascular tissues, while others such as the cecidomyiids and the adelgids induce plant tissues into becoming short distance transport systems. Cynipids can induce the cells of the inner gall layers to autolysis so that the predigested plant substances are delivered to the insect as soluble sugars, soluble proteins, and simple lipid bodies.

By examining various patterns in gall development, it becomes apparent that the most complex galls (those of cynipids) are derived from tissues that have been extensively modified and that seem under the control of the insect rather than the plant. Cells freshly incorporated into the gall are undifferentiated callus cells that have totally lost the determination provided earlier by the organ on which they originated. The gall also escapes the physiological correlations as exist between the plant organs. Polarity of gall tissues has no relation to the host organ. The gall instead develops its own polarity and symmetry, focused on the larval chamber and especially on the nutritive tissue cells. This polarization generates the different gradients discussed by Bronner (see Chapter 8) and is responsible for the appearance of the successive layers such as starch, sclerenchyma, cortex, peripheral vascular tissue, and epidermis. Apart from drawing nutrients and water from the host plant, the gall becomes a separate entity.

Galls following the cecidomyiid model result from dedifferentiation of normal plant tissue. The activity of the larva is necessary to keep nutritive cells under stimulation and to inhibit cell growth and tannin synthesis in young developing galls. In maturing galls the presence of the larva is necessary for differentiation of the sclerenchyma sheath. Cecidomyiid larvae usually derive their food from the vascular tissue of the host plant through a specialized nutritive tissue acting as a short distance transport tissue. Gradients are not as strong as they are in the cynipid model. The gall usually does not appear as a separate entity, but is grafted onto the normal plant organ along the early nutritive tissue which is composed of dedifferentiated epidermal and mesophyll cells.

Galls following the sawfly model are essentially local overgrowths of one or all tissues of the attacked organ. Normal tissue proliferates without real dedifferentiation, or sometimes gives rise to a simpler parenchyma. In these galls often the original layers of cells can be recognized. There are weak cytological or histological gradients or successive cell layers that appear around the larva. If lignification occurs, the whole gall tissue becomes lignified. Cell proliferation and lignification may even appear if the larva is killed, indicating that they appear in a passive way and the larva has little control over these processes.

As has been shown in this chapter and in others, some gall types are relatively primitive having originated long ago (sawflies) or they may be of more recent origin (moths) with their galls resulting from primitive interrelationships between plant and insect. In contrast, the formation of cynipid galls illustrates a sophisticated type of interrelationship that appears rigid and uniform. The range of cecidomyiid galls is found between the two extremes as they represent all degrees of evolution of insect–gall–plant interrelationships. The cecidomyiid model is the most common but also the most plastic. It is the model that best demonstrates the variability in gall induction. Better than any other insect group, the cecidomyiids illustrate the developmental potentials not normally expressed in plants and that can be solicited by externally applied factors.

Gall insects modify minute areas of their host plants by soliciting unusual gene expression from adjoining cells in such a way that new developmental events result. Thus the insect, through its activities, is able to reveal new morphogenetic potentialities of the host plant, which initially were directed against it, but which the insect now uses to ensure food and shelter.

LITERATURE CITED

Ananthakrishnan, T. N. 1980. On some aspects of thrips galls. Bull. Soc. Bot. Fr. 127. Actual. Bot. 1:31–34.

Arnold-Rinehart, H. 1989. Histologie und Morphogenese von *Urophora* und *Myopites* -Blütenkopfgallen (Diptera: Tephritidae) in Asteraceae. Ph.D. Thesis, University of Bayreuth, Bayreuth, FRG.

Askew, R. R. 1984. The biology of gall wasps. In T. N. Ananthakrishnan (ed.), Biology of Gall Insects, pp. 223–271. Oxford & IBH, New Delhi, India.

Barckhausen, R. 1978. Ultrastructural changes in wounded plant storage tissue cells. In G. Kahl (ed.), Biochemistry of Wounded Plant Tissues, pp. 1–42. de Gruyter, New York.

Beck, E. G. 1954. The nature of the stimulus in the *Solidago* gall induced by the larva of *Gnorimoschema gallaesolidaginis.* In Abnormal and Pathological Plant Growth, pp. 235–251. Report of Symposium 1953, Brookhaven National Laboratory, Upton, New York.

Beijerinck, M. W. 1882. Beobachtungen über die ersten Entwicklungsphasen einiger Cynipidengallen. Naturk. Verh. Köninkl. Akad. Wetenschap. 22:1–198.

Beijerinck, M. W. 1888. Uber das Cecidium von *Nematus capreae* auf *Salix amygdalina.* Bot. Ztg. 46:1–17.

Birch, M. L. 1974. Studies on gall formation in *Viola odorata* by *Dasyneura affinis* Kieffer (Diptera: Cecidomyiidae) with emphasis on the experimental modification of this process. Ph.D. Thesis, Univ. of Reading, Reading, U.K.

Bloch, R. 1965. Abnormal development in plants: A survey. Handb. Pflanzenphysiol. 15:156–183.

Boysen Jensen, P. 1948. Formation of galls by *Mikiola fagi.* Physiol. Plant. 1:95–108.

Bronner, R. 1976. Contribution à l'étude histochimique des tissus nourriciers des Zoocécidies. Marcellia 40:1–134.

Bronner, R. 1983. Adaptation Insect-Plant in Cynipid Galls. In N. S. Margaris and M. Arianoutsou-Faraggitaki (eds.), Adaptations to Terrestrial Environments, pp. 61–68. Plenum, New York.

Bronner, R. 1985. Anatomy of the ovipositor and ovipositional behavior of the gall wasp *Diplolepis rosae* (Cynipidae: Hymenoptera). Can. Entomol. 117:849–858.

Bronner, R., E. Westphal, and F. Dreger. 1989. Chitosan, a component of the compatible interaction between *Solanum dulcamara* L. and the gall mite *Eriophyes cladophthirus* Nal. Physiol. Mol. Plant Pathol. 34:117–130.

Desbiez, M. O., P. Champagnat, N. Boyer, J. M. Frachisse, Th. Gaspar, and M. Thellier. 1983. Inhibition correlative de la croissance de l'hypocotyle de *Bidens pilosus* L. par des traumatismes cotylédonaires légers. Bull. Soc. Bot. Fr. 130. Actual. Bot. 314:67–77.

Deubert, K. H. 1955. Beiträge zu den Beziehungen zwischen *Ceutorrhynchus napi* Gyll. und Winterraps hinsichtlich der Gallenbildung mit Ovarien Untersuchungen an verschiedenen *Ceutorrhynchus* Arten. Wiss. Ztschr. Martin Luther Univ. Halle 4:909–932.

Docters Van Leeuwen, W. M. 1909. Beiträge zur Kenntnis der Gallen von Java. Ueber die Anatomie und Entwicklung der Galle auf *Erythrina lithosperma* Miquel von einer Fliege, *Agromyza erythrinae* de Meyere gebildet. Recueil travaux bot. Neerl. VI:1–32.

Docters Van Leeuwen, W. M. 1910. Uber die Anatomie der Luftwurzeln von *Ficus pilosa* Reinw. und *F. nitida* L. var. *retusa* King und der von Chalcididen auf denselben gebildeten Gallen. Ber. Deutsch. Bot. Ges. XXVIII, 5:169–181.

Docters Van Leeuwen, W. M., and J. Doctors Van Leeuwen-Reijnvaan. 1907. Uber die Anatomie und die Entwicklung einiger *Isosoma*-Gallen auf *Triticum repens* und *Junceum* und über die Biologie der Gallformen. Marcellia. 6:68–101.

Docters Van Leeuwen-Reijnvaan, J., and W. M. Docters Van Leeuwen. 1926. The Zoocecidia of the Netherlands East Indies. Batavia, The Netherlands.

Docters Van Leeuwen, W. M. 1956. The aetiology of some Thrips galls found on leaves of Malaysian *Scheffleria* species. Acta Bot. Neerl. 5:80–89.

Engelbrecht, L. 1970. Cytokinin activity in larval infected leaves. Biochem. Physiol. Pflanzen 162:9–24.

Folliot, R. 1964. Contribution à l'étude de la biologie des Cynipides gallicoles (Hyménoptères: Cynipoidea). Thèse, Masson, Paris.

Fourcroy, M., and C. Braun. 1967. Observations sur la galle de l' *Aulax glechomae* L. sur *Glechoma hederacea* L. II. Histologie et rôle physiologique de la coque sclérifiée. Marcellia 34:3–30.

Freidberg, A. 1984. Gall Tephritidae. In T. N. Ananthakrishnan (ed.), Biology of Gall Insects, pp. 129–167. Oxford & IBHN, New Delhi, India.

Hatchett, J. H., and G. L. Kreitner. 1988. Mouthparts and feeding mechanism of the larva of the Hessian fly *Mayetiola destructor* (Say). 31. In Proceedings XVIII Int. Cong. Entomol. Vancouver, B.C.

Hill, R. L. 1982. Seasonal patterns of phytophagous activity on gorse *(Ulex europaeus),* and host plant quality. Proc. 5th Int. Symp. Insect-Plant Relationships, pp. 225–235. Pudoc, Wageningen.

Houard, C. 1903. Recherches anatomiques sur les galles de tiges: Pleurocécidies. Bull. Sci. France Belgique 38:140–419.

Houard, C. 1922. Les zoocécidies des plantes d'Afrique, d'Asie et d'Océanie, 2 Vols. Herrmann et Cie, Paris.

Hough, J. S. 1953. Studies on the common spangle gall of oak. I. The developmental history. New Phytol. 52:149–177.

Hovanitz, W. 1959. Insects and plant galls. Sci. Am. 201(5):151–162.

Jacobs, W. and M. Renner. 1974. Taschenlexikon zur Biologie der Insekten. Gustav Fischer Verlag, Stuttgart.

Jankiewicz, L. S., H. Plich, and R. Antoszewski. 1969. Preliminary studies on the translocation of ^{14}C labelled assimilates and $^{32}PO_4^{3-}$ towards the gall evoked by *Cynips (Diplolepis) quercusfolii* L. on oak leaves. Marcellia 36:163–174.

Kirst, G. O., and H. Rapp. 1974. Zur Physiologie der Galle von *Mikiola fagi* Htg. auf Blättern von

Fagus silvatica L.2. Transport ^{14}C markierter Assimilate aus dem befallenen Blatt und aus Nachbarblättern in die Galle. Biochem. Physiol. Pflanzen 165:445–455.

Küster, E. 1911. Die Gallen der Pflanzen. S. Hirzel, Leipzig.

Lalonde, R. G. 1983. Insect-host plant interaction in the *Urophora cardui* (Diptera: Tephritidae) *Cirsium arvense* (L.) Scop. (Compositae) gall system. M.Sc. Thesis. Laurentian University, Sudbury, Canada.

Lalonde, R. G., and J. D. Shorthouse. 1982. Exit strategy of *Urophora cardui* (Diptera: Tephritidae) from its gall on Canada thistle. Can. Entomol. 114:873–878.

Lalonde, R. G., and J. D. Shorthouse. 1984. Developmental morphology of the gall of *Urophora cardui* (Diptera: Tephritidae) in the stems of Canada thistle *(Cirsium arvense).* Can. J. Bot. 62:1372–1384.

Le Pape, H., and R. Bronner. 1987. The effects of *Ceutorrhynchus napi* (Curculionidae, Coleoptera) on the stem tissues of *Brassica napus* var. *oleifera.* In. V. Labeyrie, G. Fabres, and D. Lachaise. (eds.), Insect-Plants, pp. 207–211. W. Junk, The Hague.

Lewis, T. 1973. Thrips, Their Biology, Ecology and Economic Importance. Academic Press, London.

Loux, E. 1967. Observations biologiques sur *Rhabdophaga ramicola* Rübs. Marcellia 34:93–102.

Lutz, A., and J. Meyer. 1965. Croissance in vitro de portions de *Craneiobia cornii* (Giraud) sur *Cornus sanguinea* L. Marcellia 32:79–88.

Magnus, W. 1914. Die Entstehung der Pflanzengallen verursacht durch Hymenopteren. G. Fischer, Jena.

Mani, M. S. 1964. Ecology of Plant Galls. W. Junk, The Hague.

Maresquelle, H. J., and J. Meyer. 1965. Physiologie et morphogenèse des galles d'origine animale (Zoocécidies). Hdb. Pflanzenphysiol. 15(2):280–329.

McCalla, D. R., M. K. Genthe, and W. Hovanitz. 1962. Chemical nature of an insect gall growth factor. Plant Physiol. 37:98–103.

NcNeil, S., and R. A. Prestige. 1982. Plant nutritional strategies and insect herbivore community dynamics. Proc. 5th Int. Symp. Insect-Plant Relationships, pp. 225–235. Pudoc, Wageningen, The Netherlands.

Meyer, J. 1957. Cécidogenèse comparée de quelques galles d'Arthropodes et évolution cytologique des tissus nourriciers. Ph.D. Thèse, Université Louis Pasteur, Strasbourg, France.

Meyer, J. 1969. Irrigation vasculaire dans les galles. Bull. Soc. Bot. Fr. Mém. 75–97.

Meyer, J., and H. J. Maresquelle. 1983. Anatomie des galles. In Handbuch der Pflanzenanatomie. Gebrüder Borntraeger, Stuttgart.

Meyer, J. 1987. Plant Galls and Gall Inducers. Gebrüder Borntraeger, Berlin.

Prota, R. 1963. Sulla formazione del cecidio di *Heliozella stanneella* F.V.R. (Lepidoptera, Heliozelidae). AHI Acad. Sci., Torino 98:235–251.

Raman, A., and T. N. Ananthakrishnan. 1983. Studies on some Thrips (Thysanoptera, Insecta) induced galls. 2. Fine structure of the nutritive zone. Proc. Indian Natl. Sci. Acad. 49:525–561.

Rey, L. 1976. Recherches comparées sur trois types d'Hymenoptèrocécidies (Cynipidae et Tenthrèdinidae). Etude morphologique histologique et ultrastructurale du développement et de la differenciation. Ph.D. Thèse, Nantes, France.

Rohfritsch, O. 1967. Ejection de la coque cécidienne d'*Oligotrophus reaumurianus* de la feuille de Tilleul. Marcellia 37:139–149.

Rohfritsch, O. 1971a. Culture *in vitro* de jeunes galles d'*Aulax glechomae* L. sur *Glechoma hederacea* L. Marcellia 37:151–161.

Rohfritsch, O. 1971b. Développement cécidien et rôle due parasite dans quelques galles d'Arthropodes. Marcellia 37:233–339.

Rohfritsch, O. 1974. Infrastructure du tissu nourricier de la galle de l'*Aulax glechomae* sur *Glechoma hederacea* L. Protoplasma 81:205–230.

Rohfritsch, O. 1975a. Etude comparative de cellules du tissu nourricier de la jeune galle de l'*Aulax glechomae* L. sur *Glechoma hederacea* L. et de cellules du tissu nourricier abandonné par le parasite. Marcellia 3:185–196.

Rohfritsch, O. 1975b. Etude du développement de la galle de *Giraudiella inclusa* Fr. sur *Phragmites communis* Trin. Marcellia 38:251–270.

Rohfritsch, O. 1978. Three-dimensional study of cell-organelles in the nutritive tissue of a gall (*Liposthenes glechomae* L. on *Glechoma hederacea* L.). Protoplasma 95:297–307.

Rohfritsch, O. 1980. Relations hôte-parasite au début de la cécidogenèse de *Hartigiola annulipes* Hartig sur le Hêtre. Bull. Soc. Bot. Fr. 127 Actual. Bot. 1:199–207.

Rohfritsch, O. 1987a. Different food supply strategies in midge induced plant galls. In. V. Labeyrie, G. Fabres, and D. Lachaise (eds.), Insects-Plants, pp. 195–200. W. Junk. The Hague.

Rohfritsch, O. 1987b. Gall development and feeding strategy of *Didymomyia reaumuriana* Lw. (Cecidomyiidae) on *Tilia* (*T. cordata* Mill. × *T. platyphyllos* Scop.). Phytophaga 1:1–20.

Rohfritsch, O. 1988a. Food supply mechanism related to gall structure with the example of *Geocrypta galii* Lw. (Cecidomyiidae, Oligotrophini) on *Galium mollugo.* Phytophaga 2:1–17.

Rohfritsch, O. 1988b. The early stages of gall morphogenesis and larval behaviour. 31. In Proceedings XVIII Int. Congr. Entomol., Vancouver, B.C.

Rohfritsch, O. 1990. Two gall midges (Diptera: Cecidomyiidae) on *Tilia:* Two different patterns of interaction with the host tree. Phytophaga. 3:13–21.

Rohfritsch, O., and J. D. Shorthouse. 1982. Insect Galls. In G. Kahl and J. Schell (eds.), Molecular Biology of Plant Tumors, pp. 131–152. Academic Press, New York.

Rosenstock, G., and G. Kahl. 1978. Phytohormones and the regulation of cellular processes in aging storage tissue. In. G. Kahl (ed.), Biochemistry of Wounded Plant Tissues, pp. 623–671. de Gruyter, New York.

Roth, P. 1949. Beiträge zur Biologie der Gallwespen. Vehr. Naturf. Ges., Basel. 60:104–178.

Skuhravá, M., and V. Skuhravý. 1973. Gallmücken und ihre Gallen auf Wildpflanzen. Die neue Brehm Bücherei, Wittenberg-Lutherstadt, A. Ziemsen.

Skuhravá, M., and V. Skuhravý. 1981. Die Gallmücken (Cecidomyiidae, Diptera) des Schilfes (*Phragmites communis* Trin.). Studie C.S.A.V. Praha Academia.

Skuhravý, V. 1975. Die Entwicklung der Gallmücke *Lasioptera arundinis* (Shiner)(Diptera, Cecidomyidae) an Schilf (*Phragmites communis* Trin.) Marcellia 38:287–298.

Shorthouse, J. D. 1975. The role of insect inhabitants in six *Diplolepis* (Cynipidae: Hymenoptera) rose leaf galls of western Canada. Ph.D. Thesis, University of Saskatchewan, Saskatoon, Canada.

Shorthouse, J. D. 1980. Modifications of galls of *Diplolepis polita* by the inquiline *Periclistus pirata*. Bull. Soc. Bot. Fr. 1. Actual. Bot. 1:79–84.

Shorthouse, J. D., and R. G. Lalonde. 1984. Structural damage by *Rhinocyllus conicus* (Coleoptera: Curculionidae) within the flowerheads of nodding thistle. Can. Entomol. 116:1335–1343.

Smart, M. G., and T. P. O'Brien. 1983. The development of the wheat embryo in relation to the neighbouring tissues. Protoplasma 114:1–13.

Smith, E. L. 1970. Biosystematics and morphology of Symphyta. II. Biology of gall-making Nematine sawflies in the Californian Region. Ann. Entomol. Soc. Am. 63:36–51.

Solinas, M., and M. Bucci. 1982. An ecological investigation into flower galls on *Diplotaxis muralis* D. C. caused by the gall midge *Paragephyraulus diplotaxis* Solinas. Entomologica 17:1–22.

Sziráki, I., E. Balazs, and Z. Kiraly. 1980. Role of different stresses in inducing systemic acquired resistance to TMV and increasing cytokinin level in tobacco. Physiol. Plant Pathol. 16:277–284.

Taylor, S. H. 1949. Initiation and development of the gall of *Aylax glechomae* on *Nepeta hederacea*. Am. J. Bot. 36:222–284.

Thomas. 1893. Cecidologische Notizen, I. Entomol. Nachricht. 19:287–304.

Tscharntke, T. 1988. Variability of the grass of *Phragmites australis* in relation to the behaviour and mortality of the gall-inducing midge *Giraudiella inclusa* (Diptera, Cecidomyiidae). Oecologia 76:504–512.

Van Staden, J., J. E. Davey and A.R.A. Noel. 1977. Gall formation in *Erythrina latissima*. Zeit. Pflanzenphys. 84:283–294.

Wells, B. W. 1921. Evolution of zoocecidia. Bot. Gaz. 71:358–371.

West, A., and J. D. Shorthouse. 1989. Initiation and development of the stem gall induced by *Hemadas nubilipennis* (Hymenoptera: Pteromalidae) on lowbush blueberry, *Vaccinium angustifolium* (Ericaceae). Can. J. Bot. 67:2187–2198.

Westphal, E. 1977. Morphogenèse, ultrastructure et étiologie de quelques galles d'Eriophyes (Acariens). Marcellia 39:193–375.

Yukawa, J. 1984. Life history strategies of gall midges, with special reference to univoltine gall-making species. XVII Intern. Congress of Entomology, Hamburg.

Yukawa, J. 1987. Life history strategies of univoltine gall making Cecidomyiidae (Diptera) in Japan. Phytophaga 1:121–139.

Zweigelt, F. 1931. Blattlausgallen. Parey, Berlin.

6

Developmental Morphology of Two Types of Hymenopterous Galls

L. A. REY

Galls induced by members of the Order Hymenoptera are among the most spectacular of the insect galls because of their diversity in size, shape, and structure. The abundance of hymenopterous galls in many parts of the world has made them popular candidates for botanical studies and thus they are the subject of a rich literature (see Mani 1964; Meyer 1987; Meyer and Maresquelle 1983).

The developmental morphology of hymenopterous galls is of particular interest to gall workers because they range in complexity from simple callus-like swellings induced by certain species of sawflies to the multilayered, differentiated galls induced by cynipid wasps (see Chapters 5 and 8). Development of cynipid galls is further complicated by the occurrence of heterogeny (see Chapters 2 and 16) where alternating generations of the same species induce two distinct types of galls, each of which is usually on different organs of the same host plant.

The purpose of this chapter is to illustrate the structural complexity of hymenopterous galls by comparing in detail the developmental morphology of two completely different types of galls found in Europe; those of the heterogenous cynipid wasp *Biorhiza pallida* 01. on the buds and roots of the oak *Quercus pedunculata* Ehrh. and those of the sawfly *Pontania proxima* Lep. on the leaves of *Salix triandra* L. Particular care is taken in comparing the processes associated with initiation and the various stages in gall development, along with a comparison of the cytology of the cells fed on by the larvae. The material in this chapter is from Rey (1976).

GALLS OF THE CYNIPID WASP *BIORHIZA PALLIDA* ON *QUERCUS PEDUNCULATA*

Biorhiza pallida induces the well-known bud gall referred to as the "oak apple" whose development was observed as early as 1883 by Beyerinck. It also causes a root gall on the same tree; however, little has been published on this form of the gall owing to its underground location.

Adler (1881) was the first to show that *B. pallida,* which belongs to the Cynipini group of the subfamily Cynipinae (Kinsey 1920), had a life cycle consisting of two alternating generations. Folliot (1964) made a detailed study of the complete life cycle, which can be summarized as follows: Individuals of the bisexual generation emerge from the mature bud gall (Fig. 6-1) in late June; the females crawl into the soil after copulation and lay their eggs inside thin roots of oak; globular galls (Fig. 6-11) develop on the roots; the larval stage here lasts about 18 months; agamous, wingless individuals emerge in November or December of the following year from the root galls and climb up the tree to the oak twigs where they lay their eggs in the buds and induce new oak apples.

INITIATION AND DEVELOPMENT OF THE BUD GALL

Wounding, Oviposition, and Gall Initiation

Agamous females lay their eggs inside oak buds (Fig. 6-2). The female uses her ovipositor to saw into the inner bud axis and split the bud into two regions, an upper part that would have lengthened at the beginning of spring and a lower part from which the gall develops (Fig. 6-3). Up to 300 eggs, which remain well protected by the bud scales, are laid within the split between the two regions. A crust of dead cells forms on the plant surfaces cut by the ovipositor. The upper part of the bud withers and dies.

Tissues beneath the crust of cells killed during oviposition undergo the renewed activity referred to as "metaplasy" by Meyer (1957), at the beginning of January. The cytoplasm of these cells becomes more abundant, the vacuoles become fragmented, and there is hypertrophy of the nucleus and nucleolus. The cells remain in this condition until early March and just as nearby unattacked buds begin to flush, the metaplasied cells enter a phase of intense hyperplasy. The hyperplasy is accompanied by a dedifferentiation, which becomes more pronounced as divisions continue. The quantity of cytoplasm in the cells continues to increase, the vacuole is further reduced, tannin is lost, and plastidial starch disappears. The proliferation of these cells forms a zone characterized by dedifferentiation and homogeneity called the "gall plastem" (Fig. 6-4). The plastem grows inside the envelope of bud scales and presses the eggs against the upper surface of the proliferating mass. This early stage of egg envelopment occurs by late March.

Cells of the plastem beneath each egg undergo lysis while other nearby cells continue to divide. By the time the larvae hatch, each is situated in a small cavity on the surface of the plastem (Fig. 6-5). Bronner (1973) and Rey (1976) have shown that the eggs and larvae of *B. pallida* exude hydrolytic enzymes that are believed to be responsible for this envelopment process. As lysis and proliferation continue the larvae are progressively surrounded until they are enclosed in individual cavities within the plastem. The cavities are tear shaped at first, but they then become spherical.

Organization of Gall Structures

Anatomical features characteristic of the mature *B. pallida* bud gall begin to appear in April (Fig. 6-6). Vascular tissue differentiates within the plastem forming strips of elongated cells as the intense hyperplasy continues. Cells between these strips become vacuolated and accumulate starch. A cambial zone appears near each larval chamber where cell divisions will continue for several weeks. Cells along the margin of the larval chamber are transformed into nutritive cells (see Chapter 8), which are fed on by the larvae. The cambial zone at this stage ensures the renewal of the nutritive cells toward the inside and the development of the starch parenchyma toward the outside. The central part of the gall containing the larvae then becomes surrounded by a region where the cells undergo hypertrophy, separate from each other, and form an aeriferous parenchyma. The aeriferous parenchyma is restricted on the outside by a layer of closely joined cells. The young gall then pushes through the upper part of the bud, forces the scales aside, and becomes visible from the outside. The gall is particularly noticeable at this time for it is often red (Fig. 6-7A–D).

The nutritive cells that line the larval chamber are cytologically very interesting (Fig. 6-9). Unlike parenchyma cells more distant from the larval chamber, they retain abundant cytoplasm and do not accumulate starch. They also become rich in structural proteins and accumulate numerous lipid droplets. The nucleus, which does not divide further, is prominently enlarged, as is the nucleolus. Plastids and mitochondria become more numerous, but are undifferentiated. Autophagic vacuoles are abundant and the smooth form of the endoplasmic reticulum becomes markedly developed. As described by Roth (1949), Rohfritsch (Chapter 5), and Bronner (Chapter 8), the larvae pierce these cells with their mandibles and suck their contents.

The gall grows quickly throughout April and early May increasing from 2 to 20 mm in diameter (Fig. 6-7). Growth during this period is first due to hyperplasy of the cambial zone

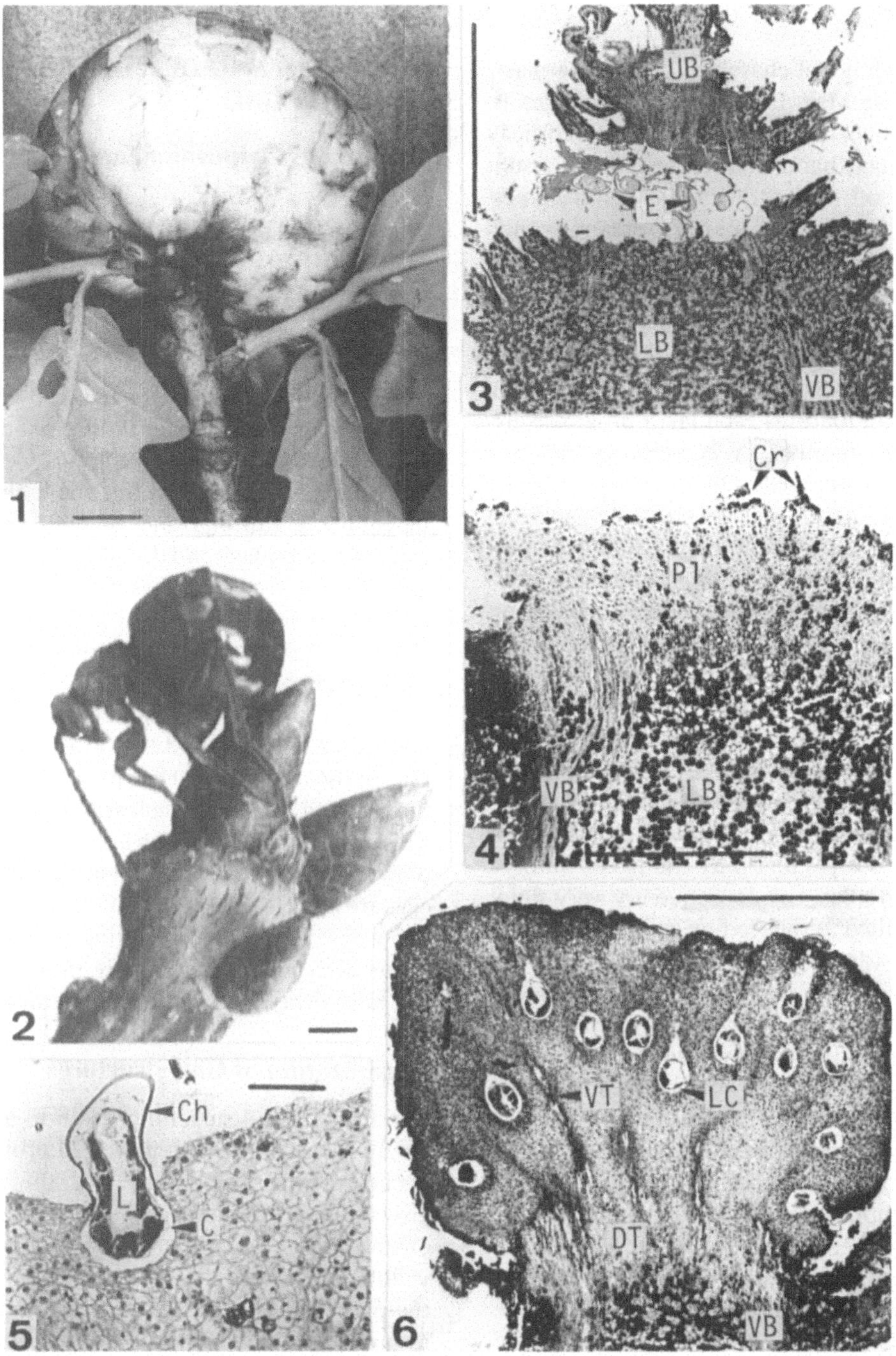

Figures 6-1–6-6. Bud galls of *Biorhiza pallida* on *Quercus pedunculata*. **Figure 6-1.** Mature gall collected at the end of May. Bar = 1 cm. **Figure 6-2.** Parthenogenetic female laying eggs in an oak bud in December. Bar = 1 mm. **Figure 6-3.** Longitudinal section of a bud just after oviposition. Bud scales were removed for better fixation. Note that the bud axis is sawn horizontally. The eggs are laid between the upper part of the bud, which will wither, and the lower part, which will give rise to the gall. Bar = 1 mm. **Figure 6-4.** Gall plastem beginning to develop in early March from tissues of the lower part of the bud axis. Note the crust of dead cells at the top of the plastem that have been crushed during oviposition. Bar = 0.5 mm. **Figure 6-5.** Freshy hatched larva on surface of gall plastem. When the larva hatches, it leaves the chorion of the egg outside and inserts itself into the cavity created by hydrolytic enzymes. Bar = 100 μm. **Figure 6-6.** Longitudinal section of a maturing bud gall collected in early April. Enclosure of the larvae is now complete. Strands of vascular tissue have differentiated within the plastem. These strands develop from a sort of distribution tray connected to the vascular bundles of the lower part of the bud. Bar = 1 mm. C, Cavity; Ch, egg chorion; Cr, crust; DT, distribution tray; E, eggs; L, larva; LB, lower part of bud; LC, larval chamber; Pl, plastem; UB, upper part of bud; VB, vascular bundle; VT, vascular tissue.

around each larval chamber. Once hyperplasy ceases in late May, the gall further enlarges as these cells hypertrophy. The aeriferous cortical parenchyma expands markedly and as a result the gall is soft to the touch. The gall becomes globular and changes from red to creamy white. Although the shape of the gall does not change as it matures, a zone of sclerenchyma cells differentiates around each larval chamber and tangentially on the inside of the vascular bundles (Fig. 6-8).

As the gall matures each larva progressively enlarges its chamber by consuming the surrounding nutritive cells. Once the cambial zone disappears, new nutritive tissue appears through the transformation of starch parenchyma cells. Nutritive cells in maturing galls continue to show the same conspicuous cytological characteristics but emphasized even more. Bronner (1977) and Rey (1976) have shown that these cells also exhibit an intense enzymatic activity, especially so for the enzymes acid phosphatase and aminopeptidase. These activities are associated with an active transport of soluble carbohydrates and with autophagic processes that may end in cell autolysis. The cytoplasm is consequently very fluid and well suited to the feeding larvae (see Chapter 8).

The gall reaches its final size (3–5 cm in diameter) by the end of May. Its surface becomes pale brown and begins to wrinkle. Each larva consumes all tissue transformed into nutritive cells up to the sclerenchyma zone. The larvae pupate once all nutritive cells are gone. The last nutritive cells undergo pathological-like alterations developing atypical formations of smooth endoplasmic reticulum (Fig. 6-10). They accumulate lipid droplets and the nuclei and nucleoli become very enlarged. These nuclei never divide and their chromatin exhibit particular structures similar to those already described by Garrigues (1951) in the nutritive cells of other cynipid galls.

The adults emerge from the galls by late June and the gall surface becomes riddled with their exit holes. Either all males, females, or both sexes may emerge from the same gall, but morphologically and cytologically all galls are similar. The galls fall to the ground soon after the insects have escaped.

INITIATION AND DEVELOPMENT OF THE ROOT GALL

Wounding, Oviposition, and Gall Initiation

Females of the bisexual generation sink their ovipositors (Fig. 6-12) into the roots of oaks 2–5 mm in diameter and deposit eggs individually or in small groups under the cambium in the nonlignified layers of the secondary xylem (Fig. 6-13). About 2 weeks later a lignified shell, originating from the cells near those crushed during oviposition, forms around the egg. Cells of the secondary xylem around this shell, along with neighboring cambial and phloem cells, then undergo metaplasy and divide just as in the bud gall. This process progressively extends to the cortical parenchyma until a gall plastem takes shape. The plastem is undifferentiated and indistinguishable from the outside (Fig. 6-14).

The larva leaves the lignified shell immediately after hatching and crawls along the lower part of the oviposition channel to the middle of the plastem. It is then enclosed in a chamber as a result of cell proliferation and cell lysis and the oviposition channel is obliterated. When the eggs are laid close to each other the plastem may contain several larvae resulting in a polythalamous gall. When the eggs are more scattered, the galls always will be monothalamous.

Organization of Gall Structures

The internal gall structures begin to appear in August (Fig. 6-15). Extensive cell proliferation occurs during this period and the resulting growth separates the cortical parenchyma. Once this occurs, the young gall appears as a yellow protuberance. While the outer layers of the plastemic tissue become progressively rich in tannins, starch begins to accumulate within the parenchyma cells inside the plastem mass. Vascular bundles begin to differentiate within this starch parenchyma, at some distance from the larval chamber, and connect with the normal cambium of the root. A zone of cambial cells forms around the larval chamber, as in the bud galls, and one or two rows of nutritive cells are differentiated (Fig. 6-16). The cytological characteristics of these nutritive cells are similar to those of the bud gall.

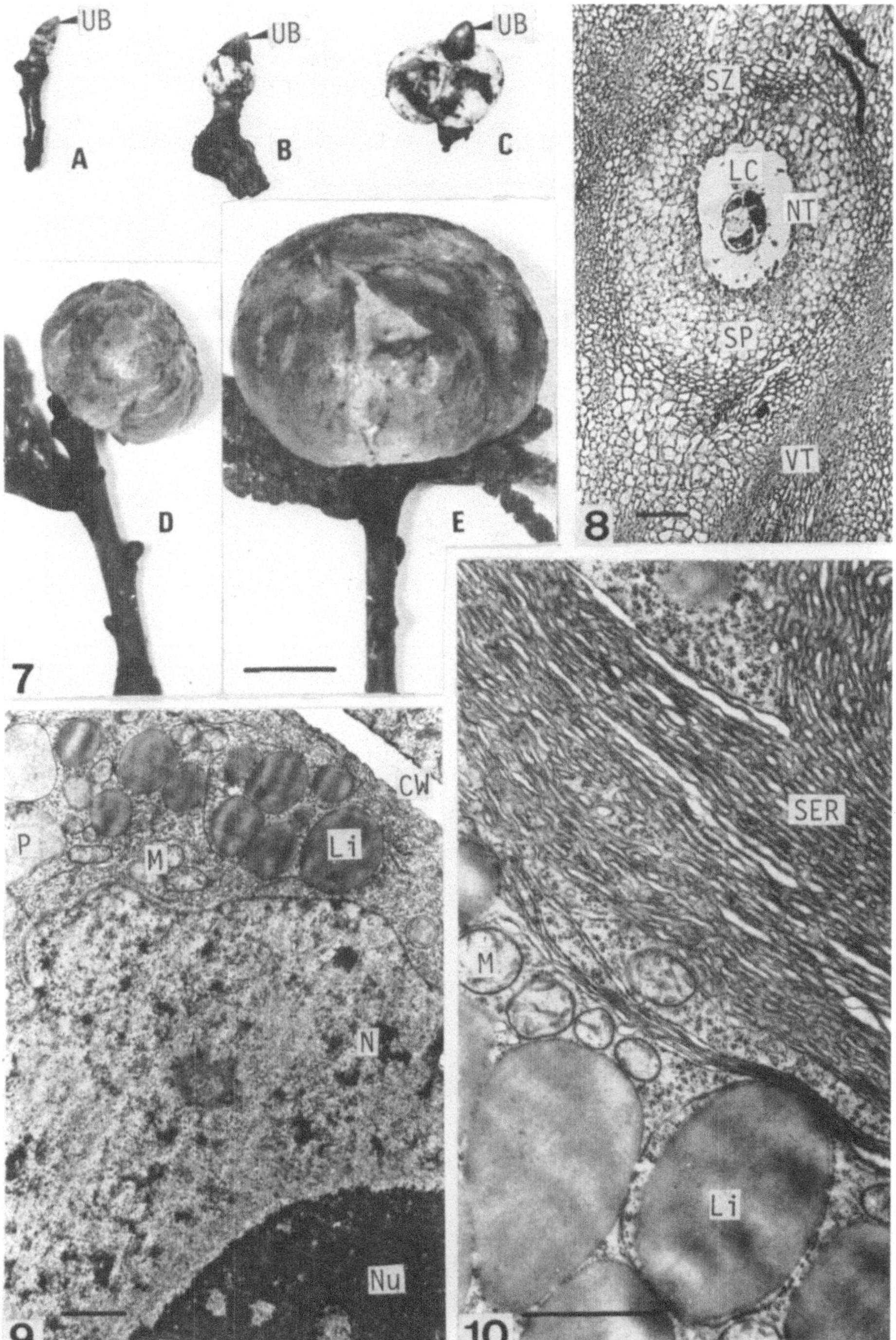

Figures 6-7–6-10. Development and structures of the bud gall of *Biorhiza pallida* on *Quercus pedunculata*. **Figure 6-7.** Gall development from late March (A) to early May (E). Bud scales have been removed in A, B, and C. The withered upper part of the bud may remain stuck at the top of the gall. Bar = 1 cm. **Figure 6-8.** Longitudinal section of a larval chamber in a gall collected in early May. Each larval chamber is surrounded by nutritive tissue, starch parenchyma, and a sclerenchyma zone that begins to differentiate at this stage. Bar = 200 μm. **Figure 6-9.** Hypertrophied nucleus and nucleolus of a nutritive cell. Note the numerous lipid droplets and undifferentiated plastids in the cytoplasm. Bar = 1 μm. **Figure 6-10.** Thick and compact bundle of smooth endoplasmic reticulum cisternae of nutritive cells. Bar = 1 μm. CW, Cell wall; LC, larval chamber; Li, lipid; M, mitochondria; N, nucleus; Nu, nucleolus; NT, nutritive tissue; P, plastid; SER, smooth endoplasmic reticulum; SP, starch parenchyma; SZ, sclerenchyma zone; UB, upper part of bud; VT, vascular tissue.

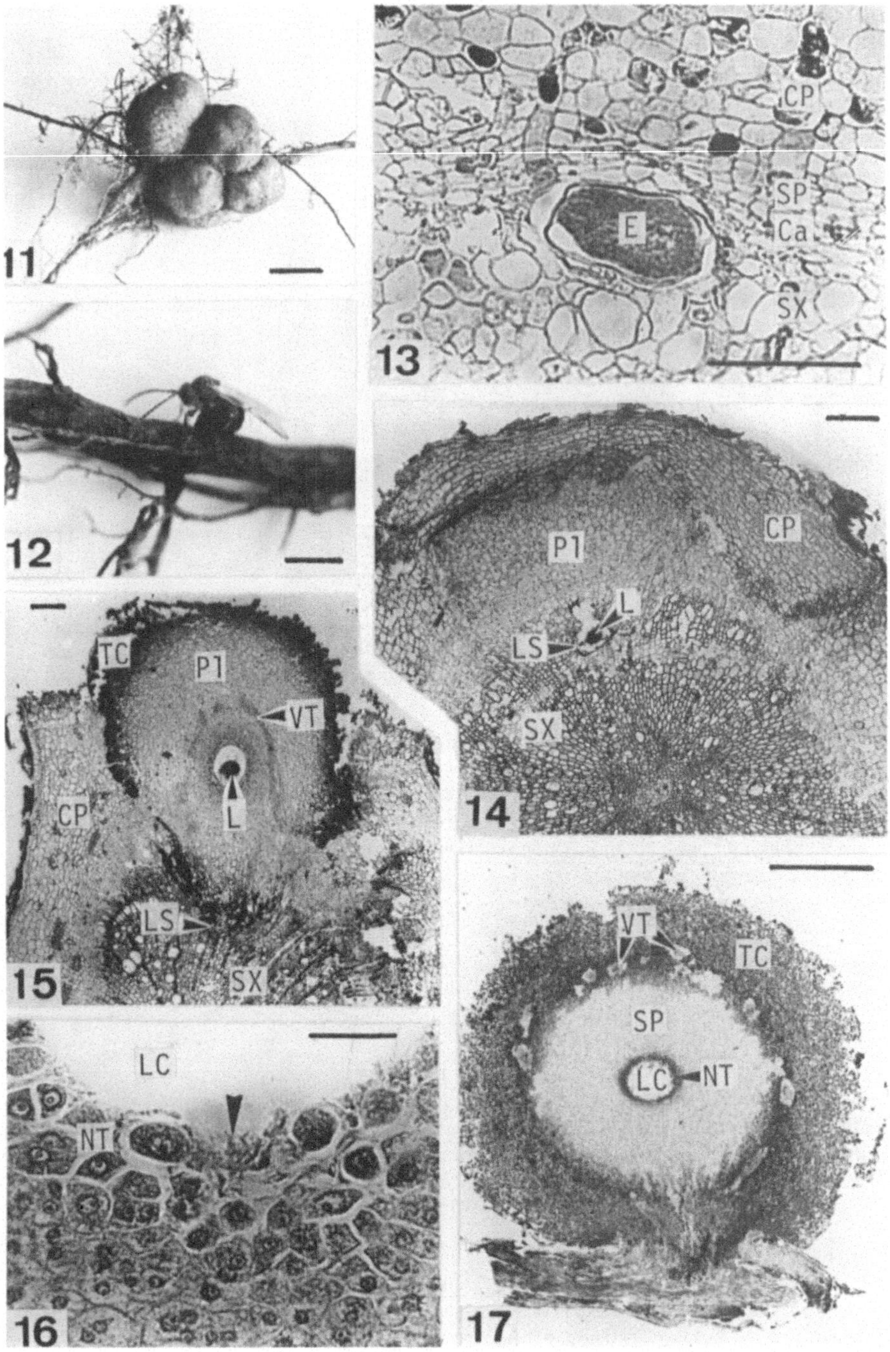

Figures 6-11–6-17. Root galls and ovipositing female of *Biorhiza pallida* on *Quercus pedunculata.* **Figure 6-11.** Mature galls collected in September of the second year of development. Bar = 2 cm. **Figure 6-12.** Female of the bisexual generation laying eggs in an oak root in late June. Bar = 2 mm. **Figure 6-13.** Cross section of a root showing the position of an egg just after oviposition. The egg is under the cambium in the nonlignified layers of secondary xylem. Bar = 100 μm. **Figure 6-14.** Cross section of a galled root in July. A gall plastem is present near the larva. The larva is surrounded by a lignified shell originating from the cells crushed during oviposition. Bar = 200 μm. **Figure 6-15.** Cross section of a galled root in August. The growing plastem has cracked the cortical parenchyma and protrudes from the root. The larva has left the lignified shell and has settled in the middle of the

Growth of the gall continues until the end of summer (3–5 mm in diameter), slows during autumn and winter, and then begins again in the spring reaching a size of 10–12 mm in diameter by April. The gall stops growing by June. The gall is spherical when single chambered (Fig. 6-17) but this shape is often distorted when the gall is multichambered (up to 4–5 larvae). The surface is light brown when the gall is immature but becomes dark brown on maturity.

Differentiation of the gall tissues is similar to the bud gall except that it is slower. The cambial zone around the larvae functions until autumn producing a thick layer of starch parenchyma (Fig. 6-17) and then the cambium ceases functioning. A layer of sclerenchyma cells is differentiated beyond the starch parenchyma. The gall continues to enlarge the following spring, but only as a result of cellular hypertrophy.

The gall is full grown by June of the second year of its development. Isolated galls are 13–15 mm in diameter; however, when they coalesce they often form masses 30–40 mm in diameter (Fig. 6-11).

Once the larvae finish consuming all starch parenchyma transformed into nutritive cells up to the layer of sclerenchyma, they pupate. The adult emerges in December, 18 months after oviposition, by chewing an exit route through the gall wall. Once the insects have left the galls become necrotic. The last nutritive cells have a pathological appearance with their smooth endoplasmic reticulum largely developed in the outer part of the cytoplasm and their nucleus and nucleolus enlarged (up to 60 and 30 μm in diameter, respectively).

GALLS OF THE SAWFLY *PONTANIA PROXIMA* ON *SALIX TRIANDRA*

Galls of sawflies are common on leaves of *Salix* both in Eurasia and North America and have been the subject of many morphological (Beijerinck 1888; Magnus 1914; Küster 1930) and ecological works (see Chapter 14). The structures and development of sawfly galls are quite different from those of cynipids and *P. proxima* serves as an ideal candidate to examine the peculiarities of this type of gall. Of special interest is that *P. proxima* is one of the species whose gall will sometimes develop without the presence of an egg in the gall chamber.

Pontania proxima belongs to the Nematinae family and has been studied by several entomologists (Carleton 1939; Benson 1954, 1960). It reproduces almost entirely parthenogenetically. The eggs are laid within young willow leaves. The larvae hatch inside the developing galls (Fig. 6-18) and feed on the inner layers of cells. During the third or fourth instar the larva bites a round exit hole in the gall wall and pushes out some of its abundant excrement. Once mature the larvae exit to the outside and drop to the ground where they pupate in the soil. Two and sometimes three generations occur between June and September. The individuals of the last generation overwinter in their cocoons in the leaf litter.

Wounding, Oviposition, and Gall Initiation

Female *P. proxima* insert their ovipositors horizontally into immature willow leaves that are composed of a few layers of tightly packed cells (Fig. 6-21) and make a cavity parallel to the plane of the leaf blade. As was shown by Magnus (1914), only one layer of median mesophyll cells is destroyed. The female then deposits one egg into the cavity (Fig. 6-19) along with a liquid secretion. Sometimes the secretion and not the egg is deposited into the cavity.

Cecidogenic action begins within 12 to 24 hours of oviposition and results in extensive hyperplasy. This hyperplasy influences all the cell

plastem. Vascular tissue has begun to differentiate at some distance from the larval chamber. Bar = 200 μm. **Figure 6-16.** Section of a larval chamber showing hypertrophied nutritive cells. Note the collapsed cell (arrow) fed on by the larva. Bar = 100 μm. **Figure 6-17.** Section of a maturing, single-chambered gall collected in November. The host root is seen in longitudinal section. Bar = 1 mm. Ca, Cambium; CP, cortical parenchyma; E, egg; L, larva; LC, larval chamber; LS, lignified shell; NT, nutritive tissue; Pl, plastem; SP, starch parenchyma; SP, secondary phloem; SX, secondary xylem; TC, tanniferous cortex; VT, vascular tissue.

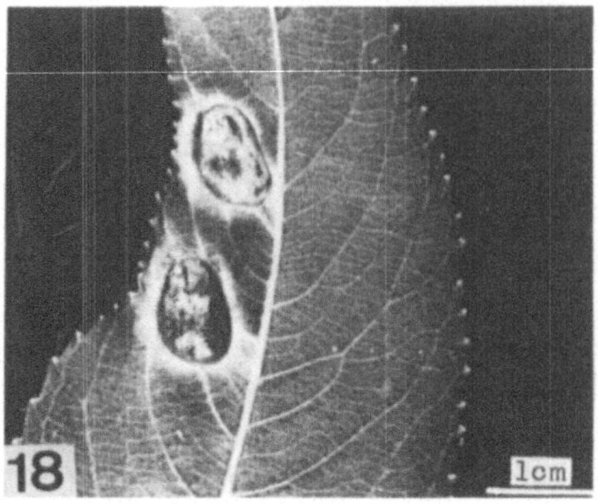

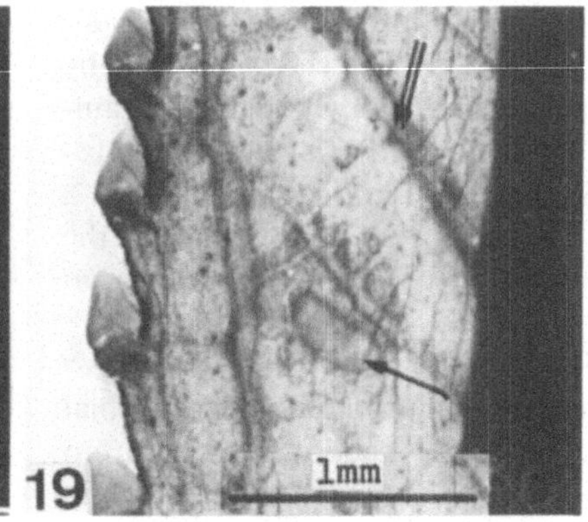

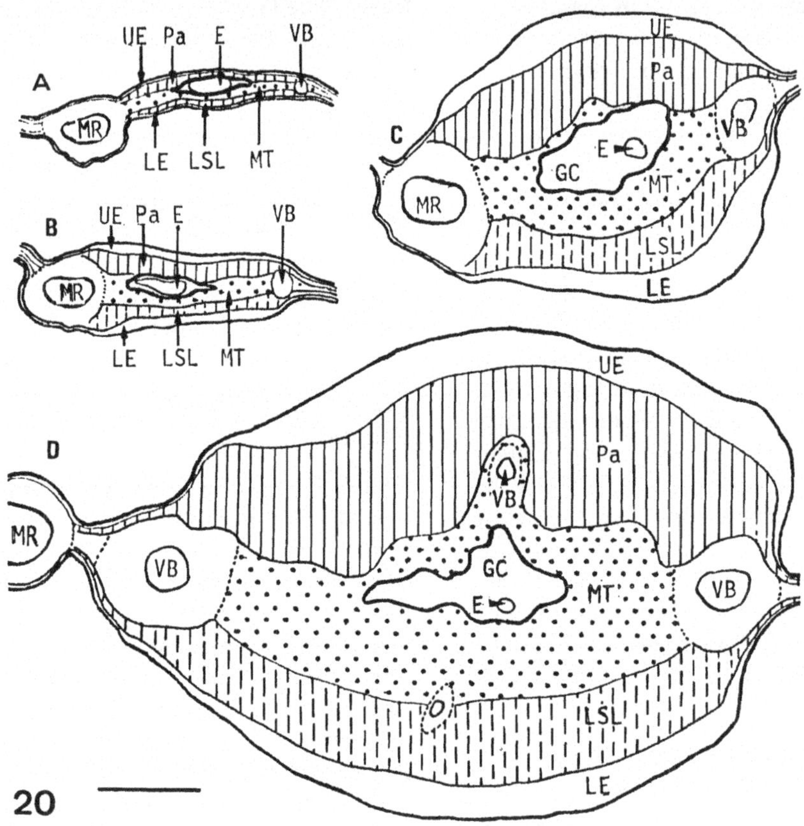

Figures 6-18–6-20. Leaf galls of *Pontania proxima* on *Salix triandra*. **Figure 6-18.** Mature galls collected in midsummer. Note that presence of the two galls causes the leaf to curl. **Figure 6-19.** Position of a freshly laid egg. The young leaf is held against an illuminated background. The egg (simple arrow) is surrounded by a darker zone, which corresponds to the part of the cavity filled with secretions from the female. The double arrow indicates the point where the ovipositor penetrated. **Figure 6-20.** Cross sections of galls at various ages post oviposition: (A) 60 hours, (B) 5 days, (C) 12 days, (D) 15 days. Symbols used for gall tissues are the same as the tissues from which they originate. Bar = 1 mm. E, Egg; GC, gall chamber; LE, lower epidermis; LSL, lower subepidermal layer; MT, median tissue; MR, midrib; Pa, palisade; UE, upper epidermis; VB, vascular bundle.

layers of the leaf blade in an area slightly larger than that of the gall cavity. Most divisions within the first 48 hours are periclinal and near the gall cavity. As a result the immature gall is composed of thick layers of cells above and below the cavity. Hyperplasy decreases toward the outside of the gall but this is partly counterbalanced by cell hypertrophy.

Thus the enclosure of *P. proxima* eggs is quite different from that of *Biorhiza pallida.* Here the egg remains passive inside the gall cavity, which is little more than the initial wound made by the ovipositing female. The cavity expands as the gall and leaf grow. Of great importance in the development of the *P. proxima* gall is that the host reacts similarly whether or not the cavity contains an egg.

Organization of Gall Structures

The gall appears as a yellow elongated spot about 2 mm wide and 4 mm long, after 60 hours of development, protruding slightly from both surfaces of the leaf blade (Fig. 6-22). On the upper surface, the gall is surrounded by a red halo due to the presence of anthocyanins.

Even as the gall matures, none of the cells is too dissimilar from those of the host leaf. In fact the layers of the normal leaf incorporated into the gall still can be identified (Fig. 6-21), although they may be thickened due to hyperplasy and hypertrophy. Thus tissues of the gall come from proliferations of the upper epidermis, palisade, mesophyll, lower subepidermal layer, and lower epidermis (Figs. 6-22 and 6-23). The egg remains unhatched during this process. Proliferation of cells around the edges of the gall is more extensive and tends to flatten the gall cavity (Fig. 6-22).

Galls developing without an egg appear little different from those containing one. The galls may be smaller, as already pointed out by Beijerinck (1888), Magnus (1914), and Küster (1930), but some reach the same size. The single peculiar point in their inner organization is that the gall cavity is obliterated and as a result galls without eggs are flattened.

However, some cytological transformations do occur in each of the tissues that proliferate to produce the gall. Tannins, for example, accumulate in tissues arising from both subepidermal layers (Fig. 6-23A, B) and chloroplasts of the median tissue surrounding the gall cavity undergo a peculiar transformation. Whereas chloroplasts of unaffected mesophyll cells exhibit normal structures (Fig. 6-24), chloroplasts of gall cells have swollen plastid lamellae (Fig. 6-25) that give rise to vesicles similar to those reported for pathological tissues (Margulies 1967; Bartels 1971; Cran and Possingham 1972). Unlike the galls of *B. pallida,* starch reserves and lipids do not accumulate in these cells and, furthermore, they contain the same amount of proteins as do cells of normal leaves.

The gall grows even more rapidly after 60 hours without changing in organization and expands in all directions (Fig. 6-20); however, there are no changes in its cellular organization. The mature gall reaches its maximum size (10 mm long, 5 mm wide, and 3–4 mm thick) in about 12 days and protrudes equally from both sides of the leaf blade. The upper surface is reddish whereas the lower surface is pale green.

Thus growth in the *P. proxima*-induced galls is due to hyperplasy and hypertrophy of various leaf tissues. Intensity of hyperplasy is reduced further from the gall cavity. Cell hypertrophy becomes important only 5–9 days after oviposition; however, the variations of hypertrophic intensity are opposite to those of hyperplasy. That is, cells of the outermost tissues are the most enlarged. The presence of enlarged cells ceases abruptly at the edges of the gall.

Cytologically, the tissues arising from the different cell layers of the normal leaf remain distinguishable (Fig. 6-20); however, those cells arising from both epidermal layers have little cytoplasm. Those arising from the palisade layer and the lower subepidermal layer are characterized by their tannins and the tissues arising from the median layers of the mesophyll, situated on both sides of the gall cavity, contain the greatest amount of chlorophyll. However, the chloroplasts in mesophyll cells appear pathological. The chloroplasts of normal mesophyll have thick grana and large starch inclusions whereas the chloroplasts of median gall tissues (Fig. 6-26) are often similar to chloroplasts in Graminae (Laetsch and Price 1969). These chloroplasts never contain starch.

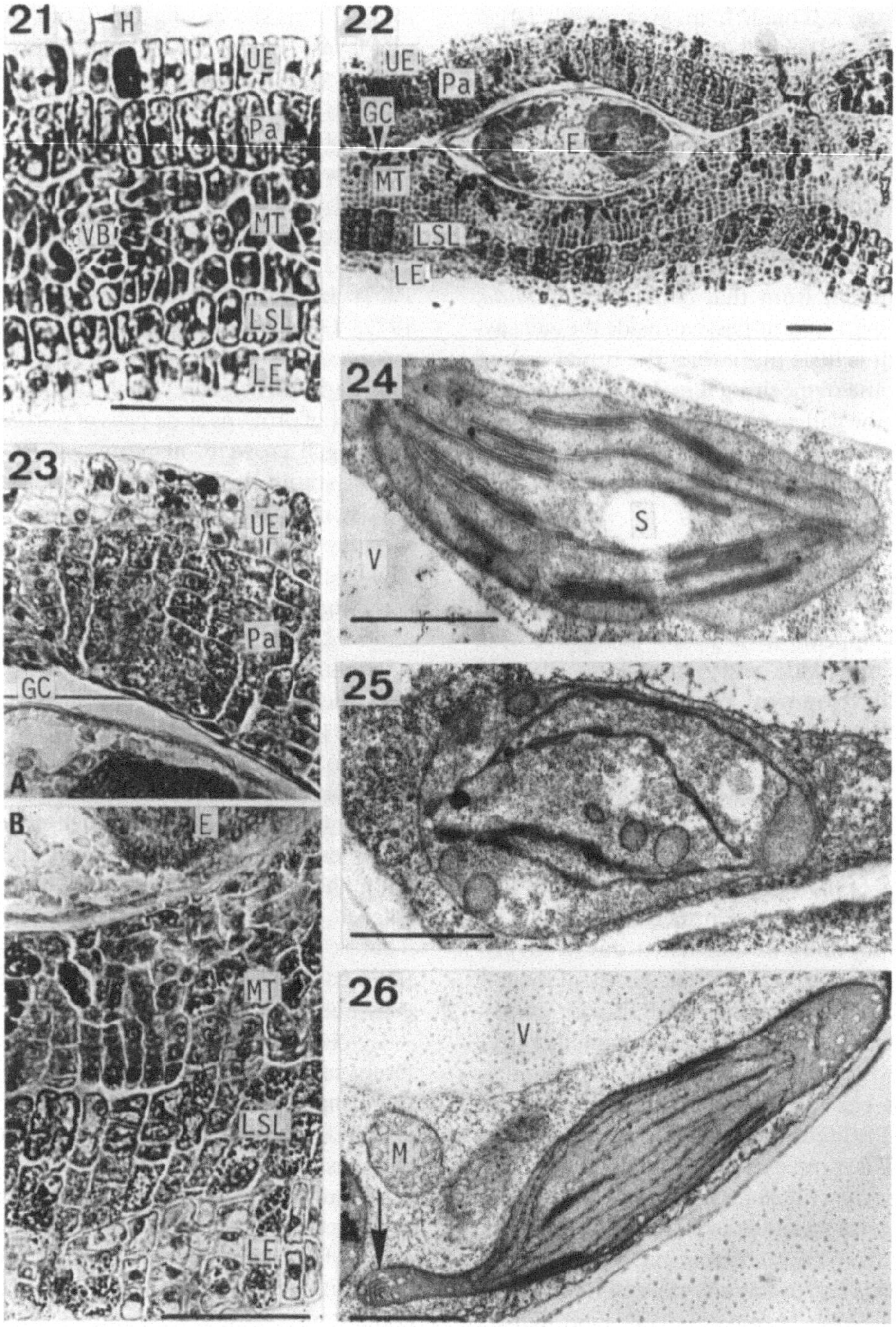

Figures 6-21–6-26. Histological and cytological features of the gall of *Pontania proxima* on *Salix triandra.* **Figure 6-21.** Cross section of a young unattacked leaf of *Salix triandra.* The ovipositor will destroy only one layer of the median tissue. Bar = 50 μm. **Figure 6-22.** Cross section of a gall 60 hours after oviposition. Cells in every layer have divided several times by periclinal mitosis. Gall walls have thickened, especially at the periphery of the egg, so that the gall cavity becomes flattened. The tissues arising from the different cell layers of the leaf remain distinguishable and symbols used here are those of the host leaf. Bar = 50 μm. **Figure 6-23.** Details of Fig. 6-22. (A) Upper wall of the gall. The tissue arising from the upper epidermis is composed of two to four cell layers. The tissue arising from the palisade is composed of six to nine cell layers. (B) Lower wall of the gall. The tissue coming from the median layers of the mesophyll is composed of about 10 cell layers. The tissue arising from the lower

Furthermore, a bundle of tubules appears in their stroma, which is very often isolated in a stretched lobe (Fig. 6-26). These curious formations cannot be compared either with a prolamellar body (Gunning 1965; Wehrmeyer 1965) or with a peripheric reticulum (Shumway and Weier 1967).

The egg hatches about 15 days after oviposition when the gall is full grown. The larva grows rapidly eating the tissue surrounding the gall cavity. The first tissues consumed are those of the median region. Some tissues formed by the leaf subepidermic layers also are consumed. Feeding by sawfly larvae induces almost no cytological modifications. The larva is full grown in about 15 days and leaves the gall to spin its cocoon.

The thin walls of the deserted gall turn yellow and then collapse. The site of the gall becomes a necrotic spot on the leaf. Galls without eggs, in contrast, remain healthy as long as the leaves that bear them.

DISCUSSION

Using the descriptions of the three galls examined in this chapter, it is interesting to compare the key steps in their development and determine how they manipulate plant growth to their own advantage. The two insects induce quite different galls; however, there are certain similarities in the early stages of cecidogenesis. Both species initiate their galls by a combination of wounding with the ovipositor, exuding secretions with the eggs, and the physical contact of the eggs with plant tissues. Solicitation of stress responses by the host plant also is important.

The females of the two species have a specialized egg-laying behavior perfectly adapted to the attacked plant organ and as a result they are able to inflict specific injuries. The wounds always occur in immature tissues or in a cambial zone, both of which are tissues amenable to manipulation. However, the eggs of *B. pallida* are separated from the living tissues by a layer of crushed cells, whereas with *P. proxima,* the fragments of destroyed cells are pushed back toward the edges of the wound cavity.

Rey (1976) has shown that the wound inflicted by *B. pallida* on the buds of winter oak is a critical event in gall initiation, but on its own has little effect. Maternal secretions applied to the eggs have no effect on the host cells; in this species it is the eggs that play the major role in gall induction. The reaction induced by the eggs is apparently due to enzymes present in the egg chorion (see Chapter 8). Some biological factors peculiar to the host plant also are important, such as the exact site where the eggs are laid inside the bud axis and the fact that the gall, initiated as early as December, starts growing only when the oak itself begins to grow in early March.

Similarly, the wound and maternal secretions in the case of root galls are not of great importance. Here as well, the eggs and young larvae play the determinant part. It also should be noted that eggs sometimes do not succeed in inducing metaplasy in the neighboring cells and are soon killed as they are enclosed by a sheath of lignified cell layers. Similarly, wounding does not play a role in the initiation of other cynipid galls. For example, hatching larvae of the spring generation of *Neuroterus quercusbaccarum* L. (Meyer 1957) sink into the young leaf tissues only through lysis of the underlying cells and induce the plastem in this region.

Since the presence of the egg is not necessary for gall development in the gall of *P. proxima,* it is more difficult to distinguish the cecidogenic importance of the wound and the maternal secretion (see Chapter 5). According to Beijerinck (1888) the secretion is the essential factor.

subepidermal layer (four to six cell layers), like the one arising from the palisade, is tanniferous. The tissue coming from the lower epidermis is now two to four cell layers. **Figure 6-24.** Normal chloroplast found in the mesophyll of a developing willow leaf. Bar = 1 μm. **Figure 6-25.** Chloroplast found in the median tissues of a gall 60 hours after oviposition. The plastid lamellae swell and give rise to vesicles. Bar = 1 μm. **Figure 6-26.** Chloroplast found in the median tissues of a mature gall 12 days after oviposition. Chloroplasts do not have vesicles, but exhibit an agranal structure. A bundle of tubules (arrow) also is present in a stretched lobe of the organelle. Bar = 1 μm. E, Egg; GC, gall cavity; LE, lower epidermis; LSL, lower subepidermal layer; M, mitochondrion; MT, median tissue; Pa, palisade; S, starch; UE, upper epidermis; V, vacuole; VB, vascular bundle.

Magnus (1914) experimented with these secretions by introducing them into young and older leaves, stems, and roots. He obtained negative results and concluded that the secretions did not play an important role. Furthermore, he thought that the gall was a reaction by the leaf to the internal wound.

Hovanitz (1959) experimented with another *Pontania* species and also found that the first part of gall development can occur without an egg in the wound cavity; however, the second part needs the presence of the larva, which hatches sooner than in the case of *P. proxima.* He showed that it is possible to obtain the second part of gall development by replacing the larva with maternal secretions. Gall formation by other genera such as *Euura* apparently is also stimulated by fluids injected with the egg during oviposition (see Chapter 14).

It also has been shown (Rey unpublished) that some *P. proxima* galls without eggs can grow very unequally. This disparity apparently is due to variation in the quantity of secretion injected into the wound. Thus it seems that in the case of *P. proxima,* the wound and the secretion are both essential to gall development. The wound is the basic event but is not sufficient to induce galls on its own. Apparently the internal wound sensitizes the cells to the action of the maternal secretion.

The first reaction of the plant tissues to the cecidozoan attack is quite different in the two galls. In the case of *Biorhiza pallida,* metaplasy is the first event followed by dedifferentiative hyperplasy, resulting in the formation of a plastem. Thus homogeneous gall tissues initially form from different plant tissues surrounding the eggs.

There is no metaplasy in the case of *Pontania proxima.* The plant cells and their nuclei enlarge slightly before dividing but their contents are not modified. Hyperplasy, which starts very quickly, is not dedifferentiative and the original leaf tissues remain discernible.

The cecidozoa in the case of bud galls of *B. pallida* make their cavities with the help of hydrolytic enzymes present in the eggs and young larvae. In the case of *P. proxima,* the cavity is preformed since it is made at the time of oviposition. For this reason, it should be called a "gall cavity" rather than a "larval cavity." The root gall of *B. pallida* is an intermediate case. The lower part of the egg-laying channel represents a preformed part of the larval cavity but the final site and shape of this cavity are determined by the larva. Although their origins are different, the cavities of the three galls are the center around which the specific gall structures are organized.

The key processes during the organization phase of most galls (see Chapter 5) are hyperplasy, hypertrophy, and differentiation, to which cell destruction by biting or lysis should be added. As most galls mature, hyperplasy decreases in intensity and then tends to disappear at a rate dependent on the amount of tissue present. Hyperplasy is then replaced by hypertrophy. The rate of hyperplasy usually decreases from the margin of the larval cavity to the gall periphery while the extent of hypertrophy increases from the larval cavities to the exterior. This trend is quite regular in the gall of *P. proxima;* however, in both galls of *B. pallida* there are important exceptions. First, cells near the vascular tissue in these two galls divide rather actively and remain small. Second, cells of the nutritive tissue are large and never divide even though they are close to the larval chamber. Vascular tissues appear in a different manner in the two systems. In the two galls of *B. pallida* vascular tissue is differentiated from new parenchymatous cells whereas in the galls of *P. proxima,* it is essentially composed of preexisting bundles.

In both galls of *Biorhiza pallida* there is an accumulation of starch in the "starch parenchyma", of lipid inclusions and of structural proteins in the nutritive tissue. There are no visible accumulations of such reserves in the gall of *P. proxima;* however, the vast quantities of *P. proxima*-induced tissues is an adequate food reserve for the hatching larva. The larvae of *B. pallida* are thought to act as physiological sinks for nutritive substances. This has been shown for other cynipid galls by using radioisotopes (Jankiewicz *et al.* 1970). However, it is probable that this action is indirect and that the flow of soluble substances toward the larval chambers is due more to a stimulation of synthesis in the nutritive cells than to larval sucking.

Both galls of *B. pallida* have a layer of nutritive tissue lining the larval chamber throughout gall development. The renewal of the consumed nutritive cells occurs first by a "cambial zone" and then later by a progressive transformation of the starch parenchyma. The peculiarities of this specialized tissue that brings rich fluid food to the larvae are characteristic of cynipid galls and have received much attention (Rohfritsch 1974; Rey 1976, 1978, 1979; Bronner 1977). Nutritive cells are the center of an active proteosynthesis and some proteolysis. The abundant lipids, mainly constituted by triacylglycerols and phospholipids (Rey *et al.* 1980), are probably synthesized, at least partly, from products arising from hydrolysis of starch. Lastly the cells closest to the larval chamber undergo pathological transformations that may lead them to lysis. There are no specialized nutritive cells in the gall of *P. proxima.*

A distinct zone of sclerenchyma cells is found in both galls of *B. pallida* and is common in most cynipid galls (see Chapter 5). No sclerenchyma forms in the gall of *P. proxima.* A gall cortex is found in galls of *B. pallida,* but it is absent in galls of *P. proxima.* Thus galls of *B. pallida* have particular structures and well-developed processes of differentiation, whereas galls of *P. proxima* remain simple in their organization. Except for some ultrastructural modifications of chloroplasts and the presence of tannin in some cells, the cecidogenic processes of *P. proxima* galls are limited to hyperplasy and hypertrophy and the normal leaf structure is not extensively modified.

Identification of the gall-inducing substances emitted by insects has been the subject of an extensive literature (Maresquelle and Meyer 1965; Bronner 1977; see Chapter 10). Studies of *B. pallida* galls contribute to this area of gall biology by showing that the larvae are the elements around which the entire central zone is organized. The hypothesis that a chemical cecidogenic gradient affects hyperplasy and hypertrophy has been developed by Garrigues (1954). Furthermore, the larvae play an important role in differentiation and this appears to be true for the galls of *B. pallida.* Arrangement of the vascular tissue is determined by larval position in the plastems and the sclerenchyma zone represents the limit of each larval "territory." The continuous renewal of the nutritive tissue shows that the larval stimulus is maintained until pupation. The cortex of *B. pallida* galls appears to be beyond direct larval influence; however, larval influence is necessary for the characteristic development of the whole gall. The larvae of *P. proxima,* in contrast, have no cecidogenic action. They simply feed on the gall tissues that have developed before they hatch.

It must be emphasized that the gall of *P. proxima* represents a borderline case among *Pontania* galls. Generally these galls are more structurally complex and the larvae play a more important role in cecidogenesis (Magnus 1914). For example, in the galls of *P. salicis* (Christ.) and *P. vesicator* (Bremi), the cavity is spherical and the original leaf tissues cannot be distinguished among gall cells. A primitive type of nutritive tissue is present in some sawfly galls; however, when it is present it has little cytoplasm and appears much less specialized than in cynipid galls. These galls clearly are initiated by the egg-laying female but are developed and organized by the larva.

Even though it seems that the cecidozoan completely controls the surrounding tissues, it must be remembered that the plant reacts such that the insect is imprisoned in an enclosed formation situated outside the host organ and attached to it through a narrow connection. In the case of *P. proxima,* gall and leaf development are intimately correlated. If the insect oviposits in leaves that are almost full grown, the resulting galls are very small. Conversely, ovipositing *P. salicis* sometimes kill their host leaves if they are too immature (Magnus 1914). This suggests a hypersensitivity reaction of the host to the maternal secretion and wounding. Furthermore, since *P. proxima* galls protrude equally from both sides of the leaf and since the different leaf tissues are still visible inside the gall wall, this gall appears more dependent on the host organ than the *B. pallida* galls.

In conclusion, there is great similarity between the bud and root galls induced by the alternating generations of *Biorhiza pallida* even though the two appear on different organs. They both show the same type of development and organization, the same differentiation processes, and, in particular, the same type of highly specialized nutritive cells. All these characteristics are influenced by the action of the eggs and then of the larvae. The cecidogenic model illustrated by the galls of *Biorhiza pallida* is typical of many other cynipid galls.

The gall induced by *Pontania proxima* illustrates a very different and unusual model of cecidogenesis for hymenopterous cecidia. Even among the different *Pontania* species, the gall of *P. proxima* is a special case. Its structure is simple and each layer of cells in the leaf blade takes part in gall formation and remains distinguishable. There are no specialized nutritive tissues. Neither the larva nor the egg is necessary for cecidogenesis. Gall formation here seems to be dependent on the initial egg-laying wound together with the maternal secretions introduced with the egg.

ACKNOWLEDGMENTS

I thank M. Boursier for his technical assistance and L. Sealy for his help with the translation of the manuscript.

LITERATURE CITED

Adler, H. 1881. Uber den Generationswechsel der Eichengallwespen. Zeit. Wiss. Zool. 35:151–246.

Bartels, F. 1971. Strukturelle Veränderungen der Chloroplasten-thylakoide in Palisadenzellen von *Peperomia metallica* im Licht-Dunkelwechsel. Protoplasma 72:27–41.

Benson, R. B. 1954. British sawfly galls of the genus *Nematus (Pontania)* on *Salix* (Hymenoptera: Tenthredinidae). J. Soc. Br. Entomol. 4:206–211.

Benson, R. B. 1960. Studies in *Pontania* (Hym.) Tenthredinidae. Bull. Br. Mus. (Nat. Hist.) Entomol. 8:369–384.

Beijerinck, M. W. 1882. Beobachtungen über die ersten Entwicklungsphasen einiger Cynipidengallen. Verhand. Könink. Akad. Wetenschappen. 22:1–98.

Beijerinck, M. W. 1888. Uber das Cecidium von *Nematus capreae* auf *Salix amygdalina*. Bot. Zeit. 46:1–28.

Bronner, R. 1973. Propriétés lytiques des oeufs de *Biorhiza pallida* Ol. C. R. Acad. Sc. Paris 276:189–192.

Bronner, R. 1977. Contribution à l'étude histochimique des tissus nourriciers des zoocécidies. Marcellia 40:1–134.

Carleton, M. 1939. The biology of *Pontania proxima*. J. Linn. Soc. Lon. Zool. 40:575–624.

Cran, D. G., and J. V. Possingham. 1972. Variation of plastid types in spinach. Protoplasma 74:345–356.

Folliot, R. 1964. Contribution à l'étude de la biologie des cynipides gallicoles (Hyménopteres Cynipoidea). Ann. Sc. Nat. Zool. 12 eme ser. 6:407–564.

Garrigues, R. 1951. Recherches sur les cécidies, le cancer et l'action des carbures cancérigenes sur les végétaux. Thése, Doct. Sc. Nat., Lyon, France.

Garrigues, R. 1954. De l'existence d'un gradient chimique, agent d'action cécidogène. Comm. Congres Intern. Bot. Paris Section B:222–226.

Günning, B. E. S. 1965. The fine structure of chloroplast stroma following aldehyde-osmium tetroxide fixation. J. Cell Biol. 24:79–93.

Hovanitz, W. 1959. Insects and plant galls. Sci. Am. 201:150–162.

Jankiewicz, L. S., H. Plich, and R. Antoszewski. 1970. Preliminary studies on the translocation of ^{14}C-labelled assimilates and $^{32}PO_4^{3-}$ towards the gall evoked by *Cynips (Diplolepis) quercusfolii* L. on oak leaves. Marcellia 36:163–172.

Kinsey, A. C. 1920. Phylogeny of cynipid genera and biological characteristics. Bull. Am. Mus. Nat. Hist. 42:357–402.

Küster, E. 1930. Anatomie der Gallen. Handbuch der Pflanzenanatomie. Börntraeger, Berlin.

Laetsch, W. M., and I. Price. 1969. Development of the dimorphic chloroplasts of sugar cane. Am. J. Bot. 56:77–87.

Magnus, W. 1914. Die Entstehung der Pflanzengallen verursacht durch Hymenopteren. G. Fischer, Jena.

Mani, M. S. 1964. The Ecology of Plant Galls. W. Junk, The Hague.

Maresquelle, H. J., and J. Meyer. 1965. Physiologie et morphogenèse des galles d'origine animale (zoocécidies). In Ruhland (ed.), Handbuch der Pflanzenphysiologie, Vol. 2, pp. 280–329. Springer, Berlin.

Margulies, M. 1967. Effect of chloramphenicol on light-dependent formation of structure and protein of chloroplasts of *Phaseolus vulgaris*. In. C. Sironval (ed.), Le Chloroplaste, pp. 192–198. Masson, Paris.

Meyer, J. 1957. Cécidogenèse comparée de quelques galles d'Arthropodes et évolution cytologique des tissus nourriciers. Thèse Doct. Sc. Nat., Strasbourg, France.

Meyer, J. 1987. Plant Galls and Gall Inducers. Gebrüder Börntraeger, Berlin.

Meyer, J., and H. J. Maresquelle. 1983. Anatomie des Galles. Borntraeger, Berlin, Stuttgart.

Rey, L. 1976. Recherches comparées sur trois types d'Hyménoptérocécidies (Cynipidae et Tenthredinidae). Etude morphologique, histologique et ultrastructurale du développement et de la différenciation. Essais experimentaux. Thèse Doct. Sc. Nat., Nantes, France.

Rey, L. 1978. Particularités ultrastructurales des cellules nourricières de la galle provoquée par *Neuroterus quercusbaccarum* L. ♀ ♂ sur *Quercus pedunculata* Ehrh. Beitr. Biol. Pflanzen. 54:315–327.

Rey, L. 1979. Observations ultrastructurales sur le tissu nourricier de la galle provoquée par *Andricus albopunctatus* Schl. ♀ ♀ sur *Quercus pedunculata* Ehrh. Bull. Soc. Bot. Fr. 126:179–190.

Rey, L., J. P. Dubacq and A. Tremolières. 1980. Lipids of oak galls. Phytochemistry 19:2569–2570.

Rohfritsch, O. 1974. Infrastructure du tissu nourricier de la galle de l' *Aulax glechomae* L. sur *Glechoma hederacea* L. Protoplasma 81:205–230.

Rohfritsch, O. 1975. Etude comparative de cellules du tissu nourricier de la jeune galle de l' *Aulax glechomae* L. sur *Glechoma hederacea* L. et de cellules du tissu nourricier abandonné par le parasite. Marcellia 38:185–196.

Roth, P. 1949. Beiträge zur Biologie der Gallwespen. Verh. Naturf. Ges. Basel. 60:104–178.

Shumway, K. L., and T. E. Weier. 1967. The chloroplast structure of iojap maize. Am. J. Bot. 54:773–780.

Wehrmeyer, W. 1965. Zur Kristallgitterstruktur der sogenannten Prolamellärkörper in Proplastiden etiolierter Bohnen. Z. Naturforsch. 20b:1270–1296.

7

Strategies in Gall Induction by Two Groups of Homopterans

O. ROHFRITSCH
AND M. ANTHONY

Insects of the Order Homoptera induce a variety of well-known galls in many parts of the world. As with other groups of gall inducers, some species are restricted to certain regions and of course the distribution of all species is influenced by that of their host plants. Most homopterans are phloem feeders; however, some gall-inducing species also feed on cortical parenchyma or on mesophyll cells. Their mouthparts are modified as stylets. All insects in the Order Homoptera secrete a viscous material that builds up a sheath around the stylets, whatever route they take in the plant tissue. Galls are initiated by the female or "fundatrix" and shelter her offspring called the "gallicolae."

Most galls induced by homopterans are relatively simple structures; however, some species induce galls that are morphologically and anatomically complex. Members of the Family Eriococcidae, for example, are the only gall insects able to induce bud galls from a leaf.

Four homopteran bud galls have been chosen as the subject of this chapter. In the first part, buds of *Picea excelsa* L. attacked by the two adelgid species *Adelges abietis* L. = *Chermes abietis* L. and *Adelges laricis* Vallot = *Chermes strobilobius* Kalt. are described. The second part describes the neoformation of buds on leaves of *Shorea curtisii* Dyer ex King attacked by the eriococcids *Gallacoccus anthonyae* Beardsley and *Gallacoccus secundus* Beardsley. The galls induced by the two species in each genera have similar forms, but in each case we show how similarities and differences are related to biology and behavior of the cecidozoans.

GALLS OF *ADELGES LARICIS* AND *ADELGES ABIETIS* (FAMILY ADELGIDAE) ON *PICEA EXCELSA*

Gall Descriptions

Galls induced by *Adelges laricis* and *A. abietis* on *Picea* (spruce) are commonly referred to as "pineapple galls" because of their external appearance (Figs. 7-1 and 7-2). Outwardly the galls seem to be composed of a number of rhomboidal "scales" or "flaps," each surmounted by a cone-shaped prolongation. The distal margin of each flap is bordered by a suture line covered with hairs. This line marks the future opening of the larval cavities.

Galls of *A. laricis* are perfectly-shaped pineapple galls (Fig. 7-1). They seem to have an axial symmetry in which the total shoot and all the needles are incorporated into the gall. The galls of *A. abietis,* in contrast, are laterally formed (Fig. 7-2); that is, only one side of the shoot is involved in the atypical growth. The unaffected side and the upper part of the shoot, along with the needles, develop normally as the gall grows.

Both galls are also different in size and color (Rohfritsch 1971). Galls of *A. laricis* are whitish-green, the gall epidermis is waxy and hairless, and there are a few white hairs along the distal margins of the flaps. Galls of *A. abietis* are

Figures 7-1–7-6. Galls induced by *Adelges laricis* and *Adelges abietis* on *Picea excelsa*. Scale represents 1 mm. **Figure 7-1.** Mature gall of *A. laricis* with needles completely modified. **Figure 7-2.** Mature gall of *A. abietis* with distal portions of the needles remaining normal. **Figure 7-3.** Bud modified by the fundatrix of *A. laricis*. Scales have been removed to show the modification of the bud. Compare the attacked bud to the normal bud (arrow). EG, Eggs. **Figure 7-4.** Small portion of a bud modified by the pseudofundatrix of *A. abietis*. Arrows indicate the modified needle base. The normal bud and the modified bud are of the same size and appear similar. **Figure 7-5.** Gall of *A. laricis* where the gallicolae have settled between the modified needles. **Figure 7-6.** Gall of *A. abietis* at the stage where the gallicolae are settling between the modified needles. The small larval cavities at the base of the needles are visible (arrow). As soon as the gallicolae are inside, the gall will grow and the larval cavities will be closed (compare with Fig. 7-2).

the same dark green of normal needles and the epidermis is covered with hairs. The hairs are denser along the distal flap margin and often are red during the first 2–3 weeks of gallicolae development. Mature galls of *A. abietis* are hard due to a zone of sclerenchyma that develops at the periphery of the vascular elements and throughout the cortical tissues. Galls of *A. laricis* remain fleshy and soft and are usually smaller than galls of *A. abietis*. The gallicolae of *A. laricis* become nymphs at the end of May while those of *A. abietis* become nymphs at the end of August. The galls of both species lose their green color at this stage and become yellowish-brown. Their tissues begin to desiccate and the cavities open. The nymphs then crawl to the twig or needles before molting.

Life Cycles of the Two Adelgids

The fundatrix nymphs of both species settle down in the fall on the current year's shoot, either near the bud *(A. abietis)* or on the bud *(A. laricis)*. Their long slender mouthparts are deeply inserted into the plant tissues. The fundatrix grows very little during the winter. Its feeding activity starts in mid-March and apparently is synchronized with the flow of sap. By the time the buds have started to swell in April (Figs. 7-3 and 7-4), the fundatrix is a mature female and lays a large cluster of eggs. The gallicolae nymphs that hatch from these eggs find easy access to the young shoot in the bud, which has been modified to a cone-like gall by the feeding activity of the fundatrix (Figs. 7-5 and 7-6). The gallicolae of *A. laricis* emerge from their galls as winged nymphs and after ecdysis fly to new hosts. *Larix* is the secondary host of *A. laricis*.

Adelges laricis follows the basic adelgid life cycle, which takes 2 years to complete and has five distinct forms. All the forms are parthenogenetic oviparous females apart from the sexuales, which are gamic males and females. Each fertilized female of this sexuales generation lays a single egg that gives rise to the fundatrix.

Reproduction is always parthenogenetic in the case of *A. abietis*. There are only two forms: the "pseudofundatrix" and the winged gallicolae, which does not migrate to a secondary host (Gaumont 1957; Carter 1971).

Like all homopterans, adelgids secrete a viscous material that forms a sheath around the stylets. One can follow the probing and feeding behavior by examining the stylet tracks. Studies at the level of the electron microscope show that the stylet tracks are always localized intercellularly (Rohfritsch 1976, 1977) and not intracellularly as was suggested earlier (Plumb 1953; Meyer 1957). Ultrastructural analysis also has shown that the very beginning of cecidogenesis does not occur at the end of winter as was previously thought, but in the autumn of the previous year on the cortex of 1-year stems as soon as the fundatrix settles down (Rohfritsch 1981, 1988).

Gall Development Related to Feeding Behavior

Once the first-instar larvae of the adelgid fundatrix settle down in the autumn on buds of *Picea*, they never move. The immatures must initiate a gall or they die during the winter. Although both *A. abietis* and *A. laricis* attack buds of the same tree, they do not choose the same type of bud; the exact feeding place on the bud also is different (Fig. 7-15).

Adelges laricis selects small lateral or terminal buds of secondary branches, which are often located more toward the inside of the tree beneath the main branches. These small buds burst later in the spring. *A. abietis* prefers vigorous terminal buds of main branches that normally are on the periphery and top of the tree. These buds will be the first to burst in the spring. Thus *A. laricis* locates buds that are not attractive to *A. abietis*. *A. laricis* locates the mid-bud scales and inserts its stylets through the first scale whereas *A. abietis* chooses the base of the bud and inserts its stylets between the pulvini or cortical leaf stalklets surrounding the bud.

Adelges laricis settles down late in the autumn (November). Its stylets cross the first scale then migrate down the scale using the free space between them. When the stylets reach the base of the scale, they enter the soft tissue and pass through the base of the more centrally located scales toward the collenchyma plate and the young leaves. They attack the buds laterally and the stylets never migrate along the vascular strands of the shoot (Fig. 7-7).

The first nutritive tissue induced by the fun-

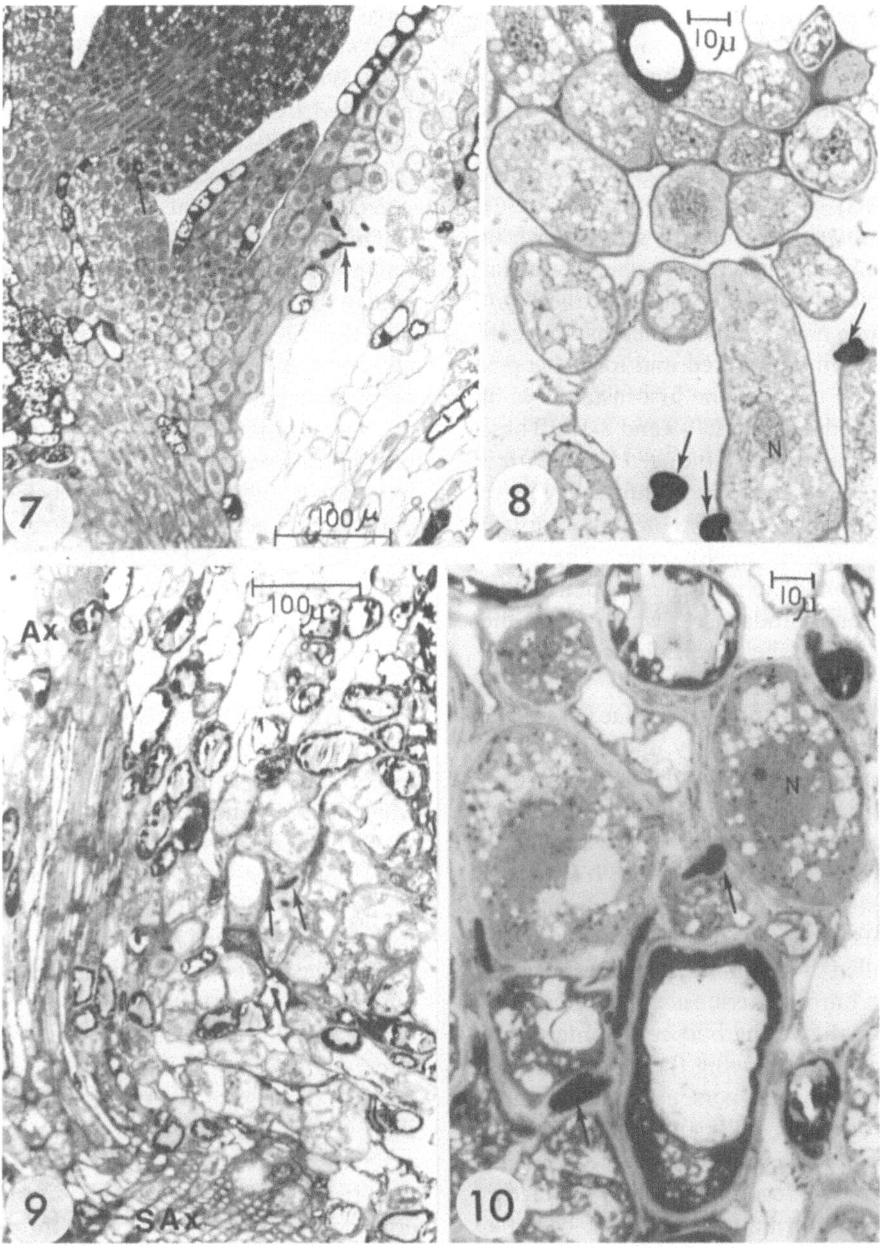

Figures 7-7–7-10. Sections of galls of *A. laricis* and *A. abietis* on *Picea excelsa.* **Figure 7-7.** Section of a bud collected in November attacked by *A. laricis.* The salivary tracks reach the second leaf of the young bud at the beginning of winter. Cells are modified around the salivary tracks in the scale base (arrow points to the salivary tracks). **Figure 7-8.** Detail of a bud attacked by *A. laricis* showing modified cells in the scale bases that constitute the first nutritive tissue of the fundatrix (NT1). Salivary tracks are identified by arrows. **Figure 7-9.** Section of 1-year-old shoot attacked by *A. abietis* collected at the end of October. The aphid's stylets are first inserted in the cortical tissue of the shoot where primary nutritive tissue develops. Ax, Bud axis. **Figure 7-10.** Modified cells of the first nutritive tissue of the pseudofundatrix larva (NT1) of *A. abietis.* The cells are rich in cytoplasm and fine droplets of lipid. They also have many small vacuoles and a large nucleus.

datrix of *A. laricis* develops in the fall along the pathway of the stylets and a rich starch tissue appears around the nutritive cells. These cells are typical of those found in galls of other insects (Chapter 8). They are hypertrophied and the nucleus and nucleolus are enlarged. They are rich in cytoplasm and the vacuoles become small; lipid droplets and starch grains are abundant (Fig. 7-8).

Adelges abietis, in contrast, settles down in October and within 1 month the first-instar nymph explores the cortical cells of the shoot leaving a large net of salivary sheaths. The cortical cells become modified and form the primary nutritive tissue of the first-instar larva of the pseudofundatrix (Figs. 7-9 and 7-10). These cells are similar to those induced by *A. laricis* and have been described by Rohfritsch (1981).

The stylets of *A. abietis* reach the cortical cells of the young bud in November. They cross the collenchyma plate near the vascular tissue. Only the pseudofundatrices, which reach the young buds before the winter, survive to spring (Thalenhorst 1972).

The role of the first nutritive tissue in the galls of both species is not clearly understood. Obviously it provides food for the first-instar larvae, but the reaction seems disproportional to the needs of this early stage. The tissue, however, may have a reserve function. It has been noted that starch disappears in the spring from galled shoots kept in water, whereas starch remains in galled shoots on the tree. This suggests that the first nutritive tissue acts as a physiological signal to the young bud cells conferring a kind of resistance allowing the larva to attack without provoking a hypersensitive reaction. It is known from other galls, such as those of sawflies, that if the attacked organ is too young, a hypersensitivity reaction occurs (Magnus 1914). From information on fungi, bacteria, and viruses, it is known that the first attack prompts a resistance in the tissue towards a second attack (Jenns and Kŭc 1980; Sziraki *et al.* 1980; Sequeira 1983).

The action of the fundatrix of *A. laricis* and the pseudofundatrix of *A. abietis* in the fall induces only physiological and cytological modifications on the 1-year-old shoot. At the end of winter the larva also induces growth modification in the bud around the feeding sites. During this period the larva shows a strong probing activity, many salivary sheaths are secreted, a large number of cells are attacked, and the stylets "sting" many cells.

Buds attacked by *A. laricis* are always more modified by spring than buds attacked by *A. abietis* (Fig. 7-15). More leaves are attacked and more cells are modified than in the case of *A. abietis.* The stylets of *A. laricis* avoid the first leaf and instead migrate to the bases of the second and third leaves such that the nutritive tissue will be localized near the center of the small bud (Fig. 7-15). Consequently, the region attacked by *A. laricis* extends early in bud development and even more when the attacked buds are very immature and would normally burst late in the season. Thus *A. laricis,* through its choice of feeding sites, behavior, and cecidogenic action, easily controls the total growth of the shoot and inhibits the normal growth of the shoot apex. *A. abietis,* in contrast, modifies only a small part of the normal shoot.

Cells withdrawn from normal patterns of development by *A. laricis* give rise, in the spring, to the main nutritive tissue (NT2) of the fundatrix (Fig. 7-15). This nutritive tissue is similar in both adelgid galls and has been described by Meyer (1957) and Rohfritsch (1976, 1977). In its early stage it is only slightly differentiated and looks like the young nutritive tissue of many zoocecidia with its nuclear and nucleolar hypertrophy, cellular hypertrophy, an increase in cytoplasm, reduced vacuoles, slightly differentiated sinuous plastids, and large numbers of droplets of lipids in the cytoplasm (Figs. 7-11 and 7-13).

As in the case of *A. laricis,* spring feeding by the pseudofundatrix of *A. abietis* begins with extensive probing. Their stylets explore the cortical tissue and the bases of the second and laterally situated leaves. The main nutritive tissue of the pseudofundatrix (Figs. 7-12 and 7-14) develops due to this aggressive behavior. In contrast to the galls of *A. laricis,* the nutritive tissue in galls of *A. abietis* is less extended and its cells are less hypertrophied. Gall tissue around the nutritive cells contains more phenolic compounds and less starch than in comparable cells induced by *A. laricis.*

Third-instar larvae and the imago of adelgids need large amounts of nutritients and thus the

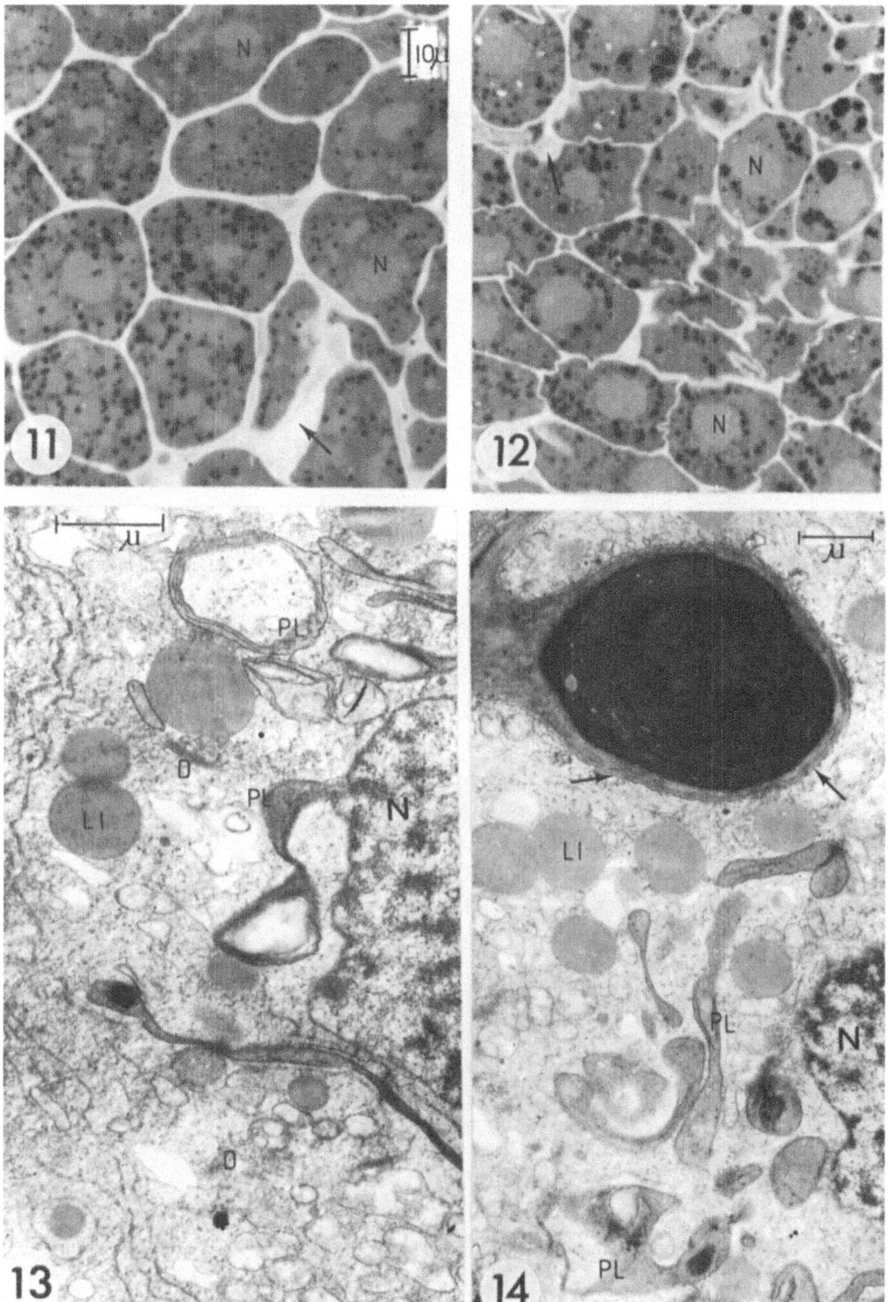

Figures 7-11–7-14. Main nutritive tissues induced by the fundatrices of *Adelges laricis* and *Adelges abietis* on *Pices excelsa* in the early spring (NT2). **Figure 7-11.** Cells of the nutritive tissue induced by *A. laricis* localized in the cortex of the bud and in the first, second, and third leaf bases. **Figure 7-12.** Cells of the nutritive tissue induced by *A. abietis* localized in the cortex and in the first and second leaf bases. Arrow indicates salivary track. N, Nucleus. **Figure 7-13.** Ultrastructure of nutritive tissue cells induced by *A. laricis.* globules. **Figure 7-14.** Ultrastructure of nutritive tissue cells induced by *A. abietis.* The host cell wall and plasma membrane (arrows) can be observed around the section of the salivary track. D, Dictyosomes; Li, lipid globules; N, nucleus; Pl, plastids.

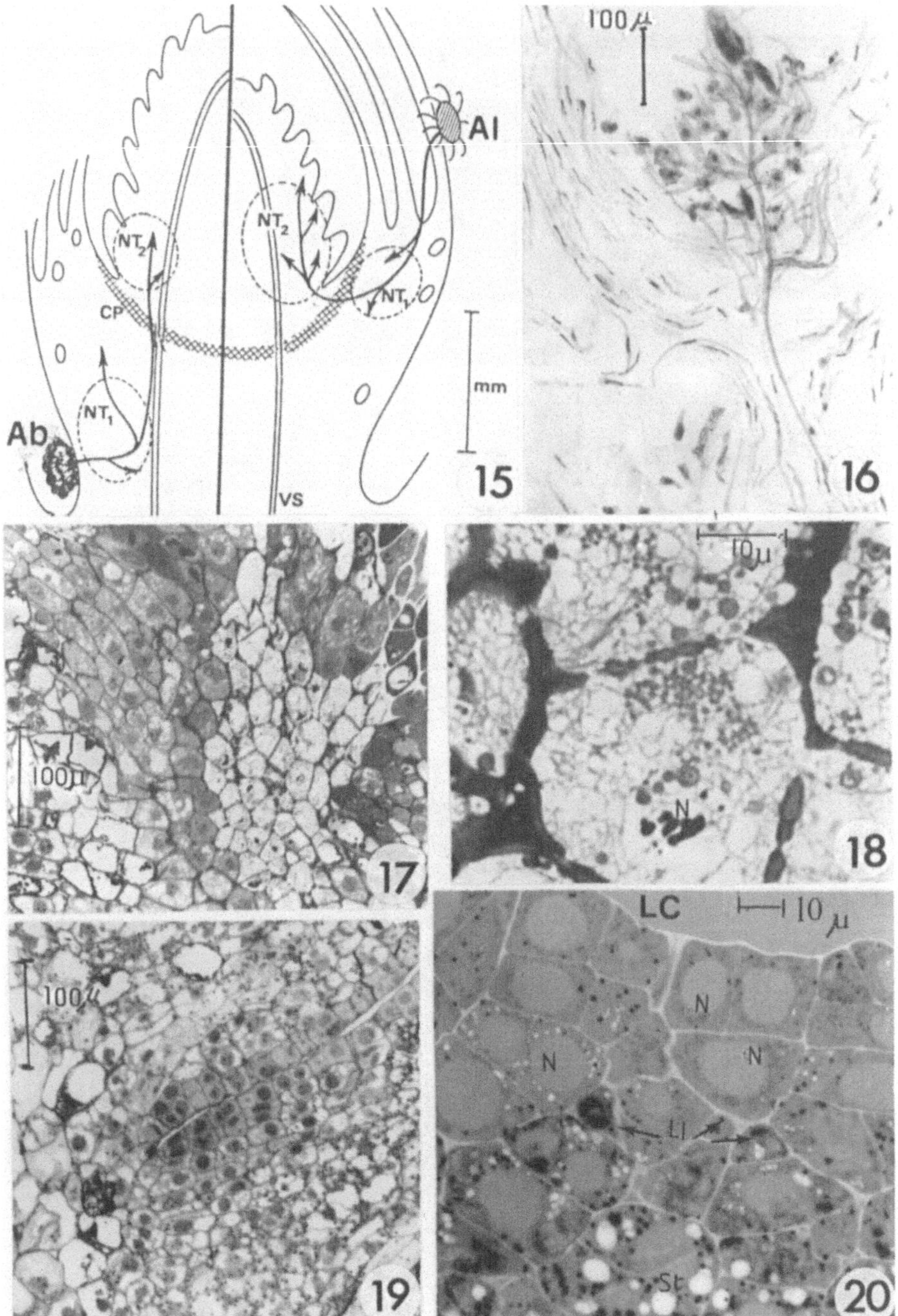

Figures 7-15–7-20. Figure 7-15. Diagram showing differences in strategy of bud attack by *Adelges abietis* (left) and *Adelges laricis* (right). The buds selected are different in size and stage of development. In both cases the fundatrix larva induces a primary nutritive tissue in the autumn in reaction to the beginning of the insect's activity. The main nutritive tissue is induced in late winter and will provide the fundatrix with food for rapid development in the spring. **Figure 7-16.** Dissected stylets and salivary sheaths left by *A. abietis* in the cortical tissue at the bases of the first and second leaves. Some plant cells remain attached to the sheath. **Figure 7-17.** Nutritive tissue of the mature fundatrix of *A. laricis.* The central area of the nutritive tissue is composed of nearly empty cells surrounded by activated cells with large nuclei and a dense cytoplasm. Tissues rich in starch are found be-

nutritive tissues are well differentiated by the time they appear (Rohfritsch 1977). Mature nutritive tissue in galls of both adelgids can extend roughly a quarter of the shoot's circumference and appears heterogeneous, being composed of two types of cells (Fig. 7-17). One type of cell, usually found in the center of the nutritive zone, is nearly devoid of cytoplasmic content (Fig. 7-18). They have necrotic nuclei, a few degraded membranes, and modified walls. These greatly modified cortical cells resemble differentiating sieve tube cells of phloem (Bouck and Cronshaw 1965; Esau 1965; Buvat 1968). The second type of cell has more typical features of nutritive tissue, with their abundant, dense cytoplasm, large lobed nucleus, numerous nucleoli, and regular cell walls. They are rich in droplets of lipids, devoid of starch, and have many small vacuoles of the lysosomal type. All modified tissues around the nutritive tissue are rich in starch.

In spite of the very aggressive probing activity of the first-instar larvae at the end of winter, the plasma membrane of the host cells never ruptures. The salivary sheath is always surrounded by a more or less thick cellulosic sheath. Cell necrosis usually does not occur along the track of the stylets, but the cortical cells near the pathway of the stylets are slightly dissociated. It is well known that the homopteran parenchyma feeders have numerous enzymes in their saliva that help the stylets progress through the tightly packed cells of the bud (Pollard 1973). This is the reason it is relatively easy to localize the salivary sheath in attacked shoots or buds (Fig. 7-16) and to completely separate it from the surrounding nutritive tissue.

By puncturing cells without rupturing the membrane and by loosening the middle lamella with salivary enzymes, the first-instar larva has a wound-like effect on a group of cells in the bud. Cells are withdrawn through this "irritating" action from the normal growth pattern and engaged in the cecidogenic field. This group of cells constitutes a "sink" that is in competition with the meristematic tissues of the shoot and whose development is independent of normal growth, but strictly related to the fundatrix development. It is easy to understand the importance of the localization of this group of cells along the normal shoot axis. Their influence will be different if localized at the base or in the center of the bud, on a vigorous and fast growing shoot or on one that is small and slowly developing (Fig. 7-15).

Maturing galls of adelgids are induced by one fundatrix and are composed of modified needles, bases of needles, and cortical tissues of the young shoot. After hatching, first-instar gallicolae crawl up to the modified needles and a negative phototropism forces them into the small holes between the modified and enlarged bases of the needles. A complex gall develops with a large number of larval chambers tightly joined.

In the case of *A. laricis,* normal growth is totally inhibited by activity of the fundatrix. When the gallicolae settle between the modified needles, only gall tissues develop around them. In the case of *A. abietis,* normal growth of nonmodified parts of shoots is continuous with gall growth.

The gallicolae begin to feed by first superficially puncturing the epidermal cells of the small cavity wall. These cells are rich in cytoplasm with a large nucleus and several nucleoli. They have a meristematic appearance (Figs. 7-19 and 7-20) with a thin wall nearly without cuticle and become the nutritive tissue of the gallicolae (Fig. 7-20). The cells are large toward the periphery of the modified leaf and contain starch and phenols. The epidermis is covered

---

yond the activated cells. **Figure 7-18.** Enlargement of near empty nutritive cells of the central area of a gall induced by the fundatrix of *A. laricis.* Note that the cell contents are highly modified as are the cell walls. **Figure 7-19.** Cells of the bases of needles modified by the pseudofundatrix of *A. abietis* on which the gallicolae will feed. **Figure 7-20.** Tissues around the larval cavity of *A. abietis* at the time when the gallicolae settle down. Cells of the first two to three layers are similar to meristematic cells. Behind these layers are cells rich in starch. Li, Lipids; N, nucleus; st, starch. Ab, *A. abietis;* Al, *A. laricis;* CP, collenchyma plate; NT_1, primary nutritive tissue; NT_2, main nutritive tissue; vs, vascular strands (Rohfritsch 1982).

with hairs, except along the nutritive cells. The gallicolae are very mobile during the first larval instar. They puncture the epidermal cells by pushing the plasma membrane into the cytoplasm. Later the stylets migrate between the cells. As the gallicolae develop, their stylets penetrate deeper into the tissues (Rohfritsch 1976).

Starch near the feeding sites of the gallicolae disappears and the cells are emptied. The gallicolae of *A. laricis* develop within 3 to 4 weeks and quickly modify the gall tissues. When they reach the third larval stage the gall tissue is nearly devoid of starch. During the last stage of the gallicolae, the nymphs take food directly from the vascular tissues of the gall. The nymphs are fixed with their heads orientated toward the back of the larval cavity.

Maturation of gall and larvae are closely correlated and occur near the end of May. At this time, the gallicolae of *A. abietis* remain in their first or second larval stage. The gall tissue becomes lignified along the shoot axis and along the leaf axis. In the distal parts of the gall the cells are filled with phenolic compounds. The gallicolae always have excess starch around the larval cavity especially near the ostiolar opening. The nymphs in this case are orientated toward the ostiole of the larval cavity, feeding only in the parenchyma. They avoid the back of the cavities where lignification has occurred. The gall and gallicolae are mature by September.

Adelges laricis has its maximum food requirement in mid-May, a period correlated with the end of the spring flush of the tree. The maximum food requirement of *A. abietis* is correlated with the second flush of the tree at the end of the summer when cones are forming (Lovett 1980). At this time, amino acids and soluble sugars are available in the vascular tissues of the tree.

Conclusions

Differences in feeding site selection and in feeding behavior allow these two species to share what superficially appears to be the same microhabitat and avoid competition for space. The actual form, color, structure, and development of each of these two galls results from the specific behavior of both insects (Fig. 7-15).

Of the two species, it appears that *A. laricis* has evolved the better strategy. Its host tree is damaged less since the main shoots are not attacked.

The gall of *A. laricis* appears as a new organ issued from a totally transformed bud. Gall tissue is parenchymatous and soft and it has a high water content. The normal characteristics of shoots or needles are no longer present. The gall of *A. abietis,* in contrast, only partly transforms the normal shoot and leaves, which explains the color, gall structure, and presence of large amounts of phenolic compounds.

GALLS OF *GALLACOCCUS ANTHONYAE* AND *GALLACOCCUS SECUNDUS* (FAMILY ERIOCOCCIDAE) ON *SHOREA CURTISII*

Gall Formation in the Superfamily Coccoidea

Many of the thousands of species of phytophagous coccids are injurious to their hosts and contain toxins in their saliva (Carter 1952); however, only a few species are capable of inducing galls. Most cecidogenic coccids induce simple galls such as bud deformation, leaf wrinkling, or pit galls (see Chapter 2); however, some species in the Brachyscelidae and Eriococcidae are able to induce complex galls.

Many of the coccids inducing simple galls are capable of feeding on other hosts without noticeably affecting plant growth (Beardsley 1984; Buhr 1964; Howell and Kosztarab 1972). However, those species inducing complex galls demonstrate a high degree of host specificity. It also is interesting to note that cecidogenic coccids specifically linked to certain plant taxa are usually restricted to certain geographic areas. For example, the Brachyscelidae, which form complex galls on Myrtaceae and Casuarinae, are restricted to Australia and New Zealand (Houard 1922) and species of *Gallacoccus* of the family Eriococcidae are restricted to inducing epiphyllous bud galls on species of *Shorea* (Family Dipterocarpaceae) in the western part of the Malaysian botanical region (Anthony 1980c).

The group of gall-inducing coccids of the

Family Eriococcidae on the leaves of *Shorea* was only recently discovered (Anthony 1980c) and it is assumed that all species belong to the genus *Gallacoccus*. This genus was described by Beardsley (1971) from two specimens discovered by Anthony (1970, 1974a, b) in Singapore. Only the females are cecidogenic in this group, inducing neoformed buds on the leaves of *Shorea*. This epiphyllous budding appears strikingly original, not only in relation to coccid galls, but to all other known types of galls.

Several aspects of the biology of *Gallacoccus anthonyae* Beardsley and *G. secundus* Beardsley on *Shorea curtisii* Dyer ex King have been studied (Anthony 1974b, 1977, 1980a, b, c); however, the following section of this chapter will focus on the life cycle of these two species and the morphogenesis of their galls. In both cases we emphasize the neoformed shoot apices.

Cecidogenesis and Life Cycle of *G. anthonyae* and *G. secundus* on *Shorea curtisii*

Galls of *G. anthonyae* (Figs. 7-21 and 7-22) develop abaxially on the midrib or petiole of leaves of *S. curtisii* and slightly inhibit leaf growth. As the galls mature, they appear like leafy shoots consisting of 25 scale-like leaves distichously arranged (Fig. 7-27). Galls of *G. secundus* (Figs. 7-28 and 7-29), in contrast, develop adaxially on leaves or stipules of *S. curtisii* and strongly inhibit growth of the host organ. As the galls mature, they appear pear shaped and consist of five scales showing a distichous arrangement. The last two scales are hypertrophied and fused at the base. Thus both galls are epiphyllous shoots produced by the functioning of a typical vegetative apex (Figs. 7-23, 7-25, and 7-31).

The epiphyllous galls of both species are initiated by a larva of a first-instar female (L1 ♀) during the first phase of nutrition. The location of this first feeding zone remains morphologically visible as a slight depression (due to inhibition of growth) situated at the base of the galls in the cortical parenchyma of a vein (Fig. 7-35a–a′). Cytologically, cells in this zone show a tanniferous and mucilaginous differentiation (Figs. 7-24 and 7-31). In the youngest galls observed, the larva had already molted and left its first feeding zone, leaving its exuvia behind. This second-instar larva (L2 ♀) shelters and feeds itself under the cover of a labial fold (1a) emanating from a basal neoformed bump on the leaf (Figs. 7-24 and 7-31). This labial fold is hemicircular in the case of *G. anthonyae* (Fig. 7-35a) and circular and bilobed in the case of *G. secundus* (Fig. 7-35a′). A young vegetative meristem has already appeared at this stage under the labial fold next to the larva.

During the second phase of insect nutrition, the neoformed apex begins to function, producing 12 scales in the case of *G. anthonyae*, but only 3 in the case of *G. secundus* (Fig. 7-35b–b′). The second-instar larva then molts into an adult female, which is subsequently fertilized by a male from a neighboring mature gall (Anthony 1980c). In *G. anthonyae* the fertilized female climbs up and establishes herself at the axil of the sixth scale whereas in *G. secundus* she is found between the two last scales (Fig. 7-35c–c′). The apex of the gall ceases to function as the insect is molting and moving, but cell hypertrophy occurs.

During the third phase of insect nutrition, the gall apex of *G. anthonyae* resumes its activity and produces 12 new leaves, before becoming parenchymatous (Figs. 7-26 and 7-35d). The apex of *G. secundus* galls ceases activity first, after the production of the first 3 leaves, while the insect is still found at the axil of the labial fold. In this case the third phase of insect nutrition is characterized by the hyperplastic development of the last two leaves that produce a gall wall with complex anatomical structures (Figs. 7-29, 7-30, and 7-35d′).

The females then give birth to first-instar male and female larvae that invade the gall bud (Fig. 7-35d–d′). The male larvae molt and transform into winged adults while the female larvae remain at their first instar. The development of the gall at this stage is only by hypertrophy. When the gall dries out, a new generation of insects is liberated to repeat the cycle. The female first-instar larvae then initiate new galls. The adult males will fertilize adult females in other more developed galls. Thus it is important for the life cycles of these insects that galls

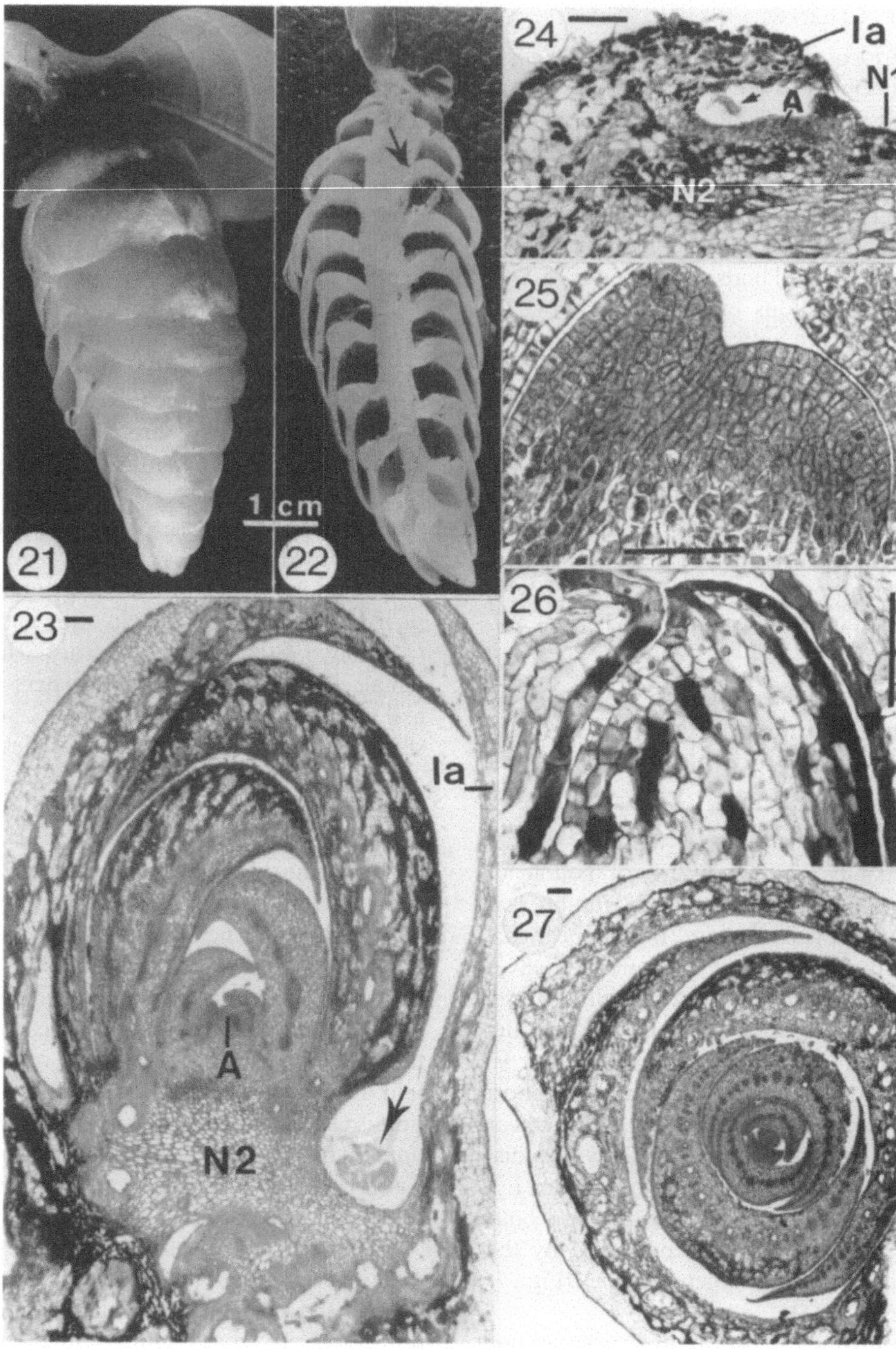

Figures 7-21–7-27. Galls of *Gallacoccus anthonyae* Beardsley on *Shorea curtisii* Dyer ex King. All scale lines indicate 100 μm unless otherwise stated. **Figure 7-21.** Mature gall. (From Anthony 1980c.) **Figure 7-22.** Longitudinal dissection of a mature gall. The adult insect (arrow) is located in the axil of the seventh scale. (From Anthony 1980c.) **Figure 7-23.** Longitudinal section of a gall halfway through its development. The second-instar larva (arrow) is located in the axil of the basal labial fold. The cecidial apex has already produced 7 scales. A, Cecidial apex; la, labial fold. (From Anthony 1977.) **Figure 7-24.** Longitudinal section of an immature gall. A newly formed apex has appeared under the labial fold next to the second-instar larva (arrow). A, Cecidial apex; la, labial fold. (From Anthony 1974b.) **Figure 7-25.** Longitudinal section of an active cecidial meristem. (From Anthony 1974b.) **Figure 7-26.** Longitudinal section of the parenchymatous apex of a mature gall. (From Anthony 1974b.) **Figure 7-27.** Transverse section near the apex of a gall halfway through its development. (From Anthony 1977.)

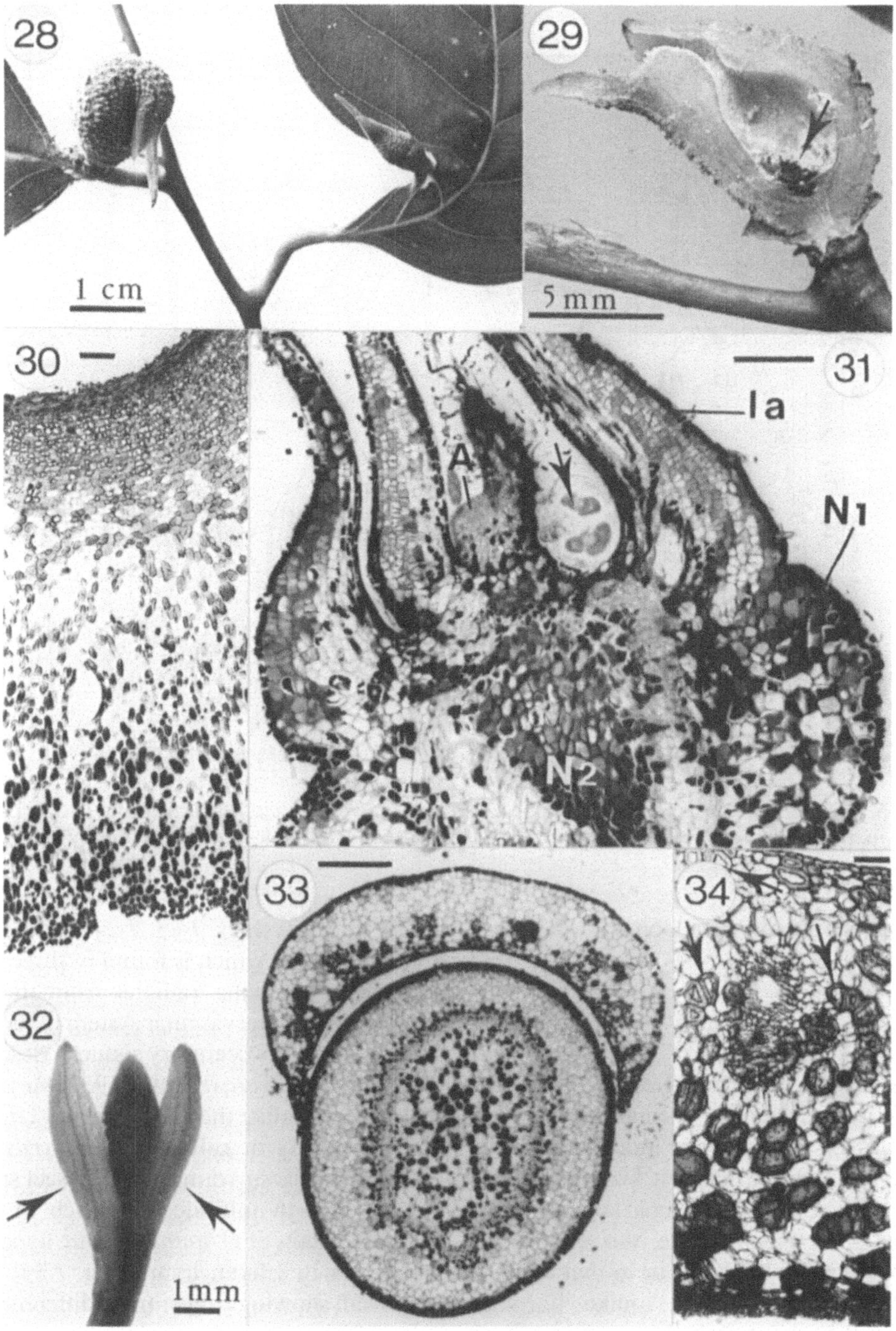

Figures 7-28–7-34. Normal shoots and shoots galled by *Gallacoccus secundus* Beardsley and *G. anthonyae* Beardsley on *Shorea curtisii* Dyer ex King. All scale lines indicate 100 μm unless otherwise stated. **Figure 7-28.** Two mature galls of *G. secundus.* (From Anthony 1974a.) **Figure 7-29.** Longitudinal dissection of a mature gall of *G. secundus.* The adult insect (arrow) is located between the last two hypertrophied bud scales that form the gall wall. (From Anthony 1974b.) **Figure 7-30.** Cross section through the gall wall of a mature gall of *G. secundus.* (From Anthony 1974b.) **Figure 7-31.** Longitudinal section of a young gall of *G. secundus.* The second-instar larva (arrow) is located in the axil of the basal labial fold. The cecidial apex has already produced two scales. A, Apex; la, labial fold. (From Anthony 1974b.) **Figure 7-32.** Young plagiotropic shoot of *S. curtisii* with its two protective bud scales (arrows). (From Anthony 1977.) **Figure 7-33.** Transverse section made at the base of a normal bud showing one of the two protective scales. **Figure 7-34.** Transverse section of a scale from the gall of *G. anthonyae.* Arrows indicate the sclerenchyma cells. (From Anthony 1974b.)

35

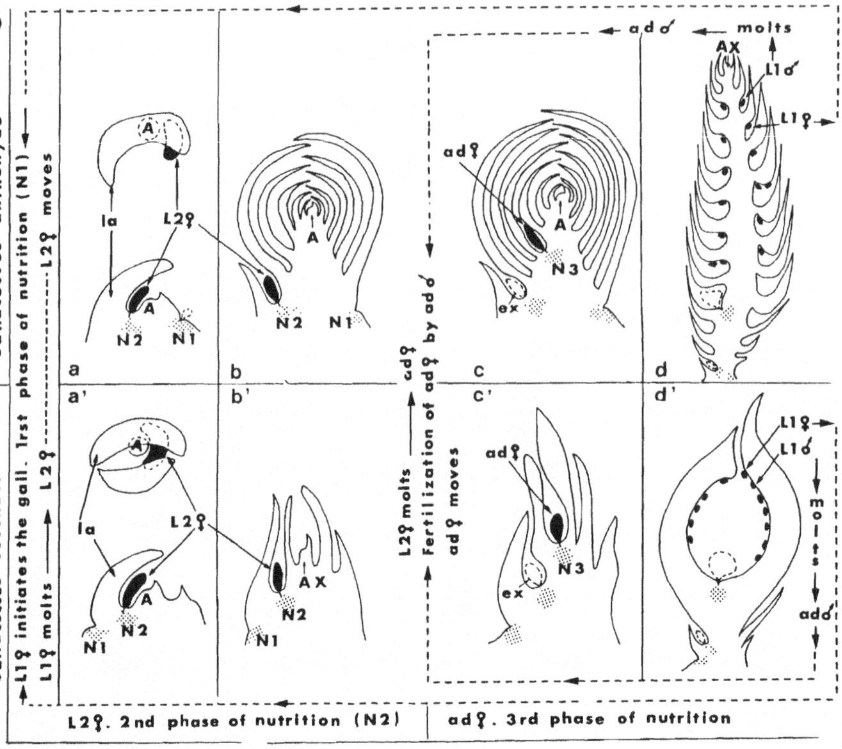

Figure 7-35. Illustration comparing the development of galls of *Gallacoccus anthonyae* and *G. secundus.* Galls are shown in longitudinal section. A polar view of the galls also is given in a–a′. (After Anthony 1974b, 1980a.)

at various stages of development simultaneously coexist.

Characteristics of Cecidial Shoots

The apices of galls of both species produce scale leaves that do not resemble normal leaves of the host. Scales of the galls are morphologically comparable to the normal first two bud scales protecting the young plagiotropic shoots (Figs. 7-32 and 7-33). Furthermore, the distichous phyllotaxy of the galls is similar to that of plagiotropic shoots. However, unlike normal shoots, cecidial shoots have no heteroblastic leaf gradient along the axis. The gall apices are capable of producing scales only. Thus gall meristems appear to have arrested juvenile development in terms of leaf production.

In addition to their juvenile characteristics, galls have specific anatomical features. For example, a specialized intermediate tissue is differentiated in both galls which is rich in tannins and mucilage (Figs. 7-23, 7-24, and 7-31). This type of tissue, which is found in three different places of the galls, radiates from the feeding zones toward the vascular system of the gall and resembles the "secretory tissues" described in galls of *Phylloxera* (Sterling 1952; Meyer 1958). However, unlike the galls of *Phylloxera,* no necrosis occurs in galls of *Gallacoccus.* In this case, the cells surrounding the insect stylets become growth inhibited and rich in tannins. Scales of galls of *G. anthonyae* are hypertrophic and rich in sclerenchyma (Fig. 7-34). A thick gall wall showing concentrical differentiation of tanniferous and mucilaginous parenchyma is found in the gall of *G. secundus* (Fig. 7-30).

Although normal and cecidial shoots share some similarities, differences are nevertheless apparent. For example, unlike normal shoots, cecidial shoots have no axillary buds and are of determinate growth; that is, the apex produces only 3 scales in *G. secundus* and 24 in *G. anthonyae* before terminating. In terms of apical anatomy, cecidial shoot development in *Shorea* appears to be linked to the life cycle of the insect, while normal shoot development appears to follow the endogenous rhythm (Anthony 1980d). Thus if the cecidozoa dies, the gall meristem ceases its activity resulting in early parenchymatization (Anthony 1974b).

Discussion

The epiphyllous bud galls of *Gallacoccus,* when compared to other coccid galls, at first appear strikingly different. However, in the pit galls of Asterolecaniidae and Diaspididae studied by Houard (1911) and Goidanich (1960), cecidogenesis was shown to result from an inhibition of growth beneath the cecidozoa and development of the surrounding rim. Such morphogenesis is reminiscent of the early developmental phenomena observed in the galls of *Gallacoccus,* which are classical morphogenetic processes normally encountered in all forms of enveloping cecidogenesis (Meyer and Maresquelle 1983).

As stated previously, inhibition of host plant growth is always more pronounced in the case of *G. secundus.* An analysis of early gall development (Anthony 1974b, 1980b) showed that inhibition of leaf growth was an early stage of induction due to the primordial inhibitory effect of the first-instar larva. Since *G. secundus,* when compared to *G. anthonyae,* establishes itself on a much younger leaf primordium, subsequent growth inhibition was enhanced. Such a phenomenon commonly appears in cecidogenesis (Maresquelle 1935, 1936, and Chapter 5). However, the development of a shoot in the axil of the primitive labial fold, as in the galls of *Gallacoccus,* appears completely original in the cecidial world.

The architecture of these two galls is quite different as they mature. The mature gall of *G. anthonyae* resembles a shoot of 6 cm in length having 24 distichous arranged scales and one basal fold. The large size of the gall is due to axis elongation and meristem functioning. In contrast, the gall of *G. secundus* is 4 cm in length, pear-shaped, and has only three distichously arranged scales which form an anatomically complex gall wall. The gall of *G. secundus* is considered to be more complex or elaborate than that of *G. anthonyae* because of special superficial structures, which make it deviate from the typical shoot structure.

At each stage of gall development, *G. secundus* are found much nearer the apex of the gall than are *G. anthonyae* (Fig. 7-35); that is, during the third period of gall growth, the insect establishes itself directly on top of the degenerated gall apex. The short life span of the gall apex in *G. secundus* appears to be linked to the insect's behavior.

CONCLUSIONS

The Adelgidae are unique among the Aphidoidea and the Eriococcidae among the Coccoidea in that they have evolved homologous types of galls in totally different ways. By distributing their feeding sites differently along the bud axis, which is the normal bud in the case of the adelgids and the neoformed bud in the case of eriococcids, the two groups of species modify in different ways the growth and development of buds. If the feeding site is too close to the terminal meristem, the shoot growth will be inhibited, as is the case with *A. laricis* and *G. secundus.* If the feeding site is kept at the bud base, as is the case with *A. abietis* and *G. anthonyae,* the shoot can develop further.

In both groups, the fundatrix changes the feeding place several times during cecidogenesis. The stylets are abandoned after each ecdysis and new stylets are introduced at a different place. All four species feed exclusively on cortical parenchyma. They all induce a kind of short distance transport tissue grafted onto the vascular tissue of the galled shoot. Cecidogenic growth in all species is initiated next to the fundatrix and not around it as occurs in most other galls (Chapter 5). The earliest site of nutrition of the first-instar larva is at the very periphery of the gall and not in the center.

We have described two gall types in each genus, but there are numerous species of Adelgidae inducing similar types of galls on *Picea* and there is a large number of *Gallacoccus* species that induce neoformed bud galls on *Shorea* leaves.

In both cases the whole insect genus expresses

the same type of cecidogenic reaction such that the gall type becomes a characteristic of the genus. Each species of the genus possesses its own specific gall features related to the specific characters of the insect's biology, and in particular its nutritional behavior. The gall features are very gradually distributed in the family Adelgidae. We may find all gradations from some species modifying only slightly the young shoot toward species that highly or totally modify the shoot. The gall types of the genus *Gallacoccus* are gathered around the two gall types described above.

We agree with Dunn (1960) working on *Pemphigus* (Homoptera) galls in that the shape of very different galls induced by different species of the same genera is determined by the probing pattern of the insect that initiates the gall. However, one must also take into account the feeding pattern of the gall insect, the plant organ attacked, and its developmental stage. What creates the specific cecidogenic structure is not one substance distributed in a specific way (Miles 1968) or a specific mixture of the insect's saliva (Dixon 1985). Thus the detailed structure of these galls is generated by the specific epigenic situation each insect induces and entertains at the plant cell level.

LITERATURE CITED

Anthony, M. 1970. Contribution à l'étude des cécidies de Singapour. Mémoire de DEA. Université de Paris.

Anthony, M. 1974a. Contribution to the knowledge of cecidia of Singapore. S'pore Gard. Bull. 27:17–65.

Anthony, M. 1974b. Cécidogenèse comparée de deux galles de Coccides (*Gallacoccus anthonyae* Beardsley et *Gallacoccus secundus* Beardsley) développées sur le *Shorea curtisii* Dyer ex King (Dipterocarpaceae). Marcellia 38:99–144.

Anthony, M. 1977. Morphological and anatomical comparison of normal and cecidial shoots in *Shorea curtisii* Dyer ex King. Marcellia 40:181–192.

Anthony, M. 1980a. Comparative study of the rates of organogenesis of normal and cecidial shoots in *Shorea curtisii* Dyer ex King (Dipterocarpaceae). Flora 169:365–375.

Anthony, M. 1980b. Effet du *Gallacoccus secundus* et de sa galle en bourgeon épiphylle sur les bourgeons de la plantule de *Shorea curtisii* Dyer ex King (Dipterocarpaceae). Bull. Soc. Bot. Actual. Bot. 127:47–56.

Anthony, M. 1980c. Etude des bourgeonnements épiphylles déterminés par l'action d'Eriococcides cécidogènes chez les Diptérocarpacées du genre *Shorea*. Thèse Dr. ès sci., Strasbourg, France.

Anthony, M. 1980d. Etude préliminaire sur le développement des bourgeons végétatifs des plantules du *Shorea curtisii* Dyer ex King (Dipterocarpaceae). Bull. Soc. Bot. Fr., Actual. Bot. 127:123–126.

Beardsley, J. W. 1971. A new genus of gall inhabiting Eriococcidae from Singapore (Homoptera: Coccoidea). Proc. Hawaiian Entomol. Soc. 21:31–39.

Beardsley, J. W. 1984. Gall-forming Coccoidea. In T. N. Ananthakrishnan (ed.), Biology of Gall Insects, pp. 79–106. Oxford & IBH, Calcutta.

Bouck, G. B., and J. Cronshaw. 1965. The fine structure of differentiating sieve tube elements. J. Cell. Biol. 25:9–95.

Buhr, H. 1964–1965. Bestimmungstabellen der Gallen (Zoo- und Phytocecidien) an Pflanzen Mittel und Nordeuropas, Band 1 & 2. Gustav Fischer, Jena.

Buvat, R. 1968. Différenciation des cellules criblées du protophloème dans les jeunes racines d'Orge *(Hordeum sativum)*. C. R. Hebd. Seanc. Acad. Sci. Paris. Sér. D 267:406–408.

Carter, C. I. 1971. Conifer woolly aphids (Adelgidae). Britain Forestry Commission Bulletin NC42, London.

Carter, W. 1952. Injuries to plants caused by insect toxins, II. Bot. Rev. 18:680–721.

Dixon, A. F. G. 1985. Aphid Ecology. Blackie and Son, Glasgow.

Dunn, J. A. 1960. The formation of galls by some species of *Pemphigus* (Homoptera: Aphididae). Marcellia 30:155–167.

Esau, K. 1965. Anatomy and cytology of *Vitis* phloem. Hilgardia 37:17–72.

Gaumont, R. 1957. Etudes sur la biologie de quelques Chermesidae. Thèse, Paris, France.

Goidanich, A. 1960. Specializzazione ecologica e nomenclatura delle *Carulaspis* del *Viscum* e della Cupressaceae (Hemiptera, Coccoidea, Diaspididae). Boll. Inst. Entomol. Bologna 24:1–38.

Houard, C. 1911. Action de cécidozoaires externes appartenant au genre *Asterolecanium* sur les tissus de quelques tiges. Marcellia. 10:3–25.

Houard, C. 1922. Les zoocécidies des plantes d'Afrique, d'Asie et d'Océanie. Vol. 1. Jules Herrmann, Paris.

Howell, J. O., and M. Kosztarab. 1972. Morphology and systematics of scale insects. Number 4. Morphology of the adult females of the genus *Lecanodiaspis* (Homoptera: Coccoidea: Lecanodiaspididae). Virginia Polytechnic Inst. and State Univ. Research Div. Bull. 70.

Jenns, A. E., and J. Kůc. 1980. Characteristics of anthracnose resistance induced by localized infection of cucumber with tobacco necrosis virus. Phys. Plant Pathol. 17:81–91.

Lovett, T. J. 1980. Some phytochemical changes in *Taxus baccata* L. shoots associated with stages in the life cycle of *Taxomyia taxi* Inch. Bull. Soc. Bot. Fr. 127. Actual. Bot. 129–136.

Magnus, W. 1914. Die Entstehung der Pflanzengallen verursacht durch Hymenopteren. Gustav Fischer, Jena.

Maresquelle, H. J. 1935. Defaut d'allongement et dépolarisation de la croissance dans les morphoses parasitaires. Rev. Gen. Bot. 47:1–56.

Maresquelle, H. J. 1936. Quelques observations sur les phénomènes d'inhibition d'origine cécidogène. In "Mélanges dédiés au Prof. Daniel," pp. 393–400.

Meyer, J. 1957. Cécidogenèse comparée de quelques galles d'arthropodes et evolution cytologique des tissus nourriciers. Thèse, Strasbourg, France.

Meyer, J. 1958. Morphogenèse comparée des galles de *Phylloxera vastatrix* Planch. sur feuilles, vrilles et tiges de Vigne. Marcellia Suppl. 30:76–91.

Meyer, J., and H. J. Maresquelle. 1983. Anatomie des Galles. Gebrüder Borntrager, Stuttgart, Berlin.

Miles, P. W. 1968. Insect secretions in plants. Rev. Phytopathol. 6:137–164.

Plumb, G. H. 1953. The formation and development of the Norway spruce gall caused by *Adelges abietis*. Bull. Conn. Agric. Exp. Sta. No. 566:1–77.

Pollard, D. G. 1973. Plant penetration by feeding aphids: a review. Bull. Entomol. Res. 62:631–714.

Rohfritsch, O. 1971. Développement cécidien et rôle du parasite dans quelques galles d'Arthropodes. Marcellia 37:233–339.

Rohfritsch, O. 1976. Traces de succion de deux Chermesidae *Chermes abietis* L. et *Chermes strobilobius* Kalt. dans les bourgeons de *Picea excelsa* L. Marcellia 39:69–84.

Rohfritsch, O. 1977. Ultrastructure of the nutritive tissue of the *Chermes abietis* L. fundatrix on *Picea excelsa* L. Marcellia 40:135–150.

Rohfritsch, O. 1981. A 'defense mechanism' of *Picea excelsa* L. against the gall former *Chermes abietis* L. (Homoptera, Adelgidae). Zeit. Angew. Entomol. 92:18–26.

Rohfritsch, O. 1982. Behavior and biology of two aphid species related to gall morphogenesis. In Proc. 5th Int. Symp. Insect-Plant Relationships. Wageningen, pp. 431–432. Pudoc, Wageningen.

Rohfritsch, O. 1988. A resistance response of *Picea excelsa* to the aphid *Adelges abietis* (Homoptera: Aphidoidea). In W. J. Mattson, J. Levieux, and C. Bernard-Dagan (eds.), Mechanisms of Woody Plant Defenses Against Insects, pp. 252–266, Springer, New York.

Sequeira, L. 1983. Mechanisms of induced resistance in plants. Annu. Rev. Microbiol. 37:51–79.

Sterling, C. 1952. Ontogeny of the *Phylloxera* gall of grape leaf. Am. J. Bot. 39:6–15.

Sziraki, I., E. Balazs, and Z. Kiraly. 1980. Role of different stresses in inducing cytokinins level in tobacco. Phys. Plant Pathol. 16:277–284.

Thalenhorst, W. 1972. Zur Frage der Resistenz der Fichte gegen die Gallenlaus *Sacchiphantes abietis* (L.) Z. Angew. Entomol. 71:225–249.

8

The Role of Nutritive Cells in the Nutrition of Cynipids and Cecidomyiids

R. BRONNER

The formation of plant galls, as illustrated throughout this volume, represents one of the most elaborately evolved and complex insect–plant relationships known to mankind. These relationships between gall insects and their host plants, as with most other types of insect–plant associations, can be defined in terms of food, shelter, and transport, and the relative importance of these factors varies according to the groups of insects and plants in question. However, nearly all galls from the simplest to the most complex are the result of many biochemical and developmental processes and in each gall it is the host plant, through the gall, that provides the insect with both shelter and food. Furthermore, some of the assimilates accumulating in the gall and fed on by the insect are transported to the gall from plant tissues some distance away.

An appreciation for the diverse methods by which cecidozoa obtain their food from gall tissues is basic to an understanding of gall biology because the specialized types of tissues that provide nourishment also play a key role in regulating gall morphogenesis (see Chapter 5) and metabolism. Feeding strategies of cecidozoa can be grouped according to the family of gall inducer and, as outlined in Chapter 2, they range in complexity from the simple punch and suck method of thrips to the subtle sucking of cecidomyiids that occurs without puncturing the cell. However, the two best studied groups of cecidozoa from the point of view of feeding and nutrition are the cynipid wasps and cecidomyiids. Consequently, it is the purpose of this chapter to examine the feeding strategies of these two groups as they relate to the nutritional value of the gall cells they consume.

GENERAL COMMENTS ON NUTRITIVE TISSUE

Lacaze-Duthiers (1853) was the first to describe a special tissue around the larval chamber of galls and referred to them as "couche ou masse alimentaire." Since then, numerous workers have described the cytological features of nutritive cells from many galls. In general terms, the nutritive cells of cynipid and cecidomyiid galls are characterized by:

1. an abundance of cytoplasm and a reduced and fragmented vacuole (Molliard 1897; Meyer 1948);
2. a hypertrophy of the nucleus and nucleolus (Molliard 1897; Meyer 1950; Garrigues 1951; Woll 1954a,b);
3. an abundance of cytoplasmic organelles (Meyer 1948; Garrigues 1951).

Ultrastructural studies also have contributed invaluable information on nutritive cells (Jauffret 1972, 1973; Meyer 1973; Rey 1976, 1979; Rohfritsch 1971a,b, 1974) and it has been shown that most nutritive cells have the following features:

1. The hypertrophied nucleus is often lobed, the nucleolus is hypertrophied and often vacuolated.

2. Cytoplasmic ribosomes are very abundant and often grouped in polysomes. Ribosomes and polysomes are often bound to a well-developed ergastoplasm.
3. Plastids and mitochondria are numerous, poorly differentiated, and often grouped around the nucleus.
4. Autophagic vacuoles are common.

These characteristics of nutritive cells serve only a basic plan since individual features vary depending on the family and species of cecidozoa. Furthermore, the cytological features of nutritive cells are not peculiar to these cells. Indeed they also can be found in normal plant cells that are physiologically and metabolically active, such as

1. secretory cells of nectary glands (Eyme 1966; Schnepf 1964; Zandonella 1970);
2. cotyledon storage cells in seedlings (Öpik 1964);
3. companion cells of phloem (Esau 1972);
4. mother cells of pollen grains (Meyer 1957).

Experimental studies by Rohfritsch (1971a) have clearly demonstrated that the active presence of the gall inducer is absolutely necessary for the harmonious development of the gall during all stages of cecidogenesis. Any cessation of larval activity brings about a rapid change in the nutritive cells to a more typical state (Meyer 1957; Rohfritsch 1966, 1967, 1968; 1975a; Jauffret 1975).

The term "nutritive tissue" as used in this chapter refers to all the tissues found between the larval chamber and the lignified sheath or the cortex. All cells in this region, even though they may initially lack the typical cytological features of nutritive cells, are destined to gain these features. Nutritive cells in cynipid galls, for example, are destroyed as the gall develops and they are systematically replaced by adjoining cells. As gall development ceases and the gall matures, no nutritive cells remain and the larva finishes its development in contact with the lignified sheath or the cortex.

It is possible to divide the nutritive tissue into two parts (Bronner 1977). The inner part is referred to as the "typical nutritive tissue" and the outer part is referred to as the "storage nutritive tissue" because it contains starch.

NUTRITIVE TISSUES OF CYNIPID GALLS

Cytological Aspects

In most cynipid galls it is possible to find three kinds of tissues surrounding the larval chamber (Fockeu 1889): a nutritive tissue lining the larval chamber, a sheath of lignified cells, and a parenchymatous zone (or cortex) with vascular tissues connected to the vascular system of the host plant. This typical succession of tissues is illustrated in Figure 8-1 for the gall induced by *Liposthenes glechomae* L. on *Glechoma hederacea* L. The nutritive tissue and the lignified sheath constitute the "inner gall"; the "outer gall" is formed by the cortical tissue and the epidermis. Observations indicate that the size of the "inner gall" is constant, whereas the size of the "outer gall" varies for a given species (Gaufilet and Fourcroy 1966; Bronner 1982). Even though there is a progressive destruction of the nutritive cells by larval feeding, the mass of plant tissues used by cynipid larvae for feeding is constant for a given species.

The importance of these different zones to larval nutrition and to gall development varies depending on the gall. However, there are certain attributes common to most galls that are readily seen in the first layer nearest the larval chamber and these attributes progressively change from the edge of the larval chamber outward. Thus a "cytological gradient" is an important feature of nutritive tissue (Rohfritsch 1974; Bronner 1977). This cytological gradient is clearly seen in Figures 8-2 and 8-4–8-9 for the gall induced by *Biorhiza pallida* (01.) ♀♂ on a *Quercus* bud.

The progressive changes in the cytological features of nutritive cells can be seen by comparing cells nearest the larval chamber (Figs. 8-4–8-6) with those of the third layer (Fig. 8-7), the fourth layer (Fig. 8-8), and the seventh layer (Fig. 8-9). Indeed, in the first nutritive layer there is nuclear and nucleolar hypertrophy, fragmentation and reduction of vacuoles, and numerous poorly differentiated organelles (plastids and mitochondria) near the nucleus. Density of the ribosomes decreases and the size of vacuoles increases in cells further away from the larval chamber. As the size of the nucleus and nucleolus decreases, the organelles become

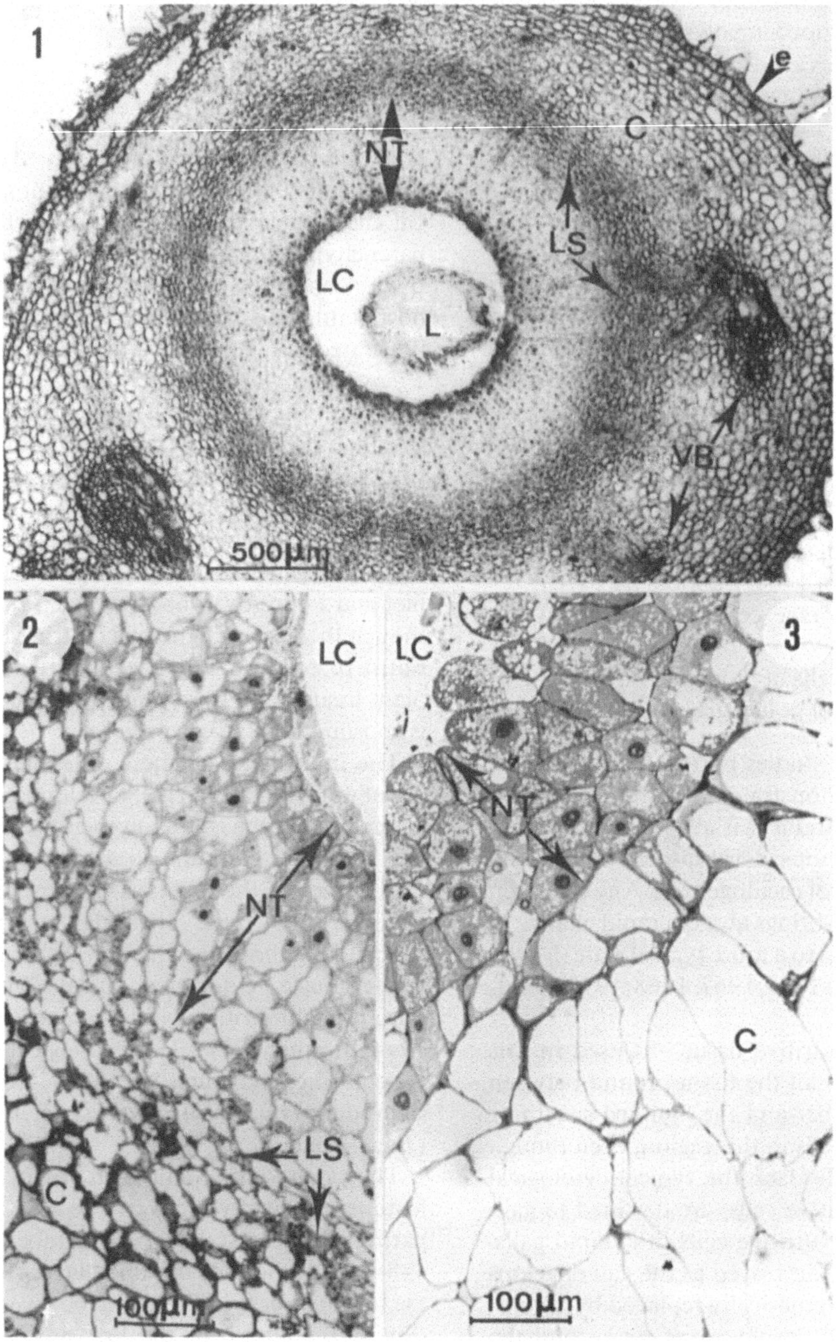

Figures 8-1–8-3. Figure 8-1. Section of a maturing gall of *Liposthenes glechomae* on a leaf of *Glechoma hederacea* showing typical tissues of many cynipid galls. Fixed with glacial acetic acid–ethanol (1:3) and stained with toluidine blue. [After Brachet (see Bronner 1977).] **Figure 8-2.** Section of a maturing gall of *Biorhiza pallida* ♀ ♂ on a bud of *Quercus.* Note the cytological gradient in the nutritive tissue increasing toward the larval chamber. Fixed with glutaraldehyde–osmium tetroxide. Stained with toluidine blue (1 percent in 5 percent borax). **Figure 8-3.** Section of a gall of *Neuroterus quercusbaccarum* ♀ ♂ on leaves of *Quercus.* Note the homogeneous nutritive tissue around the larval chamber. This gall lacks a lignified sheath. Fixed with glutaraldehyde–osmium tetroxide. Stained with toluidine blue (1 percent in 5 percent borax). C, Cortex; e, epidermis; L, larva; LC, larval chamber; LS, lignified cells; NT, nutritive tissue; VB, vascular bundles.

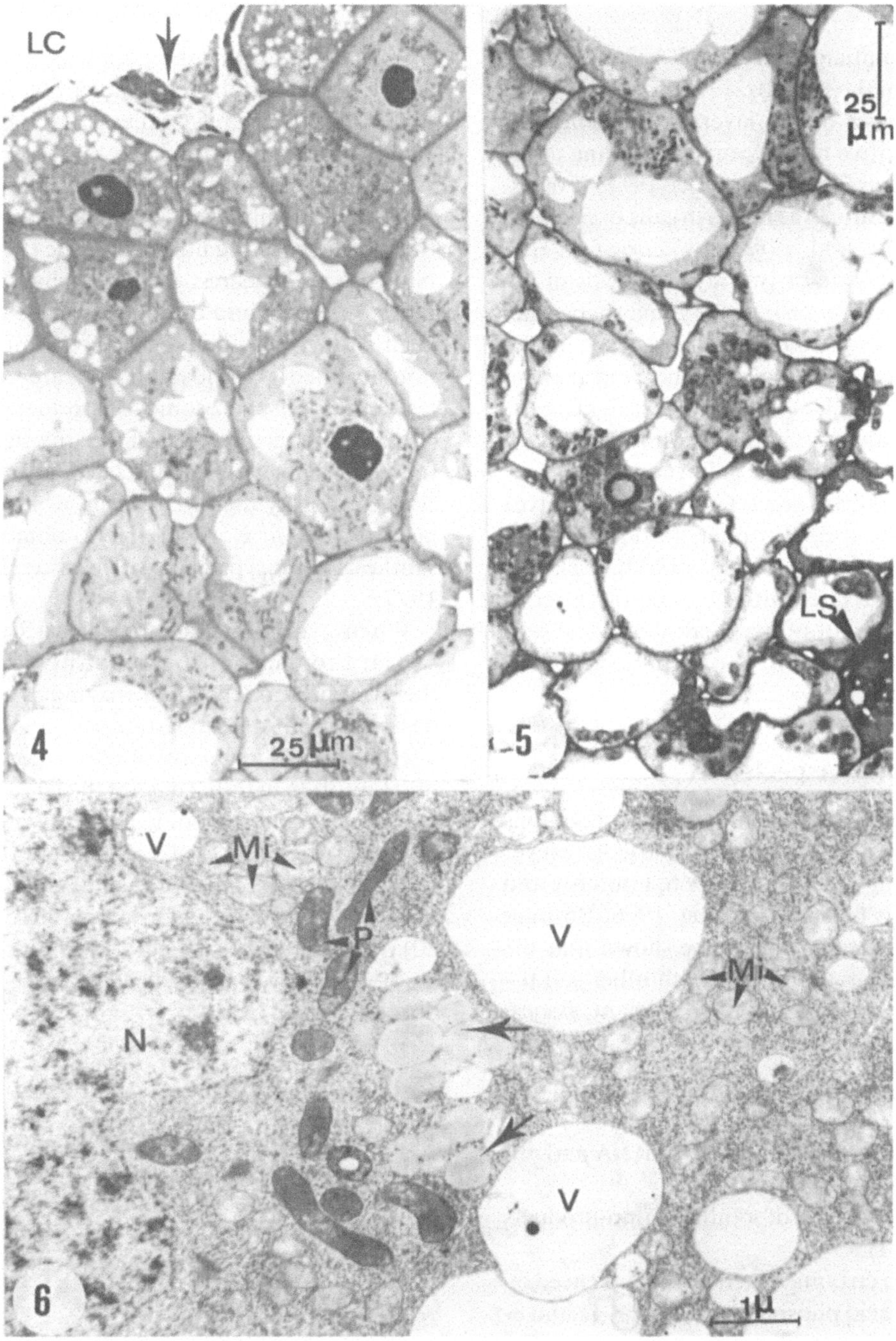

Figures 8-4–8-6. Aspects of the cytological gradient in nutritive tissues of the gall of *Biorhiza pallida* ♀ ♂ on *Quercus*. **Figure 8-4.** Nutritive tissue nearest the larval chamber. Note the remains of cells (arrow) eaten by the larva. Note the nuclear and nucleolar hypertrophy, density of the cytoplasm, fragmentation and reduction of vacuoles, and presence of numerous, poorly differentiated plastids and mitochondria near the nucleus. Fixed with glutaraldehyde–osmium tetroxide. Stained with toluidine blue (1 percent in 5 percent borax). **Figure 8-5.** Area next to Figure 8-4 showing part of the nutritive tissue nearest the lignified sheath. Note that the characteristics seen in Figure 8-4 progressively decrease away from the larval chamber. Fixed with glutaraldehyde–osmium tetroxide. Stained with toluidine blue (1 percent in 5 percent borax). **Figure 8-6.** Ultrastructural aspects of a nutritive cell of the first layer lining the larval chamber. There are numerous poorly differentiated organelles (plastids and mitochondria) around the lobed nucleus. Note the density of ribosomes in the cytoplasm. Arrows indicate lipid globules. Fixed with glutaraldehyde–osmium tetroxide. Contrasted with uranyl acetate–lead citrate. LC, Larval chamber; LS, lignified sheath; Mi, mitochondria; N, nucleus; P, plastids; V, vacuoles.

more differentiated and starch becomes more abundant in the plastids.

The number of cell layers comprising "the typical nutritive tissue" varies according to the species of gall former (Bronner 1977). For example, there are three layers in galls of *Biorhiza pallida* ♀♂ and *Andricus curvator* Htg. ♀♂ and only one or two layers in galls of *Liposthenes glechomae* l. In the grape-shaped gall of *Neuroterus quercusbaccarum* L. ♀♂, all the cells between the larval chamber and the gall cortex show similar cytological features (Fig. 8-3). It also should be noted that this gall lacks a lignified sheath. In some galls such as those induced by *Pediaspis aceris* Gmelin ♀♂ on leaves of *Acer* (Ross 1932; Bronner, 1981) and probably those by *Acraspis* (Kinsey 1930), the larval chamber is not lined with a continuous layer of nutritive cells, but rather there are several separate patches of nutritive cells projecting into the larval chamber.

Histochemical Aspects

Cytochemical and biochemical observations have been made on several cynipid galls (Meyer 1957; Turian 1958; Braun 1966; Fourcroy and Braun 1967; Gadal 1970; Rey 1976; Bronner 1977) and these authors have shown that the nutritive tissues (compared with other gall tissues and corresponding normal plant tissues) have the following features:

1. a high concentration of both structural and enzymatic proteins;
2. a richness in RNA (ribosomal RNA and nucleolar RNA) (Fig. 8-10);
3. an accumulation of soluble amino-products (Fig. 8-11);
4. very high enzymatic activity as evidenced by levels of acid phosphatases (Figs. 8-12 and 8-13), glucose-6-phosphatase, nonspecific proteases (Fig. 8-14), aminopeptidases, amylases (Fig. 8-15), invertase, and phosphorylase;
5. an accumulation of glucose in all stages of gall development.

A high concentration of starch (Fig. 8-16) is found in a gradient decreasing toward the larval chamber (Fig. 8-17). Thus the typical nutritive tissue is usually without starch whereas the nutritive storage tissue contains starch in increasing amounts toward the periphery of the gall. The observed cytological gradient corresponds to the observed starch gradient (Bronner 1977). There also is a lipid gradient, but it runs opposite to the starch gradient (Bronner 1975b). Thus in all cynipid galls lipids decrease toward the periphery of the nutritive tissue (Fig. 8-18). Nutritive cells nearest the larval chamber contain numerous lipid droplets (Fig. 8-19). It has been shown that these lipids are a mixture of nonsaturated glycerids and they are "oil bodies" and not "spherosomes" (Bronner 1980).

Experiments have shown that the starch and lipid gradients are closely related and it is possible to modify them. However, as long as the larva is alive, it is impossible to obtain starch synthesis in the typical nutritive tissue (Bronner 1977).

When cynipid larvae are killed (Rohfritsch 1975a; Bronner 1977), the nutritive cells lose their richness in RNA, the enzymatic activities decline and, at the same time, the gradients are altered. These regressive modifications occur quickly, often within 2 days of the larva's death. One of the greatest modifications observed concerns the distribution of starch. Starch is never found in the first layer of nutritive tissue when the larva is alive, but soon occurs here (Fig. 8-20) once the larva is killed. Lipids, in contrast, are not destroyed but instead coalesce in large droplets (Fig. 8-20).

Metabolical Aspects

The cytological and histochemical features of nutritive cells in cynipid galls demonstrate that there is intense metabolic activity. According to Bronner (1977) the nutritive cells are the center of the following metabolic events.

Intense Proteosynthesis

The RNA richness in nutritive tissue, the high density in ribosomes and polysomes bound to a developed ergastoplasm indicate a high proteosynthetic activity. This does not result in storage proteins, but rather in structural and enzymatic proteins.

Proteolysis

It has been reported that gall tissues contain more soluble nitrogen than ungalled plant tissues (Paris and Trotter 1911; Molliard 1913;

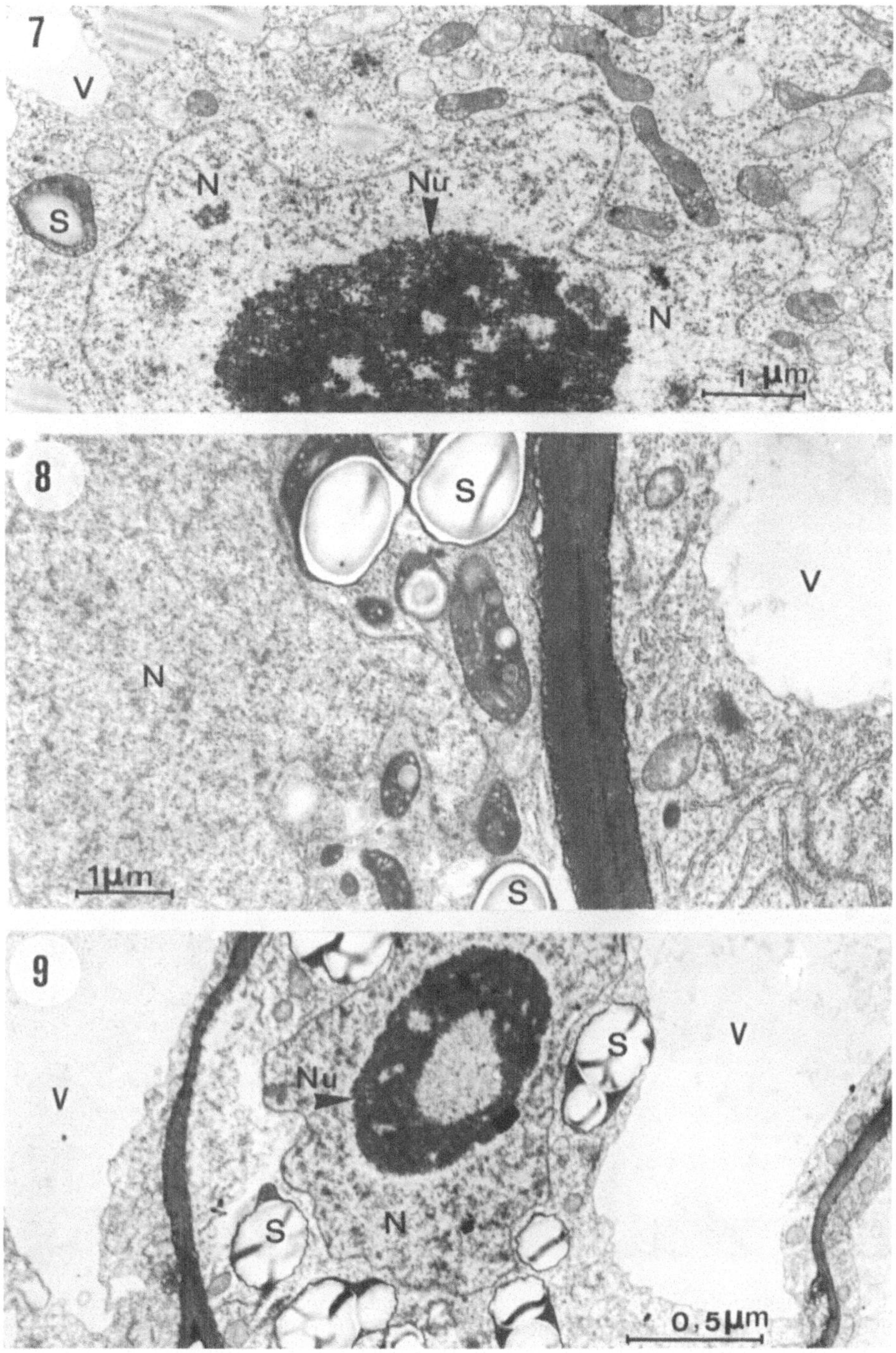

Figures 8-7–8-9. Ultrastructural aspects of the cytological gradient in nutritive cells of the gall of *Biorhiza pallida* ♀ ♂ on *Quercus.* These figures illustrate that from the larval chamber to the lignified sheath the density of ribosomes gradually decreases, the size of the vacuoles increases, starch occurs in the plastids, and sizes of the nucleus and nucleolus decrease. All sections are from material fixed with glutaraldehyde–osmium tetroxide and are contrasted with uranyl acetate–lead citrate. **Figure 8-7.** Nutritive cells of the third layer. **Figure 8-8.** Nutritive cells of the fourth layer. **Figure 8-9.** Nutritive cells of the seventh layer. N, Nucleus; Nu, nucleolus; S, starch; V, vacuole.

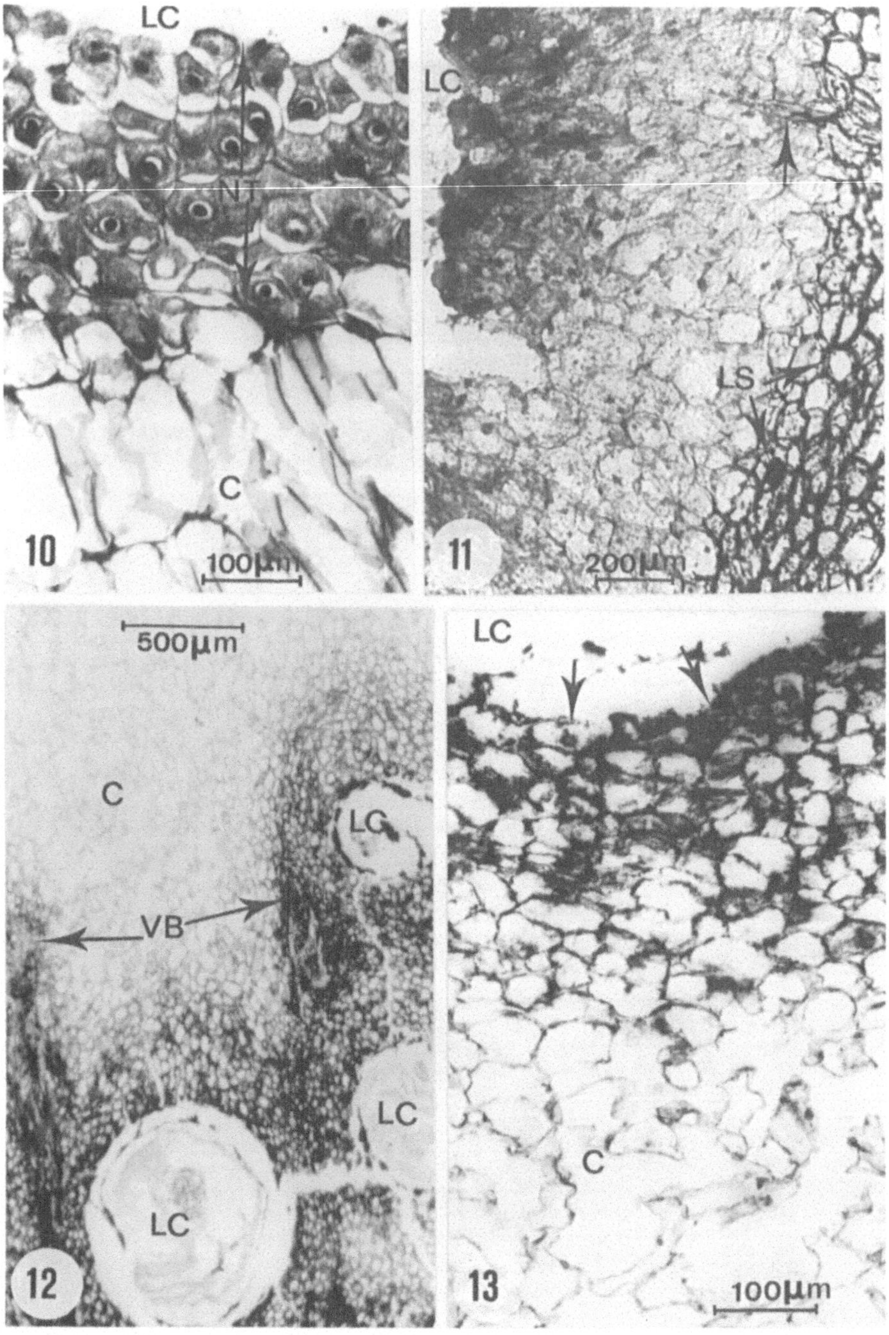

Figures 8-10–8-13. Figure 8-10. Section of a gall of *Neuroterus quercusbaccarum* ♀ ♂ showing richness of RNA in the nutritive tissue. Fixed with glacial acetic acid–ethanol (1:3). Stained with toluidine blue after the method of Brachet (1940). **Figure 8-11.** Section of the nutritive tissue of the gall of *Liposthenes glechomae* stained for soluble amino products. The amino products are particularly abundant in nutritive tissue nearest the larval chamber. Note the vascular bundle in the nutritive tissue (arrow). Fresh hand section stained with ninhydrin. (From Bronner and Meyer 1968.) **Figure 8-12.** Section of a gall of *Biorhiza pallida* ♀ ♂ showing acid phosphatases activity in the nutritive tissue around the larval chamber. Note that the vascular bundles also show high acid phosphatases activity whereas the cortical cells do not show such activity. Fixed with acetone (after Ruyter 1964). Treated by the lead nitrate method of Gomori (1956). **Figure 8-13.** Section of a gall of *Biorhiza pallida* ♀ ♂ showing detail of acid phosphatases activity in the nutritive tissue. These activities are particularly localized on cell walls, in the nucleus, and in the cytoplasm of some cells (arrows) of the first layer lining the larval chamber. These cells undergo autolysis. Fixed with acetone (after Ruyter, 1964). Treated with the lead nitrate method of Gomori (1956). C, Cortical cells; LC, larval chamber; LS, lignified sheath; NT, nutritive tissue; VB, vascular bundle.

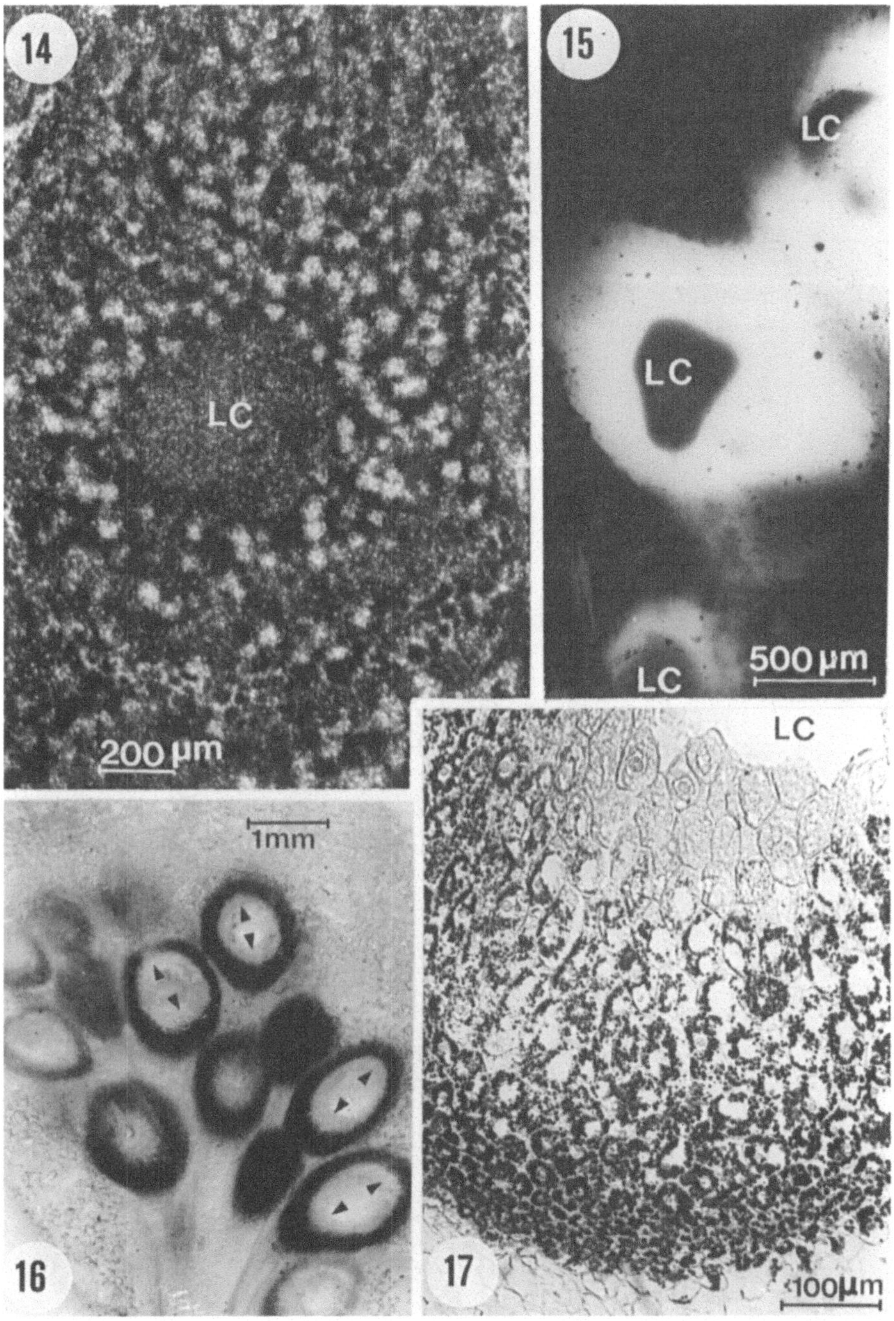

Figures 8-14–8-17. Figure 8-14. Nonspecific proteases activity (white spots) of the nutritive tissue of the gall of *Biorhiza pallida* ♀ ♂ using the modified protein substrate film method of Adams and Tuquan (1961). **Figure 8-15.** Amylasic activity (in white) of the nutritive tissue of the gall of *Biorhiza pallida* ♀ ♂ using the modified starch substrate film method of Tremblay and Charest (1968). **Figure 8-16.** Section of a gall of *Biorhiza pallida* ♀ ♂ stained to localize starch (in black). Note that starch is found around the larval chambers, but the first nutritive layers (arrows) contain no starch. Fresh hand section stained with Lugol's reagent (IK, 2 g; I_2, 1 g; H_2O, 100 ml). **Figure 8-17.** Section of a gall of *Andricus curvator* ♀ ♂ showing details of the starch gradient (black spots) around the larval chamber. Note that the typical nutritive tissue has no starch while starch increases in the storage nutritive tissue toward the periphery. Fixed with glacial acetic acid–ethanol (1:3) using the stain of Reilhes (1946). LC, Larval chamber.

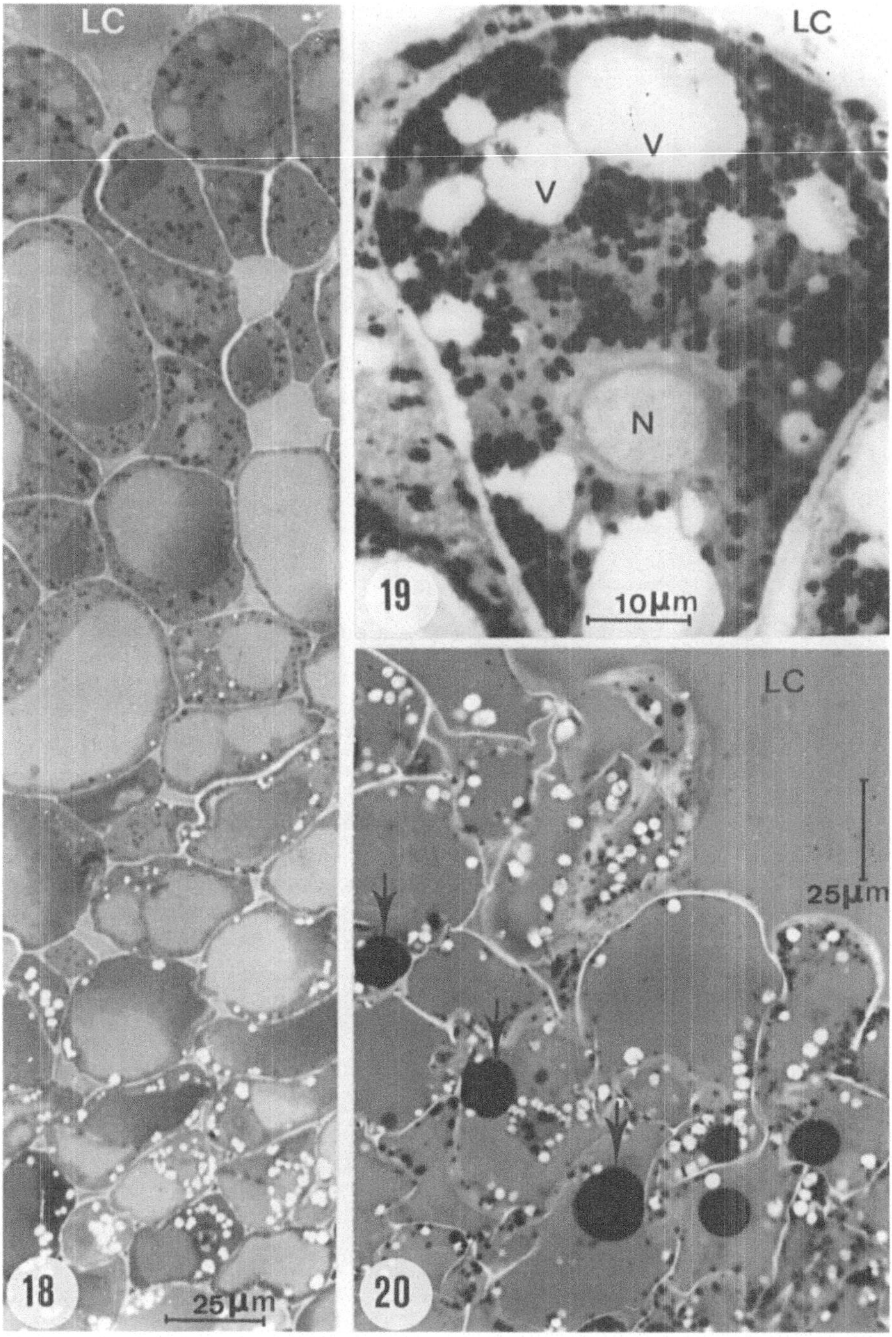

Figures 8-18–8-20. Figure 8-18. Nutritive tissue of a gall of *Biorhiza pallida* ♀ ♂ stained to show a lipid gradient (black spots) decreasing toward the gall periphery and a reverse starch gradient (white spots) decreasing toward the larval chamber. Fixed with glutaraldehyde–osmium tetroxide. Stained with Sudan Black B (from Bronner 1975a). **Figure 8-19.** Nutritive cell adjoining the larval chamber in a gall of *Biorhiza pallida* ♀ ♂. The cell is filled with numerous lipid globules (black spots). Fixed with glutaraldehyde–osmium tetroxide. Stained with Sudan Black B (from Bronner 1975a). **Figure 8-20.** Nutritive cells of the gall of *Liposthenes glechomae* in which the larva is dead showing the distribution of lipids and starch. The hypertrophied cells arising from the typical nutritive cells contain starch (white spots). The lipids (black spots) are fused together (arrows). Compare with Figure 8-18. Fixed with glutaraldehyde–osmium tetroxide. Stained with Sudan Black B (from Bronner 1975a). LC, Larval chamber; N, nucleus; V, vacuole.

Kientzler *et al.* 1962; Paclt and Hassler 1967). The increasing ratio of soluble nitrogen to insoluble nitrogen is caused by the high concentration of amides and amino acids in the nutritive tissue (Gadal 1970; Bronner 1977).

Two hypotheses can be considered to explain this accumulation of amino acids. First, there may be hydrolysis of proteins by proteolytic enzymes from either the larva (Molliard 1913) or the nutritive cells (Bronner 1977). Second, it has been hypothesized that the accumulation of amino acids comes from the host plant (Paris and Trotter 1911; Gadal 1970). The latter hypothesis is reasonable as it is well known that cynipid galls are sinks for plant assimilates (Fourcroy and Braun 1967; Jankiewicz *et al.* 1969). The vascular systems of these galls are well developed (Meyer 1969), which facilitates these sink effects. At present, there is little proof for either hypothesis; however, it is probable that there is a significant turnover of proteins in these nutritive cells. Support for this view is the appearance of "autophagic structures." It is well known that such autophagic structures contain several acid hydrolases (as proteases and acid phosphatases) and that they act in local alteration of the cytoplasm (Duve and Wattiaux 1966; Coulomb and Coulomb 1972; Matile 1975).

An Important Active Transport of Glucids Toward the Cecidozoa

Studies with the electron microscope have shown that an important fraction of the acid phosphatases activities in nutritive cells of cynipid galls is found on the cell wall and plasmalemma (Bronner 1973, 1974). It is well known that these kinds of acid phosphatases are involved in active transport of glucids (Sauter 1972; Figier 1972) and that they act in dephosphorylation of glucids (Dixon 1949; Wanner 1952).

Observations on normal vascular tissues (Sauter 1966; Sauter and Braun 1968) suggest that there is a similar active transport of sugars toward the cecidozoa in the nutritive tissue of cynipid galls. This has been confirmed by biochemical observations (Kientzler *et al.* 1962; Gadal 1970) showing that gall tissues have more sugars than normal plant tissues. Cytochemical observations indicate that sugars are concentrated in the nutritive cells, particularly in cells of the first layer (Bronner 1977). These sugars are probably transported in the form of sucrose, which is subsequently converted by the intervention of invertase and glucose-6-phosphatase enzymes, which are also found concentrated in the nutritive cells (Bronner 1977).

Simultaneous Establishment of Lipid and Starch Gradients

The starch and lipid gradients of nutritive cells are the result of interactions between the cytological and histochemical conditions of the cells and the flow of sugars from the host plant. These interactions, both of which are associated with the activities of the larva, are summarized in Figure 8-21.

Feeding by Cynipid Larvae

Cytological and histological observations on nutritive cells found in cynipid galls indicate that these cells attain the following features as they develop:

1. a richness in hydrolasic enzyme activities;
2. an active lysosomal system;
3. an accumulation of soluble compounds such as glucose and amino products;
4. an abundance of lipidic substances.

Nutritive cells supply the larva with liquid food, but in the process the larva destroys the cells. This destruction is the result of both chemical and mechanical processes. Roth (1949) stressed the importance of mouthparts of cynipid larvae that are used to tear the nutritive cells. He also showed that cynipid larvae can imbibe liquids by making unceasing sucking movements with their pharynx and its well-developed musculature. However, cytological observations (Rohfritsch 1974; Bronner 1977) show that the destruction is not solely caused by mechanical action. There also is a lysis of cells. This has been confirmed by cytochemical observations, which show cellular autolysis. Indeed, acid phosphatase is one of the markers of the lysosomal system (Duve and Wattiaux 1966) which plays an important role in autolysis and necrosis (Gahan and Maple 1966; Matile and Winkenbach 1971). It is known that the breakdown of the binding membranes of lyso-

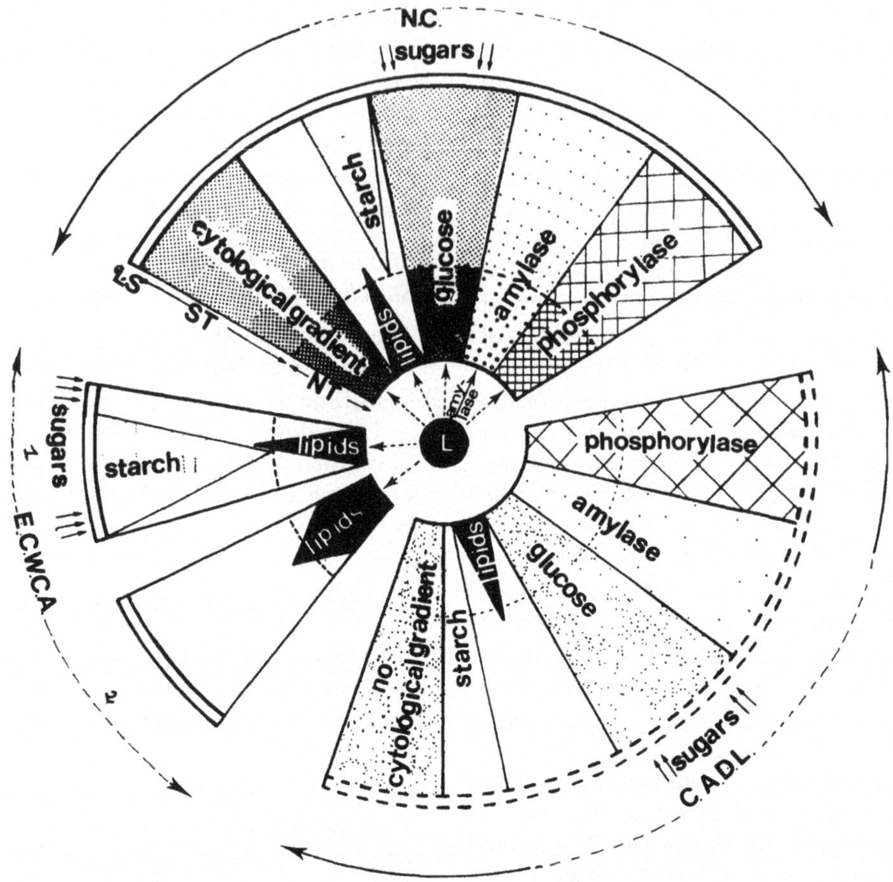

Figure 8-21. Metabolic activities within the nutritive tissues of cynipid galls. Starch and lipids are represented by triangles indicating the direction of the gradients. Other gradients are represented by a gradation in the density of the patterns. LS, Lignified sheath; NT, typical nutritive tissue; ST, storage nutritive tissue. Normal conditions (NC): Here the larva stimulates the cells and maintains the cytological and cytochemical gradients probably by enzymatic secretions. The host plant supplies sugars as well as other substances. Since galls are sinks, these sugars appear up to the first layer lining the larval chamber. Synthesis of starch occurs only in the storage nutritive tissue because of the high amylasic activity in the typical nutritive tissue. Glucose cannot be transformed into starch in the typical nutritive tissue, but instead can be transformed into lipids. Experimental conditions with cecidozoa alive (ECWCA): 1. Hypernutrition with sugars increases starch synthesis in the storage nutritive tissue; however, the typical nutritive tissue is not modified. 2. Glucidic fasting: sugars are not supplied to the gall and there is amylolysis in the storage tissue and an increase in lipids in the typical nutritive tissue. Conditions after death of the larva (CADL): In these galls the gradients soon disappear due to a cessation of larval stimulation and secretions. Amylase activity by the larva ceases and there is a reduction of proteosynthesis. The host plant continues to supply the gall with sugars for a short time period and the cells become hypertrophied. Sugars coming from the plant are not consumed by the larva and thus can be transformed into starch because there is less amylase activity in the typical nutritive tissue. Lipids remain in the nutritive cells because these cells are without structures which permit lipid hydrolysis.

somes permits acid lysosomal hydrolases to diffuse and attack the main components of the cytoplasm and nucleus (Strauss 1967). As progressive destruction of the inner layer of cells occurs, the next layer develops the cytological features of the inner layer. A key feature of this transformation is the change in the vacuoles; that is, the large vacuoles are fragmented into small vacuoles, which have lysosomal features such as a richness in acid hydrolases. After this occurs, there is a destruction of the tonoplast and acid phosphatases are found throughout the cytoplasm (Bronner 1977).

There are two hypotheses explaining this

phenomenon. The first is that the autolysis of lysosomes by destruction of their binding membranes is caused by proteases from the larvae. Bronner (1977) reported that the larva can secrete a large quantity of proteasic enzymes from their salivary glands or Malpighian system. The second hypothesis is that the autolysis is caused by mechanical slashing by the larvae. It has been shown that cell autolysis in other systems can be caused by physical action (Bitensky and Gahan 1962). Support for the idea of proteasic activity from the larvae comes from observations of the wall and plasmalemma of the nutritive cells lining the larval chamber where the walls and plasmalemma are swollen and disorganized (Bronner 1977). In fact, it is possible that the two phenomena take place simultaneously, that is, the larvae destroy the cells after autolysis has taken place. We also have to take into account another hypothesis that is related to recent results obtained in plant pathology. Indeed, it has been demonstrated that cell wall fragments can induce different kinds of plant cell reactions and that they may be involved in plant pathology. According to Darvill *et al.* (1985) some of these fragments are able to cause death of plant cells.

In phytopathology, the mechanisms whereby plant cells are killed by various parasites such as fungi, nematodes, and mites are well known (Webster and Paulson 1972; Mansfield and Sexton 1974; Westphal *et al.* 1981). During necrosis, the plant cells undergo an autolysis (Dustin 1966; Webster and Paulson 1972) and then there are the typical necrotic phenomena such as unmasking of the membrane lipids resulting in the black color of the cells (Bessis 1970; Dustin 1966). The pectocellulosic walls are not normally destroyed in necrotic cells, but instead remain perceptible for a long time even though they are tightly packed against the live cells. In nutritive cells of cynipid galls, the cells that are normally stimulated by the larvae undergo autolysis, but then the larvae burst the autolysed cells and the usual necrotic phenomena do not take place. Thus, there is a clear distinction between the modification of nutritive cells in cynipid galls and those that are diseased.

Cynipid larvae "welter" in liquids that come from the destroyed cells and as with other Apocrita, the larval midgut is not connected to the hindgut (Roth 1949; Grassé *et al.* 1961). Thus the liquid coming from the nutritive cells is not contaminated by larval excrement and the larvae have an abundant source of high-quality food at their disposal.

NUTRITIVE TISSUES OF CECIDOMYIID GALLS

Cytological Aspects

Galls induced by cecidomyiids are far more structurally heterogeneous than are the galls induced by cynipid wasps (see Chapter 5). In this group both complex and simple galls contain varying amounts of nutritive tissues. Although they may present the same cytological features as those of cynipid galls, the cellular transformations are a little less radical. For example, cytological gradients are absent, but there is a typical nutritive tissue and a storage nutritive tissue (Figs. 8-23 and 8-25).

The diversity of nutritive tissues in cecidomyiid galls can be seen by examining galls of *Dasineura urticae* Perris, *Dasineura affinis* Kffr., *Geocrypta galii* H. Lw., *Physemocecis ulmi* Kffr., and *Dasineura pustulans* Rubs. *Geocrypta galii* forms a small round swelling on the stems of *Galium mollugo* and has a nutritive tissue with several layers of cells around the larval chamber (Klein 1963; Klein and Meyer 1963; Rohfritsch 1971a,b). The gall of *Dasineura urticae* on *Urtica dioica* L. (Figs. 8-22 and 8-23) is similar to that of *G. galii*. *Dasineura affinis* provokes a rolling of *Viola* leaves (Vorms-Ravillard 1935) and in this gall the upper epidermis and some mesophyll cells are transformed into nutritive tissue (Meyer 1957; Bronner and Meyer 1976; and Chapter 11). *Physemocecis ulmi* forms a single gall on leaves of *Ulmus campestris* L. Here some leaf mesophyll cells at the top of the larval chamber are transformed into nutritive tissue (Bronner and Meyer 1972). *Dasineura pustulans* forms a gall on the leaves of *Filipendula ulmariae,* which consists of a slight arching on the upper side of the leaf. The lamina in the arching is hypertrophied and the cells are transformed into nutritive cells (Bronner 1977). Other simple galls are without nutritive cells (Bronner and Meyer

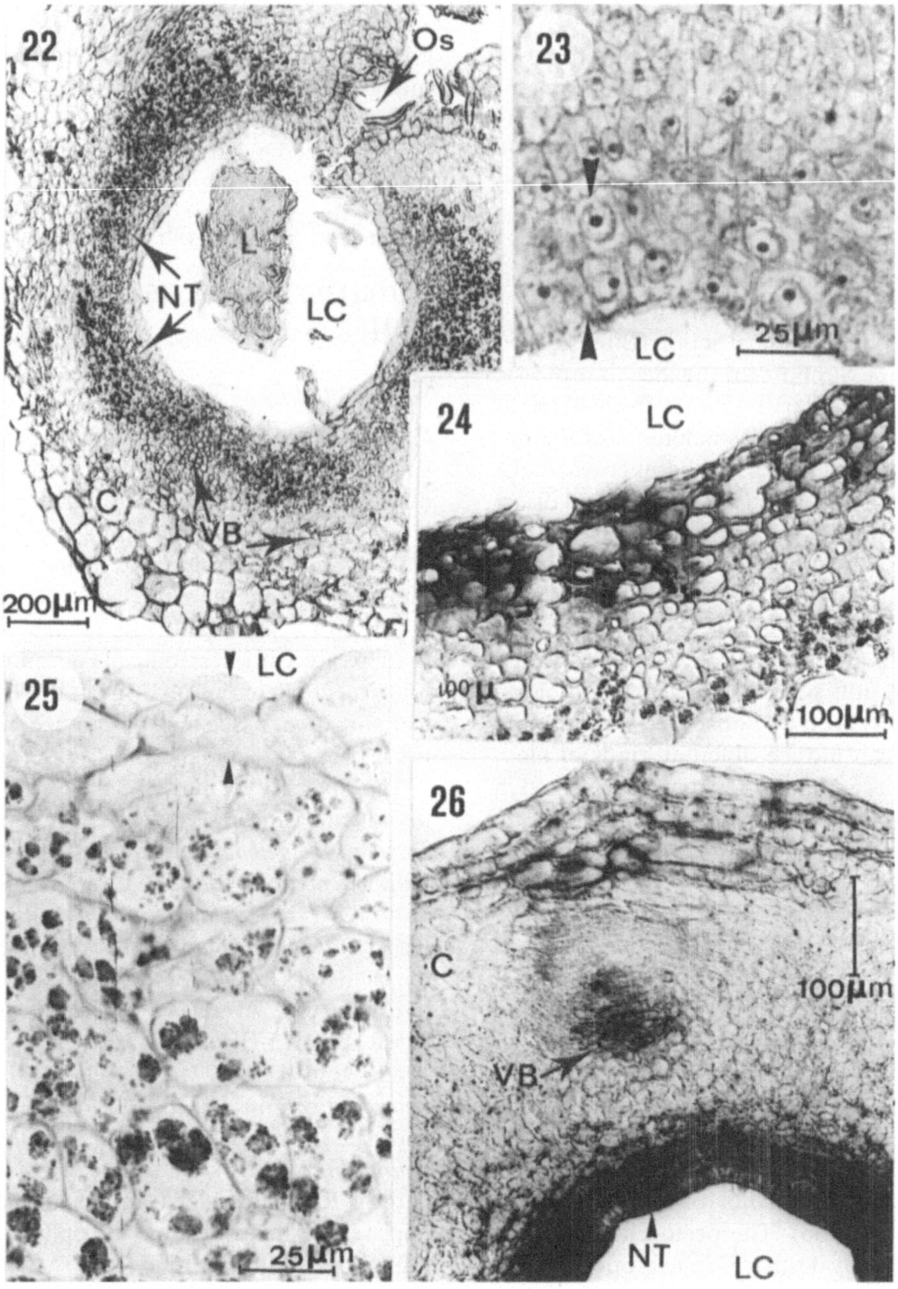

Figures 8-22–8-26. Galls of cecidomyiids with nutritive tissue. **Figure 8-22.** Section of a gall of *Dasineura urticae* on leaf of *Urtica dioica* showing the larval chamber, nutritive tissue surrounded by vascular bundles, cortical tissues, and the ostiolar opening obstructed by hairs. Fixed with glacial acetic acid-ethanol (1:3). Stained with Safranin-Fast Green. **Figure 8-23.** Nutritive tissue of the gall of *Dasineura urticae.* The typical nutritive tissue consists of two cell layers (arrows). Fixed with glacial acetic acid-ethanol (1:3). Stained with Safranin-Fast Green. **Figure 8-24.** Nutritive tissue of the gall of *Dasineura urticae* stained for soluble amino-products. Fresh hand section stained with ninhydrin (from Bronner and Meyer 1968). **Figure 8-25.** Nutritive tissue of the gall of *Dasineura urticae* stained for starch. Note the starch gradient. The typical nutritive tissue is without starch whereas starch is concentrated toward the periphery. Fixed with glacial acetic acid-ethanol (1:3). Stained with Lugol's reagent. **Figure 8-26.** Gall of *Dasineura affinis* on leaf of *Viola.* Acid phosphatases activities are very intense in the nutritive tissue and in the vascular bundles, but less so in the cortical area. Fixed with acetone using the method of Ruyter (1964). Treated by the lead nitrate method of Gomori (1956). C, Cortex; LC, larval chamber; NT, nutritive tissue; Os, ostiole; VB, vascular bundle.

1972) and the feeding of their larvae will be discussed later.

An important difference between the galls of cecidomyiids and cynipids is associated with larval feeding. Unlike cynipid galls, cecidomyiid-induced nutritive cells are never destroyed by larval feeding and only some sparse necroses are attributed to the larvae (Raymond 1927; Meyer 1957). It also is important to recognize that as in the galls of cynipids, there is an extensive development of vascular tissues joined to those of the host organ (Meyer 1969).

Histochemical Aspects

Nutritive tissues of cecidomyiid galls, compared to unaltered host tissues and to other gall tissues, have a high concentration of RNA (nucleolar and ribosomal), an accumulation of soluble amino-products (Fig. 8-24), high enzymatic activities as evidenced by the presence of nonspecific proteases, aminopeptidases (Bronner 1969), acid phosphatases (Fig. 8-26), amylases, glucose-6-phosphatases, invertase (Bronner 1970), and phosphorylase and a starch gradient, which decreases toward the larval chamber (Bronner 1977). It is always possible to identify typical nutritive tissue without starch in cecidomyiid galls (Figs. 8-22 and 8-25) and a storage nutritive tissue where starch is increasing toward the gall periphery. Glucose is accumulated in the nutritive tissue of old galls only (Bronner 1970); it never occurs in young stages as it does in cynipid galls. Glycerids do not accumulate in nutritive cells of cecidomyiid galls as they do in cynipid galls.

The active presence of cecidomyiid larvae, as in cynipid galls, is absolutely necessary to initiate and maintain these histochemical features (Bronner 1977). If the larva is removed or killed, the nutritive cells lose their richness in RNA and in enzymatic activities and the starch gradient is altered.

Metabolical Aspects

As in cynipid galls, the nutritive cells of cecidomyiid galls have a high metabolic activity controlled by the larvae (Bronner 1977). There is a high proteosynthetic activity and at the same time there is a proteolysis resulting in an accumulation of soluble amino-products (Bronner 1969). These products may also come directly from the host plant through the vascular system.

The autophagic system in cecidomyiid galls is highly developed and is directly related to the intense enzyme activities of acid phosphatases, proteases, and aminopeptidases that are present. This system and its enzymes act in controlled lysis (autophagy) of the nutritive cells, but the system does not have the same action that it does in cynipid galls. Here there is "autophagy" in contrast to "autolysis" in cynipid galls. The two systems are thought to be derived differently as they occur in conjunction with different secretions. For example, cynipid larvae can secrete proteases whereas cecidomyiid larvae do not (Bronner 1977).

There are some common metabolic features in the nutritive cells of cynipid and cecidomyiid galls such as richness in amylases, invertase, glucose-6-phosphatases, phosphorylase, acid phosphatases, and a starch gradient decreasing toward the larval chamber. But in spite of these similarities, there are important differences between the two kinds of tissues. In contrast to cynipid galls, the "typical nutritive tissue" of cecidomyiid galls can be provoked to synthesize starch, even when the larva is alive, by feeding the gall by glucidic hypernutrition. It is thought that the starch gradient is also the result of interactions between the larva, host plant, and cytological and cytochemical conditions of the nutritive cells (Bronner 1977).

In summary, the typical nutritive tissue of cynipid galls (Fig. 8-21) cannot synthesize starch when the cecidozoa is alive because amylase activity inhibits the synthetic activity of phosphorylase. Glucose is in excess and will be transformed into lipids. Bronner (1977) suggested that starch can be synthesized in the typical nutritive tissues of cecidomyiid galls because there is less amylase activity than in cynipid galls.

The lack of glycerid accumulation in nutritive tissues of cecidomyiid galls seems to be a consequence of glucidic metabolism. Normally glucose and other sugars are not in excess, but when they do occur, they are converted to starch. It is interesting to note that some cecidomyiid galls (e.g., those of *Taxomyia taxi*

Inch. on the apex of *Taxus baccata* L.) have accumulations of lipids (Jauffret 1972; Jauffret and Le Ret 1975).

Feeding by Cecidomyiid Larvae

Cytological observations show that the nutritive cells are not destroyed by cecidomyiid larvae and only very sparse necroses can be found (Raymond 1927; Meyer 1957). Apparently cecidomyiid larvae do not destroy cells because their mouthparts are poorly developed (Painter 1930) and the musculature is rudimentary (Metcalfe 1933). The mouthparts of cecidomyiids are adapted for sucking (Grassé 1951; Solinas 1968).

Ultrastructural observations do not show feeding punctures in the nutritive cells of cecidomyiid galls as they do, for example, in mite galls (Westphal 1972a,b) and aphid galls (Rohfritsch 1975b). These findings make it difficult to understand how cecidomyiids feed. Heath (1961) has suggested that the liquid food passes through the cell walls. After carefully observing many feeding cecidomyiid larvae, Rohfritsch (1980) was only able to find larvae with their heads and mouthparts close to the cell walls and she concluded that Heath's suggestion was probably correct. If this hypothesis is correct, then cecidomyiid larvae must have a way of making cell walls permeable. Even though it has been shown that cecidomyiids reduce the thickness of cuticle and cell walls beneath the larvae (Meyer 1957; Rohfritsch 1971a, 1988), there still is little information on how they accomplish this.

FEEDING IN GALLS WITHOUT NUTRITIVE TISSUE

Some types of insect galls induced by dipterans are without nutritive tissue. According to Bronner (1977) two kinds of galls can be placed in this group: "rudimentary galls" and "ambrosia galls." Rudimentary galls have a simple morphology and the larvae feed on cell sap. Ambrosia galls have a fungus lining the larval chamber and it is thought that the mycelium serves as food.

Rudimentary Galls

Rudimentary galls are induced on leaves. They have simple structures and exhibit a pock-like or arch-like shape. Although some rudimentary galls such as those of *Physemocecis ulmi* or *Dasineura pustulans* have nutritive tissues equivalent to those of more complex cecidomyiid galls, others such as the galls of *Physemocecis hartigi* Liebel on *Tilia intermedia D.C.*, *Drisina glutinosa* Giard on *Acer pseudoplatanus* L., and *Monarthropalpus buxi* Laboulb. (Geoffr.) on *Buxus sempervirens* L. have no nutritive cells (Bronner 1977). The lack of nutritive cells in these galls poses a problem in explaining how the larvae feed. Areas with nutritive cells have not been identified histochemically in these galls and thus it has been concluded that the larvae obtain their food directly from the plant sap by different means.

The gall induced by *Monarthropalpus buxi* consists of a slight swelling on the leaf that is barely perceptible on either side of the lamina. The larva is surrounded by callus tissue (Fig. 8-27) derived from the mesophyll cells. These cells have no features in common with typical nutritive cells. Of interest, the vascular bundles of the leaf can be seen in direct contact with the larval chamber (Fig. 8-28). It has been shown that females cut the vascular tissues of the leaf at oviposition (Meyer 1957) and that the buccal apparatus of the larva is frequently found toward the vascular bundles (Bronner and Meyer 1972). Thus in this gall it has been suggested that the larva feeds directly on vascular tissue. Similar observations have been made on the gall of *Physemocecis hartigi* (Figs. 8-29 and 8-30) where the gall is frequently formed on a leaf vein. Here the vascular bundles of the leaf are found directly adjacent to the larval chamber (Bronner and Meyer 1972).

The gall of *Drisina glutinosa* is even more interesting. This gall is a simple arch only slightly perceptible on the upper surface of the leaf. The larva is found on the lower side covered in a small drop of liquid identified as plant sap (Bronner 1977). How the sap is exuded has not been explained. Indeed, there are no visible changes in the cells beneath the larva (Fig. 8-31) except for a starch gradient (Fig. 8-32).

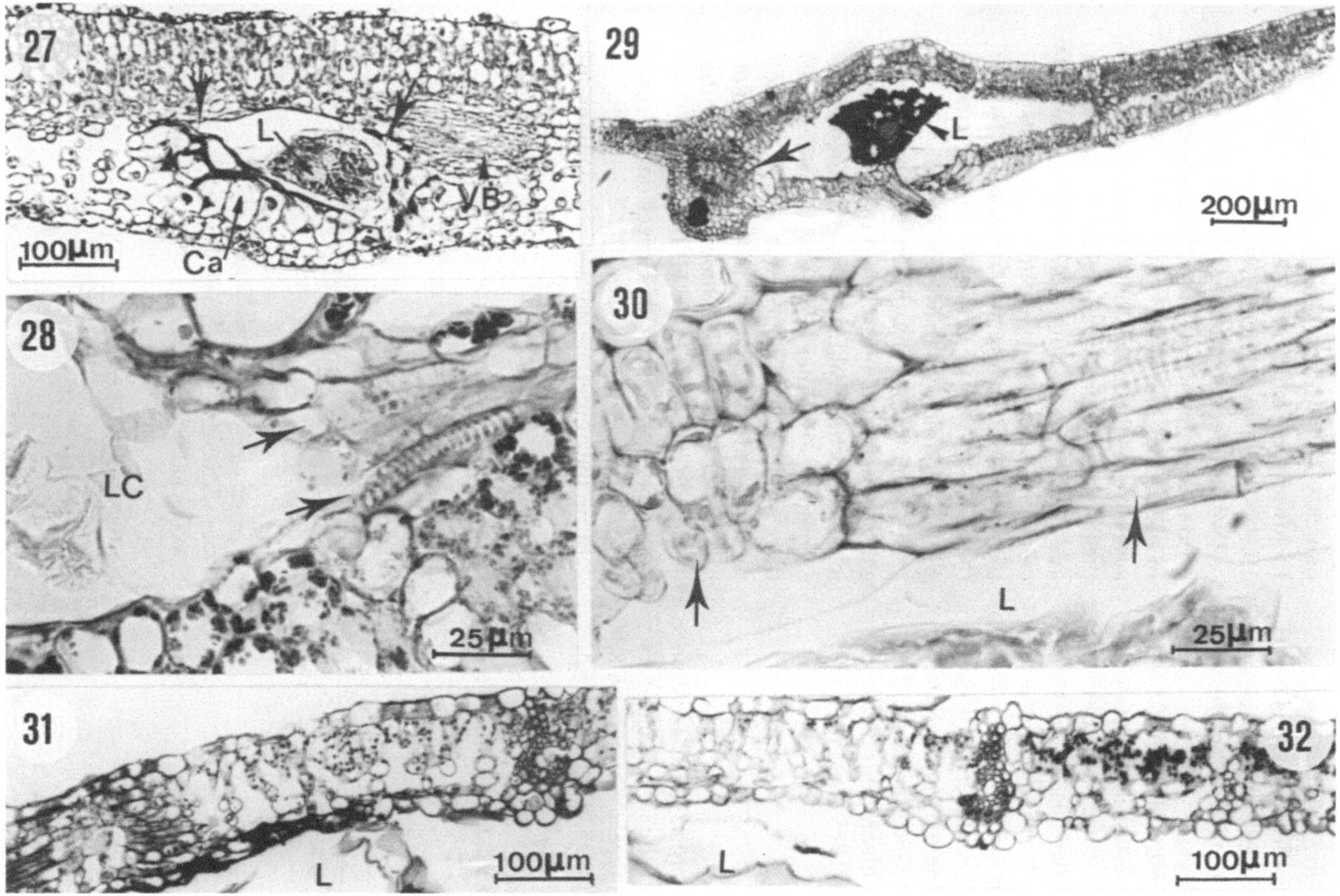

Figures 8-27 and 8-28. Galls of *Monarthropalpus buxi* on leaf of *Buxus sempervirens.* **Figure 8-27.** Section of a gall showing callus tissue under the larva and vascular bundles adjoining the larval chamber. Fixed and stained using the method of Regaud (see Meyer 1957). **Figure 8-28.** Details of a gall showing vascular bundles extending to the larval chamber. Fixed with glacial acetic acid–ethanol (1:3). Stained with the method of Reilhes (1946) for starch (black grains). **Figures 8-29 and 8-30.** Galls of *Physemocecis hartigi* on leaf of *Tilia.* **Figure 8-29.** Section of a gall showing the larva in the larval chamber on a leaf vein (arrow). Fixed and stained using the method of Regaud (see Meyer 1957). **Figure 8-30.** Details of a gall showing exposed vascular bundles (arrows) in larval chamber. Fixed and stained using the method of Regaud (see Meyer 1957). **Figures 8-31 and 8-32.** Galls of *Drisina glutinosa* on leaf of *Acer.* **Figure 8-31.** Section of a gall showing the apparently normal leaf beneath the larva. Fixed with glacial acetic acid–ethanol (1:3). Stained with aqueous toluidine blue. **Figure 8-32.** Starch gradient (in black) increasing toward the outside of the gall. Note that there is no starch in the lamina under the larva. Fixed with glacial acetic acid–ethanol (1:3) and stained with Lugol's reagent. Ca, Callus; L, larva; LC, larval chamber; VB, vascular bundle.

The feeding strategies of cecidomyiids in rudimentary galls are schematically compared in Figure 8-33. *Physemocecis ulmi* and *Dasineura pustulans* induce typical nutritive tissue, *Physemocecis hartigi* and *Monarthropalpus buxi* feed on plant sap arriving directly in the larval chamber by the vascular bundles and do not induce nutritive cells, whereas *Drisina glutinosa* feeds on exuded plant sap.

It is important to note that larvae of rudimentary galls feed without destruction of plant tissue (Bronner 1977) as is the case in complex dipterous galls with nutritive cells. The fact that larvae of rudimentary galls without nutritive cells feed directly on plant sap by several means is not unique to cecidology. Indeed, it is well known that both cecidogenous and noncecidogenous hemipterans (Aphidae and Psyllidae) thrust their stylets into phloem cells (Schmidt and Meyer 1967; Miles 1968; Saxena 1970; Carter 1973).

Ambrosia Galls

Galls referred to as "ambrosia galls" have no nutritive cells but instead the larval chambers are lined with fungal mycelia (Neger 1908, 1910; Ross 1914, 1922, 1932). They are induced by cecidomyiids belonging to four genera: *Asphondylia, Lasioptera, Kiefferia,* and *Schizomyia* (see also papers by Puzanova-Malyseva 1935; Docters Van Leeuwen 1939; Goidanich 1940; Meyer 1952, 1957; Batra and Lichtward 1963; Batra 1964; Richter-Vollert 1964; Skuhravá and Skuhravý 1981).

Two different hypotheses have been developed to explain the role of the fungal mycelium in these galls. The first hypothesis (Ross 1922) suggests that the fungus does not play a role in the nutrition of the larva and instead is parasitic and develops only after penetration of the cecidozoa. The opposing hypothesis suggests that the mycelium is essential to larval feeding (Neger 1908, 1910; Docters Van Leeuwen 1939; Goidanich 1940; Meyer 1952).

In spite of the many studies made on these galls, there is little information on the cytological or cytochemical features of the fungus. An exception is Bronner (1977), who showed that there was high RNA concentration and high activity of acid phosphatases in the mycelium of two galls (*Schizomyia galiorum* Kffr. on the flower of *Galium mollugo* H. Lw. (Figs. 8-34–8-38) and *Lasioptera rubi* Heeger on the stem of *Rubus caesius* L.).

From the numerous isolated observations on

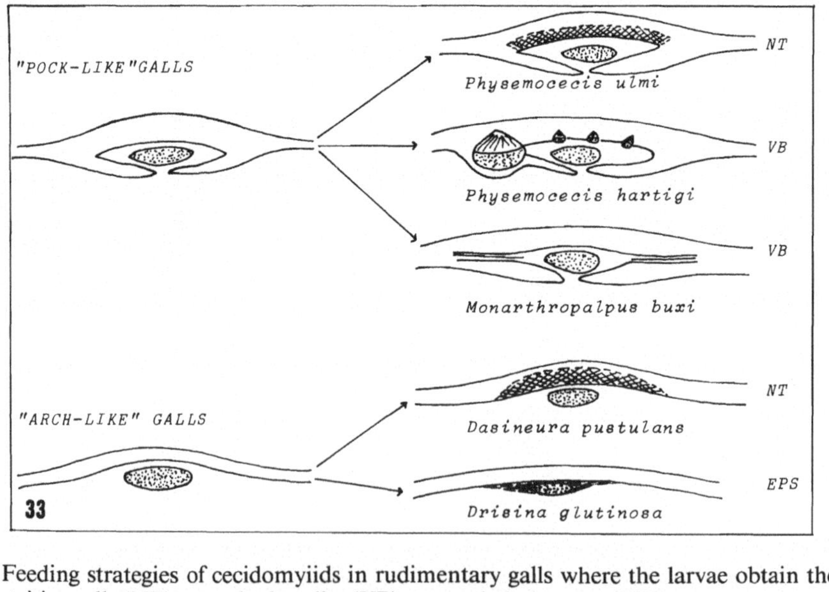

Figure 8-33. Feeding strategies of cecidomyiids in rudimentary galls where the larvae obtain their nutriment either from nutritive cells (NT), vascular bundles (VB), or exuded plant sap (EPS).

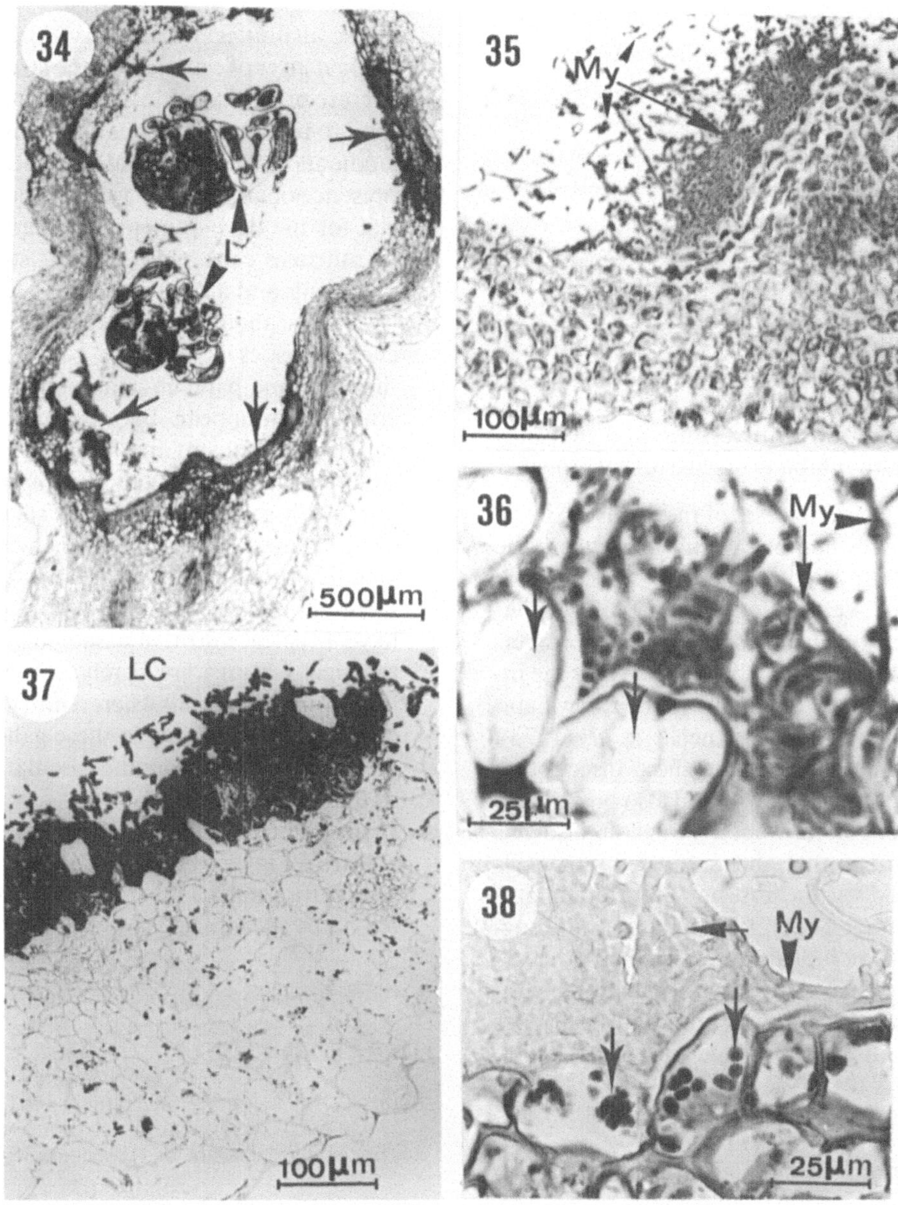

Figures 8-34–8-38. Galls of *Schizomyia galiorum* on flowers of *Galium mollugo*. **Figure 8-34.** Longitudinal section of a flower galled by two larvae. Note fungi mycelium (arrows) lining the larval cavity. **Figure 8-35.** Detail of a gall section showing mycelium. **Figure 8-36.** Detail showing mycelium. Note the lack of differentiation of nutritive tissue in the plant cells under the mycelium (arrows). **Figure 8-37.** High acid phosphatases activity (in black) in the mycelium lining the larval cavity. **Figure 8-38.** Cells beneath the mycelium stained for starch. Starch is indicated by arrows. L, Larva; LC, larval cavity; My, mycelium.

these galls, the following events apparently occur:

1. The mycelium is introduced into the gall by the larva. Skuhravá and Skuhravý (1981) showed the presence of mycelium around the larva in the earliest stages of *Schizomyia* and *Lasioptera* galls. Furthermore, if young larvae are removed from the gall and placed on a sterile medium, they are able to disseminate mycelium.
2. The larva influences the mycelium's growth. Mature *Schizomyia galiorum* larvae exit their galls to pupate in the soil and after their departure, the mycelium grows and fills all the larval chamber. Thus it appears that larvae of the four ambrosia gall species take food from fungal mycelium and that they are able to inhibit excessive fungal growth.

There are several cases of insects living in association with one or several fungi where the fungus is cultivated and fed on (Parkin 1942; Batra 1966; Weber 1966; Moser 1967; Amouriq 1973). Fungal spores are carried by insects in special "fungal containers" such that the insects disseminate the fungus in their new habitat. According to Schildknecht *et al.* (1973), substances are released by these insects that control the fungus growth and also prevent the development of unwanted fungi. It also has been recorded that the cultivated fungus is transformed by the insects (Batra and Batra 1967; Bronner 1977). In conclusion, it is suggested that the fungal mycelium lining the larval chamber plays a role in the nutrition of ambrosia gall insects just as important as it does with other insects associated with fungi.

CONCLUSIONS

The material presented in this chapter leads to the following general conclusions:

1. Most galls induced by cynipid wasps and cecidomyiids have a nutritive tissue that supplies food to the larvae. The nutritive tissue usually lines the larval chamber and its thickness depends on the species of cecidozoa.
2. Nutritive tissue, when compared with corresponding tissues found in other parts of the plant and with tissues in other parts of the gall, exhibit unique histological and histochemical attributes and they have high physiological and metabolic activities. These attributes, which are very distinct in the first layer of cells nearest the larval chamber, are progressively less distinct away from the larval chamber. Thus there are several gradients that reflect larval influence and the presence of an active larva is absolutely crucial for the development and maintenance of nutritive cells. Gall insects "stimulate" plant cells and any cessation of insect activity brings about a rapid regression of the characteristics of nutritive cells. The cells then become parenchymatous-like and the gradients disappear. Because of their own active metabolism and because galls are sinks for plant assimilates, nutritive cells are filled with soluble compounds such as glucose, sucrose, and amino acids, which provide nutrition for the insects.
3. Even though the majority of insect galls have a typical nutritive tissue, the usual cytological features are barely apparent in a few rudimentary galls where a nutritive layer does not exist. Insects in these galls extract their food directly from the vascular bundles of their host organs.
4. In the ambrosia galls, which also do not have a nutritive layer, the larval chamber is lined with fungal mycelia that are stimulated and fed on by the larvae.

LITERATURE CITED

Adams, C. M. V., and N. A. Tuquan. 1961. The histochemical demonstration of protease by gelatine-silver substrate. J. Histochem. Cytochem. 9:464–472.

Amouriq, L. 1973. Rapports entomolo-cryptogamiques. Elements sur less relations entre insectes et champignons. Hermann, Paris.

Batra, L. R. 1964. Insect fungus blister galls of *Solidago* and *Aster.* J. Kan. Entomol. Soc. 37:227–234.

Batra, L. R. 1966. Ambrosia fungi. Extent of specificity to ambrosia beetles. Science 153:193–195.

Batra, L. R., and R. W. Lichtward. 1963. Association of fungi with some insect galls. J. Kan. Entomol. Soc. 36:262–278.

Batra, S. W. T., and L. R. Batra. 1967. The fungus gardens of insects. Sci. Am. 217:112–120.

Bessis, M. 1970. La mort de la cellule. Triangle, Bull. Lab. Sandoz. 10:157–165.

Bitensky, L., and P. B. Gahan. 1962. The reversible activation of lysosomes in normal cells. Biochem. J. 84:13.

Brachet, J. 1940. La détection histochimique des acides pentosenucléiques. C. R. Soc. Biol. 133:88–90.

Braun, C. 1966. Localisation des phosphatases acides dans les galles de cynipides et de cécidomyides. C. R. Acad. Sci. 262:1437–1440.

Bronner, R. 1969. Localisation histochemique les produits aminés solubles et des aminopeptidases dans les tissus nourriciers des galles de cécidomyides. C. R. Acad. Sci. 268:810–812.

Bronner, R. 1970. Localisations histochimiques de l'invertase et du glucose dans le tissu nourricier d'une galle de cécidomyide: *Perrisia urticae* Perris sur *Urtica dioica* L. C. R. Acad. Sci. 270:641–643.

Bronner, R. 1973. Localisations ultrastructurales de l'activité phosphatasique acide (phosphomonoéstérase acide) dans les cellule nourricières de la galle du *Biorhiza pallida* 01. ♀♂ sur le *Quercus pedunculata* Ehrh. C. R. Acad. Sci. 276:2677–2680.

Bronner, R. 1974. Localisations histochimiques et cytochimiques de l'activité phosphatasique acide dans la galle du cynipide *Andricus curvator* Hartig ♀♂ sur le *Quercus pedunculata* Ehrh. Marcellia 38:3–10.

Bronner, R. 1975a. Simultaneous differential visualization of lipids and starch in plant tissues. Stain Technol. 50:1–4.

Bronner, R. 1975b. Répartition des lipides et de l'amidon dans les tissus nourriciers de quelques galles de cynipides. Marcellia 38:171–177.

Bronner, R. 1977. Contribution a l'étude histochimique des tissus nourriciers des zoocécidies. Marcellia 40:1–134.

Bronner, R. 1980. Ultrastructural observations on lipid globules in nutritive cells of cynipid galls (Hymenoptera). Cecidol. Int. 1:127–135.

Bronner, R. 1981. Observations on cynipid galls modified by inquiline larvae. Cecidol. Int. 2:53–61.

Bronner, R. 1982. Adaptation insect-plant in cynipid galls. In N. S. Margaris and M. Arianoutsou-Faraggitaki (eds.), Adaptations to Terrestrial Environments, pp. 61–68. Plenum Press, New York.

Bronner, R., and J. Meyer. 1968. Mise en évidence histochimique de l'abondance des produits aminés solubles dans les tissus nourriciers des galles. Marcellia 35:105–110.

Bronner, R., and J. Meyer. 1972. Observations histologiques et histochimiques comparées sur quatre galles 'rudimentaires' en pustule causées par des Cécidomyides. Marcellia 37:77–83.

Bronner, R., and J. Meyer. 1976. Cécidogenèse et aspects ultrastructuraux de la galle de *Dasyneura affinis* Kieffer sur *Viola odorata* L. Marcellia 39:141–154.

Carter, W. 1973. Insects in Relation to Plant Diseases. John Wiley, New York.

Coulomb, P. and C. Coulomb. 1972. Processus d'autophagie cellulaire dans les cellules de méristèmes radiculairès de la courge (*Curcurbita pepo* L., Cucurbitaceae) mis en état d'anoxie. C. R. Acad. Sci. 274:114–117.

Darvill, A. G., P. Albersheim, M. McNeil, J. M. Laum, W. S. York, T. T. Stevenson, J. Thomas, S. Doares, D. J. Gollin, P. Chelf, and K. Davis. 1985. Structure and function of plant cell wall polysaccharides. J. Cell. Sci. Suppl. 2:203–217.

Dixon, M. 1949. Multi-enzyme Systems. Cambridge Univ. Press, Cambridge.

Docters Van Leeuwen, W. M. 1939. An ambrosiagall on *Symplocos fasciculata* Zoll. Ann. Jard. Bot. Buitenzorg. 44:27–42.

Dustin, P. 1966. Leçons d'anatomie pathologique générale. Presses académiques européennes, S. C. Bruxelles et Librairie Maloine, S. A. Paris.

Duve, C., and R. Wattiaux. 1966. Function of lysosomes. Annu. Rev. Physiol. 28:435–492.

Esau, K. 1972. Cytology of sieve elements in minor veins of sugar beet leaves. New Phytol. 71:161–168.

Eyme, J. 1966. Infrastructure des cellules nectarigènes de *Diplotaxis erucoida* D. C., *Helleborus niger* L. et. *Helleborus foetidus* L. C. R. Acad. Sci. 262:1629–1632.

Figier, J. 1972. Localisation infrastructurale de la phosphatase acide dans les glandes pétiolaires d' *Impatiens holstii.* Rôles possibles de cette enzyme au cours des processus sécrétoires. Planta 108:103–120.

Fockeu, H. 1889. Contribution à l'histoire des galles. Étude anatomique de quelques espèces. Imprim. et Lib. Camille Robbe, Lille.

Fourcroy, M., and C. Braun. 1967. Observations sur la galle de l' *Aulax glechomae* L. sur *Glechomà hederacea* L. II. Histologie et rôle physiologique de la coque sclérifiée. Marcellia 34:23–30.

Gadal, P. 1970. Sur le métabolisme du chêne et de ses zoocécidies. Étude particulierè de la biosynthèse des acides aminés aromatique. Thèse, Doctorat es Sciences, Toulouse, France.

Gahan, P. B., and A. J. Maple. 1966. The behaviour of lysosome-like particles during cell differentiation. J. Exp. Bot. 17:151–155.

Garrigues, R. 1951. Recherches sur les cécidies, le cancer et l'action des carbures cancérigènes sur les végétaux. Thèse, Doctorat es Sciences, Lyon, France.

Gaufillet, C., and M. Fourcroy. 1966. Observations sur la galle d' *Aulax glechomae* L. sur *Glechoma hederacea* L. Marcellia 33:3–31.

Goidanich, A. 1940. Interpretazione symbiotica di

una associazione mico-entomatica gallare. Atti. R. Acad. Sci. 76:1–14.

Gomori, G. 1956. Histochemical methods for acid phosphatase. J. Histochem. Cytochem. 4:453–461.

Grassé, P. P. 1951. Traité de Zoologie. Anatomie systématique. Biologie 10:Fascicule 1.

Grassé, P. P., R. Poisson, and O. Tuzet. 1961. Zoologie. I. Invertébrés. Précis de Sciences biologiques. Masson, Paris.

Heath, G. W. 1961. An investigation into leaf deformation in *Medicago sativa* caused by the gall midge *Jaapiella medicaginis* Rübs. (Cecidomyidae). Marcellia 30 Suppl.: 185–199.

Jankiewicz, L. S., H. Plich, and R. Antoszewski. 1969. Preliminary studies on the translocation of ^{14}C-labelled assimilates and $32PO_4^{3-}$ towards the gall evoked by *Cynips (Diplolepis) quercusfolii* L. on oak leaves. Marcellia 36:163–174.

Jauffret, F. 1972. Transformations ultrastructurales du méristème apical du *Taxus baccata* L. en tissu nourricier sous l'action de *Taxomyia taxi* Inch. C. R. Acad. Sci. 74:226–229.

Jauffret, F. 1973. Etude ultrastructurale de la transformation du méristème apical de l' *Euphorbia cyparissiae* L. en tissu nourricier sous l'action du *Bayeria capitigena* (Bre.) C. R. Acad. Sci. 276:1177–1180.

Jauffret, F. 1975. Devenir de l'apex du *Taxus baccata* L. sous l'action du *Taxomyia taxi* Inch. et après mort provoquée du cécidozoaire. Marcellia 38:239–249.

Jauffret, F., and M. Le Ret. 1975. A comparative study on the evolution of lipids and starch reserves at the apex of *Taxus baccata* L. from healthy plants and plants parasitized by *Taxomyia taxi* Inch. Marcellia 38:311–318.

Kientzler, L., J. Bringay, and F. Cibille. 1962. Evolution des substances azotées, des glucides et des acides organiques des feuilles d' *Acer pseudoplatanus* L. parasitées par *Pediaspis aceris* Forst. C. R. Acad. Sci. 255:1467–1469.

Kinsey, A. C. 1930. The gall wasp genus Cynips. A study in the origin of species. Ind. Univ. Studies 16 (Studies 84, 85, 86).

Klein, C. 1963. Cécidogenèse et étude histocytologique de la galle de *Perrisia galii* Lw. sur *Galium mollugo* L. D. E. S. Strasbourg.

Klein, C., and J., Meyer. 1963. Etude du développement de la galle de *Perrisia galii* Lw. sur *Galium mollugo* L. Marcellia 31:77–94.

Lacaze-Duthiers, H. 1853. Recherches pour servir à l'histoire des galles. Ann. Sci. Nat. Bot. Ser. III. 19:273–354.

Matile, P. 1975. The lytic compartment of plant cells. Cell Biology Monographs, Vol. 1. Springer, Wien.

Matile, P., and F. Winkenbach. 1971. Function of lysosomes and lysosomal enzymes in the senecing corolla of the morning glory *(Ipomea purpurea).* J. Exp. Bot. 22:765–771.

Metcalfe, M. E. 1933. The morphology and anatomy of the larva *Dasyneura leguminicola* Lint. (Diptera). Proc. Zool. Soc. London:119–130.

Meyer, J. 1948. Observations sur les premiers effets cytologiques de la cécidogenèse chez deux cécidomyides. Rev. Gén. Bot. 55:256–290.

Meyer, J. 1950. Gigantisme nucléolaire et cécidogenèse. C. R. Acad. Sci. 231:1333–1335.

Meyer, J. 1952. Cécidogenèse de la galle de *Lasioptera rubi* Heeger et rôle nourricier du mycélium symbiotique. C. R. Acad. Sci. 234:2556–2558.

Meyer, J. 1957. Cécidogenèse comparée de quelques galles d'Arthropodes et évolution cytologique des tissus nourriciers. Thèse, Doctorat es Sciences, Strasbourg, France.

Meyer, J. 1969. Irrigation vasculaire dans les galles. Bull. Soc. Bot. Fr. Mem: 75–97.

Meyer, J. 1973. Evolution cytologique ultrastructurale des cellules nourricières de la galle du *Diastrophus rubi* Htg. sur le *Rubus caesuis* L. C. R. Acad. Sci. 276:387–389.

Meyer, J., and H. J. Maresquelle. 1983. Anatomie des Galles. Gebrüder Borntraeger, Stuttgart.

Miles, P. W. 1968. Insect secretions in plants. Annu. Rev. Phytopathol. 6:137–164.

Molliard, M. 1897. Hypertrophie pathologique des cellules végétales. Rev. Gen. Bot. 9:35–39.

Molliard, M. 1913. Recherches physiologiques sur les galles. Rev. Gen. Bot. 25:225–252; 285–307; 341–370.

Moser, J. C. 1967. Trails of the leafcutters. Nat. Hist. 76:32–35.

Neger, F. W. 1908. Ambrosiapilze. Ber. Dtsch. Bot. Ges. 26a:735–754.

Neger, F. W. 1910. Ambrosiapilze. III. Weitere Beobachtungen an Ambrosiagallen. Ber. Dtsch. Bot. Ges. 28:455–480.

Öpik, H. 1964. The form of nuclei in the storage cells of the cotyledons of germinating seeds of *Phaseolus vulgaris* L. Exp. Cell Res. 38:517–522.

Paclt, J. and J. Hassler. 1967. Concentration of nitrogen in some plant galls. Phyton 12:173–176.

Painter, R. H. 1930. Discussion of Haseman's paper (Haseman L. 1930. J. Econ. Entomol. 23:316–321). J. Econ. Entomol. 23:321.

Paris, G., and R. A. Trotter. 1911. Sui composti azotati nelle galle *Neuroterus baccarum.* Marcellia 10:150–159.

Parkin, E. A. 1942. Symbiosis and siricid woodwasps. Ann. Appl. Biol. 29:268–274.

Puzanova-Malyseva, E. W. 1935. *Asphondylia prunorum* Wachtl. (Dipt. Ceci.) und dere Pilzgallen am Pflaumenbaum. Z. angew. Entomol. 31:443–462.

Raymond, G. 1927. Contribution à l'étude des cécidomyides attaquant les violettes cultivées. Ann. Soc. Linn. 73:87–101.

Reilhes, R. 1946. Coloration durable par l'iode des celluloses simplement deshydratées, du glycogène et des amidons. Rev. Gen. Bot. 53:97–120.

Rey, L. 1976. Recherches comparées sur trois types

d'hyménoptérocécidies (Cynipidae et Tenthrèdinidae). Etude morphologique, histologique, et ultrastructurale du développement et de la différenciation. Essais expérimentaux. Thèse Doctorat es Sciences, Nantes, France.

Rey, L. 1979. Observations ultrastructurales sur le tissu nourricier de la galle provoquée par *Andricus albopunctatus* Sch. ♀♂ sur *Quercus pedunculata* Ehrh. Bull. Soc. Bot. Fr. 126:179–190.

Rey, L., and A. Moreau. 1983. Sur les inclusions paracristallines des cellules nourricières de la galle provoquée par *Diplolepis rosae* L. sur *Rosa canina* L. Ultrastruct. Res. 84:94–102.

Richter-Vollert, I. 1964. Untersuchungen zur Morphologie und Ökologie von *Asphondylia sarothamni* H. Lw. Zoologica Heft. 112, 40 Bd. 2. Schweizerbart, Stuttgart.

Rohfritsch, O. 1966. Analyse morphogénétique de la formation d'une galle (*Perrisia galii* sur *Galium mollugo*) par la méthode des ablations échelonnées du cécidozoaire. Mém Soc. Bot. Fr. 20–25.

Rohfritsch, O., 1967. Diverses actions cécidogènes mises en évidence par l'étude de galles artificiellement privées de leur cécidozoaire. 92e Congrès des Soc. Savantes 3:321–329.

Rohfritsch, O., 1968. Quelques méthodes expérimentales utilisées en cécidologie. Congrès de l'A. F. A S. 9:192–198.

Rohfritsch, O. 1971a. Développement cécidien et rôle parasite dans quelques galles d'Arthropodes. Marcellia 37:233–339.

Rohfritsch, O. 1971b. Infrastructure des cellules du tissu nourricier de la galle de *Geocrypta galii* Lw. sur *Galium mollugo* L. C. R. Acad. Sci. 272:76–78.

Rohfritsch, O. 1974. Infrastructure du tissu nourricier de la galle de l' *Aulax glechomae* L. sur *Glechoma hederacea* L. Protoplasma 81:205–230.

Rohfritsch, O. 1975a. Etude comparative des cellules du tissu nourricier de la jeune galle de l'*Aulax glechomae* L. sur le *Glechoma hederacea* L. et des cellules du tissu nourricier abandonné par le parasite. Marcellia 38:185–196.

Rohfritsch, O. 1975b. Traces de succion de deux Chermesidae *Chermes abietis* L. et *Chermes strobilobius* Kalt. dans les bourgeons de *Picea excelsa* L. Marcellia 39:69–84.

Rohfritsch, O. 1980. Relations hôte-parasite au début de la cécidogenèse du *Hartigiola annulipes* Hartig sur le Hêtre. Bull Soc. Bot. Fr. 127. Actual. Bot. 1980 (1):199–207.

Rohfritsch, O. 1988. Food supply mechanism related to gall structure with example of *Geocrypta galii* on *Galium mollugo*. Phytophaga 2:1–17.

Ross, H. 1914. Über verpilzte Tiergallen. Ber. Dtsch. Bot. Ges. 32:574–597.

Ross, H. 1922. Weitere Beiträge zur Kenntnis der verpilzten Mückengallen. Z. Pflanzenkr. 32:83–93.

Ross, H. 1932. Praktikum der Gallenkunde. Springer, Berlin.

Roth, P. 1949. Beiträge zur Biologie der Gallwespen. Verh. Naturf. Ges. 60:104–178.

Ruyter, J. H. 1964. Studies on an improved lead phosphate technique for the demonstration of non specific acid phosphatases in non deparaffinized organ and tissue sections. Histochemie 3:532–537.

Sauter, J. J. 1966. Untersuchungen zur Physiologie der Pappelholzstrahlen. II. Jahresperiodische Änderungen der Phosphataseaktivität im Holzstrahlparenchym und ihre mögliche Bedeutung für den Kohlenhydratstoff-wechsel und den aktiven Assimilattransport. Z. Pflanzenphysiol. 55:349–362.

Sauter, J. J. 1972. Respiratory and phosphatases activities in contact cells of wood rays and their possible role in sugar secretion. Z. Pflanzenphysiol. 67:135–145.

Sauter, J. J., and H. L. Braun, 1968. Enzymatic polarity in ray parenchyma cells of conifers in spring. Z. Pflanzenphysiol. 58:378–381.

Saxena, P. N. 1970. Staining salivary sheaths of feeding aphids: A simple prussian blue reaction for paraffin sections. Stain Technol. 45:241–242.

Schildknecht, H., P. B. Reed, R. D. Reed, and K. Koob. 1973. Auxin activity in the symbiosis of leaf-cutting ants and their fungus. Insect Biochem. 39:439–442.

Schmidt, E., and J. Meyer. 1967. Observations sur la structure des galles de *Livia juncorum* Latr. sur *Juncus articulatus* L. Marcellia 33:237–253.

Schnepf, E. 1964. Zur Cytologie und Physiologie pflanzlicher Drüsen. Protoplasma 58:193–219.

Skuhravá, M., and V. Skuhravý. 1981. Die Gallmücken (Cecidomyidae, Diptera) des Schilfes (*Phragmites communis* Trin. Studies C. S. A. V. 3:1–150.

Solinas, M. 1968. Morfologia, anatomia e organizzazione funzionale del capo della larva matura de *Phaenobremia aphidimyza* (Rondani). Entomologia 4:7–44.

Strauss, W. 1967. Lysosomes, phagosomes, and related particles. In D. B. Roodyn (ed.), Enzyme Cytology, pp. 239–319. Academic Press, London.

Tremblay, G., and J. Charest. 1968. Modified starch film method for the histochemical localization of amylase activity. J. Histochem. Cytochem. 11:202–206.

Turian, G. 1958. Recherches anatomo-physiologiques sur les cécidies. Rev. Gen. Bot. 65:279–293.

Vorms-Ravillard, A. 1935. Modifications anatomiques de galles arrêtées dans leur développement. Thèse. Sci. Nat. Paris, France.

Wanner, H. 1952. Phosphatase Verteilung und Kohlenhydrattransport in der Pflanze. Planta 41:190–194.

Weber, N. A. 1966. Fungus-growing ants. Science 153:587–604.

Webster, J. M., and R. E. Paulson. 1972. An interpretation of the ultrastructural response of tomato roots susceptible and resistant to *Meloidogyne incognita* (Kofoid and White). OEPP/EPPO Bull. 6:33–39.

Westphal, E. 1972a. Observations ultrastructurales et histochimiques sur les cônes de succion de *Cecidophyes psilaspis* Nal., Acarien cécidogène des bourgeons de l'If. C. R. Acad. Sci. 274:893–896.

Westphal, E. 1972b. Traces de succion parasitaire laissées par quelques Eriophyides cécidogènes. Aspect histochimique et observations ultrastructurales. Marcellia 37:53–70.

Westphal, E., R. Bronner, and M. Le Ret. 1981. Changes in leaves of susceptible and resistant *Solanum dulcamara* infested by the gall mite *Eriophyes cladophthirus* (Acarina, Eriophyoidea). Can. J. Bot. 59:879–882.

Woll, E. 1954a. Beiträge zum Differenzierungsproblem an Hand der Zytologie von Pfaanzengallen. Z. Bot. 42:1–29.

Woll, E. 1954b. Untersuchungen über die cytologische Differenzierung einiger Pflanzengallen. Planta 43:477–494.

Zandonella, P. 1970. Infrastructure des cellules du tissu nectarigène floral de quelques Caryophyllacées. C. R. Acad. Sci. 270:1310–1313.

9

Cecidogenesis and Resistance Phenomena in Mite-Induced Galls

E. WESTPHAL

Plant galls induced by mites have drawn the attention of botanists for over 300 years (Bumaldi 1671; Tournefort 1698); however, the fact that these rather small galls were induced by animals was not discovered until much later (Réaumur 1737). Early naturalists first thought that the extremely small, worm-like animals inhabiting these galls were insect larvae and it was not for another century before they identified the inhabitants as mites—also referred to as acari (Vallot 1820; Turpin 1833; Fée 1834; Siebold 1850). Mite-induced galls are today referred to as acarocecidia.

Gall mites belong to two families—the Tarsonemoidea and the Eriophyoidea. Only a few species of the family Tarsonemoidea induce galls and furthermore their galls are rudimentary and hardly noticeable, existing as curled leaves with simple hypertrophy (Meyer and Maresquelle 1983). In contrast, the ability to induce galls has developed to a great extent in the family Eriophyoidea. There are hundreds of gall-inducing eriophyoids and their galls exhibit a wide diversity in form and complexity.

Although this volume deals almost exclusively with galls induced by insects, the similarity of mite and insect galls warrants the inclusion of at least some aspects of mite gall biology. The comments in this chapter are restricted to the eriophyoid mites and their galls. A brief description also is given on the morphology, biology, and feeding behavior of eriophyoid mites followed by an examination of cecidogenesis by the group and a comparison of different host–mite relationships.

ERIOPHYOID GALL MITES

Morphology of Mites

Eriophyoid mites are difficult to observe and identify because of their small size (0.1–0.3 mm) and as a result the nomenclature of mite galls once followed that of their host plants. Binomial names were not established until the latter part of the nineteenth century (Siebold 1850; Frauenfeld 1864; Löw 1874; Kieffer 1885). Gall mites were especially well described in the taxonomic works of Nalepa (1911, 1929). The nomenclature of mites was recently reviewed by Keifer in Jeppson *et al.* (1975).

Eriophyoid mites show a great reduction in body structure (loss of the rear four legs, disappearance of nearly all body setae) that seems to be an adaptation for existence in microhabitats on plants (Jeppson *et al.* 1975). They usually exhibit a pale yellowish to nearly transparent, elongated, fusiform body with a cephalothoracic shield, numerous transverse ridges, and nearly 10 pairs of setae. They are characterized by the presence of four segmented legs. The strongly down-curved rostrum lies between the anterior pair of legs (Fig. 9-1). The rostrum is enclosed on each side by the pedipalps holding minute styliform chelicerae (15–50 μm long) in an anterior groove.

Life History

Galls are generally induced by overwintering females which deposit eggs that develop into adults (males and females) through two

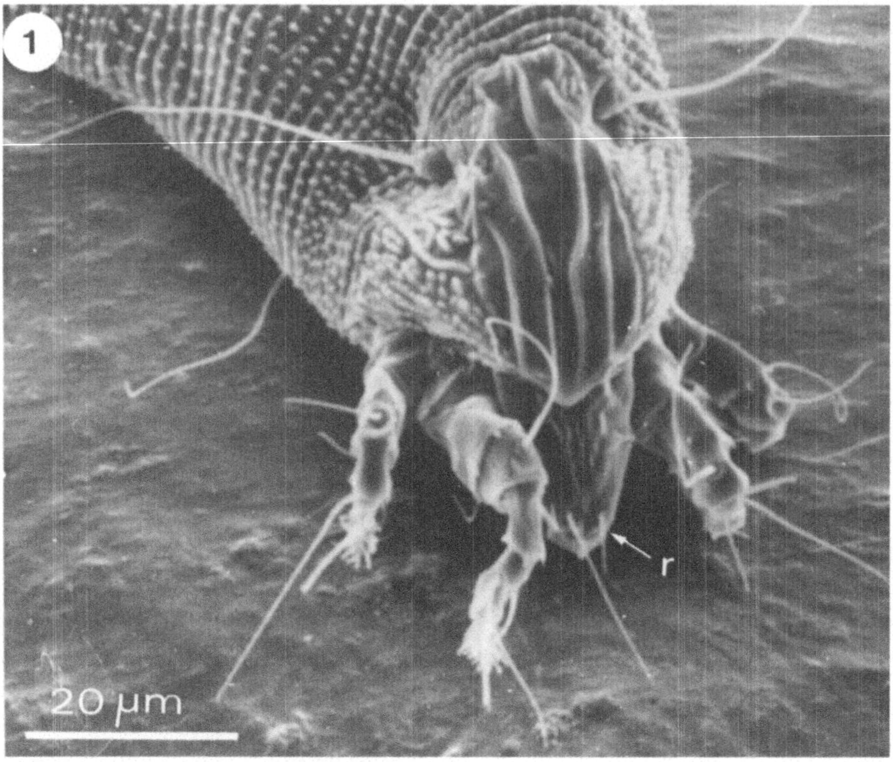

Figure 9-1. Scanning electron micrograph of *Aceria cladophthirus* Nal. (anterodorsal view). The stylet-like chelicerae are enclosed in a groove of the rostrum (r).

nymphal instars. Females breed actively, laying an average of 50 eggs each (Oldfield 1969), and populations increase rapidly. Mites of deciduous plants generally leave the galls in late summer and hibernate in crevices of the bark or under bud scales. The soft body provides a weak protection against the fluctuating weather conditions of temperate regions. To survive during unfavorable conditions, some species of gall mites undergo color and structural changes within their galls (Jeppson *et al.* 1975). The hibernating females (deutogyne) of some species are sufficiently different in appearance from breeding summer females (protogyne) that they inadvertently have been placed in separate genera. The deutogynes leave the protected areas of the host in the following spring and move rapidly over the surface of the plant to select suitable feeding sites, which are usually young leaves or buds. Eriophyoids do not penetrate the plant tissues except for species such as *Phytoptus pyri* Pgst., which induces leaf blisters or "pocken galls." All other gall mites become more or less enclosed in their galls by growth of surrounding plant tissues.

Feeding Behavior

Most species of mites attack the leaf epidermis and are more easily observed than those feeding inside buds. When a mite attacks a leaf, it comes to rest in a sucking position. It attaches itself by means of the anal sucker and arches its body before forcing stylet-like chelicerae into the plant tissues with alternate forward movements (Krantz 1973). Boring into the epidermal cell wall lasts for about 10 seconds. Feeding periods may be detected by the vibrations of the chelicerae motivator, which is visible through the transparent anterior dorsal lobe. Termination of feeding is marked by a cessation of pulsation and disengagement of the mouthparts.

The duration of feeding periods on the injured organ is an important factor in the cecidogenesis of mite galls. For example, at the beginning of pouch gall cecidogenesis, the mite remains motionless in a sucking position at the same place on the leaf for several hours to 2 days. After that, the mite begins to move about in the small newly formed leaf evagination (Westphal, 1977a,b). In contrast, other gall mites are able to change feeding sites after a short time varying from 30 seconds to 40 minutes (Westphal *et al.* 1981).

CECIDOGENESIS

Diversity of Mite Galls

Mite galls are extremely diversified. This high diversity is most striking when examining the catalogues of Houard (1908, 1922), Docters Van Leeuwen and Docters Van Leeuwen-Reijnvaan (1910, 1918), Mani (1959), and Buhr (1965). It appears that gall-inducing mites are found all over the world. They develop on all soft parts of their host plants: stems, leaves, petioles, buds, flowers, and even the fruit, except for the roots, which apparently are not suitable for mite attack.

Attempts to classify the diversity of acarocecidia have been made by Thomas (1877), Appel (1899), Cosens (1912), and Wells (1921). The diversity and gradation of representative mite galls are illustrated in Figure 9-2. These illustrations are based on morphological and anatomical data from Docters Van Leeuwen and Docters Van Leeuwen-Reijnvaan (1910–1918), Küster (1911), Ross (1932), and Mimeur (1949) and do not imply phylogenic relationships between groups. The illustrations show an increase in complexity and take into consideration the nature of the infested organ and the number of morphogenetic processes (such as inhibition of growth, correlation disturbances, labial growth, production of hairs, differentiation of nutritive tissue and lignified sheath) leading to gall formation.

Leaf Galls

Leaf galls induced by mites are generally localized malformations that are small compared to the great majority of insect galls. This size is related to the minuteness of mites, even though mite galls are inhabited by an entire colony rather than individuals.

Mites that form the simplest galls cause their hosts to produce an excess of hairs (erinea) at the site of attack, which may be uni- or multicellular, lobate, or ramified [Fig. 9-2 (1–7)]. Production of hairs also may be associated with leaf bulging [Fig. 9-2 (8–11)]. Some mites induce the formation of fleshy protuberances with or without hairs [Fig. 9-2 (12–13)], some cause pocken gàlls [Fig. 9-2 (14–15)], and others attack the midribs of the leaf resulting in fold galls [Fig. 9-2 (16–18)], whereas mites injuring the edge of leaves induce marginal leaf roll galls that may or may not be associated with thickenings of leaf tissues or hairs [Fig. 9-2 (19–24)].

The vast variety of pouch galls [Fig. 9-2 (25–43)] illustrates a gradation in the complexity of mite gall organization. Some galls protrude only through the upper side of the leaves [Fig. 9-2 (25,26,28)] whereas others are more or less prominent on the underside of leaves [Fig. 2 (31–32,34–39)]. The gall of *Eriophyes doctersi* Nal. [Fig. 9-2 (33)], for example, is almost invisible on the upper leaf surface. The growth of a labial ring on the under surface is usually associated with the appearance of ostiolar hairs which close the gall opening [Fig. 9-2 (27–30)]. Gall ostioles are usually on the under surface. However, in some cases mites attack the upper surface of the leaf and the galls become visible on both sides [Fig. 9-2 (40–42)]. One peculiar gall on some *Dicrostachys* species (Mimosae) [Fig. 9-2 (43)] results from the growth and coalescence of labial rings on two adjacent leaflets (Schnell, 1952; Jayaraman, 1988).

The gall with the most complex organization is a pouch gall caused by *Eriophyes ulmicola* Nal. on elm [Fig. 9-2 (39)]. This gall has a lignified sheath that develops in the very thick gall wall. This is considered a highly evolved feature (Westphal, 1970), but it remains an exception in mite galls.

Stem Galls

Galls induced by mites on stem tissues [Fig. 9-2 (44–46)] are far less common and diverse than leaf galls. Stem galls are often so closely packed that it becomes difficult to recognize the individual structures of each gall. Attacked twigs often swell and become knotted. The scots pine knots induced by *Trisetacus pini*

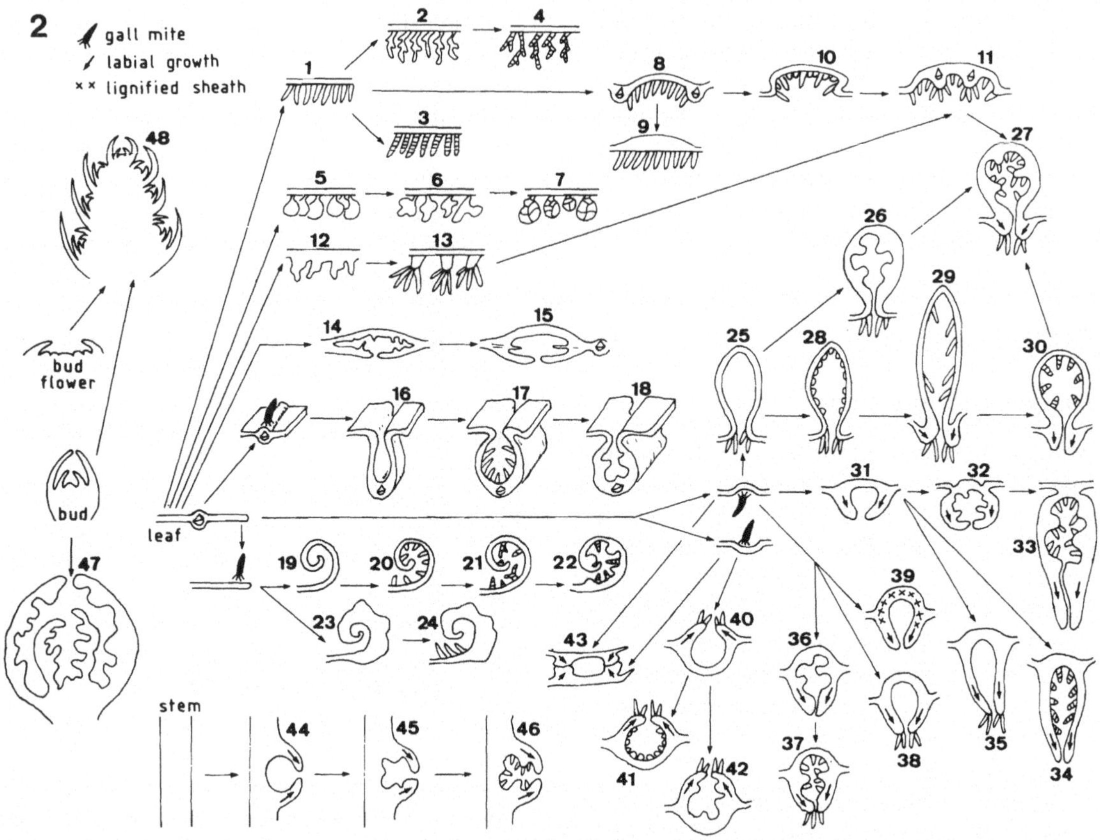

Figure 9-2. Diversity of mite-induced galls presented in order of increasing complexity. (1–43) Leaf galls. (1–7) Erineum. (1) *Phytoptus leiosoma* Nal. (2) *Eriophyjes varius* Nal. (3) *Eriophyes hippocastani* Fockeu. (4) *Eriophyes* sp. on *Uncaria glabrata* D. C. (5) *Phytoptus paderineus* Nal. (6) *Acalitus brevitarsus* Fockeu. (7) *Eriophyes* sp. on *Vitex heterophylla* Roxb. (8–11) Erineum associated with leaf bulging. (8) *Eriophyes laevis inangulis* Nal. (9) *Eriophyes vitis* Paegnst. (10) *Eriophyes* sp. on *Salvadora persica* L. (11) *Eriophyes tristriatus* var. *erinea* Nal. (12–13) Emergences. (12) *Phyllocoptes populi* Nal. (13) *Eriophyes* sp. on *Pterospermum javanicum* Jungh. (14–15) Pocken galls. (14) *Eriophyes filiformis* Nal. (15) *Eriophyes tristriatus* Nal. (16–18) Fold galls. (16) *Eriophyes macrotrichus* Nal. (17) *Phyllocoptes convolvuli* Nal. (18) *Eriophyes xylostei* Can. (19–24) Marginal leaf roll galls. (19) *Epitrimerus trilobus* Nal. (20) *Eriophyes* sp. on *Cudrania spinosa* Kds. (21) *Eriophyes* sp. on *Alyssum spinosum* L. (22) *Eriophyes laticintus* Nal. (23) *Eriophyes peucedani* Can. 24. *Phytocoptella tetratrichus* Nal. (25–43) Pouch galls. (25–30) Galls protruding through the lower surface. (31–35) Galls more visible on the lower surface than the lower surface. (36–39) Galls visible on both surfaces and with an ostiole on the upper surface. (40–42) Galls with an ostiole on the upper surface. (25) *Eriophyes macrorhynchus* Nal. (26) *Eriophyes* sp. on *Salix cinerea* L. (27) *Eriophyes viburni* Nal. (28) *Phytoptus padi* Nal. (29) *Phytoptus tiliae* Pgst. (30) *Phytoptus laevis* Nal. (31) *Eriophyes* sp. on *Haematoxylon campecheanum* L. (32) *Eriophyes* sp. on *Hypaestes floribunda* R. Br. (33) *Eriophyes doctersi* Nal. (34) *Eriophyes* sp. on *Cerbera odollam* Gaertn. (35) *Eriophyes* sp. on *Wrightia javanica* D. C. (36) *Eriophyes fraxinicola* Nal. (37) *Eriophyes* sp. on *Ficus parietalis* Bl. (38) *Eriophyes tetanothrix laevis* Nal. (39) *Eriophyes ulmicola* Nal. (40) *Eriophyes* sp. on *Toona sureni* Merr. (41) *Eriophyes similis* Nal. (42) *Eriophyes* sp. on *Pithecolobium umbellatum* Bth. (43) *Eriophyes dichrostachia* Tuck. (44–46) Stem galls. (44) *Eriophyes heteronyx* Nal. (45) *Trisetacus pini* Nal. (46) *Eriophyes* sp. on *Callicarpa arborea* Roxb. (47) Bud galls or big buds. *Phytocoptella avellanae* Nal. (48) Floral virescences. *Aceria cladophthirus* Nal.

Nal. [Fig. 9-2 (45)] persist for several years (Kruel 1963).

Bud Galls

Buds infested by mites fail to develop true leaves and may become swollen and deformed. Such galls are known as "big buds" [Fig. 9-2 (47)]. One of these big buds develops on filbert after the attack of *Phytocoptella avellanae* Nal. The apex is profoundly modified by this mite as they develop internal swollen scales and leaf primordia with thick warty protuberances. In less seriously infested buds, one or two axillary buds may develop and give rise to distorted shoots with abnormal leaves.

In other cases, considerable disturbances in plant morphogenesis may result after the attack by mites (Arnold 1967, 1968). Some plants react by brooming (bud proliferation and leaf stunting with internodal shortening), occasional stem fasciation, different degrees of floral abnormality (virescence, phyllody, sterility), and excessive budding. These galls [Fig. 9-2 (48)] may appear on most of the buds along the twig or the inflorescence and constitute voluminous malformations. The witch's broom on ash caused by *Eriophyes fraxinivorus* Nal. is of floral or vegetative origin and becomes a hanging mass of 20 cm in length or more.

All galls induced by mites, except the leaf blisters, have nutritive tissues (Westphal 1977a). This tissue is constantly fed on by the growing mite colony. In the erinea, the nutritive tissue appears discontinuous since all epidermal cells do not transform into hairs (Westphal 1975a,b). In some leaf rolls and virescences, the nutritive tissue is localized near the mites (Westphal 1977a). The inner side of other galls may be covered by one or more nutritive layers with or without fleshy protuberances, papillae, or uni- or pluricellular hairs.

Morphogenetic Phenomena in Cecidogenesis

In spite of the large morphological diversity of acarocecidia, gall formation is still governed by a number of factors as strict as those governing normal development of the plant. The first fac-

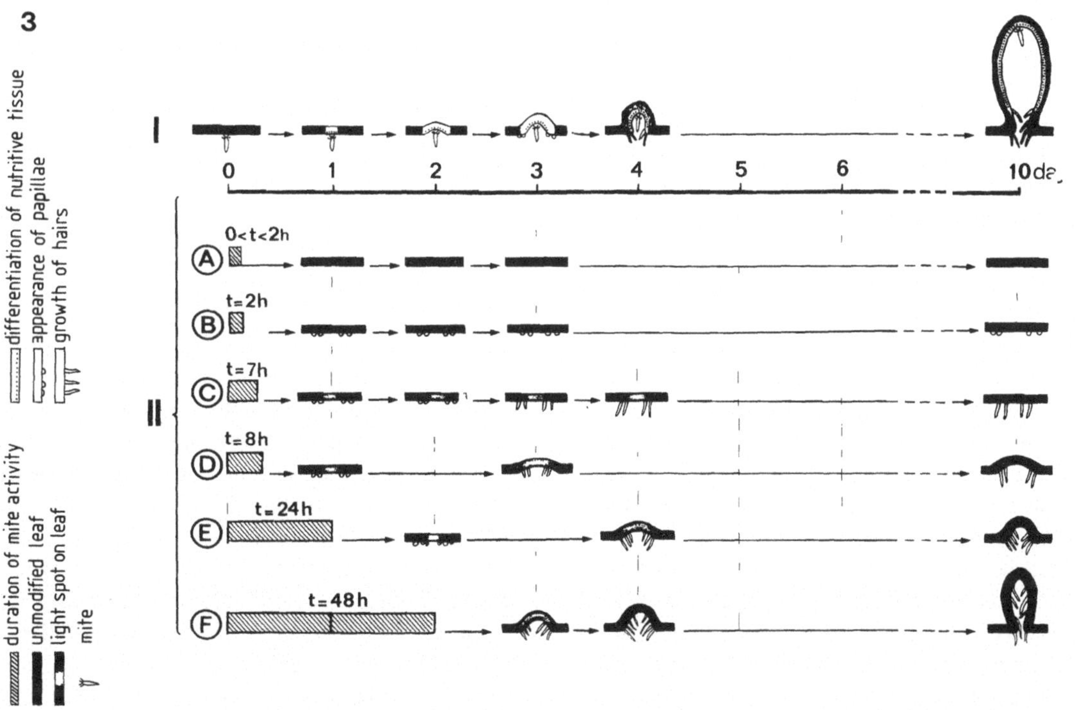

Figure 9-3. Comparison of normal cecidogenesis (I) and interrupted cecidogenesis (II) on *Prunus padus* L. over a period of 10 days of the pouch gall induced by *Phytoptus padi* Nal. When the mite is removed at the initial stage of cecidogenesis (A), there is no visible alteration of the leaf. When the mite has been allowed to feed only briefly (less than 2 hours) and then removed (B), papillae will appear the following day, but they do not develop into hairs. When the mite has been allowed to feed for 7 hours and then removed (C), papillae appear the following day that grow into elongated hairs. A faint light spot is visible on the leaf. When the mite is removed after it has fed for 8 hours (D), papillae and hairs appear, the light spot becomes more pronounced, and the upward bulge begins. When the mite is removed after 24 hours of feeding, the bulge becomes even more pronounced. Removal after 48 hours of feeding (F) results in a small pouch filled with hairs but no nutritive tissue is found. A central hairless area remains at the top of the bulge, a feature already visible in aborted galls B, C, D, and E.

tor involved in the formation of all mite galls is a mechanical action that occurs when the mite inserts its chelicerae into the cell wall for feeding. After this, the instinct and behavior characteristic to each species of mite seem to govern the succession of stimuli received by the plant, which leads to the formation of a characteristic gall. The nature of these stimuli is not known, but it is possible to understand the mechanism of any gall formation by comparing normal cecidogenesis with the development of galls from which the mites have been removed.

Experiments interrupting cecidogenesis in galls of other species (Rohfritsch 1971) clearly demonstrate how gall inducers act by successive inductions and the same can be done for the galls of mites. That is, removing the mites after various intervals following induction and then examining the stage of gall development reached permits one to identify the role of the mite in each stage of cecidogenesis. This technique was used in a study of pouch galls induced by *Phytoptus padi* Nal. (Westphal 1977a,b) and is summarized in Figure 9-3. Results of this study suggest the following sequence in gall induction: (1) inhibition of normal differentiation and/or inhibition of normal growth occurs in the attacked leaves, (2) induction of papillae which then develop into elongated hairs, (3) induction of bulging usually counteracting the previous production of papillae or hairs, (4) production of hairs restricted in the ostiolar region which has not undergone bulging, and (5) induction of the upward growth of the bulge to form the typical pouch. Added to these aspects of induction is a nutritive stimulation that ceases only with the death of the mite.

The sequence of morphological events leading to leaf roll galls is similar to that of pouch galls (Westphal 1977a) and may explain the simultaneous presence of leaf bulging and rolling after attack by *Phytocoptella tetratrichus* Nal., as was described by Schlechtendal (1916) and Buhr (1965).

Each type of gall seems to result from a given succession of stimuli related to particular feeding behaviors of the mite and from the inherent morphological characteristics of the attacked host organ. Complete gall development usually requires several months.

HOST–MITE INTERACTIONS

Research on interactions between gall mites and their host plants is a surprisingly neglected area considering the economic importance of these plant feeders (Jeppson *et al.* 1975; Keifer *et al.* 1982). Factors associated with mite-induced cecidogenesis have been well documented but until recently little was known about the resistance of plant tissues to the attack by gall mites. The only detailed report was on *Aceria cladophthirus* Nal. (previously *Eriophyes cladophthirus*) attacking *Solanum dulcamara* L. (Westphal *et al.* 1981) and this system was chosen here for an analysis of host–mite interactions.

When *A. cladophthirus* (Fig. 9-1) begins feeding on its host, it develops either a compatible or incompatible relationship (Figs. 9-4–9-6). The compatible relationship on susceptible plants leads to typical cecidogenesis (witch's brooms and virescences associated with abnormal pilosity) whereas an incompatible relationship on resistant plants is characterized by the absence of gall development and the formation of necrotic local lesions (Le Ret and Westphal 1980).

Compatible Relationships

Formation of Feeding Punctures

Before mites begin feeding on susceptible plants, they perforate the cell wall and simultaneously initiate the differentiation of nutritive tissue and gall formation. After the cell wall has been perforated, callose thickenings of the cell wall occur around the pit within 10 minutes. These feeding punctures appear cone shaped in longitudinal section (Fig. 9-7) and disc-like on epidermal imprints. Localization of callose in less contrasted areas of the cell wall thickening was confirmed by negative histochemical PAS staining (Westphal 1972). Callose deposits may occur gradually after damage of the cell wall resulting from boring with the stylet-like chelicerae. The rapid synthesis of callose associated with wall thickenings and allowing cell wall restoration seems to be a general nonspecific response of living plant cells to superficial wounding such as microneedle insertion (Nims *et al.* 1967), mechanical bending

Figures 9-4–9-6. A noninfested plant of *Solanum dulcamara* compared with shoots 3 weeks after infestation by *Aceria cladophthirus* on a susceptible plant and a resistant plant. **Figure 9-4.** Noninfested *S. dulcamara.* **Figure 9-5.** Typical symptoms of a galled, susceptible plant of *S. dulcamara.* The leaves are abnormal and there is excessive hairiness. **Figure 9-6.** Resistant *S. dulcamara* showing local necrotic lesions (arrows).

(Aist 1977), fungal infection (Aist 1976; Hinch and Clarke 1982), or eriophyoid mite feeding.

Increased Alkalinization of the Vacuoles of Injured Cells

Mite feeding provokes an early increase in the pH (≥8) of the vacuoles of susceptible epidermal cells of *S. dulcamara,* which can be detected before cytological changes become perceptible (Westphal 1982). This vacuolar alkalinization was demonstrated by the addition of neutral red, which first turns orange and then yellow within less than 1 hour. It is important to note that this physiological change is restricted to the injured cells and does not spread laterally to the adjacent cells in experimental conditions.

Vacuolar alkalinization probably expresses a modification of the permeability of the tonoplast although chelicerae of the mite never reach the vacuole. The chelicereae perforate the cell wall but whether or not the gall mite actually damages the host plasma membrane remains to be discovered. The importance of modifications in permeability has been emphasized by phytopathologists during various host–parasite interactions (Goodmann 1968; Hoppe and Heitefuss 1974; Wheeler 1975).

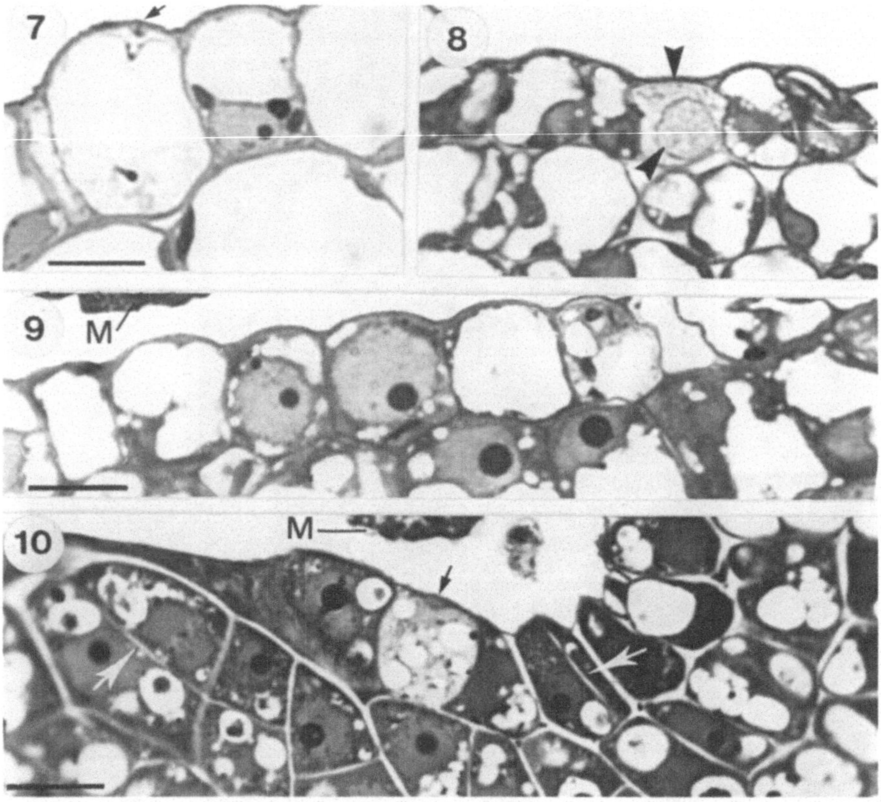

Figures 9-7–9-10. Compatible interaction between the cells of *S. dulcamara* and *Aceria cladophthirus*. The bars represent 10 μm. **Figure 9-7.** Section of epidermis showing injury (arrow) within one hour of the beginning of feeding. **Figure 9-8.** Section of epidermal cells comparing an attacked cell (upper arrow) with normal cells. Note that the nucleus (lower arrow) of the attacked cell is larger and less contrasted. **Figure 9-9.** Nutritive cells after 24 hours of feeding. Note the nuclear and nucleolar hypertrophy in the two cells near the mite (M) and in the underlying cells. **Figure 9-10.** Nutritive cells after 12 days of feeding. Some epidermal cells and underlying parenchyma cells have undergone division (white arrows). The punctured cell (arrow) has a less contrasted cytoplasm.

Cytological Changes

Gall formation occurs on immature but differentiated tissues. As with other gall inducers, mites promote a redetermination of cells. During the first 20 minutes of mite attack, cytological modifications in the injured cells are hardly noticeable except for a slight thickening at the feeding site where deposits of callose restore the cell wall. Normally the epidermal cell contains a small nucleus with a dense nucleoplasm. After mite attack, the nucleus progressively increases in size, moves within 20 minutes to a central position in the injured cell and simultaneously its chromatin becomes less contrasted (Fig. 9-8). These modifications were previously thought to prelude further cytoplasmic changes (Westphal *et al.* 1981) but the results of new investigations led to a reexamination of this interpretation. Indeed, it has been recently proven (Bronner *et al.* 1989) that the structural changes of the nucleus are concomitant with a loss of its affinity for specific DNA-binding reagents and correspond to a typical denaturation of the nuclear DNA. These severe changes occur during the first 45 minutes and appear inconsistent with prolonged survival of the injured cell. How long death of the host cell is delayed remains to be determined more precisely.

Subsequently there is a diffusion of a stimulus from the injured cell to the surrounding ones, which then undergo dedifferentiation into nutritive cells (Fig. 9-9). However, vacuolar alkalinization and nuclear DNA alteration are lacking. It therefore seems that these two modifications may be closely related and must be considered as a consequence of feeding. This implies the existence of mechanical or chemical stimuli, or the combination of both. It is possible that when the chelicerae of the mite bore into the cell wall they release some chitinous fragments into the injured cell, which are responsible for the nuclear alteration. Indeed, chitosan induces similar DNA changes when directly applied on leaves of *S. dulcamara* (Bronner *et al.* 1988). The eventual release of salivary secretions into perforated cells has not been clearly demonstrated; however, a dark granular substance has been found lining the feeding punctures. In some mite galls small vesicles with dark contents have been detected in the cytoplasm surrounding the feeding puncture (Westphal 1977a). Nothing is as yet known about the nature and origin of these dark substances.

Neighboring cells exhibit nuclear hypertrophy before cytoplasmic enrichment and fragmentation of the vacuoles. Thus dedifferentiation affecting the adjacent cells clearly indicates a progressive diffusion of stimuli responsible for gall initiation out from the cells fed on by the mites. Epidermal and parenchymatous cells start to divide later leading to the development of typical nutritive cells (Fig. 9-10).

Ultrastructural observations of nutritive cells of the gall of *A. cladophthirus* show high density of endoplasmic reticulum, ribosomes, and polysomes in the cytoplasm, prominent mitochondria and plastids showing reduced thylakoid system, and a lack of starch grains (Figs. 9-11 and 9-12). Such cytological features are common in most mite galls (Thomsen 1975; Westphal 1977a) and are similar to those of insect galls (see Chapters 5 and 8) and nematode galls (Bird 1961; Jones and Northcote 1972a,b; Wyss 1981). Mite-induced nutritive cells are also typical of cells exhibiting a high metabolic activity.

The influence of mites on susceptible plants seems to be constructive rather than destructive, resulting in cellular and growth reorganization with limited cellular damage. The differentiation of nutritive cells is essential for the maturity of the mites and supplies food for the actively breeding colony.

Incompatible Relationships

The experimental transfer of *A. cladophthirus* to certain plants of *S. dulcamara* does not result in the formation of galls thus demonstrating the existence of plants resistant to the attack of gall mites (Westphal 1980; Westphal *et al.* 1981). Such plants develop malformed leaves and thickened axes (Fig. 9-6) associated with necrotic local lesions.

Development of Necrotic Local Lesions

Macroscopic observations of injured leaves reveal that no changes appear in the first 45 minutes of mite attack. However, a tiny area (50 μm in diameter) of collapsed cells becomes perceptible after 1 hour of infestation. The necrotic lesion gradually enlarges forming a small green depression in the leaf which then browns within 4 hours. After 20 hours, the bordering zone shows radial enlargement and later produces a colorless halo around the brown lesion. However, necrosis remains barely visible (less than 0.5 mm in diameter).

Feeding Punctures

No obvious differences between the feeding behavior of mites on leaves of susceptible or resistant plants have been observed within 1 hour of mite feeding. Mites begin feeding on resistant plants by boring the cell wall but provoke a rapid death of the injured cell mostly without callose deposits. The actual presence of typical callose containing feeding punctures was at first overlooked (Westphal *et al.* 1981) because they were only found in 10 to 25 percent of the well-expanded necrotic lesions. In these cases, it seems that callose synthesis is more rapid than cell disorganization leading to cell collapse.

Vacuolar Alkalinization of Injured Cells

Vacuolar alkalinization also occurs within cells of resistant plants after the walls are perforated; however, this physiological change generally begins later (an average of 24 minutes) than in cells of susceptible plants (an average of 14 minutes) and provokes cellular death. Cell death is concomitant to irreversible plasmolysis, aggregation of cytoplasmic organelles, contraction of the small vesicles having accumulated neutral red, and a blockage of cytoplasmic streaming.

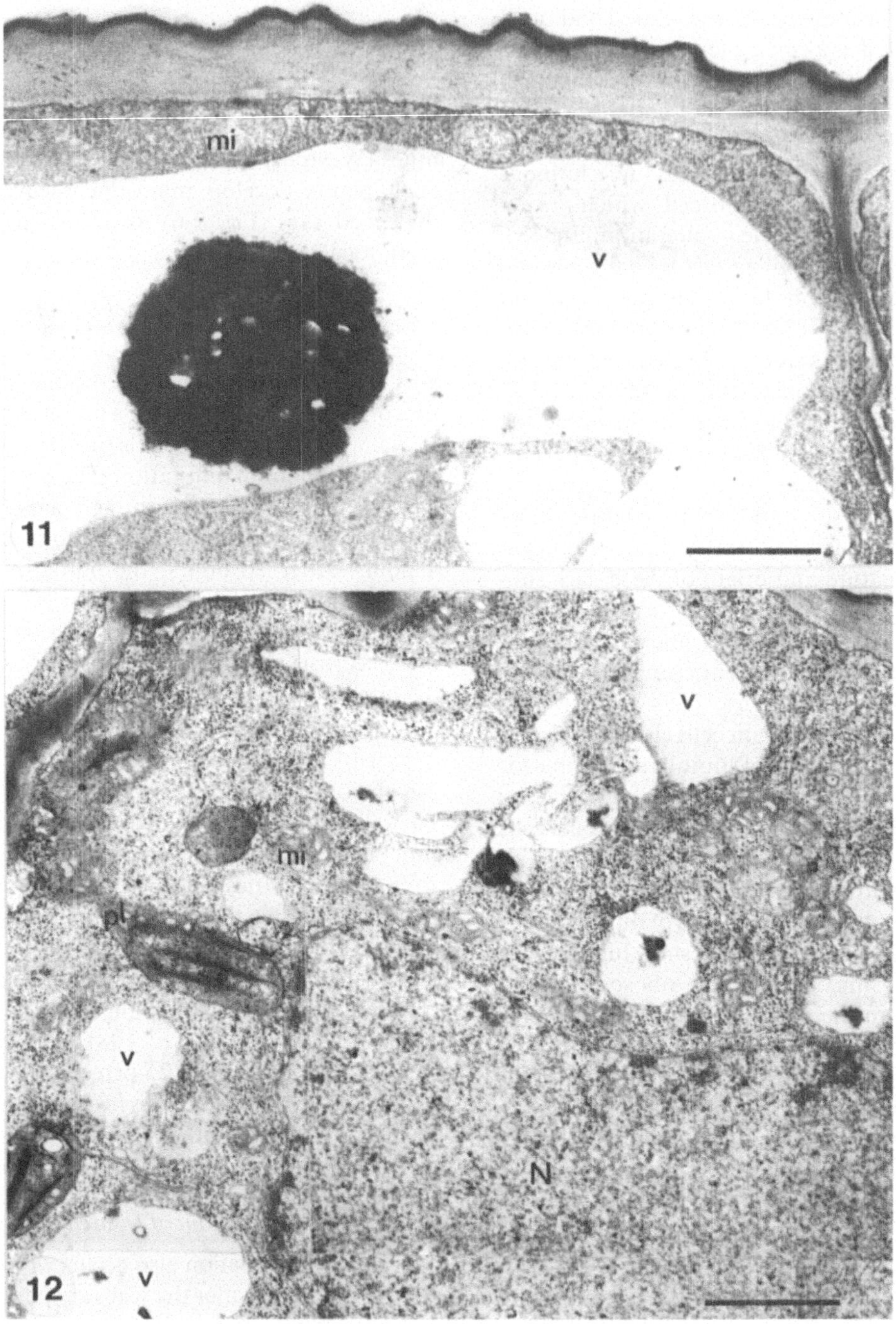

Figures 9-11 and 9-12. Comparison of the ultrastructure of a normal upper epidermal cell with a nutritive cell in a gall of *Aceria cladophthirus* on *S. dulcamara* after 12 days of feeding. **Figure 9-11.** Normal cell. Note that the cytoplasm is parietal and the main vacuole contains dark inclusions (probably alkaloids). **Figure 9-12.** Nutritive cell characterized by a large nucleus, small vacuoles, and abundant organelles in the cytoplasm. Note the reduced thylakoid system in the plastid without starch. The bars represent 1 μm. mi, Mitochondria; N, nucleus; pl, plastid; v, vacuole.

As in compatible relationships, only the punctured cells of the resistant leaf show vacuolar alkalinization, which must be correlated with direct feeding. On the other hand, the small vesicles similar to those described by Mahlberg (1972) and Wilson *et al.* (1980) during neutral red uptake undergo considerable deformation in incompatible interactions.

Cytological Changes

After cell wall perforation, the resistant cell exhibits the same rapid DNA denaturation described in susceptible injured cells (Bronner *et al.* 1989). But in this case, information on cellular death spreads quickly from the initial injured cell to the neighboring ones where pycnotic nuclei are evident within less than 1 hour (Figs. 9-13 and 9-14). After 45 minutes, some epidermal cells are collapsed and highly stained. Mesophyll cells are also affected (Fig. 9-15) leading to the formation of a small depression. Within 1 hour spongy parenchyma cells and lower epidermal cells are killed, but tissues outside the depression remain healthy (Fig. 9-16).

Necrosis is characterized by browning that appears 4 hours after infestation. Although necrosis was detected after 1 hour, the brown deposit was positively stained with phenol sensitive tests after 4 hours (Westphal *et al.* 1981). Such accumulation of products of polyphenolic oxidation is therefore a consequence of cellular

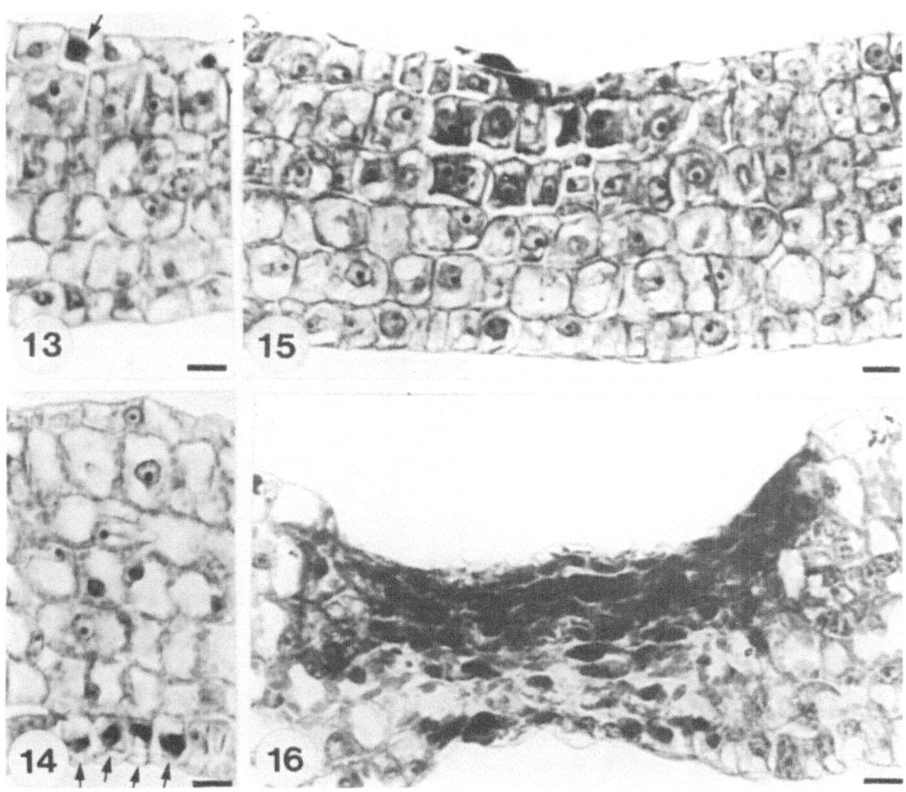

Figures 9-13–9-16. Sequence in the formation of necrotic lesions on a leaf of an incompatible *S. dulcamara* attacked either on its upper or lower side by *Aceria cladophthirus.* **Figure 9-13.** Section of a leaf showing upper epidermal cell modifications within 10 minutes of attack by the mite. Cells are slightly plasmolyzed with a pycnotic nucleus (arrow). **Figure 9-14.** Cells of the lower epidermis (arrows) showing symptoms similar to those of the upper epidermis. **Figure 9-15.** Section of a leaf 45 minutes after the beginning of an attack on the upper side. The mesophyll cells are plasmolyzed and some epidermal cells (near the depression) are necrotic. **Figure 9-16.** Section of attacked leaf 4 hours after the beginning of feeding. Note that necrosis of the epidermal cells has extended to the mesophyll and lower epidermis. The bars represent 10 μm.

death and could not be responsible for necrogenesis. The cells bordering necrotic tissues later show large amounts of callose.

Ultrastructural observations of necrotic lesions confirmed that the rapid disorganization of cell contents is characterized by a drastic detachment of the plasmalemma from the cell wall (Fig. 9-17), the appearance of small vesicles with irregular outlines (Figs. 9-17 and 9-18), the distension of the nuclear membrane with shrinkage of the darkly stained nucleoplasm (Fig. 9-17), and an increase in contrast of the cytoplasm and its organelles. Strong cell wall modifications also occur in the necrotic areas. The sharpness of cellular outlines disappears and the texture of the cuticle is modified. These observations are consistent with other descriptions of incompatible host–parasite relation-

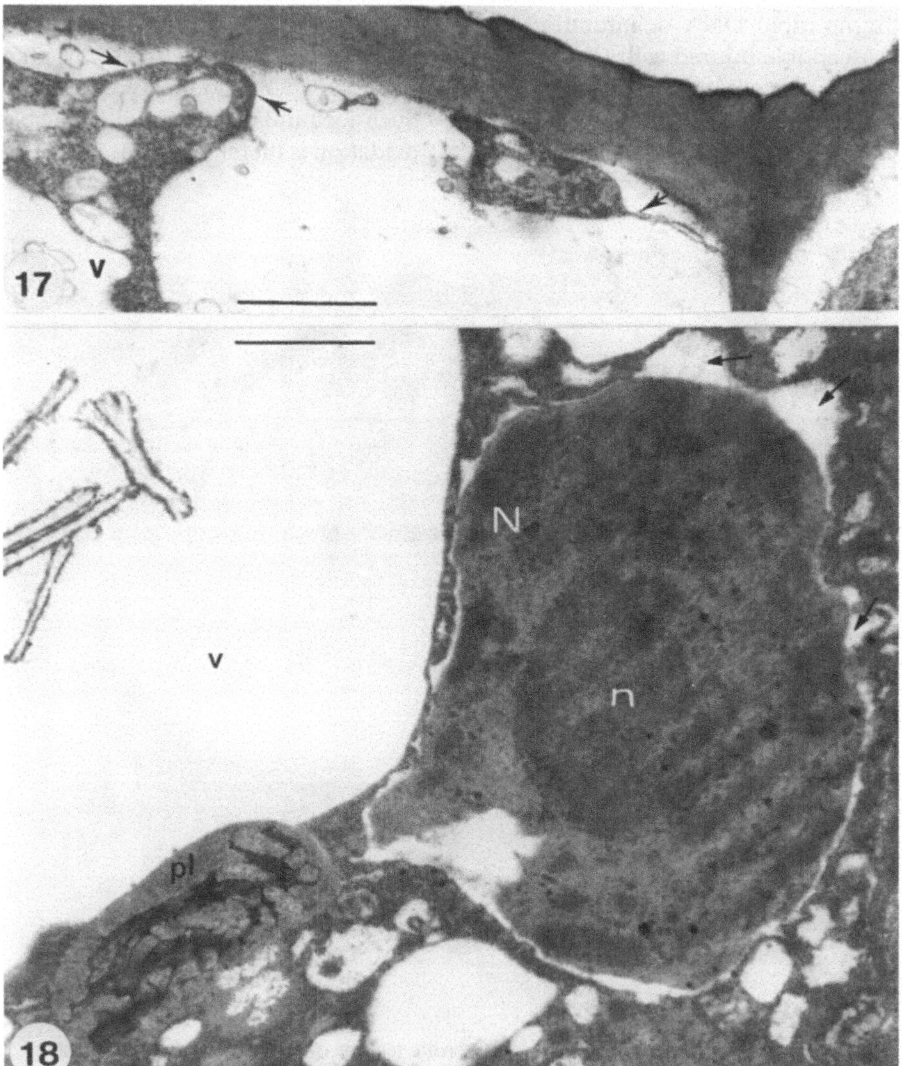

Figures 9-17 and 9-18. Ultrastructure of upper epidermal cells of a resistant *S. dulcamara* attacked by *Aceria cladophthirus.* **Figure 9-17.** Detachment of plasmalemma from the cell wall (arrows) 1 hour after the mite's attack. Note the dark stained cytoplasm and the numerous small vesicles. **Figure 9-18.** Cell after 2 hours of the mite's attack. Note the disorganization and increased staining of the cytoplasm and nucleus and the swelling of the nuclear membrane (arrows). The bars represent 1 μm. N, Nucleus; n, nucleolus; pl, plastid; v, vacuole.

ships (Paulson and Webster, 1972; Lazarovits and Higgins 1976) and confirm the rapidity of the necrotic process (Jones and Deverall 1977; MacLean and Tommerup 1979; Nozue *et al.* 1979). Early modifications of membrane permeability associated with electrolyte leakage are universally associated with plant–parasite interactions (Wheeler and Hanchey 1968; Weststeijn 1978). Changes in permeability of the plasmalemma and/or the tonoplast may explain the rapid plasmolysis observed in early stages of necrosis. This is an irreversible event and seems to play a crucial role in necrogenesis. It causes the loss of water and ions, which is a common phenomenon in phytopathology (Sigee and Epton 1976; Abawi *et al.* 1977). The cause of cell death remains unknown but the failure of nutritive cells to develop on resistant plants impedes mite development and reproduction.

Necrogenesis occurring during the incompatible interaction between *A. cladophthirus* and *S. dulcamara* is an extremely rapid process and leads to the formation of local lesions having a sharp demarcation between dead and apparently healthy tissues where callose deposits occur. Recent investigations (Westphal and Bronner 1987) demonstrated the accumulation of autofluorescent components (phytoalexine-like substances) in the lesion bordering cells and the presence of pathogenesis related proteins. These results demonstrate that the local necrotic lesions induced by *A. cladophthirus* are similar in certain aspects to the well-known hypersensitive reaction of plants to invasion by viruses, bacteria, fungi, or nematodes during incompatible plant–parasite interactions.

CONCLUSIONS

The eriophyoid mites are probably the smallest animals inducing galls. A single female induces a very large gall compared with the size of its body, but in a few weeks the gall becomes fully occupied by its offspring. However, in comparison with insect galls, mite galls are usually smaller and more discrete structures especially when they develop on host organs like leaves characterized by their limited growth. But when eriophyoid mites attack vegetative or floral buds, the galls become voluminous malformations without definite limits where they give rise to a large population of mites. Thus it seems that a good correlation exists between the final size of mite galls and their mite population.

Galls induced by mites provide microhabitats that are favorable to the development of other mites such as inquilines or predators. The best example is the well-known filbert big bud gall induced by *Phytocoptella avellanae* in which the common inquiline mite *Cecidophyopsis vermiformis* Nal. finds shelter and develops (Jeppson *et al.* 1975). On the other hand, the shape and size of some galls may be modified when predatory mites (Phytoseiidae) attack the population of gall mites.

In spite of the large diversity in shape and structure of the mite galls it is possible to disclose some common factors that govern their formation. During the succession of the inductions given by the mites, differentiation of typical nutritive cells usually occurs as long as these cells are fed upon by the mite colony. When feeding, mites provoke the formation of callose containing feeding punctures on the host cell walls. Although these structures are one of the most striking characteristics of acarocecidia (Němec 1924; Westphal 1977a), the callose deposit itself seems to be a nonspecific response of plant cells to superficial woundings.

The investigations on relationships between gall mites and their host plants have mostly focused on the compatible interactions leading to progressive and striking gall formation. The large gap in our knowledge of resistance of plants to gall-inducing mites is perhaps due to the extreme rapidity of the process and to the small size of the necrotic lesions as was observed in the *A. cladophthirus*–*S. dulcamara* interaction. In this case, preliminary results indicated that the gene for hypersensitivity of *S. dulcamara* is dominant and segregates according to simple mendelian principles. This may explain why susceptible plants are relatively scarce in a population of *S. dulcamara*. However, more investigations on incompatible interactions between other gall mites and their host plants, and the resistance processes that ensue, are badly needed to add details to the generalizations presented in this chapter.

LITERATURE CITED

Abawi, G. S., D. C. Crosier, and A. C. Cobb, 1977. Pod flecking of snap beans caused by *Alternaria alternata.* Plant Disease Reporter. 61:901–905.

Aist, J. R. 1976. Papillae and related wound plugs of plant cells. Annu. Rev. Phytopathol. 14:145–163.

Aist, J. R. 1977. Mechanically induced wall appositions of plant cells can prevent penetration by a parasitic fungus. Science 197:568–571.

Appel, O. 1899. Uber Phyto- und Zoomorphosen. Doctorat Kgl. Bayer, Julius-Maximilians Universität, Würzburg, pp. 1–58.

Arnold, B. C. 1967. A hitherto unrecorded mite gall on *Carmichaelia violacea* Kirk. Adv. Frontiers Plant Sci. 18:1–6.

Arnold, B. C. 1968. Origin and structure of mite-induced witches' brooms of *Plagianthus betulinus* A. Cunn. Phytomorphology 18:63–66.

Bird, A. F. 1961. The ultrastructure and histochemistry of nematode induced giant cells. J. Biophys. Biochem. Cytol. 11:705–715.

Bronner, R., F. Dreger, and E. Westphal, 1988. Rôle du chitosan dans les réactions de défense de *Solanum dulcamara* attaqué par un acarien. In: La paroi fongique, Proceedings of Symposium, Soc. Fr. Microbiologie, 24 Nov. 1988, Paris.

Bronner, R., E. Westphal and F. Dreger. 1989. Chitosan, a component of the compatible interaction between *Solanum dulcamara* L. and the gall mite *Eriophyes cladophthirus* Nal. (Acarina, Eriophyoidea). Physiol. Mol. Plant Pathol. 34:117–130.

Buhr, H. 1965. Bestimmungstabellen der Gallen (Zoo- und Phytocecidien) an Pflanzen Mittel- und Nordeuropas. I and II. G. Fischer, Jena.

Bumaldi, J. A. (Montallanus Ovidius). 1671. Curae analyticae aliquot naturalium observationum aldrovandicas circa historias. Bononiae.

Cosens, A. 1912. A contribution to the morphology and biology of insect galls. Trans. Can. Inst. 9:297–387.

Docters Van Leeuwen, W., and J. Docters Van. Leeuwen-Reijnvaan. 1910. Beiträge zur Kenntniss der Gallen aus Java. Ann. J. Bot. Buitenzorg. 2 e sér. VII. 2:119–183.

Docters Van Leeuwen, W., and J. Docters Van Leeuwen-Reijnvaan. 1914. Einige Gallen aus Java. Bull. J. Bot. Buitenzorg. 2e sér. XV. 1–68.

Docters Van Leeuwen, W., and J. Docters Van Leeuwen-Reijnvaan. 1916. Niederländisch ostindische Gallen No. 8. Beschreibungen von Gallen aus. Sumatra und Simaloer. Bull. J. Bot. Buitenzorg. 2e sér. XXI. 1–45.

Docters Van Leeuwen, W., and J. Docters Van Leeuwen-Reijnvaan. 1918. Niederländische ostindische Gallen. No. 10. Einige Gallen aus Java. Bull. J. Bot. Buitenzorg. 3e sér. I. 1:17–76.

Fée, A. 1834. Mémoire sur le groupe des phylleriées et notamment sur le genre *Erineum.* Paris, Strasbourg. Zeit. Naturwiss. Halle 55:486.

Frauenfeld, G. 1864. Beitrag zur Metamorphosen. Geschichte aus dem Jahre. III. Verschiedene Metamorphosen. Verhandl. d. k. k. zool. bot. Gesellschaft. Wien 13:1230–1232.

Goodmann, R. N. 1968. The hypersensitive reaction in tobacco, a reflection of changes in host cell permeability. Phytopathology 58:872–873.

Hinch, J. M., and A. E. Clarke, 1982. Callose formation in *Zea mays* as a response to infection with *Phytophtora cinnamomi.* Physiol. Plant Pathol. 21:113–124.

Hoppe, H. H., and R. Heitefuss. 1974. Permeability and membrane lipid metabolism of *Phaseolus vulgaris* infected with *Uromyces phaseoli* I. Changes in efflux of cell constituents. Physiol. Plant Pathol. 4:5–9.

Houard, C. 1908. Les zoocécidies des plantes d'Europe et du Bassin de la Méditerranée, 3 vols., Hermann, Paris.

Houard, C. 1922. Zoocécidies des plantes d'Afrique, d'Asie et d'Océanie. Hermann, Paris.

Jayaraman, P. 1988. Anatomical study on mite induced galls on *Dichrostachys cinerea.* J. Indian Bot. Soc. 67:145–148.

Jeppson, L. R., H. H. Keifer, and E. W. Baker. 1975. Mites Injurious to Economic Plants. Univ. California Press, Berkeley.

Jones, D. R., and B. J. Deverall. 1977. The effect of the Lr 20 resistance gene in wheat on the development of leaf rust *Puccinia recondita.* Physiol. Plant Pathol. 10:275–284.

Jones, M. G. K., and D. H. Northcote. 1972a. Nematode-induced syncytium: A multinucleate transfer cell. J. Cell Sci. 10:789–809.

Jones, M. G. K., and D. H. Northcote. 1972b. Multinucleate transfer cells induced in *Coleus* roots by the root-knot nematode, *Meloidogyne arenaria.* Protoplasma 15:381–395.

Keifer, H. H., E. W. Baker, T. Kono, M. Delfinado, and W. E. Styer. 1982. An illustrated guide to plant abnormalities caused by eriophyoid mites in North America. United States Department of Agriculture. Agriculture Handbook No. 573.

Kieffer, J. J. 1885. Ueber lothringische und zum Theil neue Phytoptocecidien. Z. Naturwiss. Halle 58:113–133.

Krantz, G. W. 1973. Observations on the morphology and behavior in the filbert rust mite, *Aculus comatus* (Prostigmata: Eriophyoidea) in Oregon. Ann. Entomol. Soc. Am. 66:709–717.

Kruel, W. 1963. Gallmilben an Kiefer. Beitr. Entomol. 13:566–567.

Küster, E. 1911. Die Gallen der Pflanzen. Hirzel, Leipzig.

Lazarovits, G., and V. J. Higgins. 1976. Histological comparison of *Cladosporium fulvum* race 1 on immune, resistant, and susceptible tomato varieties. Can. J. Bot. 54:224–234.

Le Ret, M., and E. Westphal. 1980. Mise en évidence

d'un phénomène de résistance à l'attaque parasitaire d' *Eriophyes cladophthirus* Nal. chez *Solanum dulcamara* L. Bull. Soc. Bot. Fr. 127:184.

Löw, F. 1894. Uber Milbengallen (Acarocecidien) der Wiener Gegend. Verhandl. Zool. Bot. Gesellschaft. 24:495–508.

MacLean, D. J., and I. C. Tommerup. 1979. Histopathology and physiology of compatibility and incompatibility between lettuce and the downy mildew fungus *Bremia lactucae* Rege. Physiol. Plant Pathol. 14:291–312.

Mahlberg, P. G. 1972. Localization of neutral red in lysosomes structures in hair cells of *Tradescantia virginiana.* Can. J. Bot. 50:851–859.

Mani, M. S. 1959. Cecidotheca Indica. Agra Univ. J. Res. Sci. 8:91–220.

Meyer, J., and H. J. Maresquelle. 1983. Anatomie Des Galles. Gebrüder Borntraeger, Stuttgart.

Mimeur, J. M. 1949. Contribution à l'étude des zoocécidies du Maroc. Encyclopédie Entomol. sér. A. 24:1–259.

Nalepa, A. 1911. Eriophyiden. Gallenmilben. Zoologica (Stuttgart) 24(61):166–293.

Nalepa, A. 1929. Neuer Katalog der bisher beschriebenen Gallmilben. Marcellia 25:67–183.

Němec, B. 1924. Untersuchung über Eriophyidengallen. Studies from the Plant Physiological Laboratory of Charles University, Prague. V. 2:47–94.

Nims, R., R. S. Halliwell, and D. W. Rosberg. 1967. Wound healing in cultured tobacco cells following microinjection. Protoplasma 64:305–314.

Nozue, M., K. Tomiyama, and N. Doke. 1979. Evidence for adherence of host plasmalemma to infecting hyphae of both compatible and incompatible races of *Phytophthora infestans.* Physiol. Plant Pathol. 15:111–115.

Oldfield, G. N. 1969. The biology and morphology of *Eriophyes emarginatae,* a *Prunus* finger gall mite and notes on *E. prunidemissae.* Ann. Entomol. Soc. Am. 62:269–277.

Paulson, R. E., and J. M. Webster. 1972. Ultrastructure of the hypersensitive reaction in roots of tomato *Lycopersicon esculentum* L., to infection by the root-knot nematode, *Meloidogyne incognita.* Physiol. Plant Pathol. 2:277–234.

Réaumur (De), R. A. 1737. Mémoires pour servir à l'histoire des insectes. III. 12*e* Mémoire Paris, pp. 413–532.

Rohfritsch, O. 1971. Développement cécidien et rôle du parasite dans quelques galles d'Arthropodes. Marcellia 37:233–339.

Ross, H. 1932. Praktikum der Gallenkunde. Springer, Berlin.

Schlechtendal (von), D. H. R. 1916. Eriophyidocecidien. Die Zoocecidien Deutschlands und ihre Bewohner. Bibliotheca Zool. 24:295–498.

Schnell, R. 1952. Contribution à l'étude des cécidies de l'Afrique occidentale. Mém. Inst. Fr. d'Afrique Noire. Mélanges Bot. 18:239–334.

Siebold (von), C. T. 1850. Uber Eriophyes. 2. Ber. über die Arb. der Entomol. Sect. Jahresbericht d. schles. Gesellschaft fur Vaterl. Cultur (Breslau). 28:88–89.

Sigee, D. C., and H. A. S. Epton. 1976. Ultrastructural changes in resistant and susceptible varieties of *Phaseolus vulgaris* following artificial inoculation with *Pseudomonas phaseolicola.* Physiol. Plant Pathol. 9:1–8.

Thomas, F. 1877. Aeltere und neue Beobachtungen über Phytoptocecidien. Zeit. ges. Naturwiss. 49:329–388.

Thomsen, J. 1975. Development and histology of galls on *Tilia platyphylla* caused by *Eriophyes tiliae tiliae.* Bot. Tidsskrift. 69:262–270.

Tournefort, J. P. 1698. Histoire des plantes qui naissent aux environs de Paris. G.A. Printzel. Thesaurus literaturea botanicae. Ed. Nova. 4e, Lipsiae, 1872.

Turpin, P. J. 1833. Sur le développement des galles corniculées de tilleul. Nouveau Bull. des Sc. par la Soc. Philomatique de Paris, 163–165.

Vallot, J. N. 1820. Note sur quelques maladies des végétaux, indépendantes de la présence des insectes et des cryptogames. Mémoires de l'Académie des Sciences, Arts et Belles Lettres de Dijon, 42–64.

Wells, B. W. 1921. Evolution of zoocecidia. Bot. Gaz. 71:358–377.

Westphal, E. 1970. Quelques aspects ultrastructuraux, cytologiques et histochimiques de la galle en bourse d' *Eriophyes ulmicola ulmicola* Nal. sur les feuilles d' *Ulmus campestris* L. Marcellia 36:199–215.

Westphal, E. 1972. Traces de succion parasitaire laissées par quelques Eriophyides cécidogènes. Aspects histochimiques et observations ultrastructurales. Marcellia 37:53–69.

Westphal, E. 1975a. Observations sur le développement et l'ultrastructure de quelques érinoses. I. Erinose provoquée par l'*Eriophyes leiosoma* Nal. sur le *Tilia intermedia* D. C. Marcellia 38:197–209.

Westphal, E. 1975b. Observations sur le développement et l'ultrastructure de quelques érinoses. II. Erinose provoquée par l' *Eriophyes paderineus* L. sur les feuilles du *Prunus padus* L. Marcellia 38:211–221.

Westphal, E. 1977a. Morphogenèse, ultrastructure et étiologie de quelques galles d'Eriophyides (Acariens). Marcellia 39:193–375.

Westphal, E. 1977b. Sequence of early changes in the leaf epidermis of *Prunus padus* L. under the influence of brief cecidogenic activity of eriophyid mites. Marcellia 40:151–157.

Westphal, E. 1980. Responses of some Solanaceae to attack by the gall mite *Eriophyes cladophthirus.* Plant Dis. 64:406–409.

Westphal, E. 1982. Modification du pH vacuolaire des cellules épidermiques foliaires de *Solanum dulcamara* L. soumises à l'action d'un acarien cécidogene. Can. J. Bot. 60:2882–2888.

Westphal, E., R. Bronner, and M. Le Ret. 1981. Changes in leaves of susceptible and resistant *Solanum dulcamara* infested by the gall mite *Eriophyes cladophthirus* (Acarina, Eriophyoidea). Can. J. Bot. 59:875–882.

Westphal, E., and R. Bronner. 1987. Quelques aspects de la résistance par hypersensibilité de *Solanum dulcamara* à l'attaque cécidogène d'un Acarien *(Eriophyes cladophthirus)*. In Proceedings ler Congrès Soc. Fr. Phytopathologie, 19–20 Nov. 1987. Rennes, France.

Weststeijn, E. A. 1978. Permeability changes in the hypersensitive reaction of *Nicotiana tabacum* cv. Xanthi nc. after infection with tobacco mosaic virus. Physiol. Plant Pathol. 13:253–258.

Wheeler, H. E. 1975. Plant Pathogenesis. Adv. Ser. in Agricultural Science. 2. Springer-Verlag, Berlin.

Wheeler, H. E., and P. Hanchey. 1968. Permeability phenomena in plant disease. Annu. Rev. Phytopathol. 6:331–350.

Wilson, C. L., G. A. Jumper, and D. L. Mason. 1980. Vacuole dynamics in fungal plant pathogens. Phytopathology 70:783–788.

Wyss, U. 1981. Ectoparasitic root nematodes: feeding behavior and plant cell responses. In B. M. Zuckerman and R. A. Rohde (eds.), Plant Parasitic Nematodes, pp. 325–351. Academic Press, New York.

10

Insect Secretions and Their Effect on Plant Growth, with Special Reference to Hemipterans

K. HORI

Galls and other symptoms of disease that develop on host plants under the influence of insects have stimulated many entomologists and plant pathologists to conduct investigations into the causes of gall formation. Up to the end of the nineteenth century, four main hypotheses had been suggested: (1) the injection of a stimulant fluid at the time of oviposition, (2) the mechanical irritation set up by the presence of a foreign body, (3) the secretion of an active saliva, and (4) the excretion of metabolic products (Carter 1962). Plumb (1953) and Carter (1962) did not support the mechanical theory, which several workers had maintained, and Felt (1936) stated that few galls were produced by mechanical injury to growing parts of plants. In the meantime, it was confirmed that artificial galls or symptoms of disease can be produced in host plants by injecting extracts of whole insects or parts of them. The results of such experiments have led many investigators to support the chemical theory.

The purpose of this chapter is to review the literature on the causes of gall formation paying particular attention to the plant-piercing-sucking gall insects in the Orders Hemiptera and Homoptera. Among these groups, Anders (1958) and Kloft (1960) maintained that amino acids injected along with the saliva of insects into plant tissues play a main role as the gall-inducing factor, whereas other investigators ascribed the induction of galls to plant-growth substances (auxin etc.) in the saliva. Schäller (1968a,b) pointed out that indole-3-acetic acid (IAA) in the insect saliva is an important factor in gall formation but that amino acids in the saliva may also play a role. Besides these, several additional hypotheses have been presented for other gall insects.

HEMIPTERANS AND HOMOPTERANS

Salivary Secretion

It is believed that plant-piercing-sucking insects and mites inject phytotoxic saliva at some time while feeding and that this saliva causes the production of galls and other malformations. Furthermore, it is known that the saliva or its components diffuse quickly from the feeding sites to surrounding tissues where they disturb metabolic and hormonal balances. Akimoto (1981) found that the fundatrix of a gall-inducing *Eriosoma* species settled on a young elm shoot and induced a gall on a leaf a little away from its feeding site. Some components in the saliva are derived from the food plants and are transmitted back via the hemolymph. The salivary glands of hemipterans and homopterans have an excretory function and contain a high level of amino acids (especially those nonessential to insects), and some other substances (Goodchild 1952; Anders 1958; Nuorteva and Laurema 1961; Schäller 1968a; Hagley 1969; Miles 1972; Rilling *et al.* 1974). Hemipteran and homopteran insects may owe their phytopathogenicity mainly to the excretory function of the salivary gland.

Amino Acids

The secretion of salivary glands of the plant-piercing-sucking insects and mites, especially of gall inducers, contains many kinds and large amounts of amino acids. These may pass directly from the food to the salivary gland through the hemolymph (Nuorteva and Laurema 1961; Miles 1968b). Up to 15 free amino acids that are either nonmetabolic or in excess in the hemolymph have been detected in the salivary gland secretion (Miles 1968b). Schäller (1968a) confirmed that in the saliva of *Viteus vitifolii* Fitch., glutamine, asparagine, serine, alanine, and aspartic acid were the main constituents and that serine, alanine, and aspartic acid seem to be contained in a higher level in the saliva of gall-inducing aphids. Moreover, Schäller (1963, 1968a) found that in aphids the composition of amino acids in the saliva was qualitatively and quantitatively specific to the species or the races. For example, leucine and valine were characteristic of fruit-feeding aphids, β-alanine of *Myzus ascalonicus* Doncaster, cystathion and cysteic acid of *Myzus cerasi* Fabr., γ-aminobutyric acid and methionine sulfoxide of *Megoura viciae* Bukton, threonine of *Viteus vitifolii,* and tyrosine of *V. vitifolii* and *Aphis sambuci* L. The concentration of total amino acids was highest in the saliva of *V. vitifolii* followed by *M. cerasi, Aphis pomi* De Geer, *Cryptomyzus ribis* L., *M. ascalonicus, A. sambuci, Sappaphis mali* Ferrari, *Eriosoma lanigerum* Hausm., *Hyalopterus pruni* Geoffr., and *M. viciae* in descending order (Schäller 1968a). On the other hand, leucine, valine, tyrosine, alanine, threonine, glycine, serine, and aspartic acid were contained in equal concentration in the saliva of all eight races of the grape phylloxera, but asparagine, glutamine, arginine, lysine, histidine, and glutamic acid differed quantitatively among the races (Schäller 1963). However, in the case of polyphagous homopterans, the chemical composition in the salivary glands may differ between individuals that have fed on different host plants and their phytopathogenicity may also vary (Nuorteva 1962).

Amino acids in the saliva of the plant-piercing-sucking insects can disturb metabolism in plant tissues to cause various injuries on plants (Eyer 1937; Kloft and Ehrhardt 1959; Kloft 1960; Marek 1961). Anders (1957) succeeded in producing nodules on vine roots by raising the seedlings in solutions of amino acids (especially tryptophan, histidine, and glutamic acid solution—singly or in combination). Anders asserted that free amino acids in the saliva were of importance as the cecidogenous agents of insects. Later Anders (1958) developed his work and confirmed that tryptophan induced swelling on the root, that a mixture of glutamic acid, histidine, and tryptophan formed galls very similar to natural galls produced by *Viteus vitifolii,* and that lysine and valine intensified this effect. In conclusion, he presumed that tryptophan was the strongest agent in cecidogenesis because it was destined to be completely excreted. Schäller (1968b) was successful in producing nodules on vine roots by the same kind of amino acid solutions used by Anders, though the form was not similar to the natural galls made by *V. vitifolii.* Moreover, Schäller could produce galls more closely resembling the natural galls on the root by treatment with an amino acid–IAA mixture. Since then, Schäller has been confident that IAA is also one of the gall inducers. On the contrary, Miles (1968a) reported that nodules could not be formed on roots developing from vine cuttings by means of the amino acid solutions of Anders but could be produced by potassium dihydrogen phosphate buffer solution alone (pH 8.0). Miles (1972) stated that perhaps the solution of other compounds that affected the permeability of the cells in the meristematic subapical region of the rootlet would cause similar growth disturbances. Schäller (1968b) did not agree with this idea and pointed out that since pure amino acid solutions without buffer induced the nodules, amino acids in the saliva might take part in gall induction. Amino acids apparently do not play a principal role in the formation, because the nodules were produced by the buffer solution alone as confirmed by Miles.

It is known, however, that amino acids greatly affect plant growth at certain concentrations. Moreover, amino acids in the saliva increase the permeability of cells of attacked plants to raise the rates of respiration, transpiration, and degree of protoplasmic streaming, and reduce the rates of photosynthesis and assimilation (Kloft 1960; Marek 1961). These conditions in the cells are typical of young leaves of plants. In general, galls are developed in young growing plant tissues by stimulation of plant cells in a meristematic or plastic condition. Thus amino acids in the saliva injected into plant tissues should make the tissues more plastic in condition and raise their sensitivity to a gall inducer. It is suspected that galls are more easily induced in such tissues. Amino acids in the saliva might therefore function as "conditioners" for gall induction by plant-piercing-sucking insects. The larvae of *Eurosta solidaginis* Fitch secrete into the host plant tissues substances that maintain a condition similar to that in the growing point of the stem (Beck 1947; Uhler *et al.* 1971). Norris (1979) stated that some substances conditioned plant cells to respond to inducers of tumorigenesis. Such results indicate that plant tissues might have to undergo some essential conditioning, possibly involving amino acids, before they become susceptible to gall induction.

Auxin (IAA etc.)

It is known that many species of aphids and some other homopterans contain auxins, IAA, or plant growth-promoting substances in their salivary glands or bodies. In some insects, auxins seem to be transfered from the food plants to the salivary glands through the hemolymph (Nuorteva 1956; Maxwell and Painter 1962b; Schäller 1965; Miles 1967; Hori *et al.* 1979), while other insects seem to produce IAA in the process of injection of the saliva into the host plant tissues. Duspiva (1954) and Miles (1968a) confirmed that in vitro IAA was produced from tryptophan in a mixture of tryptophan and a homogenate of salivary glands of an aphid or lygaeid bug. Scott (1970) suggested that lygus bugs also had the potential to produce IAA in their saliva. Miles (1968a) indicated that DOPA in the salivary gland, derived from phenylalanine or tyrosine in the hemolymph, could be converted into quinones by the function of salivary polyphenol oxidase and that, under the influence of quinones, tryptophan in the saliva could be converted into IAA during salivation or after the injection of the saliva into host plant tissues. Alternatively, IAA could be produced in the plant tissues from tryptophan injected with the saliva by the action of the phenol–phenol oxidase system of the plants (Miles 1969).

IAA in the saliva of the plant-piercing-sucking insects might be concerned in the formation of various malformations and galls on the host plants. There have been a number of successful as well as unsuccessful attempts at the production of artificial galls by IAA treatment. Successful imitations of cecidogenous processes seem to be obtained only when IAA is given to plant tissues in a correct way under strict conditions. Spectacular results are sometimes attained by the injection of synthetic auxins (Mandl 1957), probably because unnatural auxins are not under the control of the natural inhibitors of IAA activity (Kefeli and Turetskaya 1965). From these and other experimental results, many investigators have believed IAA or auxins in the saliva to be gall-inducing substances. Nysterakis (1948) reported that some aphids (*Viteus vitifolii, Anuraphis helichrysikait Kait., Pemphigus filaginis* Fonsc., and *P. bursarius* L.) secreted IAA by which galls could be produced, but that inhibition of plant growth occurred at a high concentration of IAA. Kloft (1951a) found that extracts from the San Jose scale, a nongall inducer, and from a female gall-inducing aphid, exhibited a growth promotion of the host plant at low concentration but a growth inhibition at high concentration, and stated that this relation was quite similar to the action of IAA. Schäller (1968a,b) inferred from a series of experiments that IAA and amino acids in the saliva took part in gall induction by *Viteus vitifolii* and he was successful in obtaining galls on the petioles of vines by the continuous injection of an artificial saliva including IAA and amino acids. In this case, the more the composition of the artificial saliva resembled that of the natural saliva, the more complete the induced galls were. The saliva contained a high level of IAA and amino acids and also a low level of phenolic compounds that appeared to inhibit gall induction. Thus, Schäller (1968b) concluded that IAA and amino acids in the saliva of aphids contributed to gall induction.

Miles and Lloyd (1967) found that a non-gall-inducing lygaeid, *Elasmolomus sordidus*

F., produced a gall-like swelling on the seedlings of sunflower by its feeding when a piece of agar including tryptophan and phenylalanine was inserted into the abdomen. Miles (1968a, 1969) inferred that this demonstrated the hypothesis that IAA is synthesized from tryptophan in the saliva under the influence of quinones derived from phenylalanine through DOPA. He further suggested that all phytophagous hemipterans and homopterans have the potential to induce galls.

IAA, or auxins, are known to play an important role in the syntheses of RNA and proteins in plants and influence cell enlargement. Apart from galls, the auxin (perhaps IAA) in the saliva of lygus bugs is apparently injected into carrot seeds and results in increased growth of the seedlings (Scott 1970). Since auxins raise the production of ethylene in plants, infestations of the aphid *Adelges piceae* Ratz., which contain auxin-like compounds in their saliva, increase the ethylene level in trees (Puritch 1977).

Miles (1968b, 1972) insisted that IAA was likely to be a main factor in natural cecidogenesis and stated that the final explanation of the morphogenesis of galls would probably be related to the precise placement of active substances through insect feeding activity rather than to gross concentrations of any particular substances in the tissues as a whole, though now he is not particular about this IAA theory. However, it might not be so easy to explain the role of IAA in the process of morphogenesis by this idea, since galls contain much higher amounts of IAA or auxins than homopteran saliva, either before or after its injection. Low concentrations of IAA in the saliva injected into plant tissues could hardly function in the morphogenesis of galls when IAA was already in high concentration in the gall tissues. Nevertheless, IAA in the saliva could function as an "inducer" of galls rather than as a "maturator." IAA concentration in young leaves of plants is usually very low (below 10^{-7} M). Therefore, IAA in the saliva could function in young leaves as a gall inducer. Many workers have failed to produce galls artificially by the treatment of IAA (Hough 1953; Plumb 1953). This may be due to the lack of cell conditioning prior to gall induction.

IAA Synergist

Sterling (1952), Brian (1957), and Pilet (1960) suggested that cecidogenous organisms increase IAA activity in plants by inhibiting endogenous IAA oxidase. Weidner (1950) supposed that the gall-stimulating substances of insects might be activators of the plants' own auxins. Ignoffo and Granovsky (1961) reported that a gall-inducing aphid, *Mordwilkoja vagabunda* Walsh., introduced substances that obstructed the breakdown of auxins in plant tissues and raised auxin activity. Hori (1974) detected an IAA synergist that promoted the activity of IAA and/or inhibited the activity of IAA oxidase in the salivary glands of adult *Lygus disponsi* Lin. Similarly, such an IAA synergist was found in the salivary glands of larval *Lygus disponsi* (Hori 1975) and several other hemipterans (Hori 1976; Hori and Miles 1977). The identity of the synergist revealed in these experiments has not yet been determined, nor is it known whether it is present in the saliva. The bugs *L. disponsi* produce swellings on pumpkin fruits by their feeding (Torikura 1985). Hori *et al.* (1987) found that IAA oxidase activity in the fruit tissue increased markedly immediately after the injury of the bugs but then decreased rapidly with hypertrophy of cells surrounding directly injured tissues. They suspected that this was due to IAA oxidase inhibiting substances in the saliva of the bugs injected into the tissue (IAA synergist). If it is contained in the saliva of gall insects then it could well play an important role in gall induction. As pointed out in the section on IAA, the concentration of IAA in the saliva of homopterans, or in the saliva after ejection, is probably much lower than that in the galls themselves. It is difficult, therefore, to explain how this small increment could have a significant effect on the gall tissues except possibly at the very beginning of gall induction. On the other hand, the synergist could function irrespective of the concentration of IAA in galls, because it would promote whatever IAA activity was already occurring in the gall tissues themselves. Thus the IAA synergist could behave effectively as a "maturator" of galls.

It is known that, in general, polyphenols synergize IAA-induced growth in plants. Scott

(1969) indicated that cytokinins, which might be contained in the saliva of lygus bugs, might destroy natural growth inhibitors in carrot seeds. Byers *et al.* (1976) suggested that a fly larva, *Janetiella* sp., might secrete gibberellin into the host plant, which raises the auxin level or synergizes its IAA to cause an abnormal growth.

Phenolic Compounds and Phenol Oxidases

According to Schäller (1968a), *Myzus ascalonicus* and *Megoura viciae* have seven kinds of phenol compounds in their saliva, *Sappaphis mali* six, *Viteus vitifolli*, *Eriosoma lanigerum* and *Cryptomyzus ribis* five, *Hyalopterus pruni* and *Aphis pomi* four, and *Myzus cerasi* three; those species of aphids with little ability to induce galls tended to have more kinds of phenolic compounds in their saliva than the more effective gall inducers. It is well known that monophenols tend to activate IAA oxidase and hence to inhibit IAA activity, whereas diphenols inhibit IAA oxidase and hence contribute to IAA activity.

Many plant-piercing-sucking insects are known to contain phenolases (polyphenol oxidases and peroxidases). Miles (1969) stated that polyphenol oxidase (PPO) was a component in the saliva of all phytophagous insects. On the other hand it is known that phenol oxidases and phenolic compounds increase in the plant tissues attacked by the insects and in galls.

The phenol–PPO system in the saliva of insects and in plants could take part in the production of IAA (Miles 1968b). On the other hand, it is well known that under the phenol–PPO system in plants, quinones are produced and function as a defense substance against the attack of pests and produce necrosis within the plants themselves. PPO in the saliva of insects may act to transform toxic quinones to nontoxic insoluble pigments and thus the gall inducers may succeed in their attack on the plant through the same process that produces galls (Miles 1968b). It would follow that a delicate balance exists in the interaction between the phenol–PPO systems of the insects and their host plants and that this balance determines whether the attack by insects results in necrosis (lesions) or galls. Miles (1972) suggests that evidence of this balance is to be found in the gradation from necrosis at the center of many galls to the surrounding hypertrophy. The complex roles of phenolic compounds and phenol oxidases as mentioned above may indicate that the phenol–PPO system has a more fundamental significance for gall induction than IAA itself.

Pectinases and Proteases

It is well known that various plant-piercing-sucking insects have pectinases in their salivary glands. In Hemiptera pectinase in the saliva seems to be specific to the bugs belonging to the family Miridae. Although the presence of pectinase is not correlated with gall induction as such, in aphids it serves to destroy the cell walls in the plants attacked (McAllan and Adams 1961) and is one cause of the severe injury to plants caused by mirids (Strong 1970; Varis 1972; Sapio *et al.* 1982).

In general, many plant-piercing-sucking insects, especially mesophyll feeders, have proteases in their salivary glands. Several workers have suggested that the proteases contribute to gall formation or other injuries of plants when injected into the plant tissues along with the saliva.

Other Substances

IAA or auxin inhibitors are present in the salivary glands or whole bodies of *Lygus oblineatus* Say (Fisher *et al.* 1946), female San Jose scale (Kloft 1951b), *Dolycoris baccarum* L., *Miris dolabratus* L., and *Cinara piceae* Panzer (Nuorteva 1956), *Mordwilkoja vagabunda* (Ignoffo and Granovsky 1961), *Toxoptera graminum* Rondani (Maxwell and Painter 1962b), and *Drepanosiphum platanoides* Schrank (Dixon 1971). Allen (1951) stated that phytophagous insects might inject a substance into plant tissues that increased or reduced the activity of growth substances during feeding, and that it might be a metabolic product of the insects themselves. Nuorteva (1958) reported that substances inhibiting plant growth were contained in the salivary glands of many hemipterans and at least one aphid. These substances might be phenolic compounds or substances coming from the interaction between the phenol–PPO systems of insects and plants. Moreover, Gavrilov (1964) suggested that *Pemphigus lactuarius* L. nymphs producing galls on *Populus pyramidalis* Roz. contained B group vitamins that might result in gall induction and Williams and Lindner (1965) detected kinetin and 6-methylpurine in the body extract of *Psylla pyricola* Foerster, an insect that is injurious to bean shoots.

Excreta

It has been reported that the excrement (as distinct from the saliva) of insects could produce cell division in plants and induce galls (Godan 1956; Strebler 1975; Carter 1962; Miles 1968b). However, in gall-inducing aphids the production of honeydew is usually restricted, presumably because its excretion in high amounts would constitute an unfavorable contamination of the gall cavity (Schäller 1968a), though recently Aoki (1980) found that some aphids actively push honeydew globules and exuviae out of their gall. Thus in aphids, the excrement is not likely a main factor in cecidogenesis, but it may play a minor role. However, in psyllids, it has been known that abnormal cellular activity resulting in gall induction is most probably associated with the lowering of epidermal permeability to air by the physical presence of the nymphs and their excreta (Madden and Stone 1984). Apart from galls, Pollard (1973) and Ankersmit and Rabbinage (1980) ascribed injury or physiological disturbance of plants caused by aphids partly to honeydew. It is known that in general the honeydew of aphids contains auxins (Maxwell and Painter 1962a; Eschrich 1968; Hoad *et al.* 1971; Morris and Kadir 1972; Hussain *et al.* 1974).

DIPTERA

Cecidomyiidae

Felt (1936) applied the saliva theory of gall induction by aphids to the cases of Cecidomyiidae and Cynipidae. Since then, it has been reported that the larvae of cecidomyiids may secrete plant-growth substances from their salivary glands or other organs into the host plant tissues to induce galls (Clausen 1950; Küster 1951; Boysen-Jensen 1952; Leatherdale 1955; Weidner 1957; Lee 1970; Strebler 1970). Byers *et al.* (1976) could detect a low level of gibberellin (0.7 ng/204 larvae) from the extract of the larvae of *Janetiella* sp. but auxin activity was not detected. They suggested that since auxin may be produced in the saliva and released continuously with gibberellin during feeding, these substances may be accumulating in plant tissues, even if they might not be contained in the saliva in appreciable quantity at any one time. Byers *et al.* (1976) further inferred that the gibberellin may influence gall induction; the gall-inducing larvae might secrete gibberellin and thereby induce the host plant to produce gibberellin, which would raise the auxin level in the plant or function synergistically with IAA to cause an abnormal growth. Even though the increase of gibberellin and auxin in the plant may be a result of secondary reactions during cecidogenesis, such increases might act in the morphogenesis of galls. According to Dielmann (1969), the gall midge *Wachtiella persicariae* L. has plant growth substances only at a very low level in its body, and the activity in the salivary glands is too low to explain the amount of auxin present in the galled leaf. Strebler (1975) reported that a compound either produced by action of the saliva and/or excrement, or injected in the plant tissues during egg laying, was implicated in the origin of cecidogenesis. Godan (1956) had previously denied the possibility that the eggs of *Dasyneura affinis* Kffr. exerted a gall-inducing stimulus to plant tissues.

It is quite probable that the complex cecidogenesis by cecidomyiids cannot be elucidated in such simple terms as can cases of chemically induced simple galls. However, Boysen-Jensen (1952) explained this by the feeding behavior of the insect; that is, the larva moves around being guided by instinct and secretes growth substances at exact places of the tissue in host plant leaves or galls and at exact times to induce cecidogenesis and develop the gall to a specific form. Skuhravý *et al.* (1980) stated that gall induction on *Quercus cerris* L. by *Dryomyia circinnans* Giraud was due to a relocation of most chemicals essential for the normal growth of the plant.

Larvae of the Hessian fly *Mayetiola destructor* Say destroys the hemicellulose of plants by secreting hemicellulase (Rafai *et al.* 1955). Cartwright *et al.* (1959) supposed that the larvae injected a plant-growth inhibitor into the host plant, while Byers and Gallun (1972a) detected plant-growth-promoting substances from the extract of larvae fed with a susceptible variety of wheat and stated that the larvae introduced some substances into the host plant that caused the characteristic stunting. Byers and Gallun (1972b) suggested that this stunting may not be due to deficiency of gibberellin but possibly to the inhibition of other growth-promoting metabolic systems (see Chapter 11).

Tephritidae

Mills (1969) reported that the larvae of *Eurosta solidaginis* and not their eggs elicited a response in the host plant to trigger abnormal cell division and that the larvae discharged more than one compound including auxin into the host plant, which in turn induced secretion of substances by the plant. Beck (1947) pointed out that the protease secreted into the host plant by the larvae was an important factor in gall induction and that the transformation of tryptophan to IAA in the plant tissue was a second factor. Uhler *et al.* (1971) agreed with Beck on the significance of protease secreted by the larvae in gall induction.

HYMENOPTERA

Cynipidae

Adler (1877) reported that cecidogenesis by cynipids started after the larvae had hatched and begun feeding. Shively (1950) and Viglierchio (1971) suggested the necessity of some larval secretion for the initiation of galls. However, Beijerinck (1882) and Taylor (1949) indicated that some cecidogenous effects were caused by eggs and the major or some part of the development of galls occurred after larvae appeared. Lintott (1975) suggested that the greater part of gall development occurred, and its final shape and size were determined, during the egg stage. Kostoff and Kendall (1929) stated that the wound hormones of plants produced in response to oviposition were important in the cell proliferation that occurred in the injured parts. However, the saliva also seems to be implicated in cecidogenesis by cynipids (Beijerinck 1882; Felt 1936). Therefore, the first insect-mediated disruption of plant tissue that leads to gall formation by cynipids is caused either by physical damage with the ovipositor, substances arising from the eggs, or feeding by the freshly hatched larvae (see Chapter 5 and Meyer and Maresquelle 1983).

Taylor (1949) detected a high concentration of RNA in the salivary glands of larvae of *Liposthenes glechomae* L. and took it for granted that this secretion affected the RNA metabolism of the host plant leaf to cause cell division and induce galls. Kaldewey (1965) detected auxins (IAA etc.) in an extract of *Cynips quercusfolii* L. larvae and confirmed that they came from the diet, though IAA was produced when the larvae were placed in a tryptophan solution. He concluded that tryptophan and auxin in the larvae or the host plant took part in gall induction. Matsui and Torikata (1970) reported that the larvae of *Dryocosmus kuriphilus* Yasumatsu and their galls on the chestnut tree contained little IAA but a high amount of tryptophan (a precursor of IAA) and presumed that tryptophan instead of IAA may have a function in the process of cecidogenesis because tryptophan could be gradually transformed into IAA. However, Ohkawa (1974) and Matsui *et al.* (1975) detected cytokinin-like substances in the larval extract and the latter authors, by comparing the effects of cytokinin on the metabolism of plant cells with the morphological, physiological, and biochemical conditions in galls, inferred that cytokinin may play a main role in gall induction by these insects. On the other hand, Yokota *et al.* (1973) found that the ethyl acetate-soluble neutral fraction of an extract from the larvae of *D. kuriphilus* could induce gall-like growth in plant tissue, but they assumed that the inducing factor was neither an auxin nor cytokinin even though it initiated galls and caused an increase in the activity of the endogenous plant hormones that played a role in gall development. Morover, they could detect IAA but little cytokinin in the larvae. According to Weidner (1957), cynipid galls are so complex in form that the mechanism of cecidogenesis cannot be explained merely by the function of such plant hormones. It is possible that the microfeeding behavior of the larvae may again contribute to the morphogenesis of complex galls, as was pointed out by Boysen-Jensen (1952). Shorthouse (1982) draws attention to the fact that however elaborate the morphology of cynipid galls, the individual cells are essentially normal expressions of the genetic potential of the host plant arranged under a "new" progressive, morphogenetic coordinating influence, provided by the insect (see Chapter 5). Cornell (1983) proposed that Cynipinae form mutualisms with virus or virus-like particles that work in concert with larval salivary secretions to produce galls.

Siricidae

A disease symptom appears on conifers due to the injection of mucus during oviposition by *Sirex noctilio* F. (Coutts 1969a; Spradbery 1973). Coutts (1969a) and Madden (1977) stated that mucus disturbed translocation in the plant, at first causing a stress condition. The former author supposed that the mucus might also interact with growth hormones in the tree in some unknown manner. The mucus is a glycoprotein (Coutts 1969b; Wong and Crowden 1976) and contains amylase, esterase, phenol oxidase, and protease, and it may be implicated at least in part with the natural disaggregation process (Wong and Crowden 1976).

Tenthredinidae

Küster (1930) reported that the excrement of the larvae of *Pontania* sp. could induce hyperplasia in the cells of host plants, while Felt (1936), Viglierchio (1971), and Slepyan and Gabarayeva (1978) ascribed gall formation by gall-inducing sawflies to action of the fluid accompanying the eggs rather than to any effect of the larvae. On the other hand, Hovanitz (1959) and McCalla *et al.* (1962) indicated that although the formation of galls was initiated with the injection of the ovipositional secretion, stimulation from the larvae was essential for completion of the gall. Low-molecular-weight materials (uric acid, glutamic acid, uridine, etc.) present in the ovipositional accessory gland have a gall growth-promoting activity (McCalla *et al.* 1962), but these authors were unable to clarify whether the ovipositional secretion was the same as the larval secretion, and whether the substances contributing to gall initiation also served as factors stimulating the further development of galls. They pointed out that the amount of cytokinins contained in the galls was so little that it might not contribute to the promotion of gall development (see Chapter 5).

LEPIDOPTERA

LaRue (1937) reported that the feces of many leaf-miners induced cell outgrowths on the mesophyll of plants and that the effect suggested the presence of plant-growth hormones in the feces. Unfortunately he did not describe the species of leaf-miners in his report. Beck (1949) found that the oral secretions of *Gnorimoschema gallaesolidaginis* Riley made a swelling on the stem of host plants. Engelbrecht (1971) detected large amounts of cytokinins in the green islands that developed through infestation by *Stigmella* larvae, and suggested that the larvae might produce cytokinins in their labial glands. On the other hand, Hewett (1977) pointed out that cytokinins in the host plant tissues attacked by larvae of *Anetus virescens* (Doubleday) were not identical with those in the larvae, but rather with those produced through the process of wound reaction of the plant.

COLEOPTERA

Deubert (1952) and Le Pape and Bronner (1987) reported that *Ceuthorrhynchus napi* Gyll, formed a gall-like growth on stems of the host plant, which was caused by secretions that protect the eggs. Jacquiot (1949) stated that the larvae of *Agrilus biguttatus* Fabr. injected a plant-growth substance (2,4-D-like substance) into the host plant along with their excreta and thereby induced the division of endodermis cells.

It is known that a proteinaceous extract from the larvae of the boll weevil *Anthonomus grandis* Boh. causes abscission of flower buds and debladed petioles when injected into the tissues of the cotton plant (Coakley *et al.* 1969; King and Lane 1969). This abscission agent is neither protein in the feces nor a component of the saliva, but occurs in the moulting fluid that is injected into flower bud tissue at the time of larval moulting (King and Lane 1969). King (1972) detected pectinases in the larvae and ascribed the abscission of attacked flower buds of cotton to some of these enzymes. Moreover, King (1973) confirmed that the main pectinase was endo-polymethylgalacturonase (endo-PMG) and this initiated abscission when injected into the flower buds. He suggested that imbalances in growth regulators and their degradation products in the plants, resulting from the function of endo-PMG, increased the buds' sensitivity to ethylene, and this might be the final causative agent for abscission. Levine and Hall (1978) also pointed out that the larvae of *Conotrachelus nenuphar* (Herbst.) secreted pectinases into the host plant during feeding and thereby caused tissue maceration in fruits of plums and apples. They inferred that more than one of the five pectinases detected might be the cause of premature abscission.

CONCLUDING REMARKS

It may be better to discuss the process of cecidogenesis by dividing it into the following three phases: (1) "cell conditioning" in which sensitivity of plant cells to the function of a "gall inducer" increases; (2) "gall induction" in which cecidogenesis proper is initiated; that is, cell enlargement occurs and the primary gall is constructed; and (3) "gall maturation" in which the primary gall grows and the morphogenesis of the gall is completed.

For gall induction to occur, it is necessary that the plant cells be conditioned to produce a particular physiological state (Plumb 1953; Schäller 1968b). In general, insect galls are produced on meristematic tissues or young plastic tissues. As pointed out in the section on amino acids, these may increase the sensitivity of plant cells to the gall inducer.

Felt (1936) stated that gall inducers stimulated cell production in plants at an early stage of gall development and that development of the galls depended on the adaptability of particular groups of plant cells. Hough (1953) suggested the possibility that there were two distinct stimulating agents in gall initiation and development. In the *Pontania* gall, for example, gall formation is initiated by the ovipositional secretion, but stimulation from the larvae is necessary for completion of the gall (McCalla *et al.* 1962). I should like to term the factors contributing to each process in turn as the "conditioner," "inducer," and "maturator," respectively. Of course a factor could take part in more than one process.

Schäller (1970) pointed out that IAA caused a general growth of plant tissues while amino acids contributed mainly to morphogenesis of the galls. Rohfritsch (1971) divided cecidogenesis by insects into two processes: (1) a permanent effect that is retained even if the insects in the galls are killed and (2) an effect for which continuous stimulation by the insects is necessary and which disappears if the insect dies. These statements may support the presence of two phases, "gall induction" and "gall maturation." It may be noted that such phases would also correspond to the earlier categories of "kataplasmic" galls (which do not develop in form beyond a primary stage of gall induction) and "prosoplasmic" galls (more advanced forms produced through gall maturation) (Mandl 1957).

In the "gall-maturation" phase, the feeding behavior of the insects and an increasing concentration of some plant hormones in the galls (and especially production of a hormonal imbalance in the plant tissues), in addition to production of a maturator by the insects, may be important for the morphogenesis of galls, as is indicated or implied in the publications of many workers (Nysterakis 1945; Nicolas and Aggery 1949; Boysen-Jensen 1952; Hough 1953; Wartenberg 1953; Mandl 1957; Balch *et al.* 1964; Miles 1968a; Schäller 1970; Byers *et al.* 1976; Norris 1979; Madden and Stone 1984). Livingstone (1978) considered that cecidogenesis constituted a special kind of interaction between animal and plant metabolites. Recently, Tandon and Arya (1980) isolated three auxin protectors from the stem gall tissues of *Zizyphus jujuba* Lamarck produced by a mite *Eriophyes cernuus* Massee and presumed that these disturbed IAA oxidation thereby causing high auxin activity and auxin autotrophy in the gall. Such IAA protectors could well take part in gall formation (see Chapter 11). Indeed the one unifying possibility that emerges from the many studies of gall induction by arthropods is that the definitive process may be one of protection of the plant's own IAA or other plant growth controlling hormones from natural breakdown.

ACKNOWLEDGMENTS

The author wishes to express sincere thanks to Dr. P. W. Miles of Waite Agricultural Research Institute (Australia) for his useful comments and suggestions for improving the manuscript, to Dr. Y. Nishijima, Emeritus Professor of Obihiro University, for helpful discussion, and to Mr. Y. Sugimori of the Obihiro University for assistance in surveying the literature.

LITERATURE CITED

Adler, H. 1877. Beiträge zur Naturgeschichte der Cynipiden. Deutsch. Entomol. Z. 21:202–248.

Akimoto, S. 1981. Gall formation by *Eriosoma* fundatrices and gall parasitism in *Eriosoma yangi* (Homoptera, Pemphigidae). Kontyû 49:426–436.

Allen, T. C. 1951. Deformities caused by insects. In F. Skoog (ed.), Plant Growth Substances, pp. 411–515. Univ. of Wisconsin Press, Madison.

Anders, F. 1957. Uber die gallenerregenden Agenzien der Reblaus (*Viteus* [*Phylloxera*] *vitifolii* Shimer). Vitis 1:121–124.

Anders, F. 1958. Aminosäuren als gallenerregende Stoffe der Reblaus (*Viteus* [*Phylloxera*] *vitifolii* Shimer). Experientia 14:62–63.

Ankersmit, G. W., and R. Rabbinage. 1980. Epidemiology of the cereal aphid, *Sitobion avenae.* In Integrated Control of Insect Pests in the Netherlands, pp. 291–295. Pudoc, Wageningen.

Aoki, S. 1980. Occurrence of a simple labor in a gall aphid, *Pemphigus dorocola* (Homoptera, Pemphigidae) Kontyû 48:71–73.

Balch, R. E., J. Clark, and J. M. Bonga. 1964. Hormonal action in production of tumors and compression wood by an aphid. Nature (London) 202:721–722.

Beck, E. G. 1947. Some studies on the *Solidago* gall caused by *Eurosta solidaginis* (Fitch). Ph.D. thesis, Univ. of Michigan, Ann Arbor.

Beck, E. G. 1949. The histology and biology of the *Solidago* gall caused by the larva of *Gnorimoschema gallaesolidaginis.* Am. J. Bot. 36:794.

Beijerinck, M. W. 1882. Über die ersten Entwicklungsphasen einiger Cynipidengallen. Veröff. König. Akad. Wiss. Amsterdam 22:1–198.

Boysen-Jensen, P. 1952. Untersuchungen uber die Bildung der Galle von *Mikiola fagi.* Dan. Biol. Medd. 18:1–18.

Brian, P. W. 1957. The effect of some microbial metabolic products on plant growth. Symp. Soc. Exp. Biol. 11:166–182.

Byers, R. A., and R. L. Gallun. 1972a. Ability of the Hessian fly to stunt winter wheat. 2. Paper chromatography of extracts of freeze-dried larvae and wheat plants. J. Econ. Entomol. 65:997–1001.

Byers, R. A., and R. L. Gallun. 1972b. Ability of the Hessian fly to stunt winter wheat. 3. Gibberellic acid and elongation of second and third leaves of infested plants. J. Econ. Entomol. 65:1001–1004.

Byers, J. A., J. W. Brewer, and D. W. Denna. 1976. Plant growth hormones in pinyon insect galls. Marcellia 39:125–134.

Carter, W. 1962. Insects in Relation to Plant Disease. Interscience, New York.

Cartwright, W. B., R. M. Caldwell, and L. E. Compton. 1959. Response of resistant and susceptible wheat to Hessian fly attack. Argon. J. 51:529–531.

Clausen, R. L. 1950. Observations sur la Cecidomyie de la violette *Dasyneura affinis* Kieffer. Mitt. Schweiz Entomol. Ges. 23:200–206.

Coakley, J. M., F. G. Maxwell, and J. N. Jenkins. 1969. Influence of feeding, oviposition, and egg and larval development of the boll weevil on abscission of cotton squares. J. Econ. Entomol. 62:244.

Cornell, H. V. 1983. The secondary chemistry and complex morphology of galls formed by the Cynipinae (Hymenoptera): Why and how? Am. Mid. Nat. 110:225–234.

Coutts, M. P. 1969a. The mechanism of pathogenicity of *Sirex noctilio* on *Pinus radiata.* II. Effects of *S. noctilio* mucus. Aust. J. Biol. Sci. 22:1153–1161.

Coutts, M. P. 1969b. The physiological effects of the mucus secretion of *Sirex noctilio* on *Pinus radiata* Aust. Forest Res. 4:23–26.

Deubert, K. H. 1952. Uber das durch die Eiablage von *Ceuthorrynchus napi* Gyll. (Col., Curc.) verursachte histologische Schadbild an Winterraps. Wiss. Z. Univ. Halle 2:203–205.

Dieleman, F. L. 1969. Effects of gall midge infestation on plant growth and growth regulating substances. Entomol. Exp. Appl. 12:745–749.

Dixon, A. F. G. 1971. The role of aphids in wood formation. I. The effect of the sycamore aphid, *Drepanosiphum platanoides* (Schr.) (Aphididae), on the growth of sycamore, *Acer pseudoplatanus* (L). J. Appl. Ecol. 8:165–179.

Duspiva, F. 1954. Weitere Untersuchungen über Stoffwechsel-physiologische Beziehungen zwischen Rhynchoten und ihren Wirtspflanzen. Mitt. biol. Bund Anst. Land-u.-Forstw. 80:155–162.

Engelbrecht, L. 1971. Cytokinin activity in larval infected leaves. Biochem. Physiol. Pflanzen. 162:9–27.

Eschrich, W. 1968. Translokation radioaktiv markierter Indolyl-3-essigsäure in Siebröhren von *Vicia faba.* Planta 78:144–157.

Eyer, J. R. 1937. Physiology of psyllid yellows of potatoes. J. Econ. Entomol. 30:891–898.

Felt, E. P. 1936. The relations of insects and plants in gall production. Ann. Entomol. Soc. Am. 24:694–700.

Fisher, E. H., A. J. Riker, and T. C. Allen. 1946. Bud, blossom and pod drop of canning string beans reduced by plant hormones. Phytopathology 36:504–523.

Gavrilov, K. I. 1964. On the causes of the formation of galls, tumor-like excrescences, on plants due to insect parasitism. Zh. Obsc. Biol. 26:475–478.

Godan, D. 1956. Beiträge zur Autökologie der Veilchenrollmucke (*Dasyneura affinis* Kieff.). Z. Angew. Entomol. 39:1–19.

Goodchild, A. J. P. 1952. A study of the digestive system of the West African cacao capsid bugs (Hemiptera, Miridae). Proc. Zool. Soc. London 122:543–572.

Hagley, E. A. C. 1969. Free amino acids and sugars of the adult sugar cane froghopper *(Aeneolamina varia saccharina)* (Homoptera: Cercopidae) in relation to those of its host. Entomol. Exp. Appl. 12:235–239.

Hewett, E. W. 1977. Some effects of infestation on plants: A physiological viewpoint. N. Z. Entomol. 6:235–243.

Hoad, G. V., S. K. Hillman, and P. F. Wareing. 1971. Studies on the movement of indole auxins in willow (*Salix viminalis* L.). Planta 99:73–88.

Hori, K. 1974. Plant growth-promoting factor in the salivary gland of the bug, *Lygus disponsi.* J. Insect Physiol. 20:1623–1627.

Hori, K. 1975. Pectinase and plant growth-promoting factors in the salivary glands of the larva of the bug, *Lygus disponsi.* J. Insect Physiol. 21:1271–1274.

Hori, K. 1976. Plant growth-regulating factor in the salivary gland of several heteropterous insects. Comp. Biochem. Physiol. 53B:435–438.

Hori, K., and P. W. Miles. 1977. Multiple plant growth-promoting factors in the salivary glands of plant bugs. Marcellia 39:399–400.

Hori, K., D. R. Singh, and A. Sugitani. 1979. Metabolism of ingested indole compounds in the gut of three species of Heteroptera. Comp. Biochem. Physiol. 64C:217–222.

Hori, K., H. Torikura, and M. Kumagai. 1987. Histological and biochemical changes in the tissue of pumpkin fruit injured by *Lygus disponsi* Linnavuori (Hemiptera: Miridae) Appl. Entomol. Zool. 22:259–265.

Hough, J. S. 1953. Studies on the common spangle gall of oak. II. A general consideration of past work on gall induction. New Phytol. 52:218–228.

Hovanitz, W. 1959. Insects and plant galls. Sci. Am. 201:151–162.

Hussain, A., J. M. S. Forrest, and A. F. G. Dixon. 1974. Sugar, organic acid, phenolic acid and plant growth regulator content of extracts of honeydew of the aphid *Myzus persicae* and of its host plant, *Raphanus sativus.* Ann. Appl. Biol. 78:65–73.

Ignoffo, C. M., and A. A. Granovsky. 1961. Life history and gall development of *Mordwilkoja vagabunda* (Homoptera: Aphididae) on *Populus deltoides.* Part II. Gall development. Ann. Entomol. Soc. Am. 54:635–641.

Jacquiot, C. 1949. De quelques particularités anatomiques et cytologiques observeés dans l'aubier de chênes dépérissants (*Quercus sessiliflora* Sm. et *Q. pedunculata* Erh.). Compt. Rend. Acad. Sci. 229:241–243.

Kaldewey, H. 1965. Wachstumsregulatoren aus Pflanzengallen und Larven der Gallenbewohner. Berich. Deutsch. Bot. Ges. 78:73–84.

Kefeli, V. I., and R. K. Turetskaya. 1965. Participation of phenolic compounds in the inhibition of auxin activity and in the growth of willow-shoots. Plant Physiol. 12:554–560.

King, E. E. 1972. Pectinases from boll weevil larvae, *Anthonomus grandis.* J. Insect Physiol. 18:1259–1301.

King, E. E. 1973. Endo-polymethylgalacturonase of boll weevil larvae, *Anthonomus grandis:* An inhibitor of cotton flower bud abscission. J. Insect Physiol. 19:2433–2437.

King, E. E., and H. C. Lane. 1969. Abscission of cotton flower buds and petioles caused by protein from boll weevil larvae. Plant Physiol. 44:903–906.

Kloft, W. 1951a. Insekten steuern das Wachsen von Pflanzen. Umschau Wissenschaft Technik. 51:40–42.

Kloft, W. 1951b. Vergleichende Untersuchungen an einigen Cocciden und Aphiden. Verh. Deutsch. Zool. Marburg. 1950:290–296.

Kloft, W. 1960. Wechselwirkungen zwischen pflanzensaugenden Insekten und den von ihnen besogenen Pflanzengeweben. Z. Angew. Entomol. 45:337–381 and 46:42–70.

Kloft, W., and P. Ehrhardt. 1959. Untersuchungen uber Saugtätigkeit und Schadwirkung der Sitkafichtenlaus, *Lisomaphis abietina* (Walk.). (*Neomyzaphis abietina* Walk.). Phytopathol. Z. 35:401–410.

Kostoff, D., and J. Kendall. 1929. Studies on the structure and development of certain cynipid galls. Biol. Bull. Woods Hole 56:402–458.

Küster, E. 1930. Anatomie der Gallen. Handb. Pfl. anat. 5:1–197.

Küster, E. 1951. Uber die morphologischen Kennzeichen der Gallen von *Dasyneura crataegi.* Ber. d.D. Bot. Ges. 64:283–287.

LaRue, C. D. 1937. The part played by auxin in the formation of internal intumescences in the tunnels of leaf miners. Bull. Torrey Bot. C 64:97–102.

Leatherdale, D. 1955. Plant hyperplasia induced with a cell-free insect extract. Nature (London) 175:553–554.

Lee, D. K. 1970. Physiological investigation on the damage of pine gall midge (*Thecodiplosis japonensis* Uchida et Inouye) in Korean red pine (*Pinus densiflora* S. et Z.) and pitch pine (*Pinus rigida* Mill). Res. Rep. Inst. Forest Cen. 8:33–47.

Le Pape, H., and R. Bronner. 1987. The effects of *Ceuthorrhynchus napi* on the stem tissues of *Brassica napus* var. *oleifera.* In V. Labeyrie, G. Fabres, and D. Lachaise. (eds.), Insect-Plants, pp. 207–211. W. Junk, The Hague.

Levine, E., and F. R. Hall. 1978. Pectinases and cellulases from plum curculio larvae: Possible causes of apple and plum fruit abscission. Entomol. Exp. Appl. 23:259–268.

Lintott, W. H. 1975. The gall of *Phanacis hypochoeridis* (Kief.). Eady and Quinlan on *Hypochaeris radicata* L. Marcellia 38:227–233.

Livingstone, D. 1978. Phytosuccivorous bugs and cecidogenesis. J. Ind. Acad. Wood Sci. 9:39–45.

Madden, J. L. 1977. Physiological reactions of *Pinus radiata* to attack by woodwasp, *Sirex noctilio* F. (Hymenoptera: Siricidae). Bull. Entomol. Res. 67:405–426.

Madden, J. L., and C. Stone. 1984. Induction and formation of pouch and emergence galls in *Eucalyptus pulchella* leaves. Aust. J. Bot. 32:33–42.

Mandl, L. 1957. Wachstum von Pflanzengallen durch synthetische Wuchsstoffe. Osterr. Bot. Z. 104:185–208.

Marek J. 1961. Die Wirkung von Aphidenstichen auf pflanzliche Zellen. Entomol. Exp. Appl. 4:20–34.

Matsui, S., and H. Torikata. 1970. Studies on the resistance of chestnut trees to chestnut gall wasps. III. Plant growth regulators contained in chestnut gall wasps and host gall tissues. J. Jpn. Soc. Hort. Sci. 39:115–123.

Matsui, S., H. Torikata, and K. Munakata. 1975. Studies on the resistance of chestnut trees (*Castanea* spp.) to chestnut gall wasps (*Dryocosmus kuriphilus* Yasumatsu). V. Cytokinin activity in leaves of gall wasps and callus formation of chestnut stem sections by larval extracts. J. Jpn. Soc. Hort. Sci. 43:415–422.

Maxwell, F. G., and R. H. Painter. 1962a. Auxins in honeydew of *Toxoptera graminum, Therioaphis maculata,* and *Macrosiphum pisi,* and their relation to degree of tolerance in host plants. Ann. Entomol. Soc. Am. 55:229–233.

Maxwell, F. G., and R. H. Painter. 1962b. Plant growth hormones in ether extracts of the greenbug, *Toxoptera graminum* and the pea aphid, *Macrosiphum pisi,* fed on selected tolerant and susceptible host plant. J. Econ. Entomol. 55:57–62.

McAllan, J. W., and J. B. Adams. 1961. The significance of pectinase in plant penetration by aphids. Can. J. Zool. 39:305–310.

McCalla, D. R., M. K. Genthe, and W. Hovanitz. 1962. Chemical nature of an insect gall growth-factor. Plant Physiol. 37:98–103.

Meyer, J., and H. J. Maresquelle. 1983. Anatomie des galles. Borntraeger, Stuttgart.

Miles, P. W. 1967. Studies on the salivary physiology of plant-bugs: Transport from haemolymph to saliva. J. Insect Physiol. 13:1787–1801.

Miles, P. W. 1968a. Studies on the salivary physiology of plant-bugs: Experimental induction of galls. J. Insect Physiol. 14:97–106.

Miles, P. W. 1968b. Insect secretion in plants. Ann. Rev. Phytopathol. 6:137–164.

Miles, P. W. 1969. Interaction of plant phenols and salivary phenolases in the relationship between plants and Hemiptera. Entomol. Exp. Appl. 12:736–744.

Miles, P. W. 1972. The saliva of Hemiptera. Adv. Insect Physiol. 9:183–255.

Miles, P. W., and J. Lloyd. 1967. Synthesis of a plant hormone by the salivary apparatus of plant-sucking Hemiptera. Nature (London) 213:801–802.

Mills, R. R. 1969. Effect of plant and insect hormones on the formation of the goldenrod gall. Natl. Cancer Inst. Monogr. 31:487–491.

Morris, D. A., and G. O. Kadir. 1972. Pathways of auxin transport in the intact pea seedling (*Pisum sativum* L.). Planta 107:171–181.

Nicolas, G., and B. Aggery. 1949. Nouvelles observations sur une cécidie foliaire accompagnée d'une courbure de la tige d'*Artemisia vulgaris* L. Compt. Rend. Acad. Sci. 229:1098–1100.

Norris, D. M. 1979. How insects induce disease. Plant Dis. 4:239–255.

Nuorteva, P. 1956. Studies on the effect of the salivary secretion of some Heteroptera and Homoptera on plant growth. Ann. Entomol. Fenn. 20:102–124.

Nuorteva, P. 1958. Die Rolle der Speichelsekrete im Wechselverhältnis zwischen Tier und Nahrungspflanze bei Homopteren und Heteropteren. Entomol. Exp. Appl. 1:41–49.

Nuorteva, P. 1962. Studies on the causes of the phytopathogenicity of *Calligipona pellucida* (F.) (Hom., Araeopidae). Ann. Zool. Soc. Vanamo 23:1–58.

Nuorteva, P., and S. Laurema. 1961. The effect of diet on the amino acids in the haemolymph and salivary glands of Heteroptera. Ann. Entomol. Fenn. 27:57–65.

Nysterakis, F. 1945. Phytohormones et court-noué de la vigne. Compt. Rend. Acad. Sci. 221:54–56.

Nysterakis, F. 1948. Phytohormones et inhibition de la croissance des organes végétaux attaqués par des Aphides. Compt. Rend. Hebd. Séances Acad. Sci. 226:746–747.

Ohkawa, M. 1974. Isolation of zeatin from larvae of

Dryocosmus kuriphilus Yasumatsu. Hortscience 9:458–459.

Pilet, P. E. 1960. Auxin content and auxin catabolism of the stems of *Euphorbia cyparissias* L. infected by *Uromyces pisi* (Pers.). Phytopathol. Z. 40:75–90.

Plumb, G. H. 1953. The formation and development of the Norway spruce gall caused by *Adelges abietis* L. Bull. Connecticut Agric. Exp. Sta. 566:1–77.

Pollard, D. G. 1973. Plant penetration by feeding aphids (Hemiptera, Aphidoidea): A review. Bull. Entomol. Res. 62:631–714.

Puritch, G. S. 1977. Distribution and phenolic composition of sapwood and heartwood in *Abies grandis* and effects of the balsam wooly aphid. Can. J. For. Res. 7:54–62.

Rafai, F. Y., E. T. Jones, and B. S. Miller. 1955. Some biochemical factors involved in the resistance of the wheat plant to attack by the Hessian fly. Cereal Chem. 32:437–451.

Rilling, G., A. Rapp, H. Steffan and K. H. Reuther. 1974. Freie und gebundene Aminosäuren der Reblaus (*Dastylosphaera vitifolii* Shimer) und Möglichkeiten ihrer Biosynthese aus Saccharose-^{14}C(U). Z. Angew. Entomol. 77:195–210.

Rohfritsch, O. 1971. Développement cécidien et rôle du parasite dans quelques galles d'arthropodes. Marcellia 37:233–339.

Sapio, F. J., L. F. Wilson, and M. E. Ostry. 1982. A split-stem lesion on young hybrid *Populus* trees caused by the tarnished plant bug, *Lygus lineolaris* (Hemiptera (Heteroptera): Miridae). The Great Lakes Entomol. 15:237–246.

Schäller, G. 1963. Biochemische Rassentrennung bei der Reblaus (*Viteus vitifolii* Shimer) durch Speichelanalysen. Zool. Jb. Physiol. 70:278–283.

Schäller, G. 1965. Untersuchungen über den β-Indolylessigsaüregehalt des Speichels von Aphidenarten mit unterschiedlicher Phytopathogenität. Zool. Jb. Physiol. 71:385–392.

Schäller, G. 1968a. Biochemische Analyse des Aphidenspeichels und seine Bedeutung fur die Gallenbildung. Zool. Jb. Physiol. 74:54–87.

Schäller, G. 1968b. Untersuchungen zur Erzeugung künstlicher Pflanzengallen. Marcellia 35:131–153.

Schäller, G. 1970. Wie entstehen Pflanzengallen? Vorstellungen auf Gund neuer Untersuchungsergebnisse mit pflanzensaftsaugenden Insekten. Biol. Rundschau 8:346–350.

Scott, D. R. 1969. Lygus-bug feeding on developing carrot seed: Effect on plant growth from that seed. J. Econ. Entomol. 62:504–505.

Scott, D. R. 1970. Feeding of Lygus bugs (Hemiptera: Miridae) on developing carrot and bean seeds increased growth and yields of plants grown from that seed. Ann. Entomol. Soc. Am. 63:1604–1608.

Shively, A. W. 1950. Role of the insect larva in plant gall production. Proc. Penn. Acad. Sci. 24:17–20.

Shorthouse, J. D. 1982. Resource exploitation by gall wasps of the genus *Diplolepis*. Proc. 5th Int. Symp. Insect-Plant Relationships, Wageningen, 1982, pp. 193–198. Pudoc, Wageningen.

Skuhravý, V., M. Skuhravá, and J. W. Brewer. 1980. Evaluation of plant damage caused by three species of gall midges (Diptera: Cecidomyiidae). Z. Angew. Entomol. 90:184–190.

Slepyan, E. I., and N. I. Gabarayeva. 1978. Morphological and histological features of the gut, malpighian tubes and salivary glands of *Pontania proxima* larvae (Hymenoptera, Tenthredinidae) in connection with their mode of life. Entomol. Rev. 57:333–338.

Spradbery, J. P. 1973. A comparative study of the phytotoxic effects of siricid woodwasps on conifers. Ann. Appl. Biol. 75:309–320.

Sterling, C. 1952. Ontogeny of the phylloxera gall of grape leaf. Am. J. Bot. 39:6–15.

Strebler, G. 1970. Quelques aspects de la cécidogenèse dans le cas des galles des fleurs dues a *Contarinia medicaginis* Kieffer. Marcellia 37:3–46.

Strebler, G. 1975. Les processus de la cécidogenèse. Compt. Rend. Séances Soc. Biol. Filiales 169:773–777.

Strong, F. E. 1970. Physiology of injury caused by *Lygus hesperus*. J. Econ. Entomol. 63:808–814.

Tandon, P., and H. C. Arya. 1980. Presence of auxin protectors in *Eriophyes* induced *Zizyphus* stem galls. Experientia 36:958–959.

Taylor, S. H. 1949. Initiation and development of the gall of *Aylax glechomae* on *Nepeta hederacea*. Am. J. Bot. 36:222–230.

Torikura, H. 1985. Tubercular fruit of pumpkin caused by feeding of the mirid bug, *Lygus disponsi* Linnavuori. Ann. Rept. Plant Prot. North Jpn. 36:83–86.

Uhler, L. D., C. R. Crispen, Jr., and D. B. McCormick, 1971. Free amino acid patterns during development of *Eurosta solidaginis* (Fitch). Comp. Biochem. Physiol. 38B:87–91.

Varis, A. L. 1972. The biology of *Lygus rugulipennis* Popp. (Het., Miridae) and the damage caused by this species to sugar beet. Ann. Agric. Fenn. 11:1–56.

Viglierchio, D. R. 1971. Nematodes and other pathogens in auxin-related plant-growth disorders. Bot. Rev. 37:1–21.

Wartenberg, H. 1953. Uber pflanzenphysiologischen Ursachen des Massenwechsels der Apfelblutlaus *(Eriosoma lanigerum)* auf *Malus pumila*. Mitt. Biol. Zentralanst. Forst-u. Landw. 75:53–56.

Weidner, H. 1950. The effects of the gall insects on their host plants. Naturw. Rundschau. 3:364–368.

Weidner, H. 1957. Neuere Anschauungen über die Entstehung der Gallen durch die Einwirkung von Insekten. Z. Pflanzenkrank. Pflanzenpathol. Pflanzenschutz. 64:287–309.

Williams, M. W., and R. C. Lindner. 1965. Biochemical components of pear psylla and their relative toxicity to excised bean plants. J. Insect Physiol. 11:41–52.

Wong, L. K., and R. K. Crowden. 1976. Preliminary studies on the mucus secretion of the wood wasp, *Sirex noctilio* F.I. Physiochemical and biochemical properties. Aust. J. Biol. Sci. 29:21–32.

Yokota, T., M. Okabayashi, N. Takahashi, I. Shimura, and K. Umeya. 1973. Plant growth regulators in chestnut gall tissue and wasps. Proc. 8th Int. Conf. Plant Growth Substances, Tokyo, Japan, 1973, pp. 28–38.

11

Biology of *Dasineura affinis* (Cecidomyiidae) and Influence of Its Gall on *Viola odorata*

M. L. Birch
J. W. Brewer
and O. Rohfritsch

The earliest cecidologists knew that the morphological characteristics of galls varied greatly and were dependent on a number of factors including the species of gall insect, the species of host plant, and the particular plant organ infested. Most gall inducers are quite specific in regard to the plant and the plant organ they attack. The result of these interactions is that galls induced by different cecidogenic species are generally quite distinct in regard to size, shape, and morphology (see Chapter 2).

The growth changes produced by gall inducers include both stimulation and inhibition of plant growth and sometimes both processes occur at the same moment in different areas of the same gall (see Chapter 5). The effect of these changes may be quite noticeable, as in the well-known mossy rose gall caused by *Diplolepis rosae* (L.), or very subtle as in the slightly shortened needles of *Pinus edulis* Engelm. caused by *Janetiella* sp. near *coloradensis* Felt (Brewer 1971). Various species of gall insects attack virtually all kinds of plant organs including roots, stems, buds, leaves, flowers, fruits, and seeds. The effects of such attacks vary depending not only on the type of organ attacked, but also on the developmental state of the organ at the time the attack occurs.

The gross effects of these interactions are generally easy to observe. Less noticeable, but also important, are changes that occur in levels of plant nutrients and chemistry as a result of gall inducers. Changes in levels of nitrogen, starch, sugar, and lipid have been reported as a result of infestation by gall insects (see Chapter 8). Such changes may cause serious imbalances in plant chemistry and result in alterations of plant physiology that may be more important than the galls themselves. In some cases complete eradication of the plant may result from infestation. These effects on the plant, coupled with host specificity of gall insects, have led to the use of some species as agents in the biological control of weeds (Harris 1971, 1980; Zwölfer 1976).

In this chapter we use a model system approach to present details on how one gall insect, *Dasineura affinis* Kieffer, affects its host plant *Viola odorata* L., the sweet violet. We begin with a general discussion of the insect–host plant association. We show how plant quality can influence cecidogenesis and how the life cycle of *D. affinis* influences various characteristics of the galls it causes. We then discuss the changes in various aspects of plant morphology and chemistry that occur locally. We also present information as to how these changes influence the plant overall and, where appropriate, present results of studies on other gall insect–host plant associations.

GENERAL ASPECTS OF THE INSECT-HOST PLANT RELATIONSHIP

Violets flower mainly in the spring and fruits soon follow. The plant continues to produce new leaves until growth is suspended by winter. Thus except for the winter months, susceptible plant tissues are available to *Dasineura affinis* for an extended period and as a result it is able to produce several generations per year. Attacks may occur at different stages of plant growth; however, not all *Viola* species, or all plant structures, are equally susceptible to infestation. In addition, only the early larval stages are capable of cecidogenesis. Thus larval success is dependent on the time of attack since rapid establishment of a nutrient relationship with the new plant tissues is essential and these tissues are not always available.

Time of Attack

Successful attacks may occur at any time the plant is producing new growth. For *Viola odorata* this period extends from early spring until late autumn.

Number of Generations and Their Length in Dasineura affinis

The length in time, and hence the number of generations, appears to be governed primarily by environmental temperatures. Faure (1924) reported the occurrence of only one generation of *D. affinis* but later Raymond (1928) reared at least three, the last of which overwintered. Colizza (1928) obtained five generations in Italy and noted that the generation time depended on the season. Generations that occurred early in the season were of relatively short duration while those occurring later (in the autumn) lasted longer. In England four generations usually occur with the first adults appearing in April or early May. The second generation appears in June followed by another in July–August. There is a fourth flight in September and often a small number of adults appear at the end of October as a possible fifth generation. In England the time required to complete the life cycle ranges from about 150 days for the overwintering generation to 30–45 days for the summer generation. It appears that the number of generations essentially depends on the duration of warm summer weather. In Italy, where summers are long and warm, five generations are probably common. As one proceeds north and the summers become cooler and shorter, the number of possible generations is reduced.

Emergence

The overwintering generation emerges "en masse" with the emergence period lasting only a few days. The winter or diapause period seems to act to synchronize the emergence of the adults, since in the following generations (particularly the last one in the autumn), the adult flight period is spread over a longer time. Male emergence precedes that of the females by an hour or two, with the latter usually appearing between 8 and 11 AM.

Host Specificity

Several species of *Viola* are reported to be susceptible to attacks by *D. affinis*. Barnes (1927, 1948) showed that *D. affinis* caused galls on *V. reichenbachiana* Jord., *V. alba* Besser, *V. hirta* L., *V. silvestris* Lamk., *V. riviniana* Rchb., *V. canina* L., and *V. lutea* Smith. Stokes (1955) and Godan (1961) showed that *D. violae* F. Loew induces galls on various species of the pansy group including *V. arvensis* Murray and *V. tricolor* L. Even though *D. violae* appears morphologically identical to *D. affinis*, it will not produce galls on plant species of the violet group. Faure (1924) found that *D. affinis* showed a strong preference for *V. odorata* when both it and *V. hirta* were present. We have never found galls of *D. affinis* on *V. hirta* in field plots where both plant species occur together.

Factors Affecting Infestation

Several environmental factors affect the success of attacks by cecidomyiids and some of these factors for *D. affinis* specifically include humidity, larval age, and stage of leaf development.

Humidity

The natural habitat of *V. odorata* is in areas where conditions are cool and damp with natural humidity at about 80 percent (Godan

1956). Appropriately, *D. affinis* seems adapted to these conditions that favor its host. For example, although the eggs of *D. affinis* are very vulnerable to desiccation this is not a problem as they are normally placed between the bract and the young rolled leaf where conditions are humid. However, when we experimentally moved eggs from the normal oviposition sites to other similar sites, infestation was successful only under conditions of high humidity. Godan (1961) suggested that the wide host distribution of *D. affinis* was a result of differing levels of humidity tolerance. Thus under natural conditions, it would appear that humidity levels are a limiting factor to survival of this midge. Certainly, infestation by these cecidomyiids under hot and arid conditions is more difficult.

Larval Age

Desiccation aside, it is important that larvae quickly locate a susceptible site on the host plant because their ability to cause cecidogenesis appears to decline with age. For example, we found that larvae of *D. affinis* could only induce gall formation during the very early period of their development. Older (1–2 weeks) larvae could cause necrotic patches or increased hairiness but no galls. Young first-instar larvae (2–3 days) could induce gall formation if moved from their location on a plant to a new site (on a young unfurled leaf). This may be unusual, since Rohfritsch (1980) has shown that once young larvae of *Hartigiola annulipes* (Htg.) begin to feed they are no longer able to initiate another gall elsewhere on the leaf of *Fagus silvatica* L.

Stage of Leaf Development

The stage of leaf development is also an important variable in cecidogenesis. In general, only young growing leaves can be induced to form galls. We used a number system to indicate the stage of leaf development because age was found not to be an accurate indicator of developmental stage. In this system, Class I leaves are completely furled, Classes II and III are progressively less furled, Class IV leaves are completely unfurled but have not reached full size, and Class V are fully expanded mature leaves. We found that eggs, newly hatched larvae, and larvae up to 3–4 days could not cause galls to form on leaves of Classes III through V. It is not known whether this resistance on the part of the plant is due to physical conditions (increased cuticle thickness for example) or physiological changes that occur as the leaves age. However, it appears that the extent of leaf curling is an important factor in the survival of the young larvae. It also seems likely that larvae at this stage, like the eggs, are subject to desiccation and that the furled leaf provides protection against drying.

Factors Affecting Gall Size

Several factors contribute to the variation in gall size. These include the number of larvae within a gall, the density of galls on the individual plant, the position of the infested leaf, the infested plants' particular physiological state (which in turn is governed by seasonality), its environment, and the host genes. Some of these factors will be considered here.

Number of Larvae per Gall

The most influential factor affecting variation in the size of galls of *D. affinis* is the number of larvae within the gall; that is, the final gall size and weight are generally greater for those galls containing larger numbers of larvae (Birch 1974). This seems to be a general rule for midge galls since Meyer (1957), Strebler (1971), and Houseweart and Brewer (1972) reported similar results with other species.

Pupal Weight per Gall

Pupal weight was more closely correlated with gall production than number of pupae, and was used for assessing production from galls with variable pupal numbers. It was determined that the mean weight of female pupae decreased in correlation with an increase in the numbers of insects per individual gall. Competition may provide a partial explanation for this. There also was some variation in pupal weights over the year. For example, pupal weights decreased for both males and females over the summer generation.

Gall Growth Increment

Gall weights per pupal unit decreased rapidly with an increased population within the gall until a relatively constant level was reached above 5 mg. This represents 10–15 individuals which suggests that the individual effectiveness of gall midges in gall formation is reduced in the presence of more than one individual, imposing an eventual finite gall. Apparently the host does not respond proportionally to the quantity of cecidogenic stimulus provided after a certain threshold is reached.

Larval Development and Leaf Position

We were not able to confirm earlier indications that galls appearing on the most central leaves of the plant provide a better substrate for the larvae than those on the periphery. It has been observed with *D. urticae* Perris on *Urtica dioica* L. (O. Rohfritsch, unpublished) that the best substrate for larval development is the small leaves in the stage of fastest extension growth, rather than the youngest leaves on the plant.

Host Physiology and Gall Development

Preliminary trials aimed at modifying the vigor of the host *Viola odorata* by altering light intensity and nitrogen levels and determining the effect on *D. affinis* gall production indicate that vigorously growing hosts support a greater number of galls. However, greater gall biomass per insect was produced on plants of low vigor suggesting that a vigorous host is better able to compete with the midge for its resources. Similarly Tscharntke (1988) observed that *Giraudiella inclusa* Fr. first-instar larvae have less chance of survival on vigorous shoots than on damaged or stressed shoots of *Phragmites australis* (Cav.) Trin.

Life History

Dasineura affinis lays batches of up to 60 eggs between the bracts of young, furled leaves. The larvae hatch in four days (at 18–20°C) and move to the still furled leaf margins of the youngest leaves of the plant. The larvae undergo four molts, changing from opaque white to a bright peach color. They have a bifurcated sternal spatula. The larvae are gregarious: a gall generally contains about 14 larvae though numbers range from 1 to 60. Larval development takes about 20 days (at 20°C). The larvae then cease feeding but remain inside the gall where they spin a white silken cocoon within which they pupate. Pupation lasts 10–14 days during the summer and about 200 days over the winter. Adults apparently do not feed and the life expectancy is about 1 day for males and from 2–4 days for females.

Parasitism may be very important in some years. Ghesquière (1950) reported eight species of *Platygaster* and an ectoparasitic chalcid as parasites of *D. affinis*. Birch (1974), working in England, noted that these combined to produce a rate of parasitism approaching 60 percent of the field population studied.

LOCAL EFFECT OF GALL FORMATION ON THE HOST PLANT

Morphogenesis of Gall Formation

The first externally visible sign of gall formation is not apparent until about a week after the larvae hatch from the eggs. However, careful observation reveals an inhibition of leaf unfurling about 2 days after egg hatch. By the fifth day the furled leaf becomes hypertrophied, especially in the area immediately surrounding the larva (Vorms-Ravillard 1935).

A 7-day-old gall is more easily recognizable; the rolled leaf has become harder and the leaf margins are slightly swollen and remain furled. The color of galled leaves is often paler green than normal ones. Godan (1956, 1961) observed that the galls at this stage become red, apparently caused by anthocyanins appearing in the galled tissues. The enrolled, swollen leaf tissue encloses the larva and constitutes the gall, which with time becomes progressively larger and more hairy (Fig. 11-1). The supporting petiole of the infested leaf is stunted and thickened.

The galls are generally formed on a leaf, one on each side of the midrib (Figs. 11-1 and 11-2); however, variations do occur. Occasionally one gall is found on a leaf, and more rarely, three. Flower buds also may be attacked and when they are, the infested buds twist upward on stunted stems. The time of gall initiation and the intensity of the cecidological stimulus determine the size of the laminal area displayed between the two generally symmetrical galls on the leaf margins (Figs. 11-3 and 11-4). The lamina and petiole are most thickened and reduced

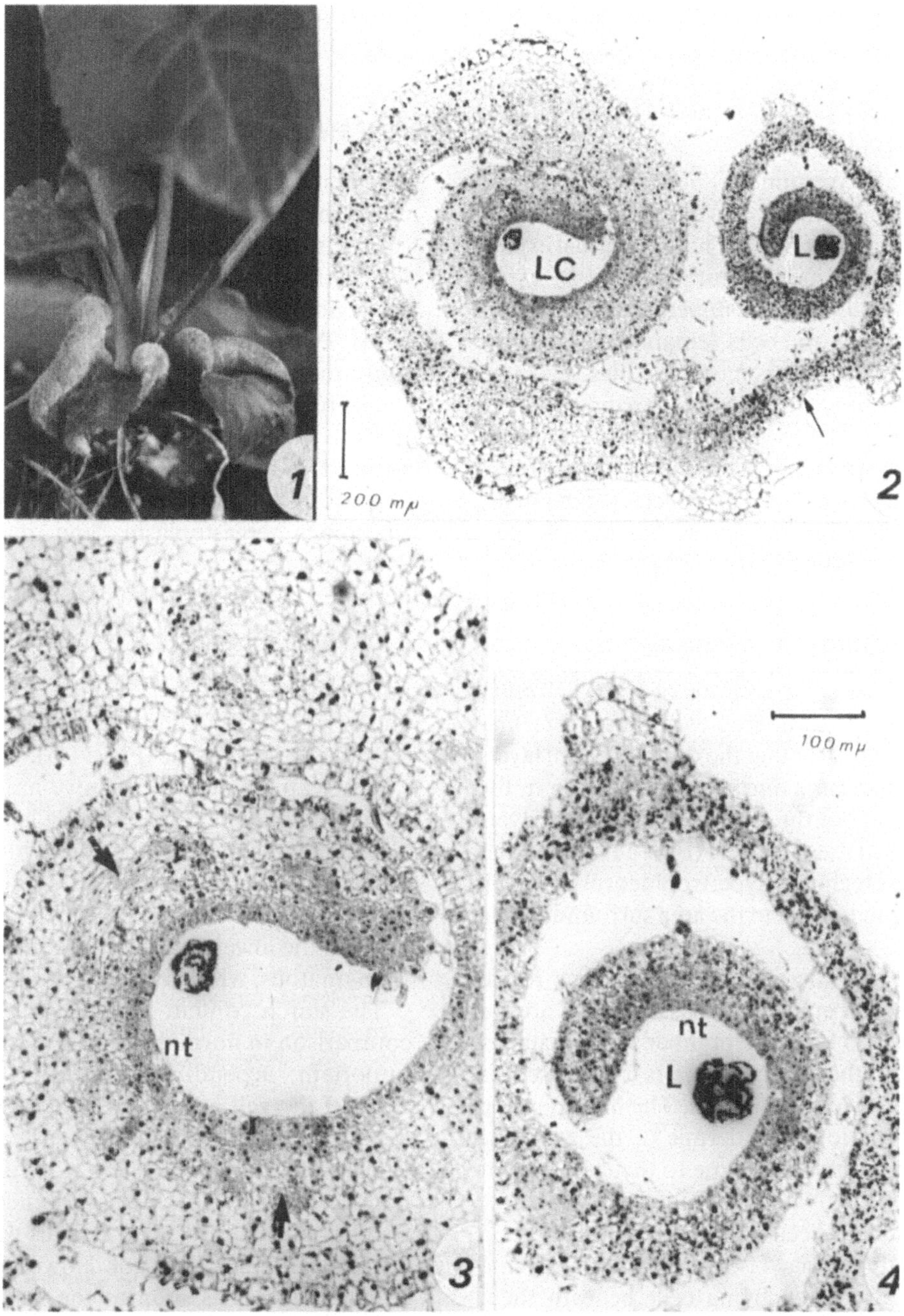

Figures 11-1–11-4. Figure 11-1. Galls of *Dasineura affinis* on *Viola odorata*. The petioles of the attacked leaves are very stunted. **Figure 11-2.** Section of a young leaf attacked by *D. affinis*. A very early stage in gall formation is shown to the right. This gall, which is 2–3 days old, is at a stage in which the different cell layers show a similar type of reaction. At the left, in a later stage, gall tissues such as the nutritive tissue (thick arrow) are beginning to differentiate. The thin arrow indicates a normal area of the leaf. (Photograph courtesy of R. Bronner and J. Meyer.) **Figure 11-3.** Detail of galled leaf shown to the left in Figure 11-2. Note the nutritive tissue, which is in contact with the larva, around the larval cavity. Hairs develop on both upper and lower epidermal layers. Vascular tissue (arrow) is found beneath the nutritive tissue. There are numerous cell divisions around the larval cavity. **Figure 11-4.** Detail of galled leaf shown to the right of Figure 11-2. Growth (cell division) of this very young gall is just starting. The lamina of the galled leaf appears relatively homogeneous and undifferentiated. (Photograph courtesy of J. Meyer.) L, Larva; LC, larval cavity; nt, nutritive tissue.

in length in a leaf galled at the youngest stage, and such leaves remain close to the crown. Leaves that are attacked slightly later in their development grow out further from the main part of the plant.

Normal leaves unfold due to differential cellular expansion between the upper and lower parts of the leaf lamina. This differential expansion is altered in leaves attacked by *D. affinis* with the result that the upper layers do not expand. Instead, the cells remain small and the number of cell layers increases (Figs. 11-3 and 11-5). There is an increased expansion in cell size towards the lower epidermis. The lack of lamina expansion in the upper epidermis keeps the gall tightly rolled. This effect is greater in the middle of the gall and diminishes laterally, parellel to the mid-rib (Birch 1974).

Gall Structure—Anatomical

When microscopic sections of young *V. odorata* leaves are compared with those from mature galls it is clear that the number of cell layers remains the same during differentiation in the normal leaf but three times as many cell layers are found in the galled leaf (Figs. 11-2–11-4 and 11-7). The region of hypertrophic growth is well defined and limited to the area surrounding the larvae.

Studies of microscopic leaf sections reveal that the lower epidermal cells of the gall are enlarged (Fig. 11-3), the number of stomata reduced, and the number of hairs is increased, as compared to normal leaves. The hairs develop all over the lower epidermis of the gall. The upper epidermal cells close to the larva appear to be highly activated to proteosynthesis. These are the nutritive cells and are the source of food for the larvae (Fig. 11-5). Epidermal cells and the palisade parenchyma cells beneath them undergo numerous divisions and the cells remain small and tightly packed (Figs. 11-5 and 11-6). The vascular tissue at the center of the modified leaf has become hypertrophied and modified. Beyond the vascular tissue there is a parenchyma tissue composed of large cells (Figs. 11-5 and 11-7). In the outer three layers of this parenchyma the cells contain chlorophyll (Bronner and Meyer 1976; Meyer and Maresquelle 1983). The most intense cell proliferation always occurs around the larva, but at a little distance away rather than immediately around it (Vorms-Ravillard 1935).

Gall Structure—Cytological Modifications

The first changes observed in young leaves attacked by *D. affinis* are modifications of the cell contents. All the cells of the leaf lamina are affected and they all react in a similar manner in the region of the larva when first attacked (Fig. 11-4). Two days after the larva has begun its activity the plant cell nucleus, with its enlarged nucleolus, becomes hypertrophied and centrally located in the cells. The vacuole becomes reduced and numerous small vacuoles appear. The cytoplasm becomes especially enriched in ribosomes, ergastoplasm, and dictyosomes. The cuticle does not develop on the upper epidermis. The cells present a pseudomeristematic aspect and are called "metaplasied" cells by Meyer (1957). Such cells have been observed in many midge galls (see Chapter 5). Cell division and new tissue differentiation will begin later in the attacked leaf tissue (Fig. 11-5).

The features described above are typical of epidermal cells in contact with a larval gall insect, though features of the young nutritive cell are less pronounced in the leaf lamina toward the vascular tissues. The other cell layers of the leaf and the lower epidermis become more parenchymatous, with large vacuoles.

The starch content of gall tissues varies in comparison to normal cells and is more or less important, depending on the developmental stage of the gall and the type of gall. In some cases starch content is increased (Shannon and Brewer 1980) and in others it is reduced (Ignoffo 1958). The normal leaf parenchyma cells also have calcium oxalate crystals whereas the normal gall tissue does not. However, this material reappears as soon as the larva is killed (Vorms-Ravillard 1935).

All cells in the mature gall become hypertrophied (parenchymatous), the nutritive tissue cells become enlarged and stretched around the larval cavity (Fig. 11-7).

Biochemical Modifications of Galled Tissue

Galls appear to be physiologically similar to fruit or meristematic tissues. In the case of the

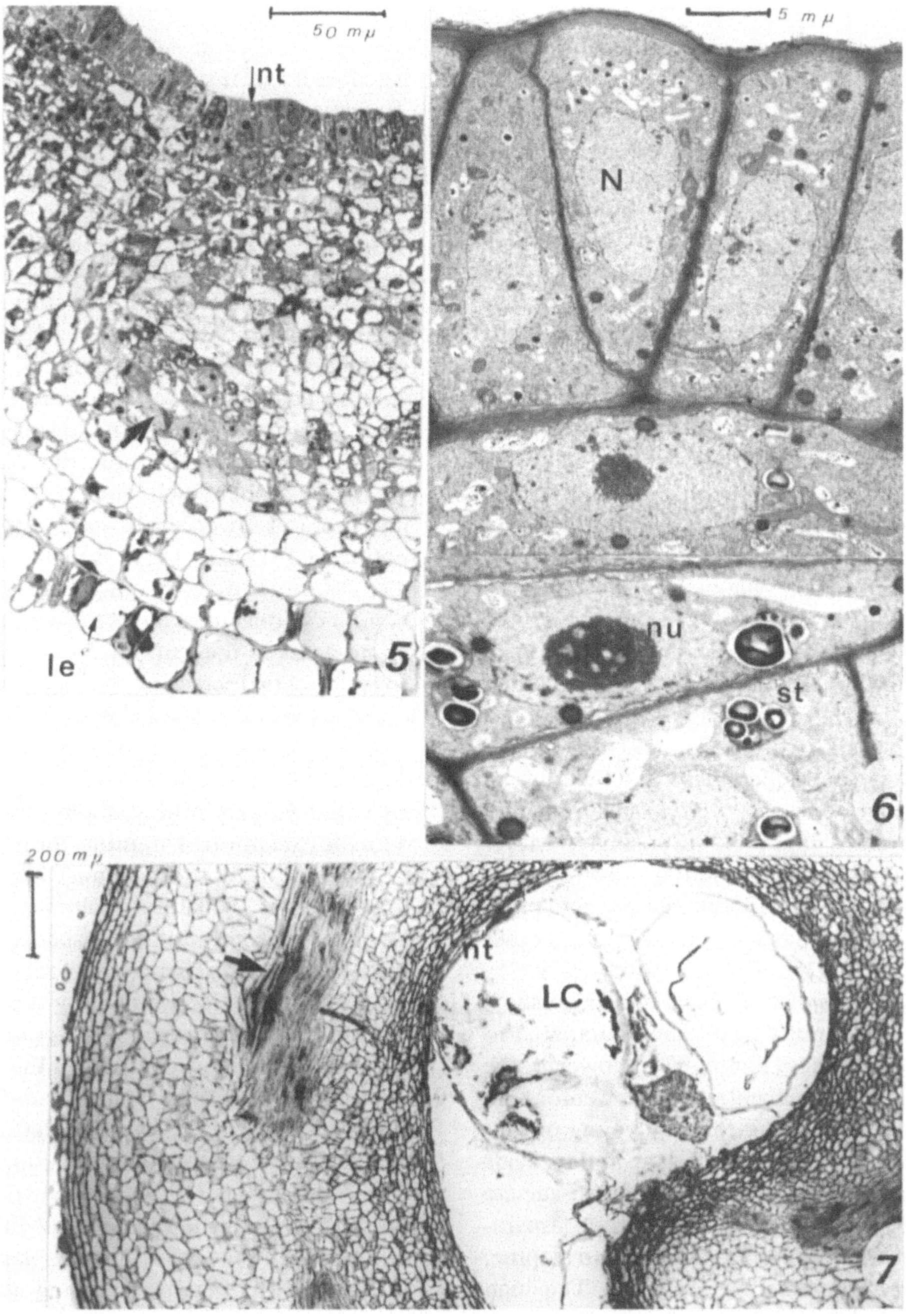

Figures 11-5–11-7. Figure 11-5. Further detail of a section of a young gall in Figure 11-3 showing upper epidermis, which has been differentiated into nutritive tissue. The nutritive tissue is six to eight cells thick and the cells become progressively more parenchymatous toward the vascular tissue, whereas the inner part of the lamina (upper part of the leaf) is a region of numerous cell divisions. The lower part consists of large parenchymatous cells. (Photograph courtesy of R. Bronner and J. Meyer.) **Figure 11-6.** High magnification of young nutritive cells 2 days after gall initiation. The epidermal cells are palisadic, rich in cytoplasm, and have a very large nucleus and nucleolus. The vacuoles are very small. Undifferentiated plastids are centrally located around the nucleus. Starch grains appear only in the second cell layer. Starch becomes more abundant toward the vascular tissue and around it. (Photograph courtesy of R. Bronner.). **Figure 11-7.** Mature gall (compare with Fig. 11-2, which is at the same magnification). The leaf lamina is now the gall wall and is very thick. The larval cavity is enlarged by cell hypertrophy. The vascular tissue (arrow) envelops the now parenchymatous nutritive tissue cells. (Photograph courtesy of J. Meyer.). LC, Larval cavity; le, lower epidermis; N, nucleus; nu, nucleolus; nt, nutritive tissue; st, starch.

leaf gall of *Aulax glechomae* L. nutrient flow is reversed so that the attacked leaf imports assimilates (Fourcroy and Braun 1967). Similarly, various authors (e.g., Jankiewicz *et al.* 1969; Kirst and Rapp 1974), working with other plant species, have demonstrated the strong drawing power of the gall for nutrients, even against the plant's own meristem. The nutrients that have been studied in various gall systems, as well as *D. affinis* on *V. odorata,* include nitrogenous materials, amino acids, starches, lipids, and others.

Nitrogenous Nutrients

Elevated levels of soluble nitrogen and protein occur in gall tissues at the expense of total nitrogen content (Ignoffo 1958; Warwick and Hildebrandt 1966; Villeret 1967; Gadal *et al.* 1969; Birch 1974; Dhingra 1981). Nitrogen levels vary considerably depending on the species and age of the gall inducer involved (Paclt and Hassler 1967). Functionally related to this, there is strong hydrolyzing activity for acid phosphatase, aminopeptidase, and invertase in the nutritive zone of *D. affinis* galls (Bronner and Meyer 1976). Not surprisingly, Meyer (1957) noted an accumulation of nitrogenous products in this layer. Soluble amino acid products also are found in the nutritive zone of other galls caused by *Dasineura* sp. (Bronner 1977).

Galled tissues of *V. odorata* of all ages show elevated total amino acid values compared to normal leaf tissues (Birch 1974); however, the total amino acid concentration is lower for 8- to 15-day-old leaves than for very young or mature leaves. The most abundant amino acids found in both normal and galled leaf tissue are amides (a complex of glutamine and asparagine), serine, α-aminobutyric acid, and alanine, in decreasing order of abundance. The high level of amides in preoviposition leaf tissues is elevated by up to three times by *D. affinis* and accounts for most of the increase in amino acid content in galled leaves. In normal leaf tissues, the amide content is reduced until growth is completed and then it returns to its original level. Molliard (1913) also demonstrated a high quantity of amides in gall tissues. He hypothesized that the amides, especially asparagine, are responsible for the hypertrophic growth and the loss of chlorophyll in gall tissues.

Serine is initially increased 9-fold in young galls of *V. odorata.* Although this level is reduced with age, it still remains higher than that of normal leaf tissues. Aspartic acid increases with age in leaf tissues and is maintained at the elevated mature leaf level throughout gall development. Threonine, alanine, proline, glycine, methionine, tyrosine, histidine, arginine, and β-aminobutyric acid are maintained at similar levels in both galled and normal tissues.

Young galled tissues contain greater concentrations of aspartic acid, serine, amides, and glutamic acid, and lower concentrations of leucine than normal tissues. In old (mature) galls, amides and serine are more abundant while glutamic acid, leucine, valine, and phenylalanine occur at lower levels than in comparable mature leaf tissues. There is usually a lack of phenylalanine in very young gall tissue. In *V. odorata* cecidogenesis alters the composition of amino acids in the leaf, raising the soluble nitrogen level by increasing the amide and serine levels. In other studies Gadal *et al.* (1969) found aspartic acid, glutamic acid, serine, amides, and alanine to be abundant both in oak leaves and gall tissue with arginine abundant only in galls. With the exception of arginine, the amino acids are the same as those which may accumulate in normal leaves. Schäller (1960) suggested that the high concentrations of asparagine, glutamine, serine, and aspartic acid found in the saliva of cecidogenic aphids may represent the end products of nitrogenous excretion in this insect and may have come from the gall tissue rich in these materials. Amino acids have been shown to be involved in the initiation of cecidogenesis of *Viteus vitifolii* Fitch on *Vitis vinifera* L. by Anders (1958). However, the main point here is that the galls provide nutrients to the cecidogen that are lost to the plant host for its own growth, thereby increasing stress on the host.

The variability in quantity of nitrogen and amino acids present, drawn to the gall, will depend on the plant species, the particular tissue involved and its age, the environmental conditions, and the species of cecidogen (Gadal 1970). The gall insect appears to increase levels of soluble amino acids, particularly amides and serine (Birch 1974). It is interesting to note that increased amides is a characteristic of sinks, since soluble nitrogen is transported mainly in this form, and that serine promotes proteolysis, which would aid in the maintenance of these elevated soluble nitrogen levels (Martin and Thiman 1972).

Sugar Nutrients

Bronner (1977) studied sugar levels in two different cecidomyiid galls, *Dasineura urticae* and *Geocrypta galii* Loew. She found that sugar (glucose) was not very abundant in tissues of young growing galls but that it appeared in high amounts in the nutritive tissue of nearly mature galls. Galls of *Dasineura affinis* have not been examined for sugars but we expect that the results might be similar to those obtained for *D. urticae.*

Starch

When starch is present in young plant organs at the time of attack, it disappears from the area near the attack as has been demonstrated by Meyer (1957), Rohfritsch (1971), and Bronner (1977) for a great number of midge galls. This is also the case with *D. affinis.* Starch appears around the vascular bundles in the vicinity of the nutritive tissue cells; however, the cells in contact with the larva remain devoid of starch (Bronner 1977).

Lipids

Lipids are known to be abundant in galls caused by cynipids but they do not normally occur in midge galls. No lipids were observed in galls caused by *D. affinis.*

Other Products

Oxalates are very common in plant tissues and they also appear often in midge galls. Galls of *Dasineura urticae* are very rich in calcium oxalate (Meyer 1957) whereas this material is absent from galls caused by *D. viola* (Vorms-Ravillard 1935). Birch (1974) found oxalic acid to be reduced 20-fold in the galled tissue of *Viola odorata.* The absence of oxalates in gall tissue is normally a characteristic of cynipid galls (Gadal *et al.* 1969).

Physiological Modifications of Gall Tissue

The formation of a gall by cecidogenic insects causes the disruption of normal ontogenic processes and produces results that are similar to the activity of plant growth regulators. Therefore, it is not suprising that growth regulators are believed to be intimately involved with gall formation (Boysen-Jensen 1948). See also Chapter 5 of this volume. These modifications in hormonal imbalances are responsible for new growth patterns and maintain the gall tissue as an importing sink. The substances believed to be involved are all plant growth regulators. It is thought that they act by altering normal growth patterns and for changing cell permeability and cell polarity.

Plant Growth Substances

It is thought that gall insects in some way maintain a high level of auxin activity in the plant for a longer period than normal. The gall is thus maintained as a sink for nutrients and the effects of hyperplasia and hypertrophy, compared with normal tissue, are produced. Galled violets and normal leaf tissues contain surprisingly similar indole promoters, apart from an extension of the initial acidic promoter profile in galled tissue. However, there may be a difference in quantity. The main disparity between the two developing tissues is the immediate loss of the neutral inhibitor and the absence of acidic inhibitors in the galled tissues. Thus the main cecidogenic effect revealed here appears to be the temporary extension of young leaf growth by increasing the auxin effectiveness (Birch 1974).

The early disappearance of neutral inhibitor in cecidogenic tissues supports Lipetz's (1959) view that although auxins may not be limiting, their inhibitors are. Thus the presence of growth regulators in galled tissue may not be a direct initiator of cecidogenesis, but a reaction to it. Once functionally active auxin levels are increased, cecidological growth may begin.

Growth enhancement and inhibition are often complementary. Although galls represent the results of growth, the effects on the plant at some distance from the gall, such as stunting, are often inhibitory. According to Byers and Gallun (1972a,b), stunting of wheat by the Hessian fly *Mayetiola destructor* (Say) results from larval feeding. The phenomenon is not just a removal of nutrients or growth promoter, but an increase in inhibitor content in some plant tissues. A similar growth inhibition is observed on hypocotyl growth of young *Bidens pilosus* L. plants a distance away from cotyledons that have been wounded by needles. The hypocotyl growth inhibition is correlated with a rapid increase of activity of basic peroxidases mediating in indole acetic acid breakdown (Desbiez *et al.* 1983).

Birch (1974) demonstrated that plants treated with growth regulators that promote vigorous growth supported higher infestations of *D. affinis* than nontreated plants. However, the cecidogenic effect was reduced, apparently as a result of competition for plant nutrients between the host plant and the galls. A growth regulator that depressed plant growth caused the opposite effect. Thus it appears that gall insects utilize the growth regulator system of their host plants by affecting both promoters and inhibitors. Since regulators are so intertwined in the control of plant growth, it also may be assumed to be the same in the growth of gall tissue.

Alteration of Normal Growth Patterns

As in other galls described by Maresquelle and Schnell (1936), Rohfritsch (1971), Rohfritsch and Shorthouse (1982), and Meyer and Maresquelle (1983), *D. affinis* very early asserts an inhibitory effect on normal leaf growth of *V. odorata.* This effect is permanent even if the cecidogen is killed during the early stages of its development. Vorms-Ravillard (1935) removed 2- to 3-day-old larvae from infested leaves and observed that they always retained their unfurled characteristics, even though the gall does not develop further. This indicates that the gall cells rapidly escape from the normal growth pattern (by the second day of attack) and transfer rapidly to the new midge-induced growth pattern (see Chapter 5).

Modifications of Cell Permeability

The galls of *D. affinis* have a higher water content than normal leaf tissue. Similar observations have been made by other workers on different types of galls (Gadal *et al.* 1969; Strebler 1971). The tissues, with high levels of soluble sugar, elevate the water content of the cells around the larval cavity; thus these cells are always turgid.

From the very beginning of gall formation, the gall tissue becomes a sink for water and nutrients. Godan (1956) demonstrated that the entire tissue implicated in cecidogenesis has a modified cell permeability. Godan (1956) used a red dye added to water in which a galled branch (*Hartigiola annulipes* on *Fagus silvatica* L.) was kept to demonstrate that the dye does not reach the gall inducer via the vascular tissue, as is normal, but rather via cell to cell transport. All of the cells around the larva transported the colored water as readily as the vascular tissue.

Kirst and Rapp (1974) demonstrated the sink effect of the midge gall induced by *Mikiola fagi* (Htg.) on *Fagus silvatica.* Nutrients (sugar) for the gall midge not only came from the galled leaf but also from the surrounding leaves. Thus, cecidogenesis affects the plant not only locally but also distally (Strebler 1971; Meyer 1957).

Effect of Galls on Different Plant Organs

The plant organ most commonly attacked by gall inducers is the leaf. However, other organs also are attacked, and the effect on the plant varies depending upon which plant organ is infested.

Leaves

The most obvious effect of galls on violet leaves is the drastic reductions in the area available for photosynthesis and transpiration. Leaf area is reduced in two ways: first, as a result of the rolling (prevention of leaf expansion) or other deformation induced by the gall maker, and second by a reduction in the number of leaves. There also is a reduced number of chloroplasts in galled tissue (Birch 1974). In addition, petioles supporting galled leaves are shorter, often thicker, and weigh less than those of normal leaves. The moisture content of galled leaves is also significantly increased over that of normal ones. The result is that photosynthetic ability of galled plants is substantially reduced in comparison with that of healthy plants. In addition, the galls maintain, and in fact increase, the nutrient sink usually associated with young leaves.

Stems

In the case of *D. affinis* on *V. odorata,* stem weights, particularly petiole weights, were also reduced in galled plants resulting in a significant reduction in the total dry weight of aerial parts, primarily leaves and petioles, although total plant weight was not affected.

Flowers and Fruits

As noted, the flowers of *V. odorata* are also attacked by *D. affinis,* if the infestation is heavy. The flower is susceptible during various stages of development with the form of the galls dependent upon the time of attack, and the stage of development of the flower. The fruits of *V. odorata* develop from flowers which are produced mainly in the early spring and ripen by the autumn. Birch (1974) reported that the weights of fruits of *V. odorata* were reduced by the presence of *D. affinis* galls and suggested that this was probably a result of competition for available nutrients in a plant of reduced photosynthetic potential. Strebler (1969) found that uninfested pods in a galled inflorescence of *Contarinia medicaginis* Kieffer on alfalfa exhibited a lower weight than usual. This reduction could be due either to a direct "hormonal" response or to nutrient diversion caused by the cecidogen. Seed production was also reduced in flower heads of *Centaurea maculosa* Lamk. and *C. diffusa* Lamk. infested by the gall flies, *Urophora affinis* Frfld. and *U. quadrifasciata* (Meig.) compared to uninfested flower heads (Harris 1980). Thus some damage by gall midges may be sufficient to reduce the reproductive potential of infested plants.

Roots

Infestations of *D. affinis* had no effect on the weight of violet roots. This is surprising since infestation was expected to direct nutrients to the gall insect, which otherwise would be available for storage in the root.

OVERALL EFFECTS ON THE PLANT

Infestation of *V. odorata* by *D. affinis* appears to affect the entire plant by causing a reduction in photosynthesis, a depletion of water and nutrients through competition between the gall and normal leaf development, and seed production.

In general, galled plants are slower growing and the leaves are thicker. The reduction in leaf number indicates a diminution of the meristematic activity while the gall insect is active on the host. The level of new leaf production increases on an infested *V. odorata* plant as soon as pupation occurs, and decreases when new cecidogenesis begins. This reduction in the formation of new leaves is most marked on heavily infested plants where many of the galls occur close to the crown, as opposed to those on longer petioles (Birch 1974). Many cecidogen-infested hosts exhibit this "inhibitive" or stunting effect, which is generally a result of reduced internode length. Successive heavy infestations of individual plants may reduce photosynthetic ability and lead to the eventual death of the plant through the repeated drain on its reserves (Godan 1956).

These effects of infestation on the overall plant may be a result of two different factors: disruption and depletion. These may affect the nutritional and/or hormonal mechanisms of the plant upsetting normal plant processes, resulting in abnormal plant growth and development. This reduced photosynthetic area of the infested plant, together with the cecidogenic sink, depletes the plant's reserves and promotes this inhibitory effect of the gall insect. Inhibition occurs very early during cecidogenesis and may be caused by growth inhibitors or by a reduction of promoters, or other factors. It is interesting that root development does not appear to be affected, either by hormonal disruption or resource depletion. It appears that the galled plants require a higher level of moisture than do normal plants. Perhaps this is why the root systems of infested plants remain well developed. At least it appears that root and gall development are not competitive. A high level of gall development seems to be accompanied by a well-developed root system just as in a normal situation a highly developed leaf system is accompanied by a well-developed root system.

DISCUSSION

The morphological and chemical changes caused by *Dasineura affinis* on *Viola odorata* have been used to illustrate the effects of cecidogenesis on host plants. The morphological changes, both local and those affecting the plant as a whole, are fairly evident. However, the biochemical changes, particularly those concerned with nutritional aspects, are only just beginning to be understood. Very few data exist on the physiology of midge galls; however, the observations and results of the effect of *Dasineura affinis* on *Viola odorata* are similar to those of Strebler (1969, 1971) on the alfalfa flower gall induced by *Contarinia medicaginis.* This suggests that our observations on the *D. affinis* gall may be generally indicative of the relations between many cecidomyiid gall inducers and their host plants. However, it is important to remember that cecidogenesis is polyphyletic (Chapter 3) and that different groups of insects may have evolved different methods for manipulating host plant tissues leading to the production of galls (see Chapter 5).

In this very specific reaction between the plant and insect, the mechanisms and importance of both growth promoters and inhibitors have yet to be fully elucidated nor indeed is the initial reaction known. The results depend on the plant's susceptibility, which is governed externally by the environment and internally by hormones, and specific recognition systems under genetic control.

The violet gall is possibly a result of an auxin-induced nutrient diversion, which reduces apical dominance and prevents leaf growth and petiole extension, perhaps with the aid of inhibitors. This inhibitory phenomenon appears early and is more or less important in nearly all cecidomyiid-induced galls. It is not related to nutrient uptake at the beginning of cecidogenesis. Larvae begin stimulating and feeding in tissues that are already importing nutrients and then increase, maintain, and extend the activity of this normal sink in competition with the rest of the plant. The cecidogen appears able to temporarily extend young leaf growth by increasing the auxin effectiveness. This induced sink will draw in nutrients that the cecidogen maintains selectively by continuous "wounding" and by increased serine levels that promote proteolysis and maintain the increased soluble nitrogen level.

Environmental conditions and plant growth regulation treatments that reduce the plant's vigor, and hence apical strength, will promote the production of gall biomass. Vigorous plants are able to support more galls with less damage to the plant but since the competition between the plant and insect is greater, gall biomass and pupal weight are reduced.

At the beginning of cecidogenesis, vigorous plant growth can be detrimental to growth of the gall. However, once the gall is established, it becomes a stronger nutrient sink than the normal water and nutrient attractor in the plant. Galls always have a higher water content than normal tissue and this characteristic becomes especially important in a dry environment. If the amount of water available to the plant is reduced, the gall will usurp most of the reserves because of its greater attractiveness. Thus the gall will be the most resistant part of the plant to water stress and may help to explain why the root system is not affected by cecidogenesis.

The effects of gall formation on the plant as a whole depend on the specific tissues being galled and on the number of cecidogens present, since attacks result in relocation and loss of chemicals important for normal plant growth. When leaves are attacked the photosynthetic capabilities, and thus nutrient production of the plant, are reduced. However, the effect on the host plant, is not usually so detrimental that it is killed.

Like many cecidomyiids, *Contarinia medicaginis* (observed by Strebler) and *Dasineura affinis* are strictly monophagous, and appear to have evolved regulatory mechanisms to control their populations. In both cases, there is an optimum number of larvae per larval cavity; above this number the pupae become small and the number of eggs per female diminishes. If the population is too high, some larvae may not mature. Competition between host and insect results in reduced plant growth, vigor, and seed production. At least in these two cases it appears that a balance has evolved between the competing processes of maximum plant growth and gall production.

LITERATURE CITED

Anders, F. 1958. Aminosäuren als gallenerregende Stoffe der Reblaus (*Viteus (Phylloxera) vitifolii* Shimor). Experientia 14:62–67.

Barnes, H. F. 1927. Material for a monograph of the British Cecidomyiidae. British gall midges of economic importance. I-V. J. S. E. Agric. Coll. Wye 24:65–146.

Barnes, H. F. 1948. Gall Midges of Economic Importance. IV: Ornamental Plants and Shrubs. Crosby Lockwood, London.

Birch, M. L. 1974. Studies on gall formation in *Viola odorata* by *Dasyneura affinis* Kieffer (Diptera: Cecidomyiidae) with emphasis on the experimental modification of this process. Ph.D thesis, University of Reading, England.

Boysen-Jensen, P. 1948. Formation of galls by *Mikiola fagi*. Physiol. Plant 1:95–108.

Brewer, J. W. 1971. Biology of the pinyon stunt needle midge. Ann. Entomol. Soc. Am. 64:1099–1102.

Bronner, R. 1977. Contribution a l'étude histochimique des tissus nourriciers des zoocécidies. Marcellia 40:1–134.

Bronner, R., and J. Meyer. 1976. Cécidogènese et aspects ultrastructuraux de la galle de *Dasyneura affinis* Kieffer sur *Viola odorata* L. Marcellia 39:141–154.

Byers, R. A., and R. L. Gallun. 1972a. The ability of the hessian fly to stunt winter wheat. 1. The effect of larval feeding on elongation of leaves. J. Econ. Entomol. 65:955–958.

Byers, R. A., and R. L. Gallun. 1972b. 3. Gibberellic acid and elongation of second and third leaves of infested plants. J. Econ. Entomol 65:1001–1004.

Couderc, J. 1933. Etude monographique de la cécidomyie de la violette. Bull. Soc. Hist. Nat. Toulouse LXV:193–279.

Colizza, C. 1928. Il moscerina delle viole (*Dasyneura affinis* Kieffer, Diptera, Cecidomyiidae) nell Italia meridionale. Boll. Lab. Zool. Portici 21:130–148.

Desbiez, M. O., P. Champagnat, N. Boyer, J. M. Frachisee, Th. Gaspar, and M. Thellier. 1983. Inhibition correlative de la croissance de l'hypocotyle de *Bidens pilosus* L. par des traumatismes cotylédonaires légers. Bull. Soc. Bot. Fr. 130. Actual. bot. 314:67–77.

Dhingra, S. 1981. Studies on some insect and mite induced plant galls of Rajasthan. Ph.D. Thesis, University of Rajasthan, India.

Faure, J. C. 1924. Sur la cécidomyie des violettes (*Perrisia affinis* Kieffer). Feuilles jeun. Nat. 45:150.

Fourcroy, M., and C. Braun. 1967. Observations sur la galle de l'*Aulax glechomae* L. sur *Glechoma hederacea* L. 2. Histologie et rôle physiologique de la coque sclérifiée. Marcellia 34:3–30.

Gadal, P. 1970. Sur le métabolisme du chêne et de ses zoocecidiés-Etude particulière de la biosynthèse des acides aminés aromatiques. Thèse D. E., Toulouse, France.

Gadal, P., S. Laurens, A. Moillo, and A. Pareilleux. 1969. Les zoocécidies du genre *Quercus*. Etude de leur composition en substances azotées et en acides organiques. Marcellia 35:225–245.

Ghesquière, J. 1950. Parasites de la cécidomyie des violettes en Belgique. Bull. Soc. Entomol. Belg. 86:64–65.

Godan, D. 1956. Beiträge zur Autökologie der Veilchenblattrollmücke *Dasyneura affinis* Kieff. Z. angew. Entomol. 39:1–19.

Godan, D. 1961. Untersuchungen über die Beziehung zwischen Parasit und Wirtspflanze bei *Dasyneura affinis* Kieffer (Diptera, Cecidomyiidae). Z. Parasitenk. 21:290–300.

Harris, P. 1971. Weed vulnerability to damage by biological control agents. Miscell. Pub. #6 Commonwealth Inst. Bio. Con., Dec. 73:29–39.

Harris, P. 1980. Effects of *Urophora affinis* Frfld. and *U. quadrifasciata* (Meig.) (Diptera: Tephritidae) on *Centaurea diffusa* Lam. and *C. maculosa* Lam. (Compositae) Z. angew. Entomol. 90:190–201.

Houseweart, M., and J. W. Brewer. 1972. Biology of a pinyon spindle gall midge (Diptera: Cecidomyiidae). Ann. Entomol Soc. Am. 65:331–336.

Ignoffo, G. M. 1958. The life history and gall development of *Mordwilkoja vagabunda* Walsh. (Homoptera: Aphidae) on *Populus deltoides*. Ph.D. Thesis, University of Minnesota.

Jankiewicz, L. S., H. Plich, and H. Antosxewski. 1969. Preliminary studies on the translocation of ^{14}C labelled assimilates and $^{32}PO_4^{3-}$ towards the gall, evoked by *Cynips (Diplolepis) quercusfolii* L. on oak leaves. Marcellia 36:163–172.

Kirst, G. O., and H. Rapp. 1974. Zur Physiologie der Galle von *Mikiola fagi* Htg. auf Blättern von *Fagus silvatica* L. 2. Transport ^{14}C markierter Assimilate aus dem befallenen Blatt und aus Nachbarblättern in die Galle. Biochem. Physiol. Pflazen. 165:445–455.

Lipetz, J. 1959. A possible role of indole acetic oxidase in crown gall tumor induction. Nature (London) 184:233–292.

Maresquelle, H. J., and R. Schnell. 1936. Etudes experimentales des phases de l'action cécidogène dans une galle. C. R. Acad. Sc. Paris. 203:270–272.

Martin, C., and K. V. Thiman. 1972. Role of protein synthesis in senescence of leaves. Plant Physiol. 50:432–437.

Meyer, J. 1957. Cécidogenèse comparée de quelques galles d'Arthropodes et évolution cytologique des tissus nourriciers. Thèse, Strasbourg.

Meyer, J., and H. J. Maresquelle. 1983. Anatomie Des Galles. Gerbrüder Borntraeger, Stuttgart.

Molliard, M. 1913. Recherches physiologiques sur les galles. Rev. Gén. Bot. Fr. 25:225–252, 285–307, 341–370.

Paclt, J., and J. Hässler. 1967. Concentration of nitrogen in some plant galls. Phyton (Australia) 12:173–176.

Raymond, G. 1928. Contribution à l'étude des Cécidomyies attaquant les violettes cultivées. Ann. Soc. Linn. 63:87–101.

Rohfritsch, O. 1971. Développement cécidien et rôle

du parasite dans quelques galles d'Arthropodes. Marcellia 37:233–339.

Rohfritsch, O. 1980. Relations hôte-parasite au début de la cécidogenèse. Soc. Bot. Fr. (Actual Bot.). 127:199–207.

Rohfritsch, O., and J. D. Shorthouse. 1982. Insect galls. In G. Kahl and J. S. Schell (eds.), Molecular Biology of Plant Tumors, pp. 131–152. Academic Press, New York.

Schäller, G. 1960. Untersuchungen über den Aminosäurengehalt des Speicheldrüsensekretes der Reblaus (*Viteus vitifolii* Shimer) (Homoptera). Entomol. Exp. Appl. 3:128–136.

Shannon, R. E., and J. W. Brewer. 1980. Starch and sugar levels in 3 coniferous insect galls. Z. angew. Entomol. 89:526–533.

Stokes, B. 1955. Behavior as a means of identifying two closely allied species of gall midges. Gr. J. Anim. Behav. 3:154–157.

Strebler, G. 1969. Etude d'un Itonididae cécidogène (*Contarinia medicaginis* Kieffer) et de son hôte. Marcellia 35:155–188.

Strebler, G. 1971. Quelques aspects de la cécidogènese dans le cas des galles de fleurs dues à *Contarinia medicaginis* Kieffer. Marcellia 37:3–46.

Tscharntke, T. 1988. Variability of the grass of *Phragmites australis* in relation to the behaviour and mortality of the gall-inducing midge *Giraudiella inclusa* (Diptera, Cecidomyiidae). Oecologia 76:504–512.

Villeret, S. 1967. Recherches sur le metabolisme azoté des galles et des feuilles du hêtre parasitées par *Mikiola fagi* Hart. Bull. Soc. Sci. Bretagne XLII:193–197.

Vorms-Ravillard, A. 1935. Modifications anatomiques de galles arrêtées dans leur développement. Thèse Sc. Nat. Paris, Edit. Proba.

Warwick, R. P., and A. C. Hildebrandt. 1966. Free amino acid content of stem and phylloxera gall tissues of grapes. Plant Physiol. 41:573–578.

Zwölfer, H. 1976. Investigations on *Sphenoptera* (Chilostetha) *jugoslavica* Obenb. (Col. Bupresidae), a possible biocontrol agent of the weed *Centaurea diffusa* Lam. (Compositae) in Canada. Z. angew. Entomol. 80:170–190.

12

Unique Aspects in the Biology of Thrips-Induced Galls

T. N. ANANTHAKRISHNAN

Galls induced by thrips are perhaps less well known to most biologists than are the galls of other species. Nonetheless, thrips are responsible for a vast diversity of structural abnormalities on a wide variety of plants. The diminutive size of thrips (0.5–12 mm) and the relative difficulty in culturing and curating them are compensated by the occurrence of large populations within galls, a feature equally characteristic of most thrips galls. Thrips are placed in the Order Thysanoptera (thysano—fringe) because of the fringed nature of their wings and their bladder-footed tarsi. Two suborders, the Terebrantia and the Tubulifera, are recognized. The Terebrantia (Fig. 12-1a) have wings lying parallel to each other bearing veins, cross veins, and setae. They generally have an ovipositor and only a single pupal stage. The Tubulifera (Fig. 12-1b) have wings that overlap and are without veins. They do not have an ovipositor, but have two pupal stages.

The majority of the thrips including most species of economic importance belong to the family Thripidae of the suborder Terebrantia and the family Phlaeothripidae of the suborder Tubulifera. Most of the gall-inducing species belong to the subfamily Phlaeothripinae of the family Phlaeothripidae; however, a few species of Terebrantia also are reported to induce galls. Studies of the biology of gall thrips have revealed several interesting relationships that are unique to the gall-inducing guild. Some of the more intriguing aspects of thrips–gall complexes are discussed here.

GENERAL COMMENTS ON THE BIOLOGY OF THRIPS

There are about 6000 species of thrips known the world over with a preponderance of species occurring in the tropics (Mound *et al.* 1980). Many species have received the attention of entomologists due to their importance as pests of agricultural crops. For a general review of the biosystematics, vector potential, as well as the phylogeny and biology of Thysanoptera, refer to Ananthakrishnan (1979a, 1979b, 1984a) and Mound *et al.* (1980).

The majority of thrips are phytophagous, attacking either flowers, pollen, leaves, fruits, twigs, or buds and a great many species of plants serve as hosts. Some thrips are predatory on small arthropods and a few are omnivorous (Lewis 1973). Thrips are often gregarious and are characteristically confined to a particular part of the plant. Based on habitat, thrips may be grouped as anthophilous or flower inhabiting, phyllophilous or leaf inhabiting, peophilous or grass inhabiting, phlaeophilous or bark inhabiting, and cecidogenous or gall inhabiting (Ananthakrishnan 1973). Many phytophagous thrips are oligophagous, although all gall inducers are strictly monophagous. Phytophagous species often restrict their feeding to rapidly growing tissues.

The mouthparts of thrips are of the "punch and suck" type and the proboscis, called a "mouth cone" (Fig. 12-2), is located posteriorly on the ventral surface of the head. The mouth-

Figure 12-1. (a) Typical terebrantian thrips. (b) Typical tubuliferan thrips.

parts are asymmetrical and include three stylets: the left mandible and the laciniae of the two maxillae (Heming 1978). The mouthparts also include an extrusible labial pad endowed with small palpi. The function of this palpi is unknown; it may be sensory or it may prevent the seepage of plant sap during feeding (Chisholm and Lewis 1984; Hunter and Ullman 1989). In the Terebrantia the maxillary stylets are short and rigid and unlike the Tubulifera they are not retracted far into the head. In the Tubulifera the stylets are more flexible and the length varies considerably, being retracted deep into the head region as far as the eyes and in extreme cases in *Casuarina*-feeding species, the stylets cross each other forming complex convolutions before reaching the mouth cone (Mound 1970).

When the mouth cone is short as in most Terebrantia and some Tubulifera, there is hardly any penetration of stylets and so feeding is often of the shallow type, resulting in the rupture of only the epidermal and a few superficial mesophyll cells. Species with longer mouth cones, as in the Tubulifera, have the penetrative mode of feeding, extending deeper into the leaf tissues, but rarely reaching the vascular tissues. The musculature of the head as well as those around the pharynx aid in the ingestion of cell sap by their contractile and pumping action. The maxillary stylets have a "tongue and groove" system on either side and fit into each other along their length. The two grooved maxillary stylets form a tube through which the liquid contents are sucked up (Mound 1971a).

Approximately 300 species of thrips are associated with galls, both as gall inducers and inquilines. Almost 75 percent of the gall-inducing thrips belong to the genera *Liothrips, Gynaikothrips, Liophlaeothrips, Eugynothrips, Eothrips,* and *Mesothrips.* Most gall-inducing thrips are tropical, being most common in Indo-Malaysia (Karny 1912, 1921, 1922; Ananthakrishnan 1978) and Australia (Mound 1969, 1970, 1971b). Others occur on Pacific Islands (Bagnall 1928) and North Africa. North

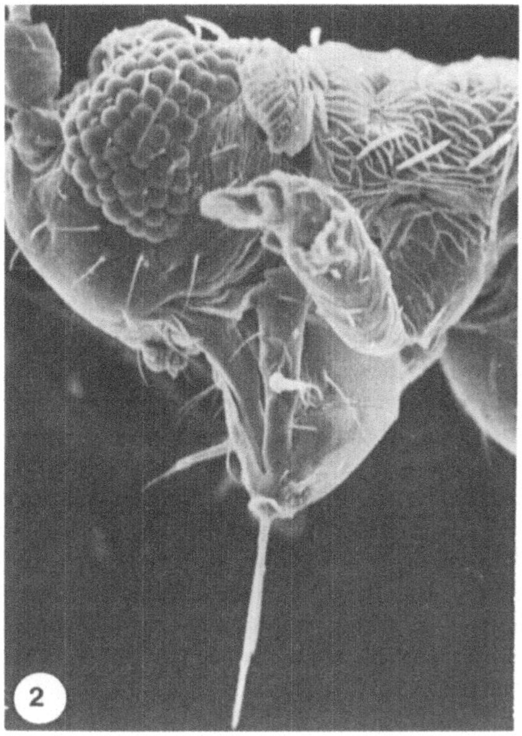

Figure 12-2. Mouth cone and extended stylet of a phytophagous terebrantian thrips.

America and Europe are poorly represented by gall-inducing thrips (Wahlgren 1945; Wood 1960).

Thrips cause about 10 percent of the galls in the Indo-Malayan region. Host plants here range from Pteridophytes such as *Asplenium nidus* L. and *Pleopeltis pteropus* T. Moore (Docters Van Leeuwen 1938) to monocotyledonous species of Araceae, Liliaceae, and Poaceae (Karny and Docters Van Leeuwen-Reijnvaan 1913). They occur on about 200 plant species in at least 60 families. The most commonly galled hosts are members of Araliaceae, Euphorbiaceae, Melastomaceae, Mimosaceae, Moraceae, Myrtaceae, Piperaceae, and Vitaceae (Ananthakrishnan 1978, 1980). There are several thrips-generic complexes associated with specific plant generic complexes. Striking instances of such relationships are *Cordia* (Boraginaceae)–*Aneurothrips* sp., *Schefflera* (Araliaceae)–*Liothrips* sp., *Maytenus* (Celastraceae)–*Alocothrips* sp., *Ficus* (Moraceae)–*Gynaikothrips* sp., and *Mallotus* (Euphorbiaceae)–*Liothrips* sp.

DIVERSITY OF THRIPS GALLS

Galls induced by thrips are usually simpler than those induced by Diptera and Hymenoptera. Many thrips galls are of the organoid type and retain close structural similarity to the attached organ; however, an endless variety of thrips galls have been described (Ananthakrishnan 1978). Most thrips galls are produced on leaves (Figs. 12-3–12-7) and they range in shape from a mere curling of the two halves of the leaf blade on either side of the midrib without the margins meeting, through more pronounced leaf rolling, folding, and swelling, to bizzare bladder, pouch, or horn galls protruding from a swollen distorted blade (Karny and Docters Van Leeuwen-Reijnvaan 1913; Ananthakrishnan 1973, 1978; Ananthakrishnan and Jagadish 1969; Raman 1981). Well-organized galls with definitive tissues are initiated on immature leaves whereas more simple galls consist of parenchymatous cells (kataplasmic galls). The more complex galls are composed of differentiated cell types (prosoplasmic galls).

Ananthakrishnan (1973, 1978) classified galls into 10 distinct categories ranging from the simple marginal leaf roll and fold galls to more complicated pouch, rosette, and horn galls (Table 12-1). About 50 percent of thrips galls are leaf rolls, 25–30 percent are leaf folds, and 10–12 percent are leaf curls. Only about 8–10 percent are of the histioid-prosoplasmic type where the gall tissues are markedly different from the host leaf.

Leaf roll galls induced by species of *Liothrips* and *Gynaikothrips* (Fig. 12-3) are especially common on species of *Ficus* and are widespread (Bagnall 1928; Priesner 1929). *Liothrips karnyi* (Bagnall) feeding on the underside of the leaf margins of *Piper nigrum* L causes a continuous marginal fold of the leaf margins (Fig. 12-4). *Psenothrips priesneri* Anan. attacking young leaves on *Walsura piscidia* Roxb. causes a laminar roll gall (Fig. 12-5) formed by inrolling of both the margins meeting together at the midrib. A similar but hypophyllous inrolling is caused by the feeding of *Liothrips ater* Anan. and Jag. and *Teuchothrips longus* (Schmutz) on a wild species of pepper and *Pavetta hispidula* L. to Hiern, respectively.

The most striking example of a pouch gall is

Table 12-1. Types of Galls Induced by Various Species of Thrips

Gall type	Species and host
Folds	
	Arrhenothrips dhumrapaksha Ramk. on *Ficus retusa* L.
	Arrhenothrips ramakrishnae Hood on *Mimusops elengi* L.
Marginal partial rolls	
	Aneurothrips punctipennis Karny on *Cordia dichotoma* Forst F.
	Aneurothrips priesneri Bhatti on *Cordia obliqua* Willd.
	Alocothrips hadrocerus (Karny) on *Maytenus senegalensis* (Lam)
Marginal complete rolls	
	Liothrips karnyi (Bagnall) on *Piper nigrum* Linn.
	Lygothrips jambuvasi (Ramk) on *Anogeissus latifolia* Wall.
Laminar complete rolls	
Simple	
	Gynaikothrips flaviantennatus Moulton on *Casearia tomentosa* Roxb.
	Eothrips coimbatorensis Ramk. on *Jasminum pubescens* Billd.
Complex	
	Crotonothrips gallarum on *Memecylon* sp.
	Crotonothrips sp. on *Memecylon* sp.
Pouches	
	Austrothrips cochinchinensis Karny on *Calycopteris floribundus* Lamk.
	Dixothrips onerosus Anan. on *Terminalia chebula* Retz.
Horns	
	Liothrips ramakrishnae Anan. and Jag. on *Schefflera racemosa* Harms
	Liothrips associatus Anan. and Jag. on *Schefflera racemosa* Harms
Rosette	
	Thilakothrips babuli Ramk. on *Acacia leucophloea* Willd.
Tubular type	
	Mesicothrips plicans Priesner on *Lonicera* sp.
Floral gall	
	Thilakothrips babuli Ramk. on *Acacia leucophloea* Willd.
Stem	
	Thaumatothrips froggati Karny on *Casuarina stricta* Miq.
	Eurhynchothrips flavicornis Bagnall. on *Microgyne microphylla* D.C.

that induced by *Austrothrips cochinchinensis* Karny on *Calycopteris floribundus* Lamk. (Fig. 12-6). This irregular, globose, unilocular gall is often about 10 cm in diameter. It is a convoluted and lobed, glabrous or pubescent hollow bag that develops from auxillary buds. The gall cavity is composed of intricated, tortuous interspaces, separated by invaginations and folds of the wall. Another interesting pouch gall is that of *Dixothrips onerosus* Anan. on leaves of *Terminalia chebula* Retz. This is an ultricular, dehiscent, pyriform, or subglobose pouch that arises on the leaf blade. The ostiole is large and epiphyllous.

Interesting horn galls are formed by several species of *Liothrips* on different species of *Schefflera*. *Liothrips heptapleuri* Karny, which is common on *Schefflera elliptica* (*S. racemosa* Harms.) causes irregular cylindrical, mostly curved horn-like structures. *Liothrips associatus* Anan. & Jag. and *L. ramakrishnae* Anan. and Jag. form similar horn galls on *Schefflera racemoca* Harms (Fig. 12-7). Feeding by large populations causes the simple, tubular galls to become converted into complex, branched, fleshy structures. The walls of the galls are succulent, enclosing a long narrow larval chamber with an ostiole at the underside of the leaf connected to the outside.

INITIATION AND DEVELOPMENT OF THRIPS GALLS

Gall-inducing thrips, as with non-gall-inducing phyllophilous species, do not have a specific pattern of infestation. Instead gall thrips invoke a variety of feeding responses depending upon site of attack, number of thrips present and age of the leaf (Gopinathan 1987). This is especially evident in galls of *Arrenothrips ramakrishnae* Hood on leaves of *Mimusops elengi* L. where feeding along the midrib prevents the laminar blades from opening. When thrips feed along the leaf margin, they either produce a leaf roll or keep juvenile leaves from unfolding.

Most thrips galls develop as a result of the cumulative feeding effects of a population, although during the initial stages of cecidogenesis, only a few individuals may be present. The most salient aspects of the interrelationships between cecidogenous thrips and their hosts are summarized by Ananthakrishnan and Raman (1989). Some species can induce galls with only one individual, but a large population is usually required to complete the gall. The horn gall induced by *Liothrips heptapleuri* is an exception. Here the females puncture the underside of the

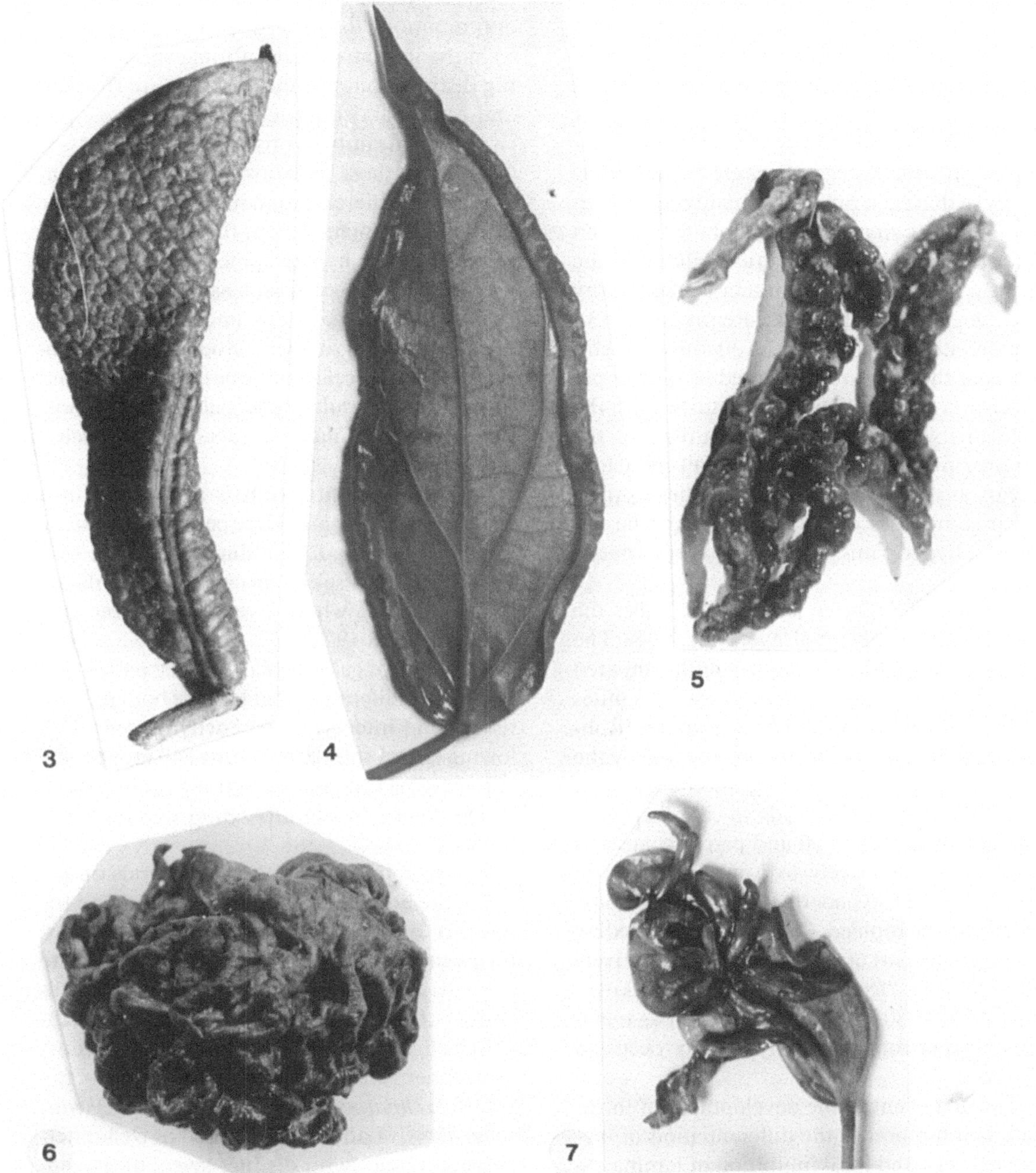

Figures 12-3–12-7. Thrips galls found on the leaves of various plants in India. **Figure 12-3.** Leaf roll gall of *Gynaikothrips uzeli* on *Ficus retusa.* **Figure 12-4.** Marginal fold gall on the underside of the leaf of *Piper nigrum.* **Figure 12-5.** Laminar roll gall of *Psenothrips priesneri* on *Walsura piscidia.* **Figure 12-6.** Pouch gall of *Austrothrips cochinchinensis* on *Calycopteris floribundus.* **Figure 12-7.** Horn galls of *Liothrips ramakrishnae* on *Schefflera racemosa.*

leaves before the tissues are fully differentiated and within a day transparent spots appear on the upper surface above each puncture. One female can initiate a large number of galls on each leaflet. The spots turn red and grow into a tubular gall 6–8 mm long in about 6 days. Female thrips then enter each hollow chamber to lay eggs. Eggs are laid intermittently so that each gall may contain eggs, larvae, and adults at any one time. Only females can induce new galls but immature galls continue to grow as long as they contain larvae (Karny and Docters Van Leeuwen-Reijnvaan 1913). Mature tubular galls are often 30 mm in length. Walls of the galls are composed of spongy parenchyma tissues about 20 cells thick, with the phloem toward the inside. Regardless of the number of individuals present, gall induction occurs as a result of wounding either a single cell in some cases, or a cluster of cells of the epidermis, hypodermis, or the first layer of mesophyll.

Feeding sites can be localized on either the upper or lower epidermal layers (Fig. 12-8). The effects of wounding are important if compared to the minute feeding punctures made by mites (Westphal 1977) or the probing of aphids (Rohfritsch 1976). The initial damage caused by the "punch and suck" feeding mechanisms of thrips (Heming 1978) leads to cell disruption and necrosis. Mesophyll and perivascular cells react more extensively to wounding than do epidermal and hypodermal cells and become greatly hypertrophied. Cells nearest the feeding sites become part of the gall and are collectively referred to as "feeding zones" (Ananthakrishnan 1978, 1980). Cell reactions such as metaplasy, hypertrophy, and hyperplasy occur in this zone.

The first event in the development of thrips galls is inhibition of the differentiation of mesophyll cells and local inhibition of lamina extension. Cell proliferation and hypertrophy follow and the thrips are enclosed in a cavity. The cellular disturbance is usually restricted to a small area, but this is sufficient to cause the host leaf to roll or fold. Irregular periclinal divisions of mesophyll cells often contribute to the thickening of these galls by 40 to 75 galls.

Feeding by some species of gall-inducing thrips affects not only the outermost epidermal layer (upper or lower) but also a few mesophyll layers up to the vascular traces. Nutritive cells appear among the hypertrophied cells (Fig. 12-8) in the immature galls of some species forming distinct zones as the galls mature (Raman and Ananthakrishnan 1983a).

As with the nutritive tissues of other galls (see Chapter 8), those of thrips galls undergo metaplasia or dedifferentiation in which the volume of cytoplasm increases and the nucleus and nucleolus become hypertrophied (Fig. 12-9). The walls of nutritive cells such as those in *Arrhenothrips*-induced galls are unevenly thickened and have deposits in their cytoplasm. Hypertrophy of affected cells continues as the galls mature and as a result, some galls consist almost exclusively of enlarged, vacuolate parenchymatous cells (Fig. 12-10).

Epidermal proliferation is also found in a number of thrips galls. Here periclinal divisions can occur adding to the thickness of the gall. However, only select groups of cells divide rather than the whole layer (Raman and Ananthakrishnan 1979).

Some thrips galls display striking patterns of vascular proliferation that involve both phloem tissues and interxylary parenchyma cells. Following lateral stretching during gall growth, the parenchyma cells show extensive proliferation, which results in the separation of xylary elements.

The structure of thrips galls depends on the age of the host leaf at the time of attack and the number of thrips at each feeding site. Most thrips galls are structurally simple compared to the galls of other insects. Leaf fold or roll galls consist of an increase in the number of constituent cell layers of laminar extensions. However, other galls such as those induced by *Alocothrips hadrocerus* Karny on *Maytenus senegalensis* Lam. are composed of well-differentiated tissues with distinct layers of sclerenchyma and tanniniferous cells (Ananthakrishnan 1978). In other galls, the sclerenchyma layer is discontinuous. In the gall of *Teuchothrips longus* on *Pavetta hispidula* the walls of some mesophyll cells near the feeding sites are thickened (Raman and Ananthakrishnan 1983b). The nutritive cells around the feeding sites are free of these wall thickenings. Some thrips galls also show proliferation of vascular

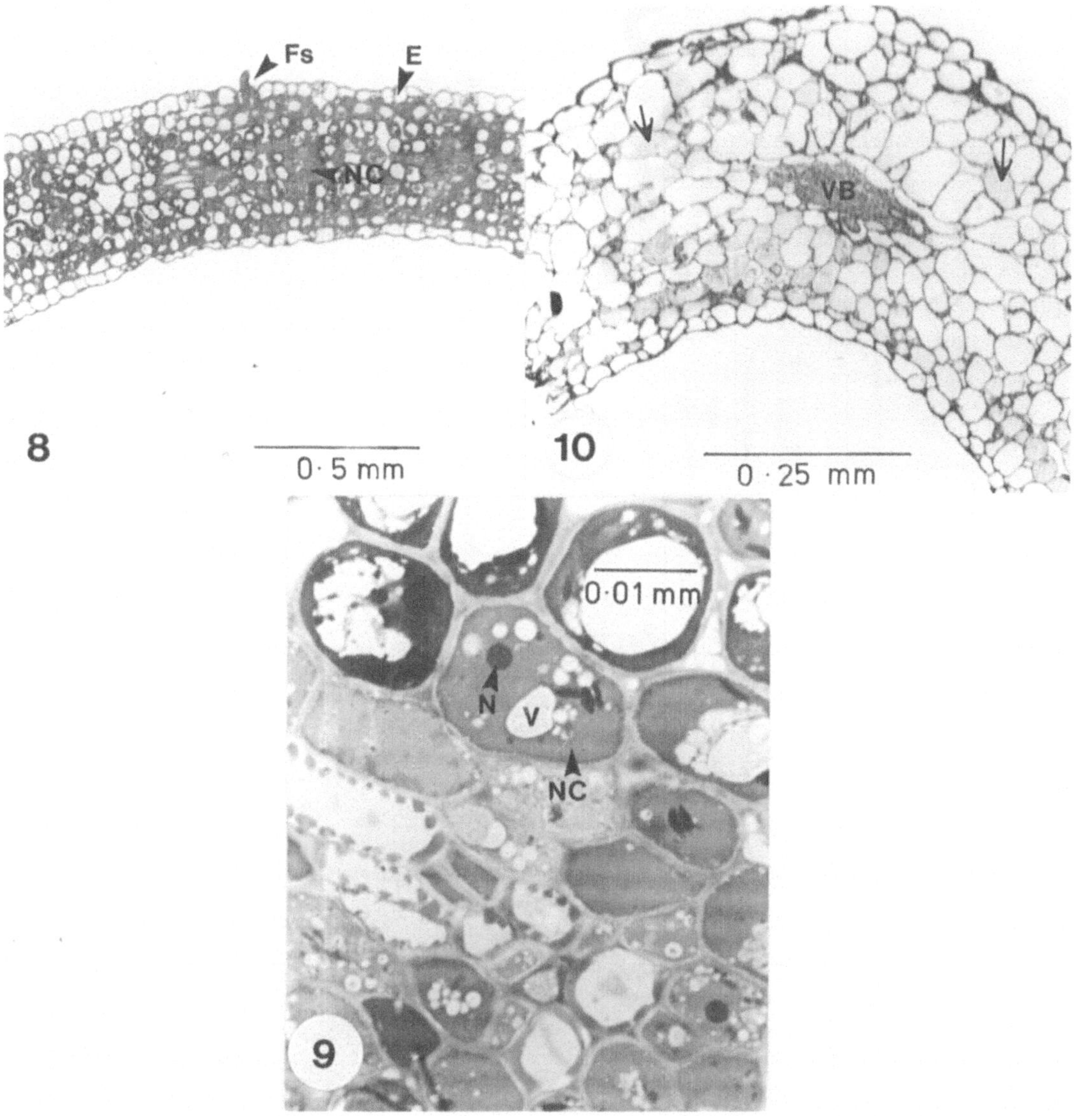

Figures 12-8–12-10. Anatomy of galls of *Gynaikothrips flaviantennatus* on the leaves of *Casearia* sp. **Figure 12-8.** Section of an immature gall showing feeding site and mesophyll cells that have begun to differentiate into cytoplasmically dense nutritive cells. **Figure 12-9.** Ultrastructure of enlarged, cytoplasmically dense nutritive cells near feeding sites. **Figure 12-10.** Section of a mature gall composed of both hypertrophied, vacuolate parenchyma cells and cytoplasmically dense nutritive cells (arrows). E, Epidermis; Fs, feeding site; NC, nutritive cells; N, nucleus; V, vacuole; VB, vascular bundle.

tissues that involve phloem tissues, interxylary parenchyma cells, and perivascular parenchyma.

IMPACT OF THRIPS POPULATION ON GALL MORPHOLOGY

Unlike the galls of most other cecidogenic insects, thrips galls are initiated as a result of the cumulative feeding effects of numerous individuals. Although only a few thrips may be present during the initial stages of cecidogenesis, mature galls often contain populations of many hundreds of individuals. Simple leaf-curl galls such as those induced by *Cercothrips nigrodentatus* (Karny) on *Planchonia valida* Blume commonly contain around 100 individuals in all developmental stages, whereas a large

Table 12-2. Population Trends in Terms of the Age of the Gall

Gall thrips (host)	Age of the gall in days (total population of gall thrips)		Population of thrips				Tissue reaction			No. of cell layers added in the gall tissue (no. of cell layers in the normal tissue)
			Immature gall		Mature gall					
	Immature gall	Mature gall	Larva	Adult	Larva	Adult	Hypoplasy	Hyperplasy	Hypertrophy	
Arrhenothrips ramakrishnae (Mimusops elengi)	6–10 (19–37)	15–21 (60–92)	2–6	2–4	14–21	18–26	+	+	Moderate	13–18 (4–6)
Schedothrips orientalis (Ventilago maderasapatana)	8–10 (7–20)	20–25 (69–93)	—	5–8	18–23	11–16	+	+	+	12–16 (5–6)
Teuchothrips longus (Pavetta hispidula)	9–10 (22–24)	23–25 (130–159)	—	9–10	45–61	19–32	+	+	+	17–60 (8–10)
Crotonothrips dantahasta (Memecylon edule)	10–13 (13–25)	20–25 (59–63)	2–4	3–6	15–19	19–21	+	+	+	22–25 (9–12)
Austrothrips cochinchinensis (Calycopteris floribundus)	— (93–136)	— (480–1400)	60–80	10–45	250–400	650–800	+	+	Moderate	47–75
Thilakothrips babuli (Acacia leucophloea)	4–8 (14–31)	20–24 (49–85)	1–2	2–3	18–32	21–28	+	+	+	30–35 (7–9)
Aneurothrips priesneri (Cordia obliqua)	5–10 (4–7)	15–30 (29–105)	2–5	1–3	14–24	7–57	Moderate	+	Moderate	10–15 (6–8)
Gynaikothrips flaviantennatus (Casearia tomentosa)	5–10 (13–22)	15–30 (46–61)	6–11	8–11	2–17	14–25	+	Moderate	+	15–20 (8–10)

gall induced by *Tolmetothrips granti* Mound from the Solomon Islands commonly harbors about 10,000 individuals (Mound 1970). Complex galls such as those of *Austrothrips cochinchinensis* on *Calycopterus floribundus* Lamk. (Fig. 12-6) often contain up to 5000 individuals, whereas galls of *Kladothrips* on *Acacia* with an internal diameter of only 8–9 mm contain up to 600 adults. On the other hand, horn galls of *Liothrips* on *Schefflera,* which appear in large numbers on single leaflets (Fig. 12-7), contain only 1 to 5 individuals each (Ananthakrishnan 1978).

The size and structure of many thrips galls are clearly dependent on the number of thrips present (Table 12-2) and unique to many species of gall thrips is the appearance of either "partial" or "complete" galls. Either type is dependent on the stage of development of the host organ at the time of attack or to the size of the thrips population. Galls induced on young, differentiating leaves become complete galls, whereas those on more mature leaves with tissues already differentiated develop into partial galls. Galls with small populations sometimes develop into partial galls whereas once a certain population is reached within the gall, a complete gall develops. *Gynaikothrips flaviantennatus* Moulton, for example, induces either a partial or complete gall depending on the number of individuals present (Raman *et al.* 1978; Raman and Ananthakrishnan 1984). Partial galls harbor an average of 12 larvae and 22 adults, while complete galls contain an average of 88 larvae and 78 adults. A similar situation occurs within the galls of *Liothrips ramakrishnae* and *L. associatus* on *Schefflera racemosa* (Krishnamurthy *et al.* 1975). This is a common feature among thrips galls and examples are known from several genera (Krishnamurthy *et al.* 1977, Ananthakrishnan and Swaminathan 1979).

Intraspecific variation in gall form and structure is a frequent feature of thrips galls with the age of the host leaf at initiation and the number of thrips at a feeding site being of particular importance. Gall initiation occurring during the differentiation phase of the leaf results in complete reorientation of the normal leaf morphogenesis, whereas feeding activities of small numbers of thrips on mature leaves results in partial or rudimentary galls.

Simple and complex galls can be produced on the same leaf depending upon the size of the thrips population. For example, both simple, tubular galls as well as fleshy, branched galls are formed by *Liothrips ramakrishnae,* side by side on leaves of *Schefflera racemosa,* either type being dependent on the number of thrips present. The walls of simple tubular galls with small numbers of thrips intermittently present have 7–12 layers of compactly arranged parenchyma cells. Complex, fleshy, and branched galls harbor large numbers of thrips for long periods and the walls of these galls have 25–35 layers.

THRIPS GALL COMMUNITIES AND COMPANY GALLS

Galls induced by thrips are commonly inhabited by numerous species besides the cecidozoan and thus can contain microcommunities. Some thrips are inquilines within the galls induced by other species, whereas other species are predacious. Sometimes galls induced by one species of thrips are inhabited by other species capable of inducing distinct galls of their own. As many as six species of thrips can be found occupying a single gall and in such instances it is difficult to determine the original gall inducer.

One of the best examples of a gall involving numerous species occurs on *Ficus retusa* L. Here the species *Arrhenothrips dhumrapaksha* Ramak., *Liothrips kannani* (Moulton), *Mesothrips bhimabahu* Ramak., *Mesothrips apatelus* Karny, *Androthrips melastomae* Zimmer., *Haplothrips aculeatus* (Haliday) *Androthrips flavipes* Schmutz, and *Mesandrothrips inquilinus* (Priesner) commonly occur together (Ramakrishna 1928). An equally interesting gall on *Geijera* sp. in Australia is commonly inhabited by the species *Sacothrips milvus* Mound, *S. ingens* Mound, *Moultonoides geijerae* (Moulton), *Chleothrips percnus* Mound, and *Chleothrips geijerae* (Moulton) (Mound 1971b). Galls such as these that develop from the activities of more than one species of thrips are called "company galls."

Some of the inquiline thrips are thought to

induce structural modifications to their host galls and consequently the transition between gall-inducing and inquilinous thrips is not well understood. Also, some of the so-called inquiline thrips seem to be predacious on the gall-inducing thrips. This is the case with *Androthrips flavipes*, one of the commonest "inquilines" in Indian thrips galls. This species has been recorded from at least 15 galls (Ananthakrishnan 1981), and has been found feeding on the eggs and immatures of gall inducers (Varadarasan and Ananthakrishnan 1981).

Besides predatory and inquilinous thrips, thrips galls are commonly inhabited by various species of coccids, aphids, anthocorids, psocids, and Lepidoptera. Thrips also are known to enter the galls induced by insects of other Orders. *Dolichothrips inquilinus* Anan., for example, has been found in psyllid galls on *Zizyphus* sp. (Ananthakrishnan 1954) and *Mallothrips indicus* Ramakrishna in the gall of the psyllid, *Trioza jambolanae. Hoplothrips atriplicis* Priesner in the gall of *Asphondylia conglomerata* Stefani on *Atriplex halimus* L. and *Liothrips devriesi* (Karny) in the gall of *Dasineura elatostemmae* Felt on *Elatostemma sesquifolium* Haask are other examples.

The size and composition of the community found within thrips galls vary with the plant bearing the galls, the gall thrips, and the other secondary species. The secondary organisms include bugs, beetles, parasitic wasps, ants, spiders, fungi, and bacteria, besides other species of thrips. There is a regular and characteristic sequence in which these organisms appear commencing with the increase in population of the gall-inducing thrips and culminating in the entry of ants and spiders in old galls that have started to senesce following the escape of the gall thrips.

LITERATURE CITED

Ananthakrishnan, T. N. 1954. New and little known Indian Thysanoptera. J. Zool. Soc. India. 6:159–166.

Ananthakrishnan, T. N. 1973. Thrips: Biology and Control. Macmillan India, New Delhi, India.

Ananthakrishnan, T. N. 1978. Thrips galls and gall thrips. Zool. Surv. Ind., Tech. Monogr. 1:1–69.

Ananthakrishnan, T. N. 1979a. Biosystematics of Thysanoptera. Annu. Rev. Entomol. 24:159–183.

Ananthakrishnan, T. N. 1979b. Thrips. In K. F. Harris and K. Maramorosch (eds.), Vectors of Plant Pathogens, pp. 149–164. Academic Press, New York.

Ananthakrishnan, T. N. 1980. On some aspects of thrips galls. Bull. Soc. bot. Fr. Actual. Bot. 127:31–34.

Ananthakrishnan, T. N. 1981. Thrips-plant gall associations with special reference to patterns of gall diversity in relation to varying thrips populations. Proc. Ind. Natl. Sci. Acad. 47:41–46.

Ananthakrishnan, T. N. 1984a. Bioecology of Thrips. Indira Publishing House, Oak Park, MI.

Ananthakrishnan, T. N. 1984b. Adaptive strategies in cecidogenous insects. In T. N. Ananthakrishnan (ed.), Biology of Gall Insects, pp. 1–10. Oxford-IBH, New Delhi, India.

Ananthakrishnan, T. N., and Jagadish, A. 1969. Studies on Indian gall thrips (Thysanoptera). Marcellia 36:7–103.

Ananthakrishnan, T. N., and A. Raman. 1989. Thrips and Gall Dynamics. E. J. Brill, Leiden, The Netherlands.

Ananthakrishnan, T. N., and S. Swaminathan. 1979. On the population trends of *Aneurothrips priesneri* Bhatti (Insecta: Thysanoptera) from the leaf galls of *Cordia obliqua.* Bull. Zool. Surv. Ind. 2:91–94.

Bagnall, R. S. 1928. Thysanoptera. On some Samoan and Tongan Thysanoptera (Tubulifera) with special reference to *Ficus* gall causers and their inquilines. Brit. Mus. Nat. Hist. London 9:55–76.

Chisholm, I. F., and T. Lewis. 1984. A new look at thrips (Thysanoptera) mouthparts, their action and effects on feeding on plant tissue. Bull. Entomol. Res. 74:663–675.

Docters Van Leeuwen, W. 1938. Zoocecidia. In Fr. Verdoorn (ed.), Manual of Pteridology, pp. 192–193. Martinus Nijhoff, The Hague.

Gopinathan, K. 1987. Morphological patterns and histochemical profile in *Mimusops - Arrhenothrips* gall system. Proc. Indian Acad. Sci. (Plant Sci.) 97:203–214.

Heming, B. S. 1978. Structure and function of the mouthparts in larvae of *Haplothrips verbasci* (Osborn) (Thysanoptera, Tubulifera, Phlaeothripidae). J. Morphol. 156:1–38.

Hunter, W. B., and D. E. Ullman. 1989. Analysis of mouthpart movements during feeding of *Frankliniella occidentalis* (Pergand.) and *F. schultzei* Tribom (Thysanoptera: Thripidae). Int. J. Insect Morphol. Embryol. 18:161–171.

Karny, H. 1912. Gallenbewohnende Thysanopteren aus Java. Marcellia 11:115–169.

Karny, H. 1921. Beiträge zur Malayischen Thysanopteren-fauna. *V.* Thysanopteren on Thee. Treubia 2(1):66–83.

Karny, H. 1922. A remarkable new gall thrips from Australia. Proc. Linn. Soc. N. S. Wales 47:266–274.

Karny, H., and W. J. Docters Van Leeuwen-Reijnvaan. 1913. Beiträge zur Kenntnis der Gallen von Java. V. Uber die Javanischen Thysanopterocecidien und deren Bewohner. Bull. Jard. bot. Buitenzorg 10:1–54.

Krishnamurthy, K. V., A. Raman, and T. N. Ananthakrishnan. 1975. On the morphology of the ceratonean thrips galls of *Schefflera racemosa* Harms. Marcellia 38:179–184.

Krishnamurthy, K. V., A. Raman, and T. N. Ananthakrishnan. 1977. Studies on plant galls from India. 2. Leaf galls of *Cordia obliqua* Willd. (= *Cordia myxa* L.) (Boraginaceae). Ceylon J. Sci. (Biol. Sci.) 12:73–85.

Lewis, T. 1973. Biology, Ecology and Economic Importance of Thrips. Academic Press, London.

Mound, L. A., 1969. Revision of three Australian genera of Phlaeothripidae (Thysanoptera) with seven new species and one new generic synonym. J. Aust. Entomol. Soc. 8:173–186.

Mound, L. A. 1970. Thysanoptera from the Solomon Islands. Bull. Br. Mus. Nat. Hist. Entomol. 24:83–125.

Mound, L. A. 1971a. The feeding apparatus of thrips. Bull. Entomol. Res. 60:547–548.

Mound, L. A. 1971b. The complex of Thysanoptera in rolled leaf galls on *Geijera*. J. Aust. Entomol. Soc. 10:83–97.

Mound, L. A., B. S. Heming, and J. M. Palmer. 1980. Phylogenetic relationships between the families of recent Thysanoptera (Insecta). Zool. J. Linn. Soc. 69:111–141.

Priesner, H. 1929. Contribution towards a knowledge of the Thysanoptera of Egypt—II. 3. A new gall making genus of the subfamily Hoplothripinae: *Megeugynothrips* gen. nov. Bull. Soc. R. Entomol. Egypt 211–216.

Raman, A. 1981. Morphological studies on some galls induced by Thrips (Thysanoptera: Insecta) from South India. Ph.D. Thesis, University of Madras, Madras, India.

Raman, A., T. N. Ananthakrishnan, and S. Swaminathan. 1978. On the simple leaf galls of *Casearia tomentosa* Roxb., (Samydaceae) induced by *Gynaikothrips flaviantennatus* Moulton (Thysanoptera: Phlaeothripidae). Proc. Ind. Acad. Sci. 87:231–242.

Raman, A., and T. N. Ananthakrishnan. 1979. On the developmental morphology of the leaf fold galls of *Maytenus senegalensis* (Lam) Excell., induced by *Alocothrips hadrocerus* (Karny) (Thysanoptera: Insecta). Proc. Ind. Acad. Sci. 88:103–107.

Raman, A., and T. N. Ananthakrishnan. 1983a. Studies on some thrips (Thysanoptera: Insecta) induced galls. 1. Developmental morphology. Proc. Ind. Natl. Sci. Acad. 49:313–358.

Raman, A., and T. N. Ananthakrishnan. 1983b. Studies on some thrips (Thysanoptera: Insecta) induced galls. 2. Fine structure of the nutritive zone. Proc. Ind. Natl. Sci. Acad. 49:525–561.

Raman, A., and T. N. Ananthakrishnan. 1984. Biology of gall thrips. In T. N. Ananthakrishnan (ed.), Biology of Gall Insects, pp. 107–128. Oxford-IBH, New Delhi, India.

Ramakrishna, T. V. 1928. A contribution to our knowledge of the Thysanoptera of India. Mem. Dept. Agr. Ind. Entomol. Ser. 10:215–316.

Rohfritsch, O. 1976. Traces de succion de deux Chermisidae: *Chermes abietis* L. et *Chermes strobilobius* Kalt. dans les bourgeons de *Picea excelsa*. Marcellia 39:69–84.

Varadarasan, S., and T. N. Ananthakrishnan. 1981. Population dynamics and prey-predator/parasite relationships of gall-forming thrips. Proc. Ind. Natl. Sci. Acad. B 47:321–340.

Wahlgren, E. 1945. Gall-bildende Thysanoptera. Opusc. Entomol. 10:119–126.

Westphal, E. 1977. Morphogenèse, ultrastructure et étiologie de quelques galles d'Eriophyides (Acariens). Marcellia 39:193–375.

Wood, G. W. 1960. Note on the occurrence of two species of thrips on low-bush blueberry in New Brunswick and Nova Scotia. Can. Entomol. 92:757–758.

13

Biology of Gall Midges on Common Reed in Czechoslovakia

M. SKUHRAVÁ
AND V. SKUHRAVÝ

The shore-line ecosystem in Czechoslovakia (Middle Europe) dominated by the common reed (*Phragmites communis* Trin.) is very old and has been little affected by the activities of humans. Eleven species of invertebrates and three species of vertebrates have been found associated with this ecosystem (Figs. 13-1 and 13-16) and the populations of each are relatively self-regulating and stable. This stability makes the system ideal for various ecological studies (Skuhravý *et al.* 1981).

Four species of gall midges (Family Cecidomyiidae) have been found on the common reed and are widely distributed throughout Europe. Three of these species *Giraudiella inclusa* (Frauenfeld), *Lasioptera arundinis* Schiner, and *Lasioptera hungarica* Möhn, are common in Czechoslovakia. A fourth species, *Microlasioptera flexuosa* (Winnertz), is rare on the same reeds.

These four gall midges are the subject of this chapter. Here we examine the morphology, bionomics, ecology, impact on the host plant, and the relationships of these midges to other insects in the common reed ecosystem. These results come from studies undertaken in Czechoslovakia from 1972 to 1976 as part of the International Biological Program and that were published in two books (Skuhravá and Skuhravý 1981; Skuhravý *et al.* 1981).

LIFE CYCLES AND GALL DESCRIPTIONS

Giraudiella inclusa

Larvae of *G. inclusa* produce galls that occur inside the stem of common reed (Fig. 13-2). Each gall contains a single white larva. This gall midge has one or two generations a year in Middle Europe. One generation develops in the spring and the second, less abundant generation, develops in the summer. Larvae of the overwintering generation hibernate in galls and pupation takes place in the spring from April 10 to May 20. Pupation lasts 2 or 3 weeks. Emergence begins about May 5 and lasts until May 30, with the maximum occurring about middle May. The adults live for only 1 or 2 days and at most 5 days. After mating, the females deposit their eggs on the earliest developed internodes on the stems of reeds.

Larvae of first generation midges are found on internodes 1–8, but occur most often on internodes 3 and 4. Larval development takes 6–7 weeks and is completed in early July. The fully developed larva is white to light yellow. Adults of the first generation start to fly early in July and females lay their eggs on soft, growing tissues of reed.

The second generation produces fewer galls than does the first. Second generation galls usu-

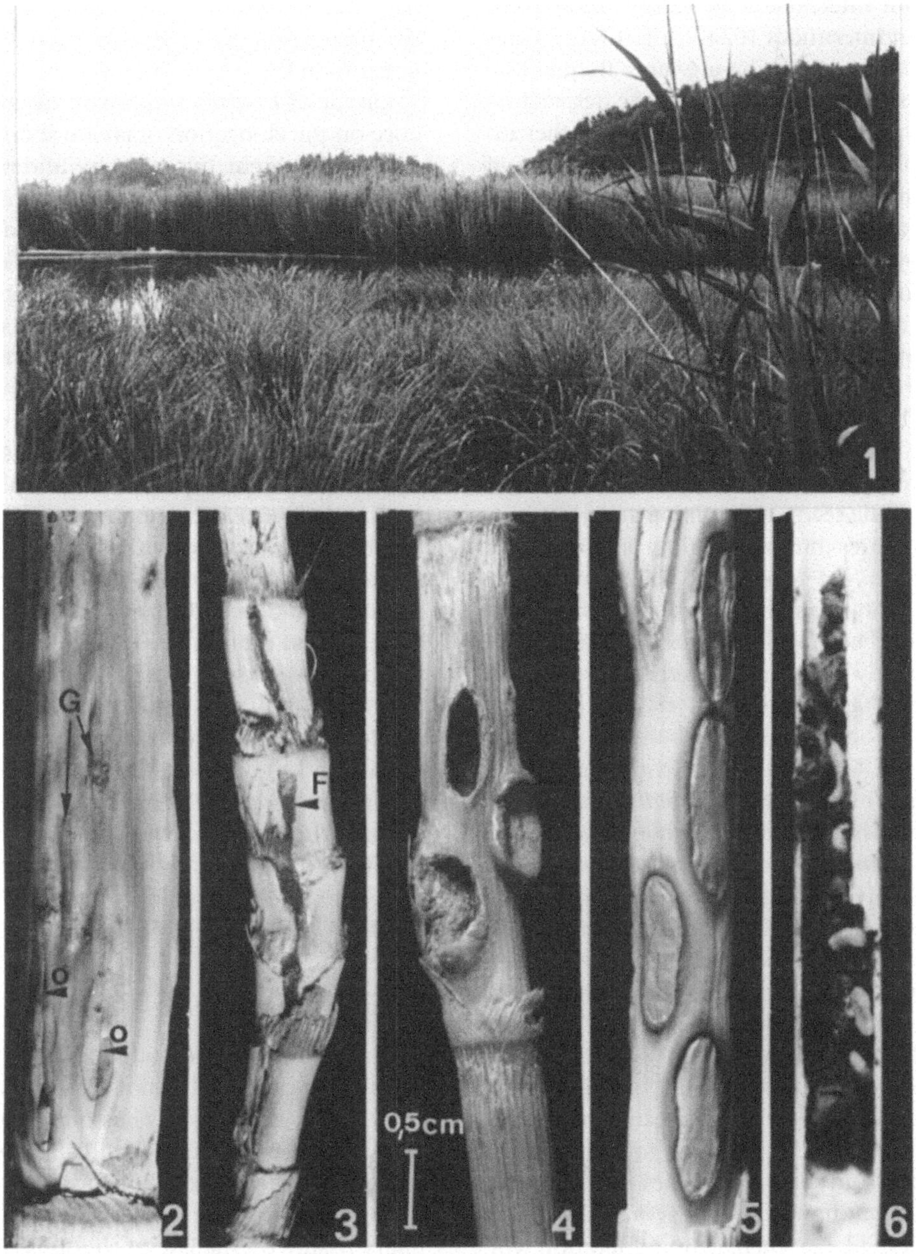

Figures 13-1–13-6. Figure 13-1. A stand of common reed near the Opatovický pond in South Bohemia, Czechoslovakia. **Figure 13-2.** Grain-like, first generation galls (G) of *Giraudiella inclusa* inside the stem. The ostiolar line (O) of each gall is visible. (From Rohfritsch 1975.) **Figure 13-3.** A lateral shoot infested by *Lasioptera arundinis* and its fungus (F). The larvae hatch at the shoot base and disseminate fungus as they move along the shoot. The fungus allows the larvae to reach the center of the stem. The adult will use the same route to exit the gall. (From Skuhravá and Skuhravý 1981.) **Figure 13-4.** Second generation galls of *Giraudiella inclusa* on a lateral shoot attacked by *Lasioptera arundinis.* The galls are incompletely closed and are externally located. **Figure 13-5.** Second generation galls of *Giraudiella inclusa* on the main stem. Galls are located internally but remain open on the stem surface. They are covered by a membrane of larval cocoon and protected by the leaf sheath. (From Skuhravá and Skuhravý 1981.) **Figure 13-6.** Dissection of a stem showing larvae of *Lasioptera hungarica* in a mass of fungi. There is no modification of normal growth of the stem. (From Skuhravý *et al.* 1981.)

ally occur on higher internodes of the main stem from internode 8 to 15, but most commonly on internodes 10–12 (Fig. 13-16). Galls of the second generation are more frequent on lateral shoots, developing after the destruction of the growing top of the stem by another invertebrate species. Development of larvae of the second generation (from the egg to the mature larva) lasts approximately 10 weeks, from early July to middle of September.

Galls of the first generation of *G. inclusa* are grain-like structures made up of cortical tissue invaginated into the stem (Fig. 13-2). The galls measure 3–10 mm in length and 1–4 mm in width. They are found only on the inner side of the stem. Generally, an infested internode contains several galls, which may often coalesce.

Galls of the second generation are also grain-like structures but are slightly smaller than those of the first generation, which develop at the base of the stem. It is unusual for these galls to coalesce; however, the outer appearance of the stem may be greatly altered by the presence of series of galls which are either slightly or completely open (Fig. 13-5).

Galls of *G. inclusa* occurring on lateral shoots, also attacked by *L. arundinis,* are different from galls on the main shoot (Fig. 13-4). Galls of *G. inclusa,* under these conditions, are only partially closed and may be externally located on the shoot. They are so different that they have been described as a separate species. For example, *Porricondyla phragmitis* (Giraud) is actually *G. inclusa.*

The development of galls of *G. inclusa* has been studied by Rohfritsch (1975). The eggs are laid on the upper part of the selected internode. After hatching, all the larvae migrate downward along one or several lines. Along these same lines but upward, each larva settles down at a specific place, where cell proliferation will be induced.

The larva does not interfere with the normal growth in length of the stem. Only lateral cell growth, of cortical origin, will cover the larva and enclose it in the modified tissue. At the same time, the larval cavity and the modified stem tissue are projected into the hollow pith by the normal growth of the stem. In young galls, the tissue is parenchymatous and at some places meristematic whereas the mature gall has a hard outer wall (cell walls become lignified) and a soft inner wall (the cells of the nutritive tissue) (Fig. 13-7). Both cell types are much larger than those of normal stems.

Larvae of *L. arundinis* have a negative influence on the elongation of lateral shoots and induce radial stem hypertrophy and when this gall is present, galls of *G. inclusa* cannot develop their usual grain-like form inside the pith of the shoot. The cortical tissue still envelopes the larva (Fig. 13-8), but the medullar layer and vascular layer push the whole gall outside of the shoot. The tissue modified by *G. inclusa* remains parenchymatous, apparently the larva has an inhibiting action on cell differentiation and lignification. All tissues of shoots attacked by *L. arundinis,* except for the cortical and sclerenchyma layers cells, are invaded by *Sporothrix* fungus associated with the larva, and the mycelium develops from the pith toward the vascular bundles. However, the cells of the gall of *G. inclusa* remain free of mycelium because the fungus is stopped by a line of resistance (Fig. 13-8). One or two layers of cells nearest the fungus develop lignified walls in the direction of the fungus and the invading mycelium is stopped by lignified papillae. This type of reaction is characteristic of nonhost tissues and is considered to be a manifestation of hypersensitivity of the tissues to the invading fungus (Vance *et al.* 1980).

Lasioptera arundinis

Larvae of this species produce a thickening and shortening of lateral shoots of common reed (Fig. 13-3). *L. arundinis* has only one generation a year. Fully developed galls occur from August to June of the following year.

The galls are made from hypertrophied medullary tissues on the lateral shoots (Fig. 13-16). Uninfested internodes of lateral shoots attain an average length of 3.8–4.8 cm; infested ones only reach a length of 1.4 cm. However, the weight of infested lateral shoots is increased by one-half in comparison with that of uninfested shoots. Larvae inside the gall are white to light orange and number from 1 to 25 per internode, with most found in the central part of infested lateral shoots. Several apical internodes of lateral shoots are usually without damage. The in-

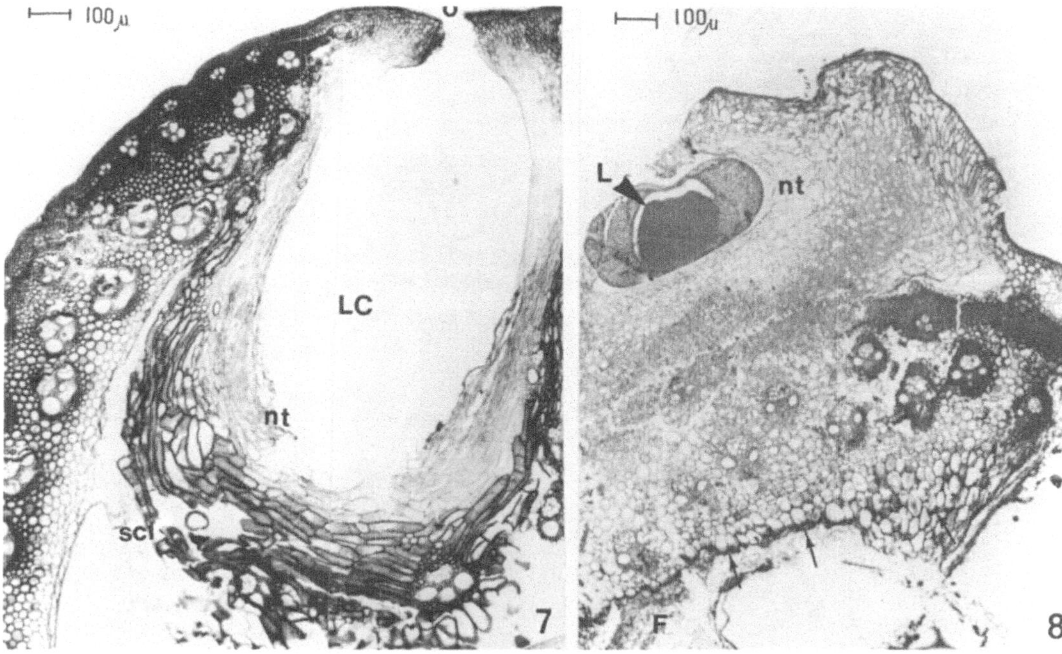

Figures 13-7–13-8. Figure 13-7. Section of a stem with a gall of *Giraudiella inclusa.* The stem surface remains perfectly round and hard. Sclerenchyma lines the nutritive tissue toward the medulla. (After Rohfritsch 1975.) **Figure 13-8.** Section of an externally located gall of *Giraudiella inclusa* developing on a lateral shoot attacked by *Lasioptera arundinis.* Most of the stem is invaded by fungus associated with *L. arundinis.* The *G. inclusa* modified tissue is strictly delimited by one or two layers of resistant cell lines (arrows). These resistant cells have lignified walls and papillae due to the invading fungus, whereas normal shoot tissue does not show such a reaction (compare with Fig. 13-12) (Rohfritsch, original). LC, Larval cavity; nt, nutritive tissue; o, ostiole; scl, sclerenchyma.

terior of the galled shoots is filled with a blackish clotted mass, which is the mycelium of a fungus of the genus *Sporothrix* (Deuteromycetes) (Fig. 13-12).

In mid-May, overwintering larvae bore a hole in the wall of the gall, through which the adults will emerge. The larva covers the opening with an operculum and starts to pupate inside its larval chamber. Pupation lasts from 14 to 25 days, or about from the middle of June to the middle of July. The pupa then pushes through the opening and the imago emerges. Emergence starts from June 26 and adults fly to the end of July.

Each female lays from 60 to 110 eggs under one of the basal leaf sheaths on the young lateral shoots (Fig. 13-9). The young shoots measure 3 to 15 cm at the time of flight of the gall midges.

This midge never attacks healthy stems, but only lateral shoots developing on the stem after the destruction of its growing top by other invertebrates such as *Archanara geminipuncta* Hb., *Platycephala planifrons* (Fabr.) or *Steneotarsonemus phragmitidis* (Schlecht.) (Fig. 13-16). Sometimes, various species of the genus *Lipara* kill the top of the reed. Thus *L. arundinis* is only a secondary inhabitant of common reed.

Larvae of *L. arundinis* hatch from eggs at the base of shoots (Fig. 13-9) and move along a narrow strip toward the apex. As they travel they disseminate fungus along the route. The larvae enter the lateral shoot through an opening formed by the fungus (Fig. 13-10). Once inside, the larvae cause a thickening of the internal tissues. The adults emerge in the following year from the galls at this site, which later appear as a longitudinal, black depression (Figs. 13-3 and 13-11).

Larvae of the first and second instar occur in July and the beginning of August. From August 7 onward, third-instar larvae are found in the galls. Larval development is complete after the

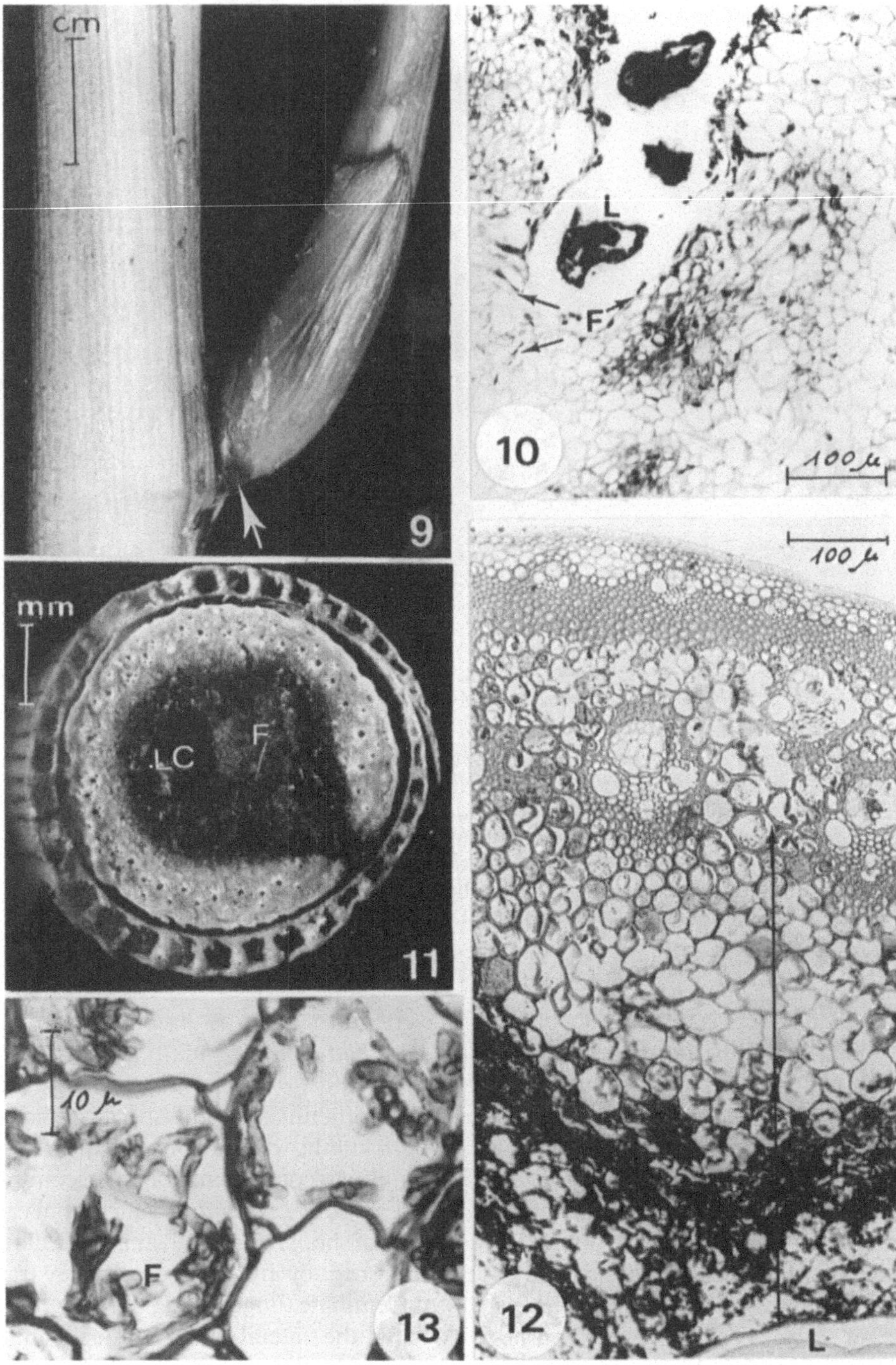

Figures 13-9–13-13. Figure 13-9. Lateral shoot attacked by *Lasioptera arundinis.* Eggs are laid at the base of the shoot (arrow). (From Skuhravá and Skuhravý 1981.) **Figure 13-10.** Section of a lateral shoot showing a larva of *Lasioptera arundinis* penetrating the stem surface with the help of *Sporothrix* fungus (F). The mycellium provokes cellular dissociation and destruction and the larva (L) enters the stem through the opening (Rohfritsch, original). **Figure 13-11.** Dissection of a lateral stem with a larval cavity (LC) of *Lasioptera arundinis.* Note that the fungus (F) penetrated the lower right side (arrow). (From Skuhravá and Skuhravý 1981.) **Figure 13-12.** Section of a lateral stem inhabited by a maturing larva (L) of *Lasioptera arundinis.* Notice the density of the fungus mycelium in the larval cavity and that the fungus has invaded cells of the medullar parenchyma and vascular bundles. All attacked tissues become hypertrophied. (line indicates the extent of the mycelium invasion) (Rohfritsch, original). **Figure 13-13.** Hypertrophied medullar cells invaded by fungus (F).

second half of September and the mature larvae remain in the galls until June of the following year.

According to O. Rohfritsch (unpublished), the fungus not only provides an entrance into the shoot for the young larva and an exit for the adult, the mycelium of the fungus also constitutes a nutritive tissue for the larva. During the early stages of gall growth, there is very little mycelium in the medullar area. However, medullar and parenchyma cells around the vascular bundles soon begin to hypertrophy. By the third instar, larvae are enveloped in a dense mycelium (Fig. 13-12), which is their only food. The mycelium also invades nearly all cells of the shoot except the cortical cells and the cells of the sclerenchyma layer (Fig. 13-12). Thus the larvae are not feeding directly on the plant tissue, but on the fungus that is exploiting the plant tissue (Fig. 13-13). Such associations have been studied by Neger (1910), Ross (1932), and Bronner (1977). These types of galls are called "ambrosia galls" and they are very common in the genus *Lasioptera* (Chapter 8).

Lasioptera hungarica

Larvae of *L. hungarica* develop inside the cavity of reed stems without the reed showing any signs of attack. This species has one generation a year and is the last of the gall midges to start its development (Fig. 13-16).

The adults fly as late as the beginning of July and first-instar larvae appear on the reed stems in about the middle of July. Larvae of *L. hungarica* penetrate the stem in a similar way to that of *L. arundinis*. However, larval development proceeds at a considerably quicker rate; less than 2 months are required for the development of first- and second-larval instars. Larvae of the third instar are completely developed after 2 months or by the second half of September. The mature, orange larvae remain in reed stems for another 9 months.

Infestations by *L. hungarica* within reed stems cannot be identified unless the stem is cut lengthwise because there are no external signs indicating the presence of the midges. Levels of stem infestation of 3–5 percent also make it difficult to locate this species. However, from the beginning of November onward, the titmouse feeds on larvae-infested stems by pecking longitudinal ridges in the stems. These ridges greatly facilitate the location of *L. hungarica* galls and larvae of this species.

The larvae of *L. hungarica* live in clusters in the cavity of the main stem. The number of larvae in an internode varies considerably from a minimum of 29 to a maximum of up to 300 (the average is 95). While 70–90 percent of the larvae occupy the lower part of an internode, there are 10–30 percent in its upper part. The larvae live in clusters in a blackish clotted mass of the mycelium of the *Sporothrix* fungus (Fig. 13-6).

Most larvae are found inside internodes 1–13 and the examination of large samples indicated that about 70 percent of the larvae were found attacking internodes 6–9. The remaining 30 percent of the larvae were in lower or higher internodes.

Larvae of *L. hungarica* do not create galls on the stem of common reed and do not damage its wall. They seem to cause no direct damage to the reed stem but they do so indirectly because birds peck on infested stems in search of food, making a longitudinal slit in the stem, which facilitates breakage at the damaged place. The higher the number of infested internodes situated in close succession on the stem, the more easily such breakage occurs.

Microlasioptera flexuosa

Larvae of this species live in the terminal parts of low and thin reed stems. Such stems may reach 20–30 cm in height and only 3–5 mm in width without any inflorescence. Infested reed stems were found on a halophytic grassland community. About 20–50 orange larvae may live in the lower part of one infested internode. The infested stem appeared to be undamaged and larvae could be detected only by splitting the stem. No fungal hyphae were observed inside the infested internodes. Under natural conditions, adults from larvae overwintering in infested reed stems emerged as late as May of the following year.

The gall midge *M. flexuosa* is a very rare species and so far has been reported only from three localities in a limited area of the Pannonian plain in Middle Europe (Skuhravá and

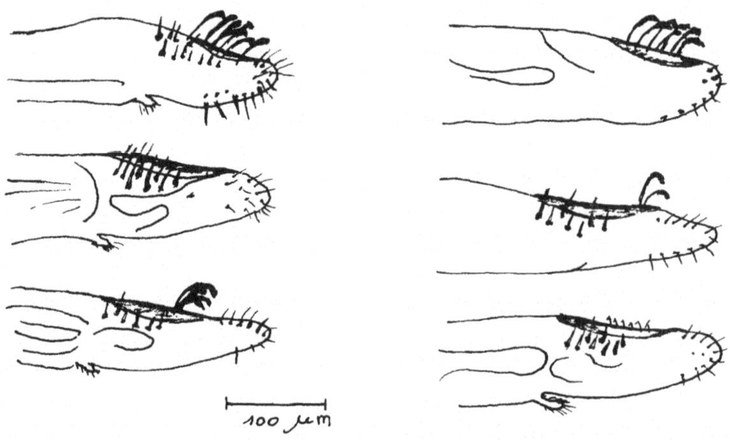

14

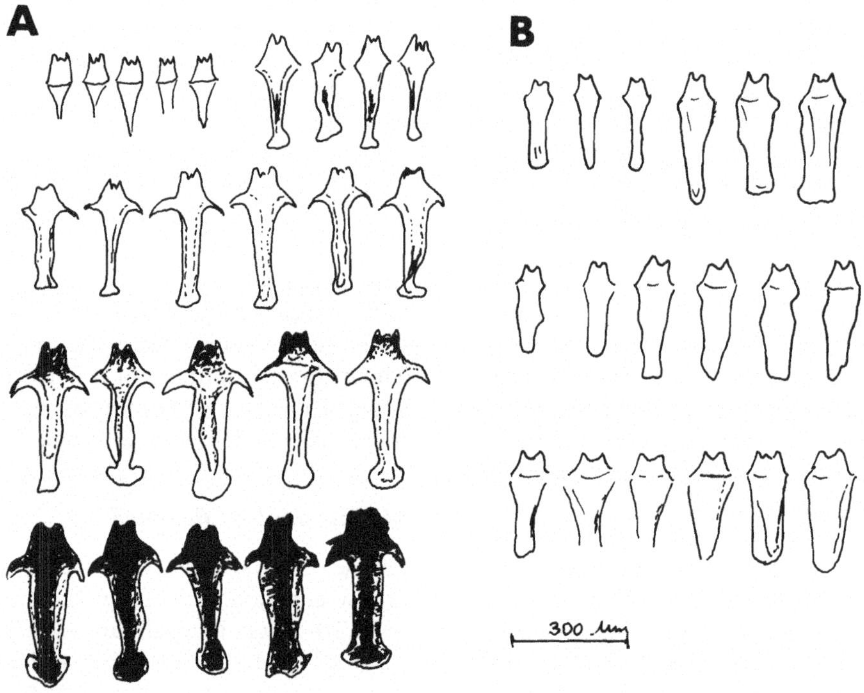

15

Figures 13-14–13-15. Figure 13-14. Terminal part of the ovipositor of *Lasioptera hungarica* showing the variability in number of spines. All insects were reared from one host plant. (From Skuhravá and Skuhravý 1977.) **Figure 13-15.** *Spatula sternalis* of two species of gall midges showing variability in shape. (A) *Lasioptera arundinis;* (B) *Lasioptera hungarica.* The size, shape, and intensity of sclerotization of the *spatula sternalis* are dependent on the function of this organ in the individual life of the larva. (From Skuhravá and Skuhravý 1977.)

Skuhravý 1981). Therefore the biology of this species is not as clear as it is for the other reed gall midges. As with *L. hungarica,* adults of *M. flexuosa* are released from the stem only after it has been broken by either biotic or abiotic factors.

MORPHOLOGY OF COMMON REED GALL MIDGES

Twenty male characters, 18 female, and 19 larval characters were analyzed and the major results were as follows:

1. There was considerable variability in the shape of the upper and middle lamellae of the hypopygium of males and also of the number of various spines on the ovipositors of females in all four species (Fig. 13-14) (Skuhravá and Skuhravý 1981).
2. There was considerable variability in morphological characters of the larval *spatula sternalis* (Fig. 13-15). Its form, generally taken as a main diagnostic character of species, depends on the age of the larval specimens and on the use of this organ during their individual life cycle.
3. There was great variability in the number of antennal segments of males and females in the ranges of the species, which depends on the size of their body.

The variability of some of these morphological structures is so high that the evaluation of only one specimen could lead to the description of a new species or a new genus. Unfortunately, this has been done many times in the past. Our detailed studies of midge morphology along with our ecological studies of common reed clearly showed that only four species of gall midges are found in Middle Europe. These results have important implications for studies of other gall midges since many species have been described from only one adult specimen or from a single larva. Thus we caution taxonomists to carefully consider the variability of morphological characters, along with ecological characters, when they are proposing new taxa.

RELATIONS BETWEEN REED DEVELOPMENT AND DEVELOPMENT OF ASSOCIATED GALL MIDGES

Growth of reed stems includes two basic components: first, the lengthwise prolongation of earlier developed tissue (extended growth), and second, the production of new internodes by the meristematic activity of the vegetative top.

In Middle Europe, common reed has its period of greatest growth in May and June. At the

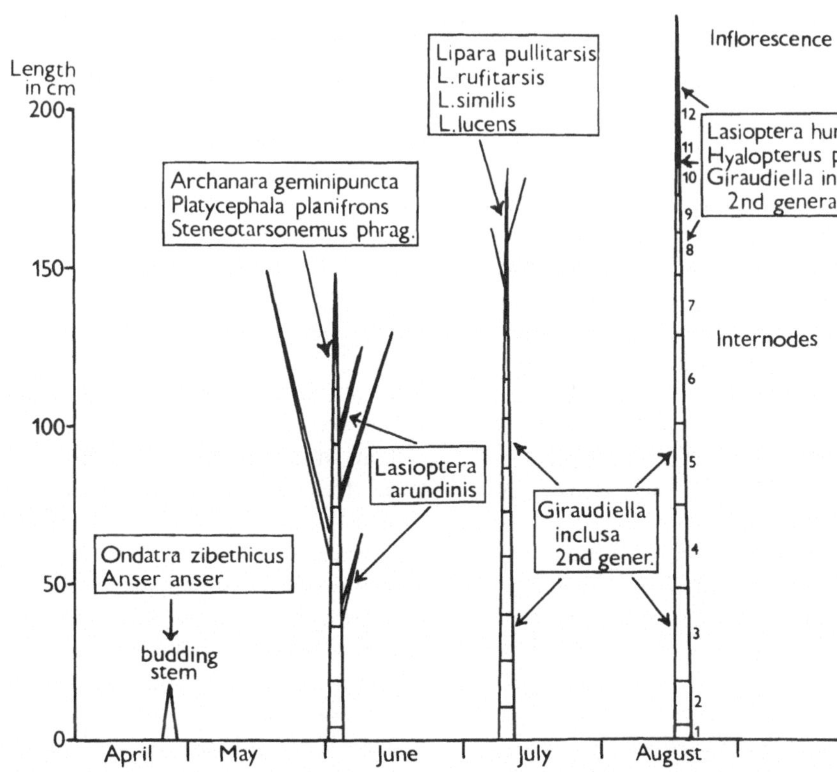

Figure 13-16. Schematic drawing of the relationships between invertebrates and vertebrates attacking the common reed occurring in the spring and early summer under Middle European conditions.

end of this period the stems (without inflorescence) reach 75–95 percent of their total length. During July the growth of reed stems slows and in August and September it ceases. When the vegetative top of the reed is destroyed or damaged during the period of extended growth, lateral shoots develop.

Development of all four gall midges is closely related to the growth and development of common reed since the larvae can develop only on young meristematic tissue. Thus the developmental state of the plant really determines which tissue of reed stem can be attacked by gall midge larvae.

Under natural conditions, there is a mutual relationship between the host plant and the gall midges. There is a simple relationship between the first generation of *G. inclusa* and *L. hungarica* when the larvae attack the main stem of common reed. A more complicated relationship exists between the host plant and *L. arundinis,* the larvae of which develop in lateral shoots. The most complicated relationship exists in the case of contemporary attack of lateral shoots by *L. arundinis* and the second generation *G. inclusa.*

Giraudiella inclusa and *Lasioptera hungarica* on the Main Stem of Reed

Female *G. inclusa* emerging from galls on the stem, where they overwintered as larvae, lay their eggs on lower internodes of the developing main stem of common reed during the second half of May. At this time the reed stems are in uneven phases of development, ranging from 40 to 150 cm in length. Therefore, the internodes that are suitable for development of larvae and galls can be found at various levels above the ground or above the water surface. This is the reason for differences in the rates of infestation, not only in various localities in Czechoslovakia and other parts of Europe, but also on reed stems in stands at the same locality.

Adults of *L. hungarica* emerge from reed stems during the month of July. They start their development later than larvae of *G. inclusa* and *L. arundinis.* When *L. hungarica* emerges, the growth of reed stems is practically finished. The places suitable for egg laying and development of larvae of this species are the 8–9 internodes of the main reed stems.

Lasioptera arundinis and *Giraudiella inclusa* on Lateral Shoots of Reed

More complicated relationships exist between the host plant and *L. arundinis* and the second generation of *G. inclusa* (Fig. 13-4). Adults of *G. inclusa* emerge at the beginning of July. The females can lay their eggs either on young developing tissue of terminal parts of the main stem or on lateral shoots budding from main stems. The emergence of *L. arundinis* continues for a long time, starting in the middle of June and continuing through the end of July with the peak flights at the end of June.

Lateral shoots on reed stems develop only after destruction of the vegetative top by invertebrates or by mechanical means. In 1973, we conducted some experiments in an attempt to understand the importance of adult midges synchronizing their emergence with the appearance of lateral shoots. We cut the vegetative tops of developing reed stems with scissors in two weekly intervals from May to August.

We discovered that lateral shoots did not develop if cutting was done in the early developmental phase at the beginning of May. Most of the stems cut at this time dried out. Lateral shoots did not develop unless the cutting was done from the second half of May until the end of July. When the vegetative top was cut in the middle of May, one to three (average 1.6) lateral shoots developed, whereas cuttings done in the middle of June produced one to eight (average 2.3) lateral shoots. During July, the number of lateral shoots produced by experimental cuttings decreased successively.

The longest lateral shoots (62 cm in length) were obtained by removing the vegetative top in the second half of June. Shorter lateral shoots (48 cm in length) developed following the cuttings done at the beginning of July and the shortest ones (average 27 cm in length) following the cuttings at the end of July.

The results of our experiments showed a relationship between the number and length of lateral shoots that developed and the time at which the vegetative top was destroyed or injured. The largest number of lateral shoots was produced after destroying or injuring the vegetative top from the middle to the end of June.

Under natural conditions similar injury and resulting growth of lateral shoots may be caused by invertebrates or other mechanical means. For example, early injury by *Archanara geminipuncta Hb., Platycephala planifrons (F.)*, and *Steneotarsonemus phragmitidis (Schlecht.)* causes the development of numerous long lateral shoots. Attacks by early developing species of flies in the genus *Lipara*, such as *L. pullitarsis* and *L. rufitarsis*, also produces growth of lateral shoots. Thus it has been shown that a secondary infestation of such lateral shoots by the gall midge *L. arundinis* is mainly dependent on primary destruction of the main stem by one of these insects. Such infestations were rare on stems infested by the species of the genus *Lipara* but did occur. The second half of June is the main emergence period of the midge *L. arundinis* suggesting that this species has adapted during the course of its evolutionary development to life on lateral shoots of common reed stems.

The lowest level of infestation of lateral shoots by *L. arundinis* was found when experimental cuttings were made after the middle of May. An average of 6.2 percent of lateral shoots were infested. When tops were cut at the beginning of June, 24.6 percent of the lateral shoots were attacked by *L. arundinis*. The highest level of 36.5 percent occurred when the tops were cut in the middle of June. At the end of June, 14.4 percent of the lateral shoots were infested. In agreement with these results, we found that the highest (76 percent) frequency of internodes of lateral shoots changed into galls of *L. arundinis* when the vegetative tops were cut in the middle of June.

Females of the second generation of *G. inclusa* also laid their eggs on lateral shoots that developed as a result of experimental or natural destruction of the vegetative tops. The precise location of the galls of *G. inclusa* on the shoot was dependent on the time in which the lateral shoots developed. Internodes of lateral shoots budding early are attacked mostly in the apical parts, while most of those budding later are attacked at the base.

Contemporary Attack of Lateral Shoots by *Lasioptera arundinis* and the Second Generation of *Giraudiella inclusa*

The most complicated relationship arises in the case of contemporary attack of lateral shoots by *L. arundinis* and the second generation of *G. inclusa*. If females of *G. inclusa* lay their eggs on lateral shoots of common reed on which females of *L. arundinis* had laid their eggs some days or weeks earlier, a complicated tissue reaction occurs under the influence of both species.

The development of all four gall midge species is closely related to the growth and development of the reed. For example, to escape from the gall, the larvae of *L. arundinis* create an exit hole for the adult insects using their *spatula sternalis*. The pupae of *G. inclusa* do the same but use their cervical armature. *L. hungarica* and *M. flexuosa* also have a *spatula sternalis*, but it is not so stout (Fig. 13-15) and they depend on natural disintegration of the reed tissue by decay, or on the stems being broken by wind, or opened by birds to escape the gall.

FACTORS INFLUENCING THE POPULATION DYNAMICS OF GALL MIDGES DEVELOPING ON COMMON REED

In the course of a year the populations of gall midges on the stems of common reed are influenced by many biotic and abiotic factors. The most important biotic factors are birds and parasitic Hymenoptera. These two factors especially affect the later phases of the developing gall midges, namely the last larval instar and pupal stage.

In addition to these factors, populations are influenced by a complex of factors which are especially important in the early developmental instars and often lead to the death of gall-midge larvae. It is not possible to determine the primary cause of mortality of those early states, but it is probably a result of abiotic factors.

Gall midges live on the common reed in a very concealed manner. Females lay their eggs under the leaf sheaths and larvae develop in their galls, or inside the cavity of the stem, without visible signs. Only damage caused by the midges can be evaluated in very early stages, even when the larvae has perished. But in most cases it is difficult to determine the real cause of mortality of each larval stadium. They may be affected by humidity or extremely dry conditions, by sunshine, high or low temperature, or by the resistance of the host plant against the larvae. Tscharntke (1988) observed that first-instar larvae of *G. inclusa* had a better chance to survive and to induce a gall on thick shoots of *Phragmites australis* (Cav.) Trin., which had been earlier damaged by a stem miner like *A. geminipunctata*. We have attempted to evaluate these factors and their influence during the course of early developmental instars by presenting the percentage losses caused by various factors (Table 13-1). By using this method it was possible to express, for the first time, how each factor affects the development of gall midges confined to one species of host plant. We also could determine the mutual relationships between each species.

The most important factors, besides those affecting early larval development, are birds and parasites. It seems that the higher the effect of birds, the lower is that of parasitic Hymenoptera. For example, birds reduce the population of *L. hungarica* by 95 percent whereas one species of parasitic Hymenoptera reduced the remaining population by only 8 percent. The population of *L. arundinis* is reduced by two species of parasitic Hymenoptera by 30–40 percent and by birds by 60 percent. Seventy percent of the population of *G. inclusa* is reduced by 6–8 species of parasitic Hymenoptera and only 27 percent by birds.

The population reduction of individual species of gall midges is related to the location of their larvae on the reed stem. Larvae of *L. hungarica* are gregarious; they are usually clustered in a small place of one internode. If a bird finds an internode full of larvae by pecking with its bill, it usually removes all larvae from that internode. Larvae of *L. arundinis* are not clustered but rather they are dispersed along their galls on lateral shoots. As a result, they have a much greater probability of surviving the attack of a bird than larvae of *L. hungarica*. Since the larvae of *G. inclusa* are solitary, birds have to peck out each larva, or the larva of a parasite, individually, and so these species also have a greater chance of survival.

These differences in gall structure and location have likely influenced the evolution of the parasitic Hymenoptera associated with the reed galls. In the case of *L. hungarica*, birds probably reduced the number of gall midges as well as the number of parasitic Hymenoptera with the result that only one species of parasitic wasp remains. The population of *L. arundinis* is attacked by two species of parasitic Hymenoptera. One, *Platygaster phragmitis* (Schrank), is very abundant and it is responsible for 95 percent of the parasitism. However, parasitic Hymenoptera apparently found the best conditions for their development in the solitary gall of *G. inclusa*. This species mainly develops inside the cavity of reeds and has two generations a year. As a result, 6–8 species of parasites are associated with this midge.

DISTRIBUTION OF GALL MIDGES ON COMMON REED

To the present, gall midges of common reed have been recorded only from Europe and mainly from the central part of this continent.

Table 13-1. Factors Influencing the Population Dynamics of Gall Midges Developing on Common Reed in Middle Europe

Factors	Gall midge species			Remarks
	Giraudiella inclusa	*Lasioptera arundinis*	*Lasioptera hungarica*	
Mortality[a] in early development instars				
Early instar	34–41	220	Not examined	Laboratory conditions
Last instar	0.2–2.5	25–35	Not examined	Laboratory conditions
Mortality by parasitic Hymenoptera	70	30–40	8	Laboratory breeding
Destroying of galls by birds	30	60	95	In natural conditions
Number of parasitic species	6–8	2	1	Ecological character
Number of generations of gall midge	1–2	1	1	Ecological character

[a]Mortality in percent of the remaining population, not of the total population.

However, common reed belongs to a plant group with a worldwide distribution, occurring also in Asia (Pakistan, India, Japan, and other lands), North America, and Africa. Undoubtedly, gall midges, along with other insects, also occur in other parts of the world. It would be most rewarding if the present results aroused an interest in the study of gall midges, and other insects attacking common reed, in other areas of the world.

LITERATURE CITED

Bronner, R. 1977. Contribution à l'étude histochimique des tissus nourriciers des zoocécidies. Marcellia 40:1–134.

Giraud, J. 1863. Mémoire sur les insectes qui vivent sur le roseau, *Phragmites communis* Trin. (*Arundo phragmites* L.) et plus spécialement sur ceux de l'ordre des Hyménoptères. Verh. zool. bot. Ges., Wien 13:1251–1296.

Neger, F. W. 1910. Ambrosiapilze. III. Weitere Beobachtungen Ber. Dtsch. Bot. Ges. 28:455–480.

Rohfritsch, O. 1975. Etude du développment de la galle de *Giraudiella inclusa* Fr. sur *Phragmites communis* Tr. Marcellia 38:251–270.

Ross, H. 1932. Praktikum der Gallenkunde. Springer, Berlin.

Skuhravý, V. 1975. Die Entwicklung der Gallmücke *Lasioptera arundinis* Schiner (Diptera, Cecidomyiidae) an Schilf (*Phragmites communis* Trin.). Marcellia 38:287–298.

Skuhravá, M., and Skuhravý, V. 1977. The development of *Giraudiella inclusa* (Frauenfeld) and the variability of its galls on *Phragmites communis* Trin. Marcellia 40:195–206.

Skuhravá, M., and Skuhravý, V., 1981. Die Gallmücken (Cecidomyiidae, Diptera) des Schilfes (*Phragmites communis* Trin.). Studie ČSAV, 1981 (3), Praha.

Skuhravý, V., V. Pokorný, J. Pelikán, M. Skuhravá, K. Hudec, and B. Rychnovský. 1981. Invertebrates and vertebrates attacking common reed stands (*Phragmites communis*) in Czechoslovakia. Studie ČSAV, 1981 (1), Academia, Praha.

Tscharntke, T. 1988. Variability of the grass *Phragmites australis* in relation to the behavior and mortality of the gall-inducing midge *Giraudiella inclusa* (Diptera, Cecidomyiidae). Oecologia 76:504–512.

Vance, C. P., T. K. Kirk, and R. T. Sherwood. 1980. Lignification as a mechanism of disease resistance. Annu. Rev. Phytopathol. 18:259–288.

14

Evolution and Ecology of Gall-Inducing Sawflies

P. W. PRICE

This chapter reviews knowledge on the gall-inducing sawflies (Hymenoptera: Tenthredinidae) with emphasis on their evolution and ecology, and presents new information based on research in Northern Arizona on the stem galler, *Euura lasiolepis* Smith.

The study of sawfly galls and gallers is very interesting and significant for many reasons: in understanding the molecular and cellular biology of gall formation, in the development of ecological understanding of plant–herbivore relationships, in evolution and speciation of specialized herbivores, and because they have an impact on plants grown for purposes of soil conservation and biomass production. These areas will be discussed in turn.

Sawfly galls are initiated by the ovipositing female in the genera *Euura* and *Pontania.* In some cases, under her influence the gall grows to full size even though an egg does not hatch (Magnus 1914; Smith 1970; Price and Craig 1984; Price and Clancy 1986a). In others, the female causes a small gall but a larva stimulates full gall development (Hovanitz 1959, and Chapters 5 and 6). This character provides an unusual opportunity for experimentally studying the initiation of gall formation. Contents of the colleterial female gland feeding into the ovipositor (or saw) can be extracted and injected into very young leaves or petioles used as oviposition sites, and consequent development can be studied. Hovanitz (1959) pioneered this approach and although his results were very promising they have not been followed up. He found that when cell cultures of willow were exposed to gland extract of the female, masses of filamentous strands were stimulated to grow. Injection of female gland secretion into willow leaves also induced small growths. Hovanitz found the female secretion rich in nucleic acids and proteins. Heating to inactivate proteins did not reduce gall-inducing activity, suggesting the essential role of nucleic acids in the process. Hovanitz noted that nucleic acids were the basis of the hereditary material, and without saying so, left the impression that sawflies may be engaged in genetic manipulation of plant tissue to form galls; they may be genetic engineers. Nucleic acids could be vectored by viruses, plasmids, or other means, although search for such vectors in *Euura lasiolepis* by Dr. B. S. Vinson at Texas A&M University has not been fruitful (personal communication). Searching for viral or plasmid genome in sawflies or in gall tissue may be very fruitful (see also Cornell, 1983, who argues for a viral role in gall formation). In addition to studies of female secretions, Hovanitz placed sawfly larvae or gall tissue on willow cell cultures and in both cases found locally stimulated cell division. These exciting results should be followed up with our better developed techniques in micromanipulation, microanalysis, tissue culture, and genetic engineering. These systems with *Euura* or *Pontania* sawflies offer great potential for elucidating a mechanism in gall formation in plants.

A second area of science to which the study of sawfly galling can contribute significantly is the ecology of plant–herbivore relationships.

Many features make gall-inducing sawflies superior to free feeders for answering questions about ecological relationships involving population dynamics, resource quality, community development, and three trophic level interactions. Galls are prominent and easily located, even at very low densities, so sawflies, parasitoids, and inquilines can be sampled accurately. The sawflies remain in one location on the plant throughout their development, so the effect of resource quality and position on sawfly survival and growth can be evaluated with precision. In addition, resources available to sawflies can be relatively easily measured since the gall is formed on a particular plant part. Survivorship of sawfly eggs and larvae can be accurately measured since the gall contains the history of activity within. Galls provide a very tangible link between the first and third trophic level (plants and the enemies of herbivores) because rate of gall growth and ultimate gall size greatly influence accessibility of herbivores to their enemies. In sawfly gall communities, as well as in others (see Chapter 16), parasitoids and inquilines are usually abundant, so data on the third trophic level are relatively easy to collect. Since a gall is static on the plant, its influence on plant growth and reproduction can be evaluated, whereas impact of free-feeding herbivores on plant reproductive success is harder to document. With the great interest in plant–herbivore relationships and the rapid development of this field (e.g., Rosenthal and Janzen 1979; Denno and McClure 1983; Price 1984a; Strong *et al.* 1984), the study of gall-inducing sawflies can make an important contribution to general ecology.

A third discipline in which studies on sawfly gall inducers can have significant impact is the evolution of plant–herbivore relationships and theories on speciation of specialized herbivores. Much interest has been generated in sympatric speciation of herbovires through host shifts (Bush 1975a,b). Heslop Harrison (1927) was a pioneer in this subject, using *Pontania salicis* (Christ.) to test experimentally for host plant shifts, and found evidence of shifting preferences that were ultimately heritable. Although he took a Lamarckian view of the mechanisms involved, his results were very provocative, but have not been followed up. Now we know that many sibling species of gall-inducing sawflies exist (Smith 1970; Kopelke 1986, 1989). "The stem gall-makers are host species-specific, and form complex sibling arrays within related host species groups" (Smith 1970). Food plant races recognized by Benson (1950) also may be sibling species. These patterns of sibling species distribution across closely related host plants fit the predictions of Bush's (1975a) sympatric speciation model. Studies on the genetic relatedness within genera of galling sawflies would therefore be very interesting. Another question arises, on the evolution of new genera, discussed below, and estimates of genetic relatedness would be very valuable on *Pontania* and *Euura* species on the same host plant.

The fourth area of interest for the study of sawfly gall inducers concerns their impact on willow growth in nurseries, in biomass production for fuel, and in plantings for soil conservation. The only sawfly on willows in New Zealand, *Pontania proxima* (Lepeletier) is of considerable interest in selection of the best willow clones for soil conservation and erosion control, all clones being rated for susceptibility to this sawfly (R. L. Hathaway, personal communication). The energy forests using willows being developed in Sweden and Finland (e.g., Larsson and Wiren 1982) are not attacked by galling sawflies although the potential for accidental introduction is present. Our studies on the impact of *Euura lasiolepis* on willow growth suggest that species in this genus could be serious pests in managed stands of willow.

Life histories of galling sawflies are all similar, and differences between genera will be discussed in more detail in the section on evolutionary relationships. Adults emerge in the spring, mate on willow plants and females begin oviposition soon after mating on the rapidly growing new plant shoots. Eggs are deposited into plant tissue after injection of the fluid secreted by the colleterial gland by the female (see also Chapters 5 and 6). The gall begins to grow and subsequently the larva hatches and feeds in the new gall tissue. When feeding is completed a cocoon is spun in the gall, or after cutting an exit hole in the gall and falling to the ground. Adults emerge to start a new generation. Species may be univoltine (e.g., *Euura lasiolepis,* Price and Craig 1984), bivoltine (e.g.,

Pontania proxima, Carleton 1939), or multivoltine (e.g., *Pontania pacifica* Marlatt in California, Caltagirone 1964).

Our ecological studies have been located around Flagstaff, Arizona where the arroyo willow, *Salix lasiolepis* Bentham, is the most common species of willow. It is a shrubby willow, growing 3–4 m tall and spreading vegetatively through layering of branches, such that one plant may develop into a large clone covering 10–30 m^2. It grows wherever water is plentiful, mainly along creek banks, such as Schultz Creek draining from the south of the San Francisco Peaks, and the main drainage of the Flagstaff area, the Rio de Flag. The willow is also found at springs, borrow pits, and drainage ditches along roads. Two willows we have studied occur at Coyote Spring (CS) on Museum of Northern Arizona property within the Flagstaff city limits. Coyote Spring produces water all year. Other willow clones we studied are located on Museum of Northern Arizona (MNA) property on Schultz Creek and about 1 km below this site at Northland Press (NP) close to the confluence of Schultz Creek and the Rio de Flag. Both of these drainages have temporary water flow during spring melt of snow, if enough snow falls in the area. Elevation of these sites is about 2132 m above sea level.

EVOLUTIONARY RELATIONSHIPS

The association between tenthredinid sawflies and plants has been a long one, with primitive plants like horsetails and ferns associated with primitive sawflies. In North America the horsetail genus *Equisetum* has 11 species of tenthredinid associated with it, and the fern genus *Pteridium* has 9 (Smith 1970). The adaptive radiation of the sawflies was probably well underway before the angiosperms appeared (see also Chapter 3).

However, it is on the angiosperms that sawflies have radiated most extensively. In the superfamily Tenthredinoidea, woody plants, especially in the Rosaceae, Betulacae, Fagaceae, and Salicaceae, provide many of the host species. In the family Tenthredinidae in North America, 95 species are recorded from *Salix,* 25 species each on *Betula* and *Quercus,* and 23 species on *Populus* (Smith 1979).

This adaptive radiation has largely retained the primitive larval habit of feeding externally on leaves with the tenthredinoid families Pergidae, Argidae, Cimbicidae, and Diprionidae composed almost exclusively of species with this habit (Table 14-1). In the Tenthredinidae about 80 percent of the species are free feeding as larvae but minor radiations have occurred involving feeding within the plant as leaf and bud miners, fruit feeders, and species that modify plant growth into leaf folds and galls.

The evolution from free-feeding species to leaf-folding and galling species can be traced by a comparative approach on extant species of nematine sawflies (Tenthredinidae: Nematinae: Nematini). All the relevant genera of sawflies are present on *Salix lasiolepis* and their feeding and galling positions are illustrated in Figure 14-1. Members of the genus *Nematus*

Table 14-1. Feeding Habits of North American Sawflies in the Superfamily Tenthredinoidea[a]

	Feeding habits						
Family	Total species	External leaf feeders	Stem borers[b]	Leaf and bud miners	Fruit feeders	Leaf folders	Gall formers
Pergidae	5	5					
Argidae	59	58		1			
Cimbicidae	12	12					
Diprionidae	48	47	1				
Tenthredinidae	824	677	1	25	21	26[c]	74[d]
Totals	948	799	2	26	21	26	74
%	100	84	<1	3	2	3	8

[a]Derived from host records in Smith (1979).
[b]Numbers do not include species that bore stems after completion of feeding.
[c]All in genus *Phyllocolpa.*
[d]Includes *Pontania* leaf gallers and *Euura* stem, petiole, and bud gallers; 42 sp. *Pontania;* 29 sp. *Euura (Euura)* stem and petiole gallers; 3 sp. *Euura (Gemmura)* bud gallers.

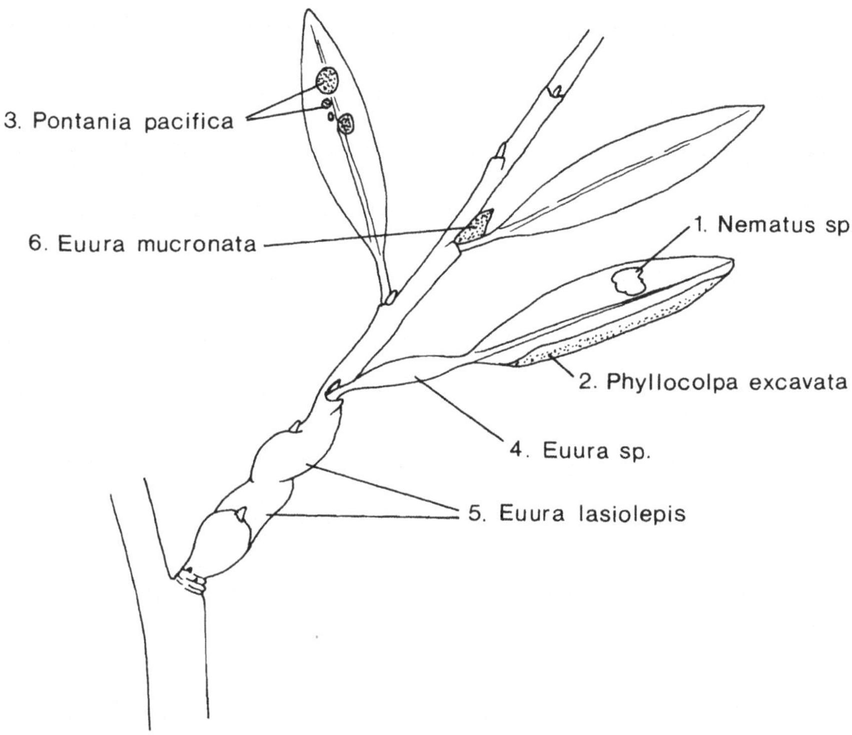

Figure 14-1. Diagrammatic view of sawfly feeding positions and galls on *Salix lasiolepis.* (1) *Nematus* sp., hole in leaf made by external feeder. (2) *Phyllocolpa excavata,* leaf fold. (3) *Pontania pacifica,* leaf gall. (4) *Euura* sp., petiole gall. (5) *Euura lasiolepis,* stem gall. (6) *Euura mucronata,* bud gall. View is of abaxial leaf surfaces. (From Price 1984b.)

oviposit below the epidermis of the leaf lamina, and the larva feeds externally on the leaf, first in a hole in the leaf and expanding this to the leaf edge. This is clearly the type of progenitor from which the galling species are derived. *Phyllocolpa* oviposits in the leaf edge and causes abnormal leaf development into a leaf fold, and the larva feeds within the fold. Once this ability to alter leaf development had evolved, a slight shift in site of oviposition in the leaf lamina, as in *Pontania proxima,* may lead to the initiation of gall development. Another small shift to oviposition in the midrib would result in the characteristic gall position in the genus *Pontania.* Indeed, the most primitive *Pontania* species form galls close to the leaf edge (Smith 1970) and represent a stage intermediate between *Phyllocolpa* and the majority of *Pontania* species whose galls are associated with the midrib of the leaf. A slight shift in oviposition site to lower on the midrib could give rise to petiole galls typical of the genus *Euura,* and even lower oviposition on the petiole would produce stem galls as in other members of *Euura.* Finally a shift in oviposition from low on the petiole to the bud could readily occur resulting in the bud gallers, also in the genus *Euura.* Thus divergence of species from a *Netamus*-type sawfly onto different parts of the host plant may have resulted by accidents in oviposition behavior leading to speciation events, as illustrated in Figure 14-2.

Other characters of these sawflies correlate with this proposed phylogenetic trend and have been studied by Smith (1968). As the common name "sawfly" indicates, the ovipositor is well developed for sawing into plant tissue, the saw being the gonapophyses VIII. The saw-like appearance is created by a series of annuli along each gonapophysis with tooth-like projections on the ventral edge, each annulus bearing complex spines or ctenidia. These characters of the saw are diagnostic, as described by Smith (1968). With the trend from *Phyllocolpa* to

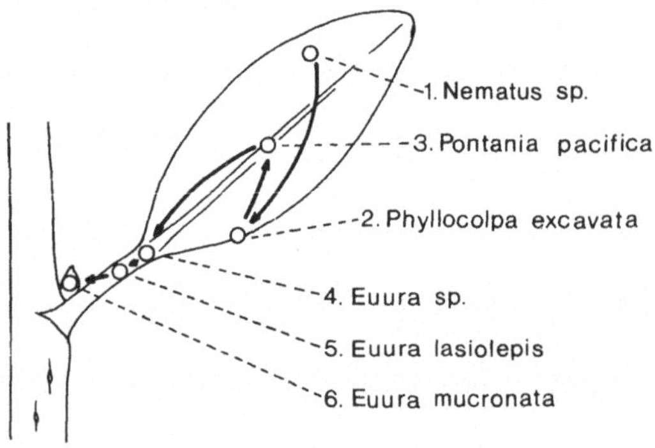

Figure 14-2. Diagram of oviposition sites of the six species of sawfly on *Salix lasiolepis*, with the possible trend in phylogeny indicated with arrows based on shifts in oviposition behavior. View is of abaxial leaf surface.

Euura described above, there is also a trend of increasing ovipositor length, increasing narrowness of the saw (length:width ratio), and increasing number of annuli (Table 14-2), all associated with deepening oviposition into plant tissues. The only departure from the scheme above is that *Euura (Gemmura)*, the bud gallers, appear to be intermediate between *Pontania* and *Euura (Euura)*.

Another revealing characteristic discussed by Smith (1968, 1970) is larval behavior. In *Pontania, Euura (Gemmura)*, and the petiole-galling *Euura* the larva cuts an exit hole in the gall and emerges to pupate in the soil, forming a silken cocoon to which soil particles adhere. In the *exigua* species group of *Euura* the larva cuts an exit hole but remains in the gall to spin a cocoon and eventually pupate. Larvae in the *lasiolepis* species group cut no emergence hole, and the adult sawfly must cut the hole through which it escapes from the gall. In this phylogeny there has been an increasingly intimate relationship between the sawfly and the host plant, with species in the *exigua* and *lasiolepis* species groups of *Euura* living in the plant throughout their larval life and adults emerging ready to oviposit, often on the same host plant. Benson (1950) and Smith (1968, 1970) note that they are the most specific of the sawflies, with each species being specific to a species of willow.

Gall-inducing sawflies of nematine stock are clearly closely related and have radiated exclusively on the family Salicaceae, with the majority of species on *Salix* and a few on *Populus* [in North America the recorded host relationships for *Phyllocolpa* are 3 species on *Salix* and 4 on *Populus;* for *Pontania* 25 species on *Salix;* for *Euura* 24 species on *Salix* (Smith 1979) (see also Chapter 3)].

The subfamily Blennocampinae in the Tenthredinidae also contains gall-inducing species. *Blennogeneris spissipes* Cresson. induces terminal bud galls in *Symphoricarpus* (Ross 1932), *Blennocampa pusilla* Klug. induces leaf rolls on *Rosa* (Darlington 1975), and *Hoplocampoides xylostei* Giraud attacks buds of *Lonicera.* However, very little is known about the gall inducers in this subfamily, and they will be discussed in this chapter no further.

It appears that gall induction has evolved at least twice in the Tenthredinidae, once in the subfamily Nematinae, and once in the Blennocampinae.

Table 14-2. Probable Phylogenetic Relationships between Nematine Sawflies on Willows, following Smith (1968), with Trend Moving from Left to Right

	Taxon genus, subgenus, and species group					
	Phyllocolpa	*Pontania*	*Euura, Gemmura*	*Euura, Euura*	*Euura, Euura, exiguae*	*Euura, Euura, lasiolepis*
Ovipositor Length (mm)	0.5–0.7	0.6–1.0	0.6–0.8	0.9–1.3	0.9–1.5	1.5–2.0
Saw length:width	3–6:1	6–8:1	6–8:1	10–13:1	10–14:1	15–20:1
Number of annuli	15–20	15–20	15–20	20–25	20–25	25–30
Oviposition site	Leaf	Leaf	Bud	Petiole	Stem	Stem
Larva cuts exit hole		Yes	Yes	Yes	Yes	No
Larva leaves gall	Yes	Yes	Yes	Yes	No	No
Site of cocoon and pupa	Soil	Soil	Soil	Soil	Gall	Gall

ECOLOGICAL RELATIONSHIPS

Food Webs

Feeding relationships within galls have been described by Carleton (1939) for *Pontania proxima* on *Salix triandra* L. and *S. fragilis* L. in Britain (Fig. 14-3), Caltagirone (1964) for *Pontania pacifica* on *Salix lasiolepis* in California (Fig. 14-4), Kopelke (1983a, 1984, 1985) for *Pontania proxima* on *Salix fragilis*, and *P. dolichura* (Thomson), *P. viminalis* L., and *P. vesicator* (Bremi-Wolf) on *Salix purpurea* L. in Germany (see also Kopelke 1986, 1987a,b), and personal observations in Arizona on *P. pacifica* and *Euura lasiolepis* on *S. lasiolepis* are included here (see also Price and Craig 1984; Clancy *et al.* 1986). Smith (1970) discussed natural enemies of *Euura* and *Pontania* species in California. *Pontania pacifica* in Arizona appear to be the same species based on morphological criteria (H. Goulet, personal communication), but galls are different, so they probably represent a sibling species pair on the same host plant. In Arizona, galls are spherical on the underside of the leaf, but in California they are "as much developed on the upper as on the lower surfaces of the leaf" (Caltagirone 1964, p. 280).

A comparison of food webs based on willows and their gall-inducing sawflies shows remarkable similarities. Both *Pontania proxima* and *P. pacifica* have the following common elements (cf. Figs. 14-3 and 14-4): ecto- and endoparasitic ichneumonids, and the unusual inquilines, weevils (Curculionidae), and moth caterpillars (Tortricidae and Cosmopterygidae). A comparison of all described food webs based on gall-inducing sawflies shows that other equivalent elements are repeatedly present (Table 14-3): primary ectoparasitoid braconids, pteromalids, and eurytomid inquilines. Thus seven taxa are commonly represented in sawfly galls, both in the Nearctic and Palaearctic realms.

The communities of parasitoids and inquilines are so similar that several elements of the community may have tracked phylogenetically the speciation events of the host sawflies. Of particular interest are the genera *Lathrostizus, Scambus, Pteromalus (Habrocytus), Eury-*

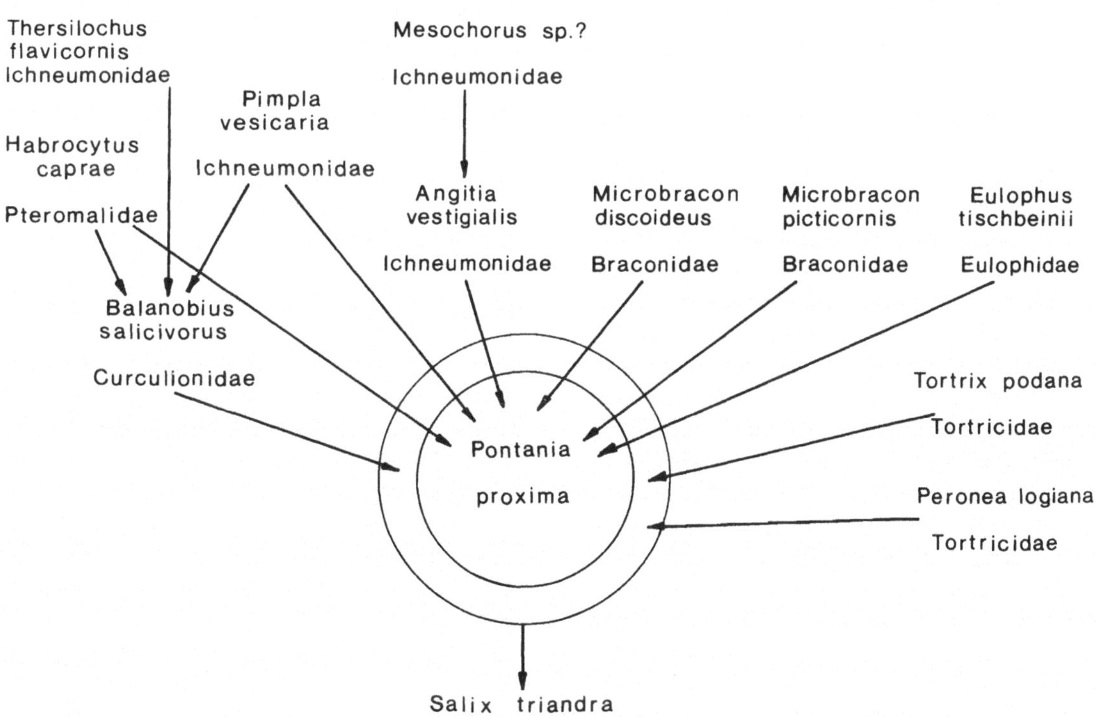

Figure 14-3. Food web of the major species associated with *Pontania proxima,* based on studies by Carleton (1939). Parasitoids have arrows to the gall chambers and inquilines have arrows indicating feeding on gall.

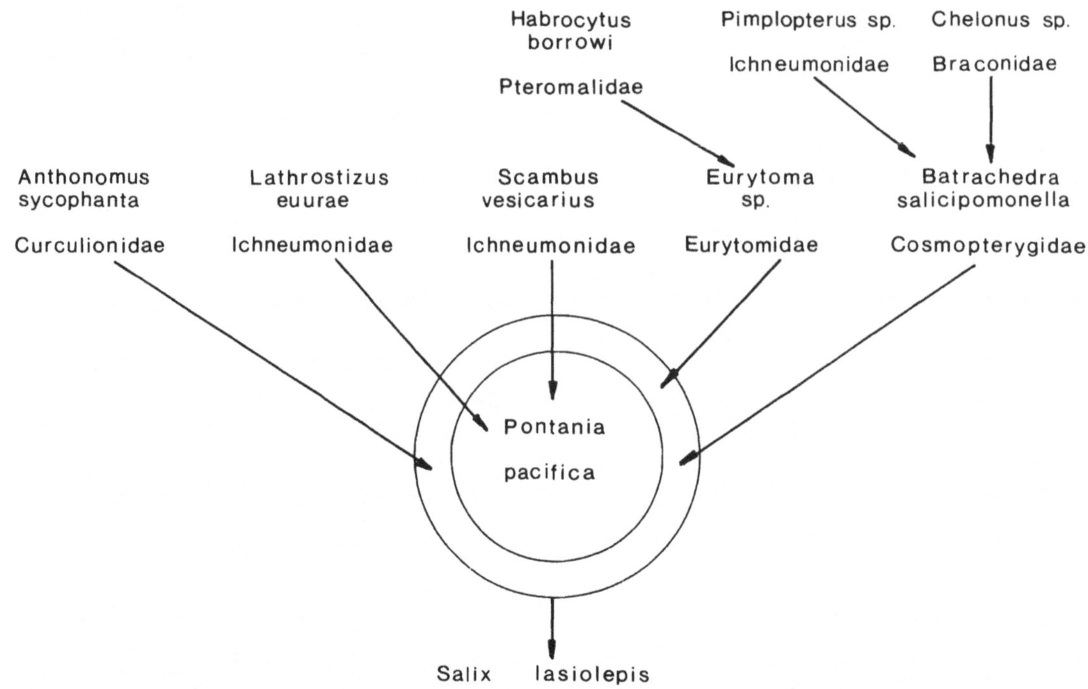

Figure 14-4. Food web of the major species associated with *Pontania pacifica,* based on studies by Caltagirone (1964). Parasitoids have arrows to the gall chamber and inquilines have arrows indicating feeding on the gall.

toma, and the braconids. Where genera differ, further study may easily recognize closer relationships. This may even be true for the curculionid and lepidopteran inquilines. Such phylogenetic tracking of parasites is thought by Brooks (1980) to be very important in parasite communities. We should predict equivalent, closely related species in the same genus on different species (vicariants), as *Lathrostizus lugens* and *L. euurae* (Ashmead) appear to be. The alternative hypothesis is that after sawflies speciated and diverged in geographic location, the community of parasitoids and inquilines developed from the newly sympatric fauna. We would predict considerable generic difference between equivalent species in this case not seen in Table 14-3. These possibilities are worth more study, but the parallels in communities on sawfly hosts are similar to the communities based on gall wasps studied by Askew (1961), in which the genera *Eurytoma, Mesopolobus, Olynx, Synergus,* and *Torymus* repeatedly occur. "The food webs in galls of corresponding generations of closely allied species are often similar" (Askew 1961). Askew and Shaw (1986) reviewed the kinds of parasitoids associated with different feeding habits of sawflies, including free feeders, gall inducers, and leaf miners. Price and Pschorn-Walcher (1988) argued that there has been a gradual loss of parasitoid species as the gall-inducing habit has evolved from free feeders to *Euura.*

Survivorship Curves

Survivorship curves of sawfly cohorts on clones are generated by dissection of galls at the end of sawfly development, but before emergence of adults. One of the great advantages of studying galling species is that so frequently the history of life and death is preserved in the gall. *Euura lasiolepis* galls can be dissected from March to early May to reveal survivorship and causes of death. The stages of development of the sawfly used in this analysis are Stage 1, egg; Stage 2, I and II instar larvae; Stage 3, III and IV instar larvae; Stage 4, V and VI instar larvae; Stage 5, larvae in cocoons [i.e., late V instar males and

Table 14-3. Comparison of Communities of Parasitoids and Inquilines in Galls of Sawflies, Showing the Common Components in Each Community[a]

Sawfly species	Location	Primary endoparasitoid ichneumonid	Primary ectoparasitoid ichneumonid	Primary ectoparasitoid braconid	Primary ectoparasitoid pteromalid	Eurytomid inquiline	Curculionid inquiline	Lepidopteran inquiline	Reference
Pontania proxima	Somerset, Glocestershire, England	*Angitia vestigialis*[b]	*Pimpla vesicaria*[c]	*Microbracon discoideus* *Microbracon picticornis*	Habrocytus *caprae*[d]	None	*Balanobius salicivorous*	*Tortrix podana* *Peronea logiana*	Carleton (1939)
Pontania proxima	Germany	*Lathrostizus lugens*[b]	*Scambus vesicarius*[c]	Braconid unnamed	Pteromalid unnamed	?	*Curculio salicivorus* *Curculio crux*	Tortricid unnamed	Kopelke (1982, 1983a,b)
Pontania pacifica	California, U.S.A.	*Lathrostizus euurae*	*Scambus vesicarius*	None	None[e] *Habrocytus borrowi* (hyper)	*Eurytoma* sp.	*Anthonomus sycophanta*	*Batrachedra salicipomonella*[f]	Caltagirone (1964)
Pontania pacifica	Arizona, U.S.A.	?	*Scambus* sp.	*Bracon angelesius*	*Pteromalus* sp.	*Eurytoma fossae*	None	*Batrachedra striolata*	This study
Euura lasiolepis	Arizona, U.S.A.	*Lathrostizus euurae*	*Scambus* sp.	None	*Pteromalus* sp.	None	None	*Batrachedra striolata*	Price and Craig (1984)

[a]Species names are as in original publications, and synonymy is indicated below where known. All species cause death of sawfly larvae. ? indicates that presence or absence is uncertain.
[b]*Lathrostizus lugens* (= *Angitia vestigialis*) (Kopelke 1983a).
[c]Probably the same species, but name is not recognized in Carlson (1979).
[d]*Pteromalus* and *Habrocytus* are so close that the genera integrade (Burks 1979).
[e]*Habrocytus borrowi* is a hyperparasitoid on *Eurytoma* sp.
[f]Caltagirone (1964) actually described *B. striolata* according to Hodges (1966).

late VI females (Smith 1970)]. The mortality factors are categorized as follows:

1. *Abortion.* The galls of *Euura* are unusual in that they are formed from stimulation by the female parent during oviposition, and not by the feeding larvae (Magnus 1914; Smith 1968, 1970). Colleterial gland fluid injected with the egg apparently stimulates gall formation. Therefore, galls indicate an oviposition event, and lack of any evidence of larval feeding within the gall indicates an aborted attempt and death of the egg or very early I instar larva. Early death through abortion is a very common cause of mortality, and is recorded as death between developmental stages 1 and 2.
2. *Death of I and II instar larvae.* A small amount of larval feeding and tunneling at the top of the gall, near the oviposition site, indicates that the egg hatched but larval mortality occurred soon afterward. This is recorded as death between developmental stages 2 and 3, and is presumably the result of a host resistance factor. Oviposition by the ichneumonid, *Lathrostizus euurae,* occurs at this time, but it is an internal parasitoid and death of the host does not occur until late in the following spring. Therefore, its effects are not recorded in the survivorship curves discussed here.
3. *Death of III and IV instar larvae.* A moderate amount of feeding so that tunneling and frass are clearly visible in the gall indicates survival to a mid-larval stage. When development has not proceeded beyond this stage it is recorded as death between stages 3 and 4. This may be caused by inadequacy of the host plant as a food source or attack by the parasitoid *Pteromalus* sp.
4. *Death of V and VI instar larvae.* Death of late instar larvae occurs for unknown reasons, but a large amount of feeding, large frass, and the dead larva are evident in the gall. This is recorded as mortality between developmental stages 4 and 5.

Once the larva has survived to stage 5, and has spun its cocoon, subsequent survival is usually 100 percent to the emerging adult stage except for mortality caused by *Lathrostizus euurae,* and predation by chickadees; estimated at 10 and 4 percent, respectively, in one population (Price and Craig 1984).

Survivorship curves of sawflies differ dramatically from year to year on the same host willow clone (Fig. 14-5). On the two clones illustrated, survivorship was relatively high in the 1979 generation (based on gall dissections in March 1980), significantly reduced in the 1980 generation, and very low in the 1981 generation. The patterns of mortality differed in the same clones from year to year and between clones. Mortality between stages 1 and 2 became increasingly severe from 1979 to 1981 in both clones. Mortality between stages 3 and 4 increased in Clone MNA 2 in 1980 but not in Clone NP 8, but it was significant on both clones in the 1981 generation.

Moreover, some clones yielded relatively high survivorship of sawflies in the 1981 generation (Fig. 14-6). Although these clones were growing side by side, with branches intermingling, the shapes of the curves differed, but in the upper canopy sample both clones yielded about 30 percent survival of sawflies.

The reasons for these kinds of differences will be discussed in the next two sections.

Host Plant Quality

The large differences between years in percentage of eggs that do not hatch (abortion) seen in Clones MNA 2 and NP 8 (Fig. 14-5), and other clones examined (Table 14-4), probably results from changes in host plant quality. The hypothesis addressed here is that host plant quality affects abortion. In three of the four clones examined each year, abortion increased from 1979 to 1981 (Table 14-4). Willow plants were becoming increasingly resistant to sawfly attack. This trend was correlated with a trend of reduced summer and winter precipitation in

Figure 14-5. Survivorship curves for *Euura lasiolepis* on Clones MNA 2 and NP 8 for the 1979, 1980, 1981, and 1982 generations. For stages of development 1 = egg, 2 = I and II instar larvae, 3 = III and IV instar larvae, 4 = V and VI instar larvae, 5 = larvae in cocoons. All estimates are based on dissections of 100 or more galls.

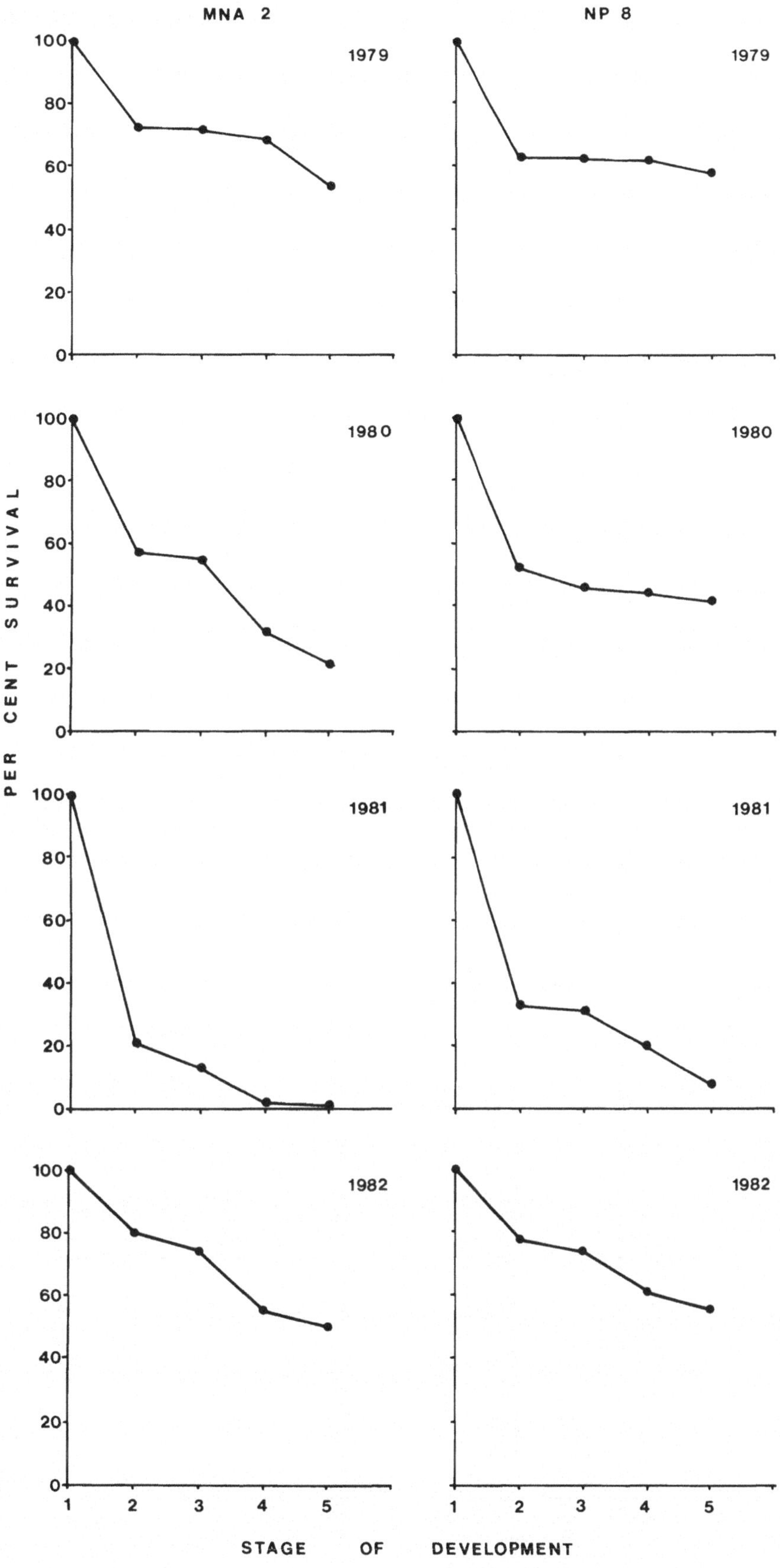

MNA 2
NP 8
1979
1980
1981
1982
100
80
60
40
20
0
PER CENT SURVIVAL
1
2
3
4
5
STAGE OF DEVELOPMENT

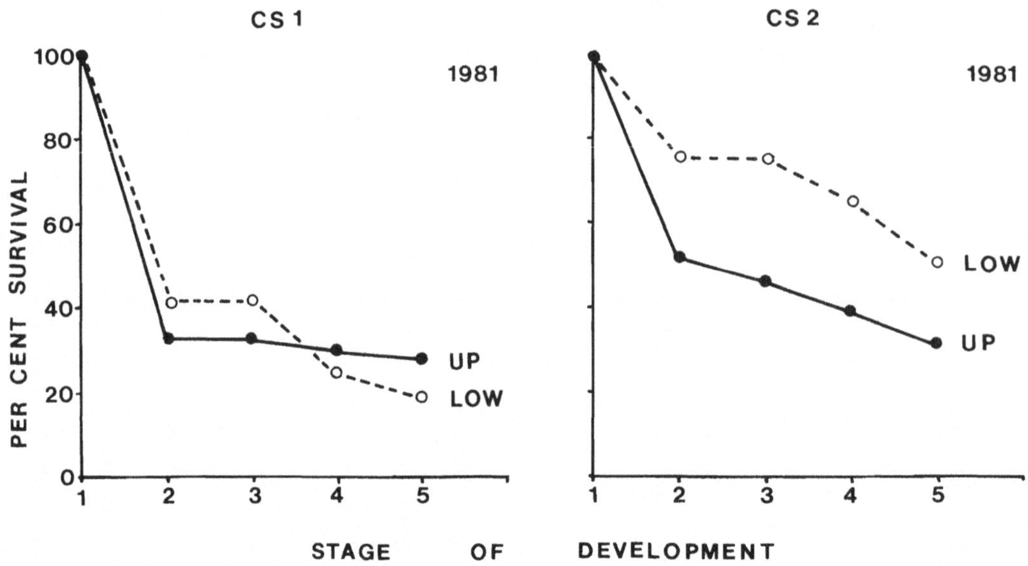

Figure 14-6. Survivorship curves for *Euura lasiolepis* on clones CS 1 and CS 2 for the 1981 generation in the upper (UP) or lower (LOW) canopy of each clone. Stages of development are as in Figure 14-5. Upper canopy estimates were made from over 100 galls. Lower canopy estimates were based on 36 galls and 63 galls, respectively, because fewer galls were present in the lower canopy.

the Flagstaff area, causing reduced plant growth, and a tendency for reduced gall growth. Mean gall diameter declined steadily in Clones MNA 2 and 4 from 1979 to 1981, and in Clones NP 8 and 9 there was a slightly increased gall size in the 1980 generation, but gall size in 1981 was lower than in 1979.

To test the hypothesis that water availability affects host plant quality we analyzed two clones growing at Coyote Spring (CS), on Museum of Northern Arizona property, which runs throughout the year. This contrasts with sites at which the other willow clones grow (MNA and NP) since they are situated on

Table 14-4. Comparisons of Survival of *Euura* Sawflies on Willow Clones, Involving the Mortality Factors Abortion and Parasitism

			Percent mortality			
				Parasitism		
Generation	Clone	Mean gall diameter (mm)	Abortion	Total	Interval	Percent survival
1979	MNA 2[a]	5.53	27	3	4	54
	MNA 4	4.74	29	23	36	27
	NP 8	6.98	37	1	2	59
	NP 9	4.90	56	17	41	11
1980	MNA 2	5.24	43	23	41	22
	MNA 4	4.54	35	17	28	26
	NP 8	7.29	48	2	4	41
	NP 9	5.34	56	9	21	24
1981	MNA 2	4.80	79	9	33	0.5
	MNA 4	3.52	58	3	30	0.8
	NP 8	6.72	67	10	28	7
	NP 9	4.75	49	2	15	4
	CS 1 up	7.53	67	3	8	28
	CS 1 low	6.99	58	14	33	19
	CS 2 up	5.96	48	2	5	31
	CS 2 low	5.48	24	10	13	51

[a]Site MNA = Museum of Northern Arizona, Flagstaff. Site NP = Northland Press, bordering MNA property. Site CS = Coyote Spring on MNA property.

Schultz Creek, which flows briefly, if at all, during snow melt in the spring. Throughout the spring and summer of 1981 this creek was dry, apparently influencing growth and quality of willows during 1981 on which the 1981 generation of sawflies developed. Survival on these two CS clones, at about 30 percent was much higher than on any clone on Schultz Creek, and gall size was relatively large (Table 14-4). However, abortion was relatively higher on CS 1 but relatively low on CS 2. Unfortunately we have no data on previous years to test if abortion is independent of precipitation at this site. These results suggest that water availability can ultimately affect survival of sawflies, and that on many clones this works through host quality changes (Price and Clancy 1986b).

A second test of the hypothesis on water availability affecting plant quality was made by comparing survival of sawflies in the upper canopy versus the lower canopy in Clones CS 1 and 2. The prediction was that water stress would affect more severely older, taller willow branches, than younger, shorter branches, and therefore plant quality should be lower in the upper canopy than in the lower canopy. This prediction was supported by the reduced abortion in both clones (Fig. 14-6), recorded as higher survivorship in the lower canopy sample at stage 2 of sawfly development.

After the 1981 generation we predicted that high winter precipitation would reduce abortion. Precipitation from October to May 1980–1981, relevant to the 1981 generation, was 23.5 cm whereas in 1981–1982 it was almost double at 41.5 cm. Survivorship of the 1982 generation was indeed much higher and abortion was lower than any other year studied on MNA 2 and NP 8 (Fig. 14-5) (see also Price and Clancy 1986b).

In general, there appears to be a change in host plant quality, at least in some clones, related to water availability. We have yet to determine what this resistance factor is.

Influence of the Parasitoid *Pteromalus* sp.

Mortality between developmental stages 3 and 4 is caused largely by the parasitoid *Pteromalus* sp. This mortality factor changes from year to year on the same clones and differs between clones in any one year (Fig. 14-5, Table 14-4). It can be expressed as parasitism on the surviving cohort at stage 3 in development (interval parasitism), which estimates parasitism on hosts actually available to the parasitoids.

We predicted that host plant quality would interact with parasitoid attack through the influence of gall size. As plant quality increased, gall size would increase, making sawflies less available to their enemies, and percentage parasitism would decline. Thus, parasitism could not be understood without understanding the role of plant quality. In these galling systems it is likely that such three trophic level interactions are inevitable, such that all levels must be studied to understand any one (Price *et al.* 1980; Price and Clancy 1986a; Price 1988).

In general, this prediction is supported. In the 1979 generation, the two clones with large galls, MNA 2 and NP 8, had very low levels of parasitism, whereas the clones with small galls had high parasitism; MNA 4, close to MNA 2, and NP 9 adjacent to NP 8 (Table 14-4). The comparison between NP 8 and NP 9 shows many differences (Table 14-5). Mean gall diameter was significantly higher on NP 8, more galls were formed per stem, and survivorship was much higher because of lower abortion, lower larval mortality, and lower parasitism. The large differences in appearance between galls on these clones are illustrated in Figure 14-7. Vegetative growth was stunted, and flowering and

Table 14-5. Comparison of 100 Galls on Each of Clones NP 8 and NP 9 Growing Side by Side in Apparently Identical Physical Conditions in 1979[a]

Characteristics	NP 8	NP 9
Sex	Male	Female
Stem color	Yellow	Red
Mean gall diameter (mm)	6.9	4.9
Range in gall diameter (mm)	3.5–9.1	2.7–6.9
Mean number of galls per stem	2.5	1.6
Total sawflies alive per 100 galls	76[b]	14[b]
Total aborted sawflies	37	59
Total sawflies dead (late instars)	3	12
Total living parasitoids plus inquilines	1	18
Total organisms	117	103

[a]Note high survivorship and low parasitism on NP 8 and the reverse on NP 9.
[b]Note that these figures differ from the 59 and 11 percent survival for these clones in Table 14-4, because 17 galls and 3 galls, respectively, had 2 sawflies per gall, whereas in Table 14-4 all numbers are expressed on a per gall basis.

Figure 14-7. Representative shoots from clones NP 8 (five shoots on left) and NP 9 (six shoots on right) galled by the 1979 generation of *Euura lasiolepis.* Note the large galls on NP 8 and the small galls on NP 9; up to five adjacent internodes were galled on NP 8 but no more than two or three on NP 9; no inflorescences were produced on illustrated stems from NP 8 but many were formed on some stems of NP 9.

fruiting were almost totally suppressed in NP 8, whereas in NP 9 vegetative shoots developed well and many inflorescences were developed. This relationship also was seen in the 1980 generation in Clones NP 8 and 9, and comparisons between upper canopy and lower canopy samples in the 1981 generation on Clones CS 1 and CS 2, which grow side by side, also show that within clones the areas with larger galls have lower parasitism (Table 14-4). Overall data in Table 14-4, there is a significant correlation between gall size and percent interval parasitism (Fig. 14-8) (Price and Clancy 1986a; Price 1988).

However, there are complexities in this relationship between quality and parasitism that we cannot as yet account for. The general picture of large galls reducing parasitism accounts for at most 40 percent of the variance, so other factors must be important. For example, the upper canopy on CS 2 has smaller galls than the lower canopy on CS 1 but smaller galls than NP 8 but lower parasitism (Table 14-4). In the 1981 generation NP 9 has smaller galls than NP 8 but lower parasitism. One major factor not considered is population size of available sawfly hosts. Galls with living sawflies on NP 9 in 1981 were so rare that they must have been very difficult to find for parasitoids. On MNA 2 in the 1981 generation only 13 percent of the original galls still had living sawflies available to parasitoids (Fig. 14-5), and 33 percent of these were parasitized (Table 14-4), a strong response considering the low density of hosts available. Thus the pure effects of gall size are complicated by interacting effects of host density in the current year, probably host density and parasitism in the previous year, and possibly individual willow clone and microhabitat differences, making individual clones more or less favorable to parasitoids. These kinds of factors will be more readily analyzed in choice experiments with parasitoids on clones in a willow garden growing under identical conditions.

The interactions between these three trophic levels, plant, herbivores, and the enemies of herbivores, are complex. Interaction is clearly important, for the combination of abortion and parasitism has pushed populations on some clones almost to extinction (e.g., MNA 2 and MNA 4 in 1981), even on a clone that seemed to be very favorable for the 1979 generation (MNA 2) (Table 14-4). On MNA 2, in 210 galls dissected, only 1 living sawfly was found at the end of the 1981 generation.

The study of gall-inducing sawflies offers

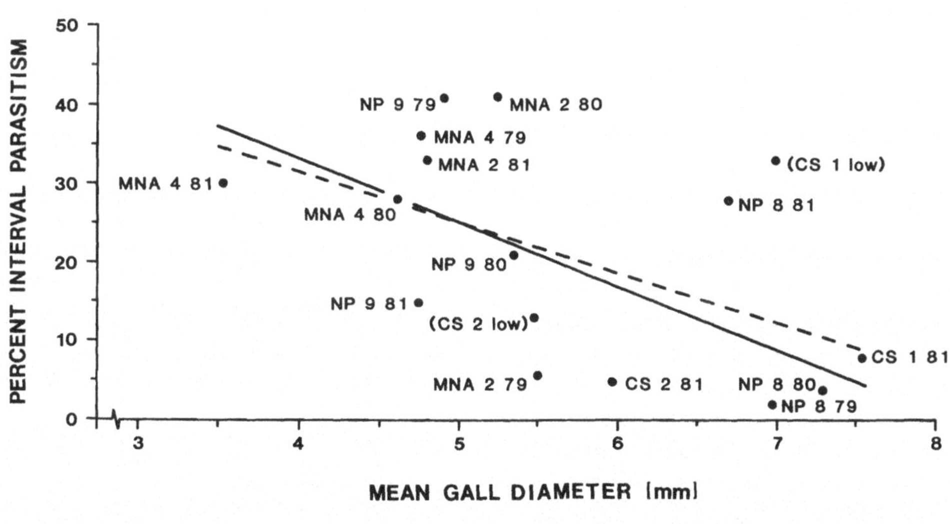

Figure 14-8. Relationship between mean gall diameter (millimeters) and percent interval parasitism for data in Table 14-4. Clone identity and year are given for each point. Except for CS 1 and 2 low (in parentheses) galls were collected in the upper canopy and the regression equation for upper canopy galls is $Y = 64.61 - 7.82X$ ($n = 14$, $R^2 = 0.40$, $p < .05$) accounting for 40 percent of the variance (solid line). For all gall collections the equation is $Y = 56.36 - 6.20X$ ($n = 16$, $R^2 = 0.26$, $p < .05$) accounting for 26 percent of the variance (dashed line.)

many interesting possibilities, particularly using experimental approaches. Nearly all species are found on poplars and willows, many species of which are particularly easy to propagate vegetatively from cuttings. Therefore, a strong experimental approach can be developed using potted plants and applying treatments of such factors as water, fertilizer, hormones, and shading, to study sawfly choice and survivorship and gall growth in different treatments (e.g., see Price and Clancy 1986a; Preszler and Price 1988; Craig *et al.* 1989). Sawflies behave naturally in arenas with high humidity, so behavioral aspects of oviposition, site choice on shoots, and choice between plants can be studied in detail (e.g., Craig *et al.* 1988, 1989, 1990). Mechanisms of host shifting and speciation may be investigated by testing for variation in populations for host choice, oviposition "errors" and gall induction in sympatric willow species, not used as host plants. Testing effects of colleterial fluid on different willows, and the ability of eggs to develop in galls initiated by other species would help unravel the mechanisms of adaptation to hosts and the differences in host specificity seen in genera. Studies on genetic similarities between host plant and sawfly species, and sawfly genera, would contribute to the understanding of the phylogeny and radiation in this group. Mechanisms of gall formation also may be studied, and emphasis may best be placed on the content and effect of colleterial fluid in *Euura* females, since this fluid stimulates gall growth, and development can be studied in the absence of larval sawflies. Experimental injection of colleterial fluid from *Nematus, Phyllocopa, Pontania,* and *Euura* into different plant parts following the species-specific wounding behavior of the female, coupled with a comparative study on gland content, may prove revealing on the evolutionary pathway to the gall-inducing habit.

Gall-inducing sawflies are widely distributed in north temperate, subarctic, and alpine environments, wherever climate is cool and wet enough for willows and poplars to grow well. They are easy to find and readily available for study to many centers where research activity is concentrated. With new techniques now available, with conceptual developments in behavior, ecology, and evolution, with increasing pressure to understand unusual growth in plants, and the growing importance of energy forests using willow, the study of gall-inducing sawflies should increase dramatically. They have the potential for contributing significantly to the understanding of many aspects of evolutionary biology, from molecular and cellular processes to macroevolution and the phylogenetic development of taxa.

ACKNOWLEDGMENTS

I am grateful to Karen Clancy for help with dissection of galls, analysis of data, and to her and Gwen Waring for comments on this chapter. The studies reported here were supported by National Science Foundation Grants DEB 80-21754 and BSR 83-14594. Genus and species identifications were made by E. Smith of the California Academy of Sciences *(Euura lasiolepis),* H. Goulet of the Biosystematics Research Institute, Ottawa (*Pontania pacifica* from Arizona), and by members of the Insect Identification and Beneficial Insect Introduction Inst. Systematic Entomology Laboratory, U.S.D.A.: E. E. Grissell *(Pteromalus),* D. R. Smith *(Euura),* and S. R. Shaw and P. M. Marsh *(Bracon angelesius).* I am very grateful for their expertise.

REFERENCES

Askew, R. R. 1961. On the biology of the inhabitants of oak galls of Cynipidae (Hymenoptera) in Britain. Trans. Soc. Br. Entomol. 14:237–258.

Askew, R. R., and M. R. Shaw. 1986. Parasitoid communities: Their size, structure and development. In J. Waage and D. Greathead (eds.), Insect Parasitoids, pp. 225–264. Academic Press, New York.

Benson, R. B. 1950. An introduction to the natural history of British sawflies. Trans. Soc. Br. Entomol. 10:45–142.

Brooks, D. R. 1980. Allopatric speciation and non-interactive parasite community structure. Syst. Zool. 29:192–203.

Burks, B. D. 1979. Family Pteromalidae. In K. V. Krombein, P. D. Hurd, Jr., D. R. Smith, and B. D. Burks (eds.), Catalog of Hymenoptera in America North of Mexico, pp. 768–835. Smithsonian Institution, Washington, D.C.

Bush, G. L. 1975a. Sympatric speciation in phytophagus parasitic insects. In P. W. Price (ed.),

Evolutionary Strategies of Parasitic Insects and Mites, pp. 187–206. Plenum, New York.

Bush, G. L. 1975b. Modes of animal speciation. Annu. Rev. Ecol. Syst. 6:339–364.

Caltagirone, L. E. 1964. Notes on the biology, parasites, and inquilines of *Pontania pacifica* (Hymenoptera: Tenthredinidae), a leaf-gall incitant on *Salix lasiolepis.* Ann. Entomol. Soc. Am. 57:279–291.

Carleton, M. 1939. The biology of *Pontania proxima* Lep., the bean gall sawfly of willows. J. Linn. Soc. London *(Zool.)* 40:575–624.

Carlson, R. W. 1979. Family Ichneumonidae. In K. V. Krombein, P. D. Hurd, Jr., D. R. Smith and B. D. Burks (eds.), Catalog of Hymenoptera in America North of Mexico, pp. 315–740. Smithsonian Institution, Washington, D.C.

Clancy, K. M., P. W. Price, and T. P. Craig. 1986. Life history and natural enemies of an undescribed sawfly near *Pontania pacifica* (Hymenoptera: Tenthredinidae) that forms leaf galls on arroyo willow, *Salix lasiolepis.* Ann. Entomol. Soc. Am. 79:884–892.

Cornell, H. V. 1983. The secondary chemistry and complex morphology of galls formed by the Cynipidae (Hymenoptera): Why and how? Am. Midl. Nat. 110:225–234.

Craig, T. P., J. K. Itami, and P. W. Price. 1988. Plant wound compounds from oviposition scars used as oviposition deterrents by a stem-galling sawfly. J. Insect. Behav. 1:343–356.

Craig, T. P., J. K. Itami, and P. W. Price. 1989. A strong relationship between oviposition preference and larval performance in a shoot-galling sawfly. Ecology 70:1691–1699.

Craig, T. P., J. K. Itami, and P. W. Price. 1990. Intraspecific competition and facilitation by a shoot-galling sawfly. J. Anim. Ecol. 59:147–159.

Darlington, A. 1975. The Pocket Encyclopedia of Plant Galls in Colour, rev. ed. Poole, Dorset Blandford Press, London.

Denno, R. F., and M. S. McClure. 1983. Variable plants and herbivores in natural and managed systems. Academic Press, New York.

Heslop Harrison, J. W. 1927. Experiments on the egg-laying instincts of the sawfly, *Pontania salicis* Christ., and their bearing on the inheritance of acquired characters; with some remarks on a new principle in evolution. Proc. R. Soc. London Ser. B 101:115–126.

Hodges, R. W. 1966. Review of the New World species of *Batrachedra* with description of three new genera (Lepidoptera: Gelechioidea). Trans. Am. Entomol. Soc. 92:585–651.

Hovanitz, W. 1959. Insects and plant galls. Sci. Am. 201(5):151–162.

Kopelke, J.-P. 1982. Die gallenbildenden *Pontania*—Arten—ihre Sonderstellung unter den Blattwespen. Teil I: Gallenbildung, Entwicklung und Phänologie. Nat. Mus. 112:356–365.

Kopelke, J.-P. 1983a. Die gallenbildenden *Pontania*—Arten—ihre Sonderstellung unter den Blattwespen. Teil II: Konkurrenz und Koexistenz, Parasitismus. Nat. Mus. 113:12–23.

Kopelke, J.-P. 1983b. Brutparasitismus mit partieller Entomophagie. Die besondere Lebensweise der in Blattwespengallen auftretenden Rüsselkäferarten. Nat. Mus. 113:333–344.

Kopelke, J.-P. 1984. Biologie und Parasiten der gallenbildende Blattwespe *Pontania proxima* (Lepeletier 1823) (Insecta: Hymenoptera: Tenthredinidae). Senck. Biol. 65:215–239.

Kopelke, J.-P. 1985. Uber die Biologie und Parasiten der gallenbildenden Blattwespenarten *Pontania dolichura* (Thoms. 1871), *P. vesicator* (Bremi 1849) und *P. riminalis* (L. 1758) (Hymenoptera: Tenthredinidae). Faunistisch-Okol. Mitt. 5:331–344.

Kopelke, J.-P. 1986. Zur Taxonomie und Biologie neuer *Pontania*—Arten der *dolichura*—Gruffe. Senk. Biol. 67:51–71.

Kopelke, J.-P. 1987a. Die Rolle der Brutparasiten in Feindkomplex der gallenbildenden *Pontania*—Arten (Hymenoptera: Tenthredinidae: Nematinae). Zool. Jahrb. Syst. 114:487–508.

Kopelke. J.-P. 1987b. *Adelognathus cubiceps* Roman. 1924. (Ichneumonidae; Adelognathinae)—ein ungewöhnlicher Parasitoid der gallenbildenden *Pontania*—Arten (Tenthredinidae: Nematinae) (Insecta: Hymenoptera). Senk. Biol. 67:253–259.

Kopelke, J.-P. 1989. Mittel- und nordeuropäische Arten der Gattung *Pontania* Costa 1859 aus der *herbaceae*- und *polaris*-Gruppe (Insecta: Hymenoptera: Tenthredinidae). Senk. Biol. 69:41–72.

Larsson, S., and A. Wiren. 1982. Leaf-eating insects in an energy forest stand of *Salix viminalis* L. in Central Sweden. Ann. Entomol. Fenn. 48:119–125.

Magnus, W. 1914. Die Entstehung der Pflanzengallen verursacht durch Hymenopteren. G. Fischer, Jena.

McCalla, D. R., M. K. Genthe, and W. Hovanitz. 1962. Chemical nature of an insect gall growth-factor. Plant Physiol. 37:98–103.

Preszler, R. W., and P. W. Price. 1988. Host quality and sawfly populations: A new approach to life table analysis. Ecology 69:2012–2020.

Price, P. W. 1984a. Insect Ecology, 2nd ed. Wiley, New York.

Price, P. W. 1984b. Alternative paradigms in community ecology. In P. W. Price, C. N. Slobodchikoff, and W. S. Gaud (eds.), A New Ecology: Novel Approaches to Interactive Systems, pp. 353–383. Wiley, New York.

Price, P. W. 1988. Inversely density-dependent parasitism: The role of plant refuges for hosts. J. Anim. Ecol. 57:89–96.

Price, P. W., C. E. Bouton, P. Gross, B. A. McPheron, J. N. Thompson, and A. E. Weiss. 1980. Interactions among three trophic levels: Influence of plants on interactions between insect herbivores and natural enemies. Annu. Rev. Ecol. Syst. 11:41–65.

Price, P. W., and K. M. Clancy. 1986a. Interactions among three trophic levels: Gall size and parasitoid attack. Ecology 667:1593–1600.

Price, P. W., and K. M. Clancy. 1986b. Multiple effects of precipitation on *Salix lasiolepis* and populations of the stem-galling sawfly, *Eurra lasiolepis.* Ecol. Res. 1:11–114.

Price, P. W., and T. P. Craig. 1984. Life history, phenology, and survivorship of a stem-galling sawfly, *Euura lasiolepsis* (Hymenoptera: Tenthredinidae), on the Arroyo Willow, *Salix lasiolepis,* in Northern Arizona. Ann. Entomol. Soc. Am. 77:712–719.

Price, P. W., and H. Pschorn-Walcher. 1988. Are galling insects better protected against parasitoids than exposed feeders?: A test using tenthredinid sawflies. Ecol. Entomol. 13:195–205.

Pschorn-Walcher, H. 1975. Massenauftreten der Blattwespe *Hoplocampoides xylostei* Giraud (Hym.: Tenthredinidae) im badischen Rheintal und ihr Vorkommen im schweizer Jura. Mitteil. Schweiz. Entomol. Ges. 48:141–145.

Rosenthal, G. A., and D. H. Janzen (eds.). 1979. Herbivores: Their Interaction with Secondary Plant Metabolites. Academic Press, New York.

Ross, H. H. 1932. The subfamily Lycaotinae in North America (Hymenoptera: Tenthredinidae). Can. Entomol. 64:40–45.

Smith, D. R. 1979. Suborder Symphyta. In K. V. Krombein, P. D. Hurd, Jr., D. R. Smith and B. D. Burks. (eds.), Catalog of Hymenoptera in America North of Mexico, pp. 3-317. Smithsonian Institution, Washington, D.C.

Smith, E. L. 1968. Biosystematics and morphology of Symphyta. I. Stem-galling *Euura* of the California region, and a new female genitalic nomenclature. Ann. Entomol. Soc. Am. 61:1389–1407.

Smith, E. L. 1970. Biosystematics and morphology of Symphyta. II. Biology of gall-making nematine sawflies in the California region. Ann. Entomol. Soc. Am. 63:36–51.

Strong, D. R., J. H. Lawton, and R. Southwood. 1984. Insects on Plants: Community Patterns and Mechanisms. Harvard Univ. Press, Cambridge, MA.

15

Ecology of *Pemphigus* Gall Aphids

T. G. WHITHAM

Many ecological principles lack critical examination because the required data for testing are difficult, if not impossible, to collect (Stearns 1976; Mitchell 1983). In many instances (e.g., adaptation and competition theory) we have relied on indirect tests or inference to examine major ecological paradigms. However, gall-inducing aphids and other galling insects show great promise for surmounting the problems that have prevented robust testing of many ecological principles. First, due to their gall-inducing behavior, they are essentially sessile for a major portion of their life cycle. This feature not only permits an accurate census of their population but allows one to follow the progress of individuals through time. Second, although censuses of most organisms are imcomplete because individuals that die leave no evidence of their presence, with *Pemphigus* gall aphids (and probably many other gall insects as well) even the aphid that dies trying to colonize the host plant leaves a scar on the leaf that can be easily identified at any later time. Knowledge of those individuals that fail to reproduce and die may be even more informative than knowledge of the survivors. Third, since the gall is induced by a single founding individual that subsequently reproduces within the gall, the reproductive success or relative fitness of individual colonizers can be quantified by counting the number of progeny. This is a powerful tool because it allows the ecologist to discriminate between alternative hypotheses by comparing differences in fitness. Fourth, resource quality can be quantified by determining how parameters of the host and aphid environment (e.g., plant defenses, nutrition, competitor density, and predators) affect aphid survival and reproduction. Lastly, the gall is essentially a community in microcosm composed of the plant, the aphid herbivores or plant parasites, and the various predators and parasites of the aphids. Considering that a single tree may support over 58,000 galls containing 4,500,000 aphids, sample size is not a problem.

When taken in combination, these five features enable the observer of gall-inducing insects to critically examine many current and important questions to ecologists and evolutionary biologists that are unanswerable in most systems. It is the aim of this chapter to examine the basic biology and ecology of *Pemphigus* gall aphids, with the goal of showing how they may be especially useful in studying coevolution, parasite–host interactions, plant defense strategies, territoriality, and host selection behavior.

LIFE CYCLE AND GALL FORMATION

Colonizing stem mothers emerge from eggs that overwintered in the deep fissures of the primary host, cottonwood trees of the genus *Populus* (Harper 1963), and migrate en masse to the developing leaves where they attempt gall induction. Although the stem mothers are wingless, dispersal from the overwintering site is rapid with as much as 83 percent of the population arriving at the developing leaves over a 3-day period that peaks just after bud burst (Whitham 1978). Considering that as many as 58,000 stem mothers may make the migration on a single tree, this period of the life cycle is analogous to a bird migration but on a microscale. After moving between numerous leaves, shoots, and branches of the host tree, the stem mother selects a leaf for colonization. Because aphids are thought to be excellent phytochemists capable of discriminating between relatively small differences in sap nutrients and/or plant defenses (Auclair 1963; van Emden 1972; Dixon 1975; Zucker 1982) their settling preferences are probably based largely on biochemical cues. With repeated probing of the leaf tissues with her stylets, the stem mother induces a hollow gall. Within the confines of this gall the stem mother (fundatrix) parthenogenetically and viviparously produces 300 or more progeny.

When the gall dehisces in midsummer, the winged offspring, the spring migrants or fundatrigenia, migrate to a secondary host where they go underground and feed on the roots. Although the primary host is always a cottonwood, depending on the aphid species, the secondary host ranges from economically important crops such as beets, carrots, spinach, lettuce, potatoes, celery, turnips, and cabbage (Palmer 1952; Harper 1958, 1959) to nonagricultural plants often associated with disturbance such as lambsquarters (*Chenopodium album* L.), goldenrod (*Solidago* sp.), umbrella plant (*Erigeron* sp.), water parsley (*Oenanthe sarmentosa* Presl.), *Polygonum aviculare* L., *Aster multiflorus* Ait., *Agropyron* sp., *Achillea* sp., *Sonchus* sp., and *Lapsana communis* L. (Palmer 1952; Harper 1959; Wool 1984). Foster (1975) also found that *Pemphigus trehernei* Foster could withstand saltwater submergence on its secondary host, *Aster trifolium* L., which is found in saltmarshes subjected to tidal submergence. Thus, although *Pemphigus* gall aphids are quite specific in their primary host preference (Family Salicaceae), their secondary hosts are diverse and span at least eight plant families (Chenopodiaceae, Compositae, Cruciferae, Euphorbiaceae, Gramineae, Polygonaceae, Solanceae, and Umbelliferae). It would be interesting to discover if there exists any biochemical similarities between the many secondary hosts and *Populus* that would account for the associations. Alternatively, the phloem feeding habit of these aphids may permit them to circumvent the diverse defenses of these plant families.

On the secondary host a wingless generation, the alienocola, are produced, which in turn produce a winged generation, the fall migrants or sexupara. These return to the primary host where the wingless sexuals are produced ($2N-1$ males and $2N$ females). This is the only time that sexual reproduction occurs and each female lays a single overwintering egg to complete the life cycle. This represents the generalized life cycle; however, variations exist (see Harper 1963; Foster 1978).

In summary, the life cycle of *Pemphigus* gall aphids is complex involving migration between a primary host, a woody perennial, and an herbaceous secondary host. The migration from a woody perennial in mid-summer to an herbaceous secondary host and then back again in the fall may be in part associated with changes in nutrient levels in the two host species. Dixon (1971) found that the nongalling bird cherry-oat aphid, *Rhopalosiphum padi* L. left bird cherry in the late spring when the percent soluble nitrogen in the leaves fell to a low level and then returned again in the fall as the leaves were senescing and the soluble nitrogen once again increased. Presumably, the herbaceous secondary host is more actively growing in the summer than the woody perennial, thereby making it profitable to shift hosts and continually feed from the nutritionally superior host plants. As will be developed later in this chapter an alternative to the above logic is that with gall-inducing aphids, host alternation may be forced on them by the counterdefenses of their primary host plant.

Although the selection of an individual leaf for galling by *Pemphigus* aphids takes place very rapidly, leaf selection is perhaps the most crucial period of the life cycle. The reproductive fate of the colonizing stem mother is closely tied to the quality of the leaf selected and once the settling decision is made the commitment is total, with no chance of moving to a better leaf. Once a leaf is selected, the stem mother probes the immature leaf tissues with her stylets to induce gall formation. Figure 15-1 shows a scanning electron micrograph of a first-instar stem mother of *Pemphigus betae* Doanne in the process of inducing a gall on a leaf of *Populus angustifolia* James. At this stage the stem mother is about 0.6 mm in length. This photograph shows the stem mother with her stylets inserted

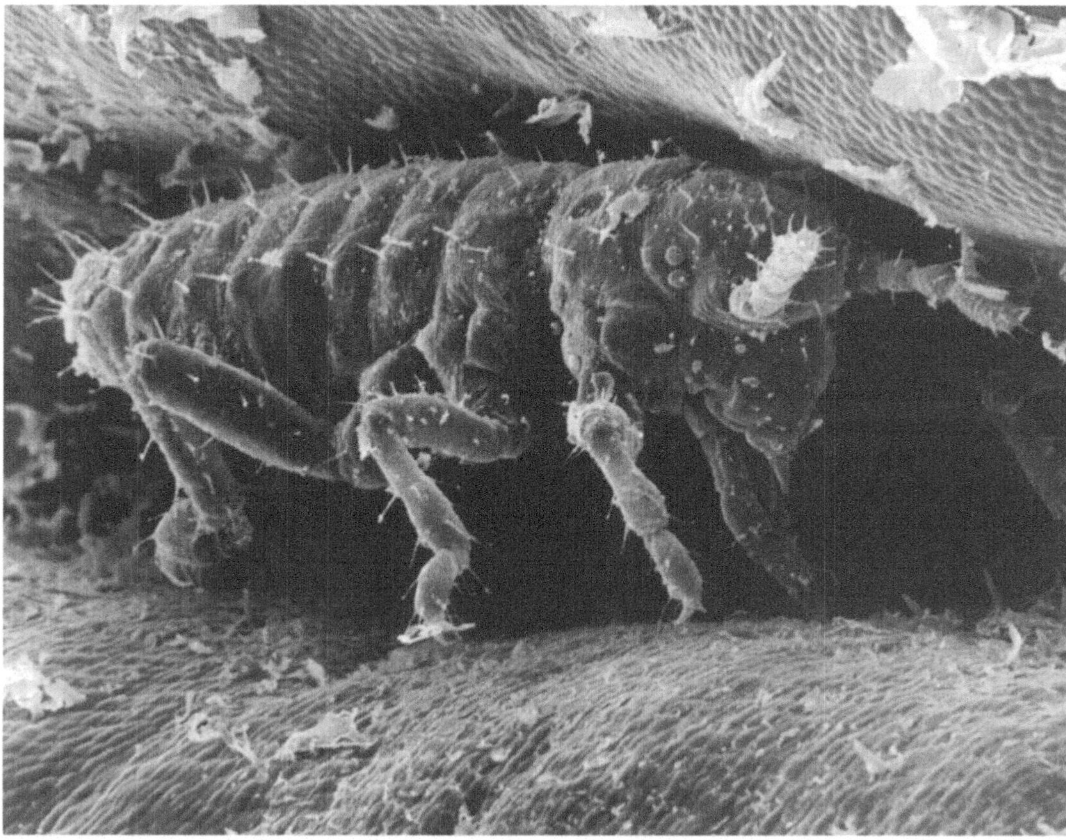

Figure 15-1. Scanning electron microscope photograph of a first-instar stem mother of *Pemphigus betae* in the process of forming a rudimentary gall on an immature leaf of narrowleaf cottonwood, *Populus angustifolia.* The stylets are inserted into the midrib of the leaf blade and a depression measuring about 1 mm deep has been formed. Within another 1–2 days the leaf tissues would have further expanded and completely enclosed the stem mother within a hollow gall. (Photo by D. L. Wright, S. Pratt, and T. G. Whitham.)

into the midrib of the leaf blade. A slight depression is evident. The leaf tissues most intensively probed seem to sink below the level of the surrounding tissue to form a gall resembling a pocket or purse. After approximately 3–4 days of probing the stem mother is surrounded by modified leaf tissues.

If the stem mother dies after the slightest depression has been formed (within the first 12–24 hours of probing) the rudimentary gall will become even more noticeable as the leaf expands, but it will fail to close over or to approach the size of a gall containing a living stem mother. Thus the continued presence of the stem mother and stimulation of leaf tissues are required for successful gall formation. These observations suggested that by censusing the number of rudimentary galls or scars and the number of successful galls containing living aphids, an accurate census of the colonizing population could be obtained. To test this, a branch containing 135 stem mothers in which half had formed rudimentary galls and the other half had not yet produced any visable depression was isolated by placing tanglefoot at the base of the branch to prevent further recruitment or interbranch movement. Nine weeks later 99 percent were accounted for either by the scar they produced before dying (30 percent died) or by their successful gall (Whitham 1978). These and other observations demonstrated an important feature of this system—the ability to accurately census the colonizing population.

The probing behavior of the colonizing stem mother on the immature leaf determines the size and shape of the gall. Dunn (1960) has shown that the striking differences in gall morphologies of three *Pemphigus* species occupying the same host plant, *Populus nigra* L., are associated with different probing behaviors in which the patterns of stylet insertion were quite different. For example, the spiral probing pattern of *Pemphigus spirothecae* Pass. on the leaf petiole results in a spiral-shaped gall. It appears that the growth of cells associated with stylet insertion is arrested while the unaffected cells on the opposite side of the petiole continue normal extension thereby causing the petiole to bend toward the affected area and produce a spiral gall. This probing behavior results in the petiole bending on itself at the rate of 90° every 12 hours until three coils are formed.

Studies of *Pemphigus betea* on narrowleaf cottonwood show that the probing behavior also affects gall formation and deviations from the basic pattern negatively affect reproductive success (Whitham 1986). Rohfritsch (1982, 1988 and Chapter 7) demonstrated in the case of two adelgids how a specific probing and feeding pattern of the immature stem mother is responsible for specific gall structures. *Pemphigus betea* probes back and forth along the midrib of the leaf blade to produce a linear pocket-shaped gall. Figure 15-2 shows the normal probing pattern obtained by recording the position of a single stem mother on the leaf blade every 10–15 minutes over a 2-day period. In this example, and other replicates, the stem mothers were observed to be continuously feeding and/or probing the leaf tissues. Competitive interactions between colonizing aphids and territorial behavior (to be discussed later) interrupt this basic probing pattern resulting in reduced probing time, higher mortality, smaller galls, and fewer progeny.

Although Dunn's (1960) studies clearly demonstrate the importance of probing behavior in producing a specific morphology, the mechanisms of how aphids induce galls are poorly understood. Prat (1955) showed that plant resistance to gall formation increased with cell differentiation and age, with meristematic tissues being the most easily affected by gall-inducing organisms. *Pemphigus* gall aphids can form galls only on immature or actively growing tissues and fail to develop when transferred to mature leaves (T. G. Whitham, personal observation). However, the first nymphal stage of the adelgid stem mother is able to induce cell dedifferentiation and to feed on the cortical tissue of 1-year-old spruce stems (Rohfritsch 1988 and Chapter 7).

It has been commonly thought that along with saliva, aphids inject something into the leaf such as amino acids or plant growth substances (Schäller 1968; Miles 1968; Hori 1974; Chapter 10). Today it is assumed that all the active substances are elaborated by the plant itself when it is wounded or attacked by the gall inducer. The plant produces growth or secondary metabolites oriented against the inducers (Berryman 1988). The inducing molecules are thought to be released by the plant when its cell walls are attacked by fungi or insects (Darvill and Albersheim 1984; Broadway *et al.* 1986). Forrest and Dixon (1975) found that for the production of galls by the apple aphid, *Dysaphis devecta* (Walker) and *D. plantaginea* Passerini the aphids produced their normal leafroll galls (which are morphologically different) even when they were forced to feed on the stem. Thus for the production of galls it is not necessary for aphids to probe at the site of gall induction. This supports the idea that substances produced at the feeding site of the insect are translocated through the plant as has been demonstrated with proteinase inhibitor-inducing factor (PIIF) (Broadway *et al.* 1986). Ryan (1988) characterized the factor as cell wall-derived galacturonic acid oligomers.

THE HOST AS A LIMITING RESOURCE

In contrast to the view that gall-inducing insects manipulate their host plant for their sole benefit, Mani (1964) suggested that gall induction is a plant adaptation to isolate its parasites in time and space. Studies of *Pemphigus betae* gall aphids support the alternative hypothesis that galling may benefit the host plant. For the aphids the primary advantages of galling include protection from predators and a constant environment. The disadvantages include hazards of gall initiation [100 percent of the colonizing population can die trying to initiate a gall on a resistant host (Whitham 1983)], highly variable host leaf quality to which the repro-

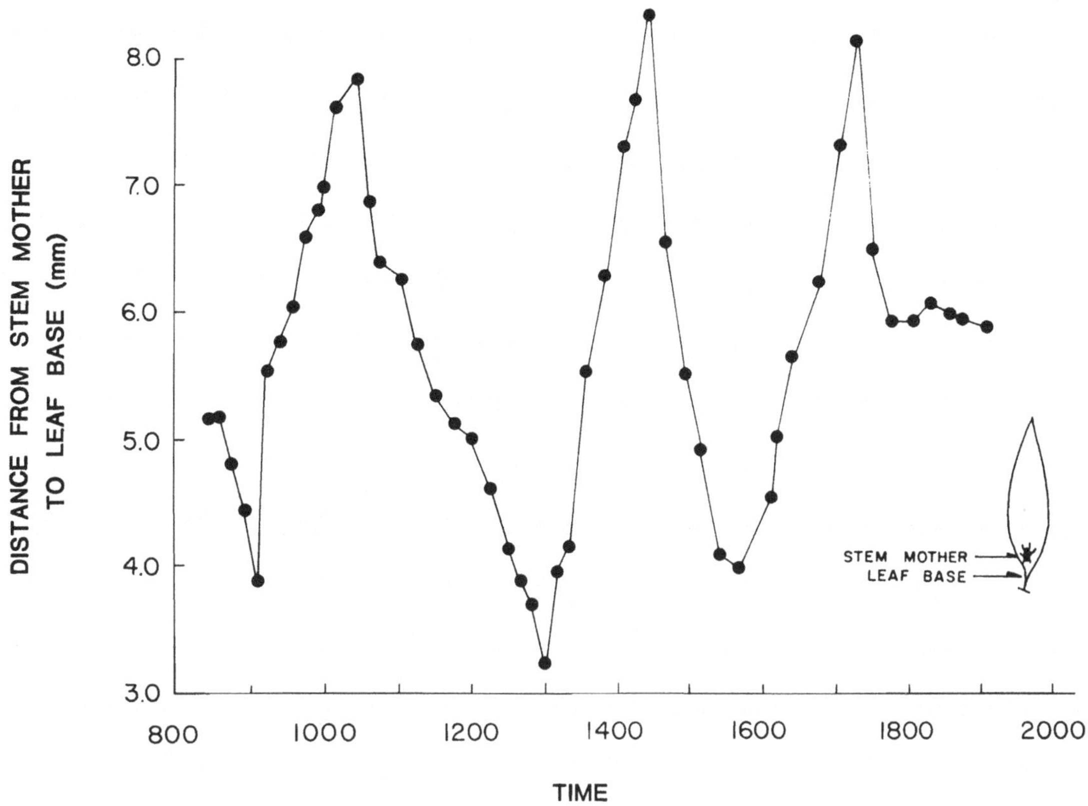

Figure 15-2. The back and forth probing behavior of a first-instar stem mother of *Pemphigus betae* induces the pocket-shaped gall typical for the species. Because a solitary stem mother continuously probes the leaf tissues with her stylets to produce a linear gall adjacent to the midrib of the leaf blade, their characteristic gall-inducing behavior can be quantified by measuring the position of the aphids on the leaf blade as a function of time (see text).

ductive fate of the colonizing aphid is tied (Whitham 1978), mortality resulting from severe competition for superior gall sites (Whitham 1978, 1986), isolation on a leaf that can be selectively abscised and dropped by the host plant before the stem mother's progeny mature (Williams and Whitham 1986), and obligatory migration to a secondary host with unknown but probably high mortality during migration. Although the advantages and disadvantages are not necessarily clear-cut, it would appear that there are definite trade-offs involved with galling.

Central to several of the points on the disadvantages of galling is the high level of variation in gall site quality. Only a relatively small portion of the host plant leaves are suitable or susceptible to galling. Figure 15-3 (solid line) shows, in terms of reproductive success, the importance of choosing the right leaf to gall. On the smallest leaves ($\leq$ 5.0 cm) the average stem mother produced only 32 progeny while on the largest leaves ($>$ 15 cm) the average stem produced 189 progeny. In part, the failure of stem mothers to thrive on small leaves is due to their high mortality during the gall-inducing process. On the smallest leaves as many as 80 percent of the colonizers died trying to initiate a gall, whereas on the largest leaves 0 percent died. Although there is a clear advantage to galling leaves that mature to large size, large leaves are relatively rare. Figure 15-3 (dashed line) shows the distribution of leaf sizes available for colonization. The smallest leaves that result in high aphid mortality and low fecundity are the most common leaves produced by the plant. The largest leaves which result in 0 percent mortality and the highest fecundity are relatively rare,

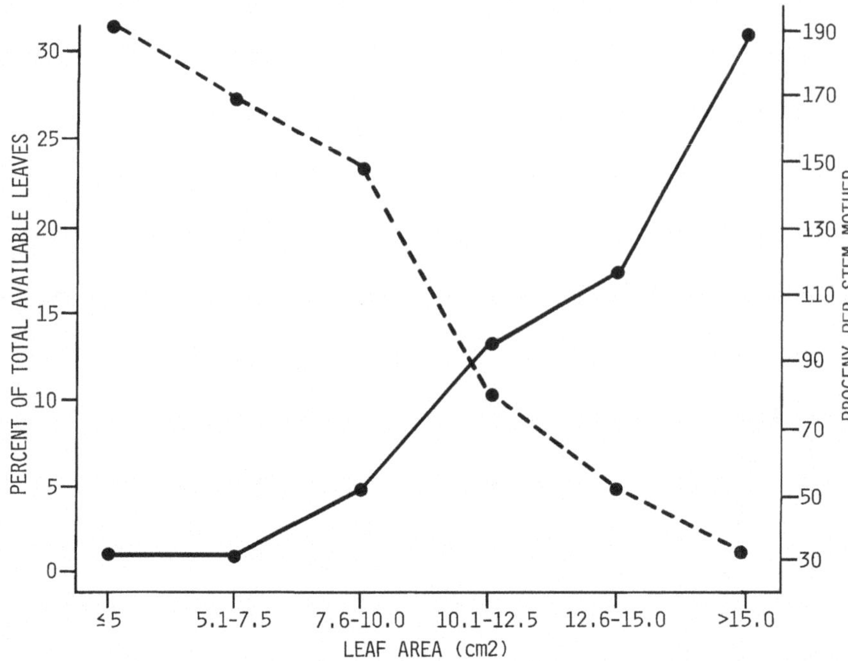

Figure 15-3. Solid line shows the number of progeny that stem mothers of *Pemphigus betae* achieved on leaves of different size. Dashed line shows the distribution of leaf sizes available for colonization. The largest leaves, which result in the highest reproductive success, are relatively rare. (Adapted from Whitham 1978.)

representing only 1.6 percent of the total leaves available. Thus the entire tree is not open to colonization; only a small portion is suitable to these galling aphids.

Several factors correlate well with the observation that large leaves are superior to small ones for *Pemphigus* aphids. Since productivity is related to photosynthetic surface area, the size of the leaf should be correlated with the quantity of resources translocated into and/or out of the leaf. The largest leaves are also positioned on terminal shoots, which are the sites of most active growth and represent natural sinks for the plant's resources. For example, Kozlowski and Ward (1961) observed that the terminal leaders of *Pinus resinosa* Ait. exhibited as much as 12 times more growth than lateral buds and were more likely to continue producing leaves after the initial flush or spring growth. In addition to colonizing natural sinks, gall aphids may be able to increase the sink capacity of their host leaves. Using ^{14}C-labeled sucrose, Way and Cammel (1970) showed that leaves with aphids received 70 to 80 times more sucrose than the adjacent uninfested leaves.

In addition to large leaves representing a superior resource for galling aphids, they may also be less defended by the host plant. The preference of *Pemphigus betae* for large leaves of *Populus angustifolia* is negatively correlated with the concentration of phenolics (Zucker 1982), which have been strongly implicated in the defensive chemistry of many plants (Levin 1971). The small leaves that colonizing stem mothers avoid have a higher concentration of phenolics than the preferred large leaves. In paired comparisons of leaves within individual shoots the smallest leaves averaged 1.7 times more total phenols per milligram of dry leaf tissue than the largest leaves (Zucker 1982).

Perhaps even more important than the quality of the resource and the concentration of plant defenses, large leaves may be less susceptible to wounding and blockage of the phloem from which most aphids feed. Milburn (1972) presented evidence suggesting plants have a protective sealing mechanism in which the sieve tubes are induced to rapidly close down once a threshold level of sap loss has been reached. Evert *et al.* (1968) found that phloem from which the aphid *Longistigma caryae* Harris had recently withdrawn its stylets was blocked by tylosoids and massive amounts of callose. The deposition of callose extended not only to the penetrated sieve elements but to the adjacent elements as well. Furthermore, the smallest sieve elements were most severely affected, indicating that large sieve elements are less likely to produce the wounding response. As noted by Dixon and Logan (1973) the wounding response of plants to aphids may account for the observation that large free-living sycamore aphids, *Drepanosiphum platanoides* Schrk., selectively feed on large leaves where large veins are most abundant. For galling aphids such as *P. betae* in which the fate of the individual aphid is tied to a single leaf, it is crucial for the vascular system to remain fully functional.

Recent evidence (Williams and Whitham 1986) demonstrates the presence of a gall induced defensive response in which galled leaves are selectively abscised or dropped from the host tree. Once the leaf falls to the ground it quickly withers, killing the trapped aphids. Galled leaves typically start showing signs of chlorosis within 1 week of gall initiation at bud break and begin falling from the tree 2 weeks later, well before the aphids mature in early July. By examining the frequency of galled and gall-free leaves on the ground and on the tree it was possible to calculate the probability of premature leaf fall. It was found that galled leaves had a much higher probability of premature leaf fall than gall-free leaves. Leaves without galls had a 1.8 percent chance of falling whereas galled leaves had a 5 percent chance of being abscised. It would appear that the host plant can discriminate between galled and gall-free leaves and can shed its parasites.

In agreement with the wounding hypothesis that vascular blockage is most likely to be induced on small leaves with smaller sieve tubes, galled small leaves are far more likely to become chlorotic than the large leaves that gall aphids prefer. It was also found that as the number of galls per leaf increased the probability of premature leaf fall also rose. For example, with one gall per leaf the probability of premature leaf fall was 10 percent; however, as gall densities increased to four or more galls per leaf the probability of leaf fall increased to 54 percent. Again, this is consistent with the hypothesis that at higher gall densities the level of sap loss exceeds the threshold at which a wounding response is triggered. Using the above example taken from a single tree, the observed level of leaf fall resulted in 25 percent of all galls being dropped from the tree before the aphids matured. A 25 percent reduction in parasite numbers on a mature tree can translate into a loss of aproximately 3 million aphids. Considering that aphids are capable of acting as resource sinks by attracting nutrients from adjacent leaves (Miles 1968; Way 1968; Way and Cammel 1970; Hori 1974), the loss of galled leaves relatively early in the season may be an effective plant mechanism of "cutting its losses" to minimize the potential impact of its parasites.

In an effort to circumvent such plant defenses, and/or a deteriorating gall environment caused by invading predators, Setzer (1980) found that immatures of *Pemphigus populitransversus* Riley may leave the gall and crawl to another gall. Using electrophoretic techniques to measure variation, the average proportion of heterotypic aphids at different localities ranged from 0.027 to 0.250. *Pemphigus populicaulis* Fitch was also found to move but at much lower levels. Although there are definite advantages to moving when the gall environment deteriorates, Williams and Whitham (1986) found that in comparison with free-living aphids, immatures of the galling aphid *Pemphigus betae* were not physiologically adapted to living outside the gall and suffered high mortality when exposed for short periods of time.

From the above discussion it would appear that the host plant and its parasites are engaged in a struggle in which the aphids can successfully circumvent the plant's defenses only on a small fraction of the total available leaves. The effectiveness of the plant's defenses by blocking the vascular system and, on some plants, by dropping the galled leaf in mid-season may have, through evolution, forced these aphids into complex life cycles involving costly migration to a secondary host. Examination of the life cycles of galling and free-living aphids (Williams and Whitham 1986) shows that galling aphids are 3.5 times more likely to have a complex life cycle involving a secondary host than are free-living aphids. Thus, rather than the aphids manipulating the host plant solely for their benefit, it would appear that galling has associated costs for the insect that may ultimately benefit the host plant.

TERRITORIAL BEHAVIOR

Because parthenogenetic reproduction and high population growth rates are not generally associated with competition for resources in short supply, it is perhaps surprising to find that aphids, the epitome of an r-selected species, can be highly territorial. Territorial behavior has probably evolved in these gall-inducing insects due to the scarcity of superior gall sites.

Once the density of colonizing *Pemphigus betae* stem mothers exceeds the availability of superior gall sites on the narrowleaf cottonwood, three options, each of which has fitness trade-offs, are available to the residents and remaining colonizers. First, once a superior gall site is occupied, one or more additional stem mothers can pile up on the same large leaf. However, a single leaf is not a uniform substrate and there is a position effect. For example, stem mothers of *P. betae* nearly always prefer to colonize the base of the leaf blade; this preference is strongly correlated with reproductive success and the lack of plant defenses. When three stem mothers colonized the same leaf it was found that the basal stem mother produced an average of 138 progeny; the intermediate stem mother, 75 progeny; and the distal one only 29 progeny. Thus, due to the position effect, stem mothers that choose to double-up on a leaf with another colonizer in the basal position are at a reproductive disadvantage (Whitham 1978, 1980). In agreement with these findings, Zucker (1982) found that the concentration of phenolics at the preferred base of the leaf was nearly half the level found more distally on the leaf blade. A second option for colonizing aphids is that once the superior leaves are occupied, successively smaller leaves can be colonized until all stem mothers have settled. However, this also results in those settling last or choosing to occupy smaller leaves suffering a reduction in fitness. A third option is for the largest stem mothers to aggressively displace and/or defend superior gall sites to effectively become "chief of the leaf."

Whitham (1979, 1986) showed that colonizing stem mothers were territorial and that they aggressively defended gall sites. The outcome of a territorial conflict is determined by kicking-shoving contests in which the largest stem mother usually wins. When competitor densities are very high, territorial behavior can result in a floater population of colonizers unable to find suitable gall sites. When this occurred, the mortality of the floater population was found to be 72 percent, as compared to only 24 percent of the resident population. With such differential mortality it is not surprising that contests can last 2 days and result in the death of the inferior competitor. It was also observed that a resident with a newly formed but still open gall could be displaced by an intruder after an intensive conflict. This observation is similar to those made by Akimoto (1981) with the gall aphid *Eriosoma yangi*.

Although territorial behavior permits large, dominant stem mothers of *P. betae* to sequester superior resources, even for the winner there are costs associated with territorial behavior. Territorial contests occur primarily during the crucial and brief gall-inducing period, which lasts about 3 days when the leaves are immature. Territorial conflicts are usually conducted at the expense of the gall-inducing behavior discussed above and in Figure 15-2. While engaged in kicking and shoving contests, the stem mothers were not observed probing the leaf tissues with their stylets, as both competitors spent most of their time at a common boundary (Whitham 1986). Due to the trade-off between territorial and gall-inducing behavior, the resulting galls of the "winners" were smaller and contained fewer progeny. The fitness of the losers, however, was substantially lower, making the aggressive behavior adaptive.

Such aggressive behavior is not unique to *Pemphigus betae*. Aoki (1978) described first instars of *Pemphigus dorocola* Matsumura attacking lepidopterous larvae by stinging them with their stylets and tearing the skin with their large hind legs and tarsal claws. Such behavior towards potential herbivores could be adaptive because some lepidopterous larvae preferentially feed on leaves galled by *P. betae* rather than gall-free leaves (A. G. Williams and T. G. Whitham, unpublished data).

Aoki et al. (1977) and Aoki (1979) describe even more elaborate behavior for the gall aphid, *Astegopteryx styracicola* Karsch., which in Taiwan forms large galls on cork leaf snowbell *Styrax suberifolia* Hook. These very small aphids form galls measuring 9 cm in diameter and contain as many as 100,000 aphids. Aoki (1979) observed what is believed to be a sterile caste of soldiers defending the gall. In this species, dimorphism occurs in the second instar in which the aggressive morph is hairy and does not continue further development. Soldiers were observed patrolling the surface of the gall and adjacent twigs and quickly attacked syrphid predators, which were soon covered and killed, or fell from the gall with soldiers attached. Aoki (1979) also observed that when the gall was disturbed from below, soldiers would drop from the gall and bite the intruder, including man. For example, when ten galls were disturbed by shaking the twigs, nearly 4000 soldiers dropped. If they landed on the skin they would readily bite, using their stylets and causing intensive irritation of the skin. Their biting has a similar effect on mice and it is proposed that such behavior is effective against vertebrate predators. That vertebrates can be effective aphid predators is supported by observations that chickadees and chipmunks will readily slice open galls of *P. betae* and eat the aphids (Whitham 1987). Furthermore, in comparison with insect predators, these vertebrate predators were equally effective in reducing the aphid population.

HOST SELECTION BEHAVIOR

Organisms that face variable environments should become evolutionarily adapted to discriminate between the mosaic patterns of variability and selectively settle where their reproductive success or fitness is highest. The decision of whether to settle should be based on the fitness to be expected in a given habitat. Fretwell and Lucas (1970) and Fretwell (1972) formalized the logic of habitat selection behavior and although originally based on bird populations, the basic ideas apply equally well to gall insects. The settling decision depends on the interaction of several factors to produce a given fitness. If habitats vary in quality, obviously the habitat producing the highest fitness should be preserved and selectively colonized. If there are negative effects of increasing density (e.g., several stem mothers on the same leaf), as colonizers continue to settle in the best habitat, a density should eventually be reached at which expected fitness in the so-called "poorer habitat" (e.g., a small leaf) with no competitors would be as high as in the best habitat. At this point colonization of the poorer habitat should begin and as population densities increase even further, successively poorer habitats should be colonized. Settling decisions should be further affected by the absence or presence of territorial behavior in which individuals through their aggressive interactions can prevent a habitat from being colonized by other competitors. Whether territorial behavior is exhibited should depend on the payoff or profitability in terms of increased fitness. Thus for colonizing gall aphids to behave in this evolutionarily appropriate and "precise" fashion requires an ability to integrate several variables that include habitat quality, competitor dignity, and one's own competitive ability in territorial interactions.

Whitham (1980) examined the ability of *Pemphigus* gall aphids to integrate these variables and make appropriate settling decisions. It was found that within a single large branch of a tree they did so with considerable precision. Colonizing stem mothers adjusted their densities on leaves of different quality such that the average fitness achieved on a small leaf with only one gall was identical to the average number of progeny produced on a large leaf with three galls. To examine the precision of their settling behavior, a computer simulation was run to determine what percentage of their potential reproductive success was being achieved. Although the simulation was unrealistically generous in allowing colonizers unlimited search time and perfect knowledge of both the distribution of leaves available for colonization and competitor densities on all leaves, it was found that the aphids in the field were achieving at least 85 percent of their potential fitness. Although such discriminatory ability may be unexpected in small insects that lack the mobility of "intelligence" of higher organisms, the reproductive advantage gained by those that make appropriate settling decisions probably accounts for the evolution of their selective settling behavior.

The ability to settle appropriately within a single branch extends to the entire tree and different trees in the host population. Figure 15-4 shows how an estimated 53,000 stem mothers of *Pemphigus betae* settled on 20 branches of a 20-m narrowleaf cottonwood tree (Whitham 1983). The galls produced by these stem mothers when mature contained approximately 2,100,000 aphids or about 1.5 aphids per cm^2 of leaf area. The pattern of settling shows that adjacent branches as well as branches at various heights can have very different parasite loads. Gall densities per branch ranged from 19 to 794 galls per 1000 leaves, a 42-fold difference in aphid preference for one branch over another.

Preference for individual branches reflects an underlying pattern of host resistance in which survival can vary greatly from one branch to another. Within a single tree survival rates on nearby branches ranged from 0 to 72 percent. The difference in survival rates between branches was strongly correlated with leaf size, which is consistent with previous data (see earlier sections of this chapter). Colonizing stem mothers responded to the high levels of variation by avoiding those branches where mortality was highest. Although their settling behavior was appropriate, superior gall sites were so limiting that the negative effects of density and territoriality on the aphid population were sufficient to suggest that within-plant variation may serve as a significant host defense against parasites (Whitham 1983).

The differences between trees growing near one another in the same grove are similar to the variation observed within a single tree. Consequently, the selection of an individual tree by the fall migrants is an important phase in the life cycle that affects the fitness of the stem mothers the following spring. In an examination of 33 mature trees, aphid preference for individual trees measured as gall densities ranged from 5 to 729 galls per 1000 leaves (Whitham 1983). Similarly, the avoidance of specific trees was correlated with survival rates, which ranged from 76 percent on large-leaved trees that were

Figure 15-4. This figure depicts how an estimated 53,000 galls of *Pemphigus betae* containing an estimated 2,100,000 aphids were distributed over 20 branches of a single 20 m narrowleaf cottonwood, *Populus angustifolia.* The drawing is to scale and reflects the leaf area of each branch. (Adapted from Whitham 1983.) Colonizing stem mothers do not settle randomly on their host trees but clump on specific branches where their survival and reproduction are highest (see text).

preferred, to as low as 0 percent on small-leaved trees that were avoided. The ability of colonizing aphids to discriminate between inferior and superior hosts at all levels of habitat selection (within a leaf, and between leaves, shoots, branches, and trees) is clearly adaptive.

Censuses of the aphid population on individual branches of the same tree and between trees in the host population show that colonizing aphids prefer to colonize the same branches and trees year after year. This is particularly interesting considering that the complex life cycle of these aphids involving migration between a primary and secondary host requires the subsequent recolonization of preferred branches and trees each year. Furthermore, such specificity continues even when the aphid population fluctuates widely from one year to the next. Whether these aphids have become genetically adapted to specific host plants or are simply responding to host trees or plant parts in various states of continual stress is presently unknown but of great interest (see Whitham and Slobodchikoff 1981; Whitham 1981, 1983; Whitham *et al.* 1984 for further discussions on the implications of within-plant variation of parasite populations).

CONCLUDING REMARKS

Because galling insects and their host plants are so closely tied together in coevolutionary interactions and due to the numerous advantages of working with galling insects (i.e., censusing, fitness estimates, and sample size), they may well represent model systems for the study of many ecological principles. In this chapter I have tried to bring together information on gall formation, plant resistance, defense characteristics, and heterogeneity, which in part explain why gall aphids have evolved territorial behavior, complex life cycles, and a high degree of discriminatory behavior at the various levels of the host selection process. It should be the goal of future research to more thoroughly integrate plant and insect behavior, biochemistry, population genetics, and other disciplines to ellucidate both the evolutionary selection pressures and the underlying mechanisms.

ACKNOWLEDGMENTS

I gratefully acknowledge the support of NSF Grants DEB7905848, DEB8005602 and BSR8303051. The use of field facilities provided by T. K. Swan, J. Begay, R. White, E. Bozniak, and D. M. Gatherum was greatly appreciated.

LITERATURE CITED

Akimoto, S. 1981. Gall formation by *Eriosoma yangi* (Homoptera, Pemphigidae). Kontyû 49:426–436.

Aoki, S. 1978. Two pemphigids with first instar larvae attacking predatory intruders (Homoptera, Aphidoidea). New Entomol. 27:7–12.

Aoki, S. 1979. Further observations on *Astegopteryx styracicola* (Homoptera: Pemphigidae), an aphid species with soldiers biting man. Kontyû 47:99–104.

Aoki, S., S. Yamane, and M. Kiuchi. 1977. On the biters of *Astegopteryx styracicola* (Homoptera, Aphidoidea). Kontyû 45:563–570.

Auclair, J. L. 1963. Aphid feeding and nutrition. Annu. Rev. Entomol. 8:439–490.

Berryman, A. A. 1988. Towards a unified theory of plant defenses. In W. J. Mattson, J. Levieux, and C. Bernard-Dagan (eds.), Mechanisms of Woody Plant Defenses Against Insects, pp. 39–53. Springer-Verlag, New York.

Broadway, R. M., S. S. Duffey, G. Pearce, and C. A. Ryan. 1986. Plant proteinase inhibitors: a defense against herbivorous insects? Entomol. Exp. Appl. 41:33–38.

Darvill, A. G., and P. Albersheim. 1984. Phytoalexins and their elicitors. A defense against microbial infection in plants. Annu. Rev. Plant Physiol. 35:243–275.

Dixon, A. F. G. 1971. The life-cycle and host preference of the bird cherry-oat aphid, *Rhopalosiphum padi* L., and their bearing on the theories of host alternation in aphids. Annu. Appl. Biol. 68:135–147.

Dixon, A. F. G. 1975. Aphids and translocation. In M. H. Zimmermann and J. A. Milburn (eds.), Transport in Plants, Vol. I, pp. 154–170. Springer-Verlag, New York.

Dixon, A. F. G., and M. Logan. 1973. Leaf size and availability of space to the sycamore aphid *Drepanosiphum platanoides*, Oikos 24:58–63.

Dunn, J. A. 1960. The formation of galls by some species of *Pemphigus* (Homoptera: Aphididae). Marcellia 30:155–167.

Evert, R. F., W. Eschrich, J. T. Medler, and F. J. Alfieri. 1968. Observations on penetration of linden branches by stylets of the aphid *Longistigma caryae.* Am. J. Bot. 55:860–874.

Forrest, J. M. S. and A. F. G. Dixon. 1975. The induction of leaf-roll galls by the apple aphid *Dysaphis devecta* and *D. plantaginea.* Ann. Appl. Biol. 81:281–288.

Foster, W. A. 1975. The life history and population biology of an intertidal aphid, *Pemphigus trehernei* Foster. Trans. R. Entomol. Soc. London 127:193–207.

Foster, W. A. 1978. Dispersal behavior of an intertidal aphid. J. Anim. Ecol. 47:653–659.

Fretwell, S. D. 1972. Populations in a Seasonal Environment. Princeton University Press, Princeton, NJ.

Fretwell, S. D., and H. L. Lucas, Jr. 1970. On territorial behavior and other factors influencing habitat distribution in birds. I: Theoretical development. Acta Biotheoret. 19:16–36.

Harper, A. M. 1958. Notes on behavior of *Pemphigus betae* Doane (Homoptera: Aphididae) infected with *Entomophthora aphidis* Hoffm. Can. Entomol. 90:439–440.

Harper, A. M. 1959. Gall aphids on poplar in Alberta. I: Descriptions of galls and distributions of aphids. Can. Entomol. 91:489–496.

Harper, A. M. 1963. Sugar-beet root aphid, *Pemphigus betae* Doane (Homoptera: Aphididae) in Southern Alberta. Can. Entomol. 95:863–873.

Hori, K. 1974. Plant growth-promoting factor in salivary glands of the bug, *Lygus disponsi.* J. Insect Physiol. 20:1623–1627.

Kozlowski, T. T., and R. C. Ward. 1961. Shoot elongation characteristics of forest trees. Forest Sci. 7:357–368.

Levin, D. A. 1971. Plant phenolics: An ecological perspective. Am. Nat. 105:157–181.

Mani, M. S. 1964. Ecology of Plant Galls. W. Junk, The Hague.

Milburn, J. A. 1972. Phloem transport in *Ricinus.* Pest. Sci. 3:653–665.

Miles, P. W. 1968. Studies on the salivary physiology of plant bugs: Experimental inductions of galls. J. Insect Physiol. 14:97–106.

Mitchell, R. 1983. Effects of host-plant variability on the fitness of sedentary herbivorous insects. In R. F. Denno and M. S. McClure (eds.), Variable Plants and Herbivores in Natural and Managed Systems, pp. 343–370. Academic Press, New York.

Palmer, M. A. 1952. Aphids of the Rocky Mountain region. Thomas Say Foundation, Volume 5.

Prat, H. 1955. Relations existant entre les notions de gradient chimique et de rayon d'activité cécidogénétique. Marcellia 30:51–53.

Rohfritsch, O. 1982. Behavior and biology of two aphid species related to gall morphogenesis. In Proceedings of the 5th Int. Symposium on Insect-Plant Relationships, Wageningen, pp. 431–432. Pudoc, Wageningen.

Rohfristch, O. 1988. A resistance response of *Piceae excelsa* to the aphid *Adelges abietis* (Homoptera, Aphidoidea). In W. J. Mattson, J. Levieux, and C. Bernard-Dagan (eds.), Mechanisms of Woody Plant Defenses Against Insects, pp. 253–266. Springer-Verlag, New York.

Ryan, C. A. 1988. Response mechanisms of plant tissues to herbivore feeding. In Proceedings of XVIII Int. Congress of Entomology, Vancouver, B. C., Abstract p. 165.

Schäller, G. 1968. Biochémische Analyse des Aphid enspeichels und seine Bedeutung für die Gallenbildung. Zool. Jahrb. Physiol. 74:54–87.

Setzer, R. W. 1980. Intergall migration in the aphid genus *Pemphigus.* Ann. Entomol. Soc. Am. 73:327–331.

Stearns, S. C. 1976. Life history tactics: A review of the ideas. Q. Rev. Biol. 51:3–47.

van Emden, H. F. 1972. Aphids as phytochemists. In J. B. Harborne (ed.), Phytochemical Ecology, pp. 25–43. Academic Press, London.

Way, M. J. 1968. Intra-specific mechanisms with special references to aphid populations. In T. R. E. Southwood (ed.), Insect Abundance, pp. 18–36. Symp. R. Entomol. Soc. London, No. 4. Blackwell, Oxford, U.K.

Way, M. J. and M. E. Cammell. 1970. Aggregation behavior in relation to food utilization by aphids. In A. Watson (ed.), Animal Populations in Relation to Their Food Resources, Symp. Brit. Ecol. Soc. No. 10. Oxford: Blackwell Scientific.

Whitham, T. G. 1978. Habitat selection by *Pemphigus* aphids in response to resource limitation and competition. Ecology 59:1164–1176.

Whitham, T. G. 1979. Territorial behavior of *Pemphigus* gall aphids. Nature (London) 279:324–325.

Whitham, T. G. 1980. The theory of habitat selection: Examined and extended using *Pemphigus* aphids. Am. Nat. 115:449–466.

Whitham, T. G. 1981. Individual trees as heterogeneous environments: Adaptation to herbivory or epigenetic noise? In R. F. Denno and H. Dingle (eds.), Insect Life History Patterns: Habitat and Geographic Variation, pp. 9–27. Springer-Verlag, New York.

Whitham, T. G. 1983. Host manipulation of parasites: Within-plant variation as a defense against rapidly evolving pests. In R. F. Denno and S. McClure (eds.), Variable Plants and Herbivores in Natural and Managed Systems, pp. 15–41. Academic Press, New York.

Whitham, T. G. 1986. Costs and benefits of territoriality: Behavioral and reproductive release by competing aphids. Ecology 67:139–147.

Whitham, T. G. 1987. Evolution of territoriality by herbivores in response to host plant defenses. Am. Zool. 27:359–369.

Whitham, T. G., and C. N. Slobodchikoff. 1981. Evolution by individuals, plant-herbivore interactions, and mosaics of genetic variability: the adaptive significance of somatic mutations in plants. Oecologia 49:287–292.

Whitham, T. G., A. G. Williams, and A. M. Robinson. 1984. The variation principle: Individual plants as temporal and spatial mosaics of resistance to rapidly evolving pests. In P. W. Price, C. N. Slobodchikoff, and W. S. Gaud (eds.), A New Ecology: Novel Approaches to Interactive Systems, pp. 15–51. Wiley, New York.

Williams, A. G., and T. G. Whitham. 1986. Premature leaf abscission: An induced plant defense against gall aphids. Ecology 67:1619–1627.

Wool, D. 1984. Gall-inducing aphids. In T. N. Ananthakrishnan (ed.), Biology of Gall Insects, pp. 11–58. Oxford and IBH, New Delhi, India.

Zucker, W. V. 1982. How aphids choose leaves: The roles of phenolics in host selection by a galling aphid. Ecology 63:972–981.

16

Ecological Relationships of Insects Inhabiting Cynipid Galls

A. A. WIEBES-RIJKS
and J. D. SHORTHOUSE

All students of insect galls undoubtedly have been struck by the diversity of insects that emerge from mature galls, especially those induced by cynipid wasps. In fact, some galls issue so many species of inhabitants that it is often difficult to identify the gall inducer. The cynipid gall of *Diastrophus kincaidii* Gillette on thimbleberry, for example, supports 14 species (Wangberg, 1976), whereas Askew (1961) found some oak galls supporting as many as 22 species. The oak gall induced by the cynipid *Biorhiza pallida* Ol. has over 75 species associated with it (Mani 1964). The ease by which large numbers of cynipid galls can be sampled and the diverse assemblage of inhabitants they contain make them attractive subjects for studies by ecologists and gall workers alike.

The purpose of this chapter is to review various aspects of the community ecology of cynipid galls. Here we explain the basic community organization in cynipid galls and outline the general feeding relationships of the most common types of inhabitants. We do this by examining the communities associated with two galls induced by cynipids of the genus *Diplolepis* found on wild rose in North America and a European gall (known as the oak apple) induced by *Cynips quercusfolii* L., which exhibits alternation of generations. In doing so we hope to draw attention to the research potential of this fascinating aspect of gall biology.

ATTRIBUTES OF GALL COMMUNITIES

Ecologists have long held a fascination with the attributes of plant and animal communities. Communities can be broadly defined as assemblages of populations living in a prescribed area that function as a unit through coupled metabolic transformations. Communities not only have a definite functional unity with characteristic trophic structures and patterns of energy flow, they also have compositional unity in that certain species will occur together from one year to the next (Anderson and Kikkawa 1986). The intensity of the interactions and the degree to which they maintain the composition of the community represent a major aspect of community organization. Areas of current study in community ecology include the predictability of natural communities, the importance of competition between component species in determining structure, and the coevolution between community members (Kikkawa and Anderson 1986).

There has developed a rich literature in the area of insect communities on plants (see Strong *et al.* 1984 for references). Plants serve as hosts to numerous small but readily identifiable communities associated with organs such as roots, stems, or flowerheads. Such assemblages of insects in particular parts of plants have been referred to as "miniature ecosystems" by Zwölfer (1979). They have measurable energy input, a fairly diverse infrastructure, distinct food chains with several trophic levels, and show numerous interactions among their insect components. Other workers have preferred to call the assemblage of species associated with such microenvironments "component communities" (Root 1973).

Cynipid galls can be considered as atypical outgrowths of plant organs (Rohfritsch and Shorthouse 1982) and the assemblage of insects associated with them as forming small and relatively simple, but distinct component or microcommunities. We define gall communities as the assemblage of insect populations interacting within galls initiated by the same species of gall inducer, in the same location and during the same season. Gall communities are closed communities where the potential number of interactions is limited and bounded in space and time.

As with all phytophagous communities, the host plant is central to all gall communities. Besides serving as the energy source for all gall inhabitants, the host plant is the site where the adults seek shelter during inclement weather, search for mates, undergo courtship, select oviposition sites, and then overwinter in the prepupal stage. The plant is the producer (or converter) that forms the various organs and directs a stream of assimilates to the gall over a lengthy period. The gall inducers and other phytophagous inhabitants constitute the primary consumers, which in turn are exploited by various groups of secondary consumers such as ectoparasites and endoparasites. Several galls are inhabited by fungi and fungivorous and detritivorous species.

Insect galls have many attributes that make them ideal for studies in the area of community ecology. In contrast to most insects that can be located in either the larval or adult stage for a short period of the season, gall insects, via their galls, are easily located and sampled over much of the year. This also means that distribution patterns for gall-inducing species can be established without ever observing adults in the field. Communities associated with some galls in northern regions, that remain attached to the host plant, are easily collected in mid-winter when the galls are contrasted against the snow. The adult stage, which takes place outside the galls, lasts for only a few weeks. Adults are easily emerged from mature galls.

The larvae of gall inhabitants are easily identified such that fluctuations of populations of all species can be followed throughout a season by periodically dissecting large random samples. Thus gall communities provide an opportunity to study how a single host species can support large numbers of entomophagous species and how several often closely related species are able to coexist within the same community. Blair (1944) in his study of parasitoids associated with galls induced by *Diplolepis rosae* (L.) pioneered research into the complex and subtle trophic relationships existing in gall communities. Askew (1961) followed Blair's approach in examining the large but seemingly structured parasitoid communities associated with a range of cynipid oak galls.

Many gall communities are remarkably consistent across the range of the gall-inducing species. For example, Schröder (1967) found that galls of *Diplolepis rosae* contained the same species of inhabitants across Europe. However, Askew and Shaw (1987) warn that some communities cannot be sharply defined because many of the entomophagous species regularly or occasionally enter the complexes of more than one species of gall.

The larvae of most species of cynipids are attacked by entomophagous species referred to as parasitoids and phytophagous species referred to as inquilines. The larvae of inquiline cynipids feed on gall tissues and may or may not directly harm the gall inducer. Some species of parasitoids are always more numerous than others and are referred to as dominant or core species whereas others, referred to as incidental or satellite species, are found so sporadically or in such low numbers that they are ignored in community studies. Parasitoids and inquilines contribute substantially to the mortality of most cynipid populations.

ROLES OF INSECTS INHABITING CYNIPID GALLS

Most of the many insects associated with cynipid galls are of the Order Hymenoptera; however, a few spiders, beetles, and moths are known to use galls as shelters. Most of the first workers to study gall communities concentrated on identifying the species of adult wasps reared from galls. This sort of information was catalogued by Fulmek (1968) for European galls and some information is in Felt (1965) for North American galls. More recent attention has shifted from merely collecting galls and taking stock of the emerging adults to more detailed studies of the ecological relationships between the participants. A few examples are given in Table 16-1. A classic study is that by Varley (1947) in which he examined the inhabitants of galls induced by the tephritid *Urophora jaceana* (Hering) and established the factors responsible for fluctuations in the numbers of each species.

The assemblage of inhabitants associated with galls of a particular species of cynipid can be determined by collecting large samples of mature galls and emerging the inhabitants in the laboratory. In northern regions this can be done only after the galls have been subjected to cold temperatures to break diapause of the inhabitants. The sequence by which the inhabitants emerge under laboratory conditions provides an approximation of the periods during the season when the habitants initiate galls or search out and oviposit in immature galls. One also should collect some maturing galls in late summer to check for fall emergents. Some species of *Eurytoma* and *Torymus* emerge from maturing galls in the same season the galls are initiated and search for other galls in which to oviposit and overwinter. However, to obtain an accurate picture of the trophic relationships of gall inhabitants one needs to collect samples throughout the season and return them to the laboratory for dissection. Most inhabitants leave characteristic signs within gall chambers facilitating their identification. Rearing mature larvae that have completed feeding, in gelatin capsules (Shorthouse 1972), is another way of associating larvae with adults.

To illustrate the ease by which the larvae of mature cynipid galls can be distinguished, we first have chosen the gall induced by *Diplolepis nebulosa* Basset (Fig. 16-1). This gall is common in Western Canada on the abaxial surface of leaves of *Rosa woodsii* Lindl. growing on dry exposed prairie (Shorthouse 1975). One usually can find all of the larvae shown in Figure 16-1 by randomly collecting 100 mature galls. Some members of gall communities begin attacking the larvae of the gall inducers within days of gall initiation and in fact there is a sequence of attacking species throughout the summer until the tissues of the gall become too brittle for the ovipositors to penetrate. An example of a typical food web, based on another species of cynipid is taken from Askew (1961) and shown in Figure 16-2.

Table 16-1. Studies on the Relationships between the Inhabitants of Galls, Listed According to the Year of Publication

Group	Gall inducer	Host	Reference
Tephritidae	*Urophora jaceana* (Hering)	*Centaurea*	Varley (1947)
Cynipidae	Various	*Quercus*	Askew (1961)
Cecidomyiidae	*Mikiola fagi* (Hartig)	*Fagus*	Dziurynski (1961)
Tenthredinidae	*Pontania pacifica* Marlatt	*Salix*	Caltagirone (1964)
Psyllidae	*Pachypsylla* sp.	*Celtis*	Moser (1965)
Cynipidae	*Diplolepis rosae* (L.)	*Rosa*	Schröder (1967)
Cynipidae	*Andricus kollari* (Hartig)	*Quercus*	Duffet (1968)
Cynipidae	Various	*Quercus*	Brookfield (1972)
Cynipidae	*Diplolepis* sp.	*Rosa*	Norlander (1973)
Cynipidae	*Diplolepis polita* (Ashmead)	*Rosa*	Shorthouse (1973)
Cecidomyiidae	*Massalongia* sp.	*Betula*	Askew and Ruse (1974)
Cynipidae	*Diastrophus kincaidii* Gill.	Rubus	Wangberg (1976)
Cynipidae	*Acraspis hirta* Bassett	Quercus	Washburn and Cornell 1979)
Cynipidae	*Diplolepis triforma* Shorthouse and Ritchie	*Rosa*	Zuchlinski (1979)
Cecidomyiidae	*Semudobia* sp.	*Betula*	Roskam (1979, 1986)
Cecidomyiidae	*Semudobia* sp.	*Betula*	Roskam and Van Uffelen (1981)

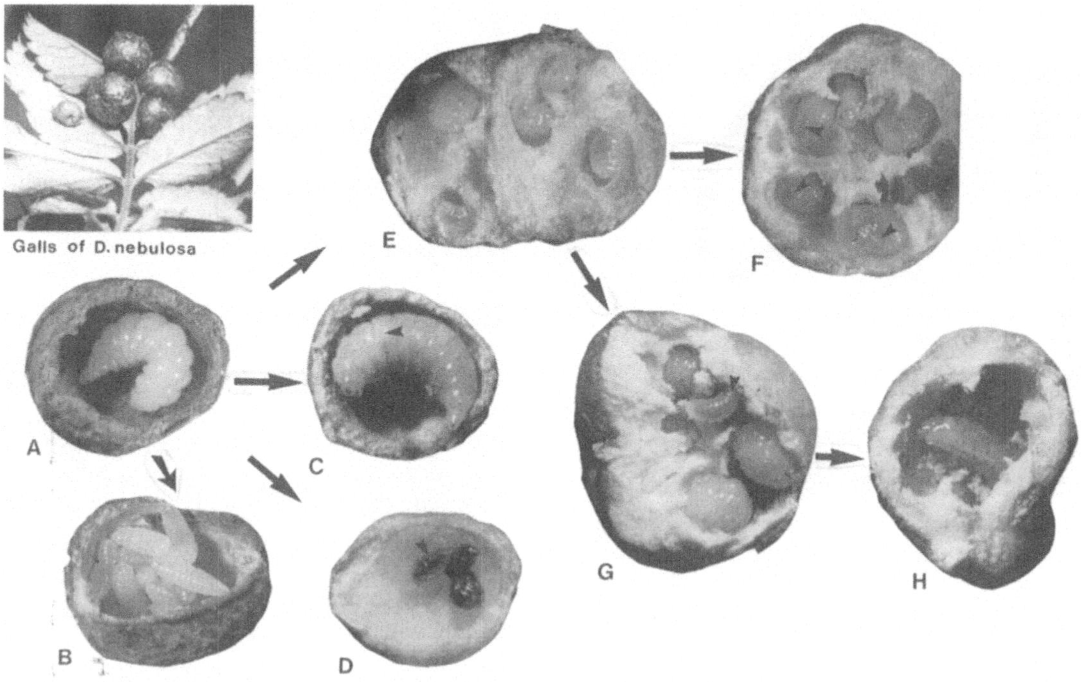

Figure 16-1. Feeding relationships of the insects inhabiting rose leaf galls of *Diplolepis nebulosa.* (A) Larva of *Diplolepis nebulosa,* the gall inducer. (B) Larvae of *Tetrastichus* after consuming *D. nebulosa.* (C) Larva of *Torymus* (arrow) feeding on larva of *Glyphomerus stigma* that has fed on *D. nebulosa.* (D) Larva of *Habrocytus* (arrow) that has fed on *D. nebulosa* (black pellet). (E) Larvae of *Periclistus* that have modified the gall of *D. nebulosa.* (F) Larva of *Habrocytus* feeding on larvae of *Periclistus.* (G) Larva of *Eurytoma* (arrow) feeding on larvae of *Periclistus.* (H) Larva of *Eurytoma* that has consumed larvae of *Periclistus* and gall tissues induced by *Periclistus.*

The Gall Inducer

The primary inhabitant of all cynipid galls is the larva (or larvae) of the gall inducer itself. Galls are induced by oviposition and by freshly hatched larvae when they begin to feed on adjoining plant tissues (see Chapters 5 and 6). The larvae are quickly surrounded by rapidly proliferating cells such that each larva becomes enclosed within its own chamber (Fig. 16-1A). The entire larval stage is spent within the chamber feeding on nutritive cells that line the chamber surface. In other words, each larva is completely confined within plant tissues making it easily located by parasitoids. Other factors that likely encourage parasitism of cynipid larvae are their inability to escape by wiggling or dropping from aerial vegetation, their low levels of predation and infection by disease, and their low mortality due to abiotic factors. Most temperate cynipid wasps overwinter within their galls in a prepupal stage, pupate the following season, emerge to the adult stage within the gall, then chew an exit channel to the outside, mate, and immediately begin searching for susceptible host tissues in which to oviposit.

Inquilines

Inquilines are gall inhabitants that feed on gall tissues but are unable to induce galls of their own. Most inquilines associated with cynipid galls are themselves cynipids. Inquilines of the genus *Periclistus* are exclusively associated with galls of *Diplolepis* whereas inquilines of the genera *Synergus, Ceroptres,* and *Saphonecrus* are inquilines in galls on *Quercus,* whereas *Synophromorpha* are inquilines on *Rubus* (Ritchie and Shorthouse 1987). Most inquilines have

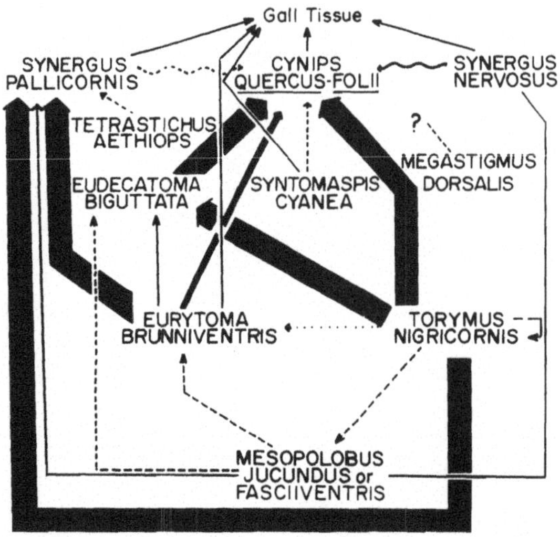

Figure 16-2. Food web of inhabitants associated with galls of *Cynips quercusfolii*. Thickness of the arrow gives an approximate indication of the frequency of the relationship. (From Askew 1961.) *Synergus pallicornis* and *Synergus nervosus* are cynipid inquilines. All the remaining species are parasitoids.

similar life histories. Adults emerge from galls soon after the gall inducers and oviposit in freshly initiated galls. Some species of inquilines develop on tissues of the gall without harming the inducer whereas other species kill the inducer at oviposition. An example of the first group is *Synergus pallicornis* Htg. (Fig. 16-2) whereas all *Periclistus* belong to the second group. *Periclistus* kill the larva of *Diplolepis* at oviposition and freshly hatched *Periclistus* not only feed on tissues induced by the inducer but also modify attacked galls (Shorthouse 1980). Each inquiline becomes enclosed in its own larval chamber (Fig. 16-1E) and feeds on nutritive cells lining the surface. Such inquilines commonly increase the size of inhabited galls. They also increase the insect biomass per gall and consequently serve as an important source of food for many entomophagous species. Convincing quantitative data on the increase of biomass are available for *Synergus clandestinus* Eady in the gall of *Andricus legitimus* Wiebes Rijks (Wiebes Rijks 1980).

Entomophagous Inhabitants

Entomophagous inhabitants feed either on larvae of the inducers, inquilines or primary parasitoids (Figs. 16-2, 16-4, and 16-7). Most belong to the Superfamily Chalcidoidea and are found in the families Eurytomidae, Torymidae, Pteromalidae, Eulophidae, Eupelmidae, and Ormyridae, but others occur in the Superfamily Ichneumonoidea, family Ichneumonidae. These inhabitants feed either as endoparasites, developing within the bodies of their hosts, or as ectoparasites, feeding externally on the host larvae. Most entomophagous inhabitants are considered parasitoids as they eventually kill their hosts and thus resemble predators (parasitoids are insects with a life style intermediate between that of predators and true parasites). Only a few species of parasitoids are gall specific; most attack a limited range of galls.

Female parasitoids usually lay one or more eggs on hosts and the ensuing larvae consume the host, killing it in the process. Endoparasitoids tend to enter the community early in the season whereas ectoparasitoids tend to enter later when the host larvae are more mature. Some entomophagous species are host specific, whereas others feed on any larvae inhabiting a chamber. *Eudecatoma biguttata* (Swed.) of the family Eurytomidae, for example, is a host-specific endoparasitoid feeding only on the gall inducer (Fig. 16-2). Although its eggs are laid while the gall and larvae are immature, the larva does not emerge from the host until the gall is fully developed. *Syntomaspis cyanea* Boh., of the family Torymidae, is a delayed entomophagous ectoparasitoid (Fig. 16-2) that feeds on gall tissues before it attacks the host *Cynips*. Species of *Torymus*, also of the family Torymidae (Fig. 16-2), are polyphagous ectoparasitoids that attack any larva (Fig. 16-1C) in the gall (even those of its own species). Some *Torymus* have more than one generation per year. The various categories of host relationships are shown in Table 16-2. These categories were mainly distinguished by Askew for cynipid communities found within galls on oak, but they also are suitable for food webs of other galls, many of which are much simpler.

Parasitoids associated with galls of cynipids oviposit during a temporal oviposition window bounded on one side by development of the cynipid larva and on the other by development of the gall. Structural and behavioral limitations also restrict attack by particular species of parasitoid to certain periods of the gall's develop-

Table 16-2. Categories of Inhabitants of Galls Induced by *Cynips quercusfolii*[a]

Agamic generation		
Phytophagous		
Gall inducer	*Cynips quercusfolii* L.	Cynipidae
Harmless inquiline	*Synergus pallicornis* Htg.	Cynipidae
Harmful inquiline	*Synergus nervosus* Htg.	Cynipidae
Entomophagous parasitoids		
Polyphagous ectoparasitoid	*Eurytoma brunniventris* Ratz.	Eurytomidae
Monophagous ectoparasitoid	*Syntomaspis cyanea* Boh.	Eurytomidae
Entomophagous parasitoids		
Monophagous endoparasitoid	*Eudecatoma biguttata* Swed.	Eurytomidae
Polyphagous ectoparasitoid	*Torymus nigricornis* Boh.	Torymidae
Polyphagous ectoparasitoid	*Torymus geranii* Wlk.	Torymidae
Sexual generation		
Phytophagous		
Gall inducer	*Cynips quercusfolii* L. *forma taschenbergi* Von Schl.	Cynipidae
Harmful inquiline	*Synergus gallaepomiformis* Boyer de Fonscolombe	Cynipidae
Entomophagous parasitoids		
Polyphagous ectoparasitoid	*Torymus auratus* Fourc.	Torymidae
Polyphagous ectoparasitoid	*Torymus geranii* Wlk.	Torymidae
Polyphagous ectoparasitoid	*Mesopolobus fasciiventris* Ww.	Pteromalidae
Monophagous ecto (mainly)	*Mesopolobus tibialis* Westw.	Pteromalidae
Monophagous ecto (mainly)	*Mesopolobus fuscipes* Wlk.	Pteromalidae
Monophagous ecto (mainly)	*Mesopolobus dubius* Wlk.	Pteromalidae

[a]The term "monophagous" refers only to this gall community. Some species referred to as monophagous are also associated with communities of other cynipids.

ment. Thus some species attack small developing galls, whereas others attack only mature galls. Such differences in capability lead to a temporal succession of parasitoids throughout the season of gall growth. This temporal succession is often indicated by the length of their ovipositors, that is, species attacking early in gall development have shorter ovipositors than late-attacking species. Several workers have shown how the degree of parasitization declines with increasing gall size (Stille 1984; Jones 1983; Weis *et al.* 1985). Jones (1983) found that the most deeply set larvae of *Diastrophus kincaidii* are the least parasitized and suggested that communal oviposition of this species has evolved as a means of increasing gall size.

Temporal succession of the parasitoids on a host does not preclude interaction between the species because early attacking parasitoids can be killed by those attacking the same host individual later in the season. Evidence for such competitive interactions in gall parasitoid communities is presented by Askew (1975). He found that many parasitoids attacking cynipids could be placed in two broad categories: those that were monophagous (feed exclusively on the gall inducer or inquilines) and those that were polyphagous (feed opportunistically on either the gall inducer or any other parasitoid larva found in the galls). The polyphagous species tend to attack late in gall development thereby causing mortality among the early monophagous species.

Washburn and Cornell (1981) studied the influence of parasitoids and inquilines on galls of *Xanthoteras politum* (Bassett) and found them to be the major mortality factor of galls found on newly regenerated patches of host oaks after fire. They found that in each successive year after the galls became established, a higher frequency of the host population was killed. Galls apparently disappeared from the study sites because phenological changes in host plants eventually eliminated the gall inducer. Therefore the apparent overexploitation of the host population by the inquilines and parasitoids may be to the selective advantage of the attackers because the gall inducers will be eventually eliminated from the postfire sites. According to Price (1980), parasitoids exist in nonequilibrium conditions exploiting small, discontinuous environments. The instability of the local patch makes colonization of new patches necessary and parasitoids have qualities such as high fecundity and the ability of single females to start new colonies, enabling them to successfully do so.

It also has been suggested that parasitoids are partly responsible for the extraordinary diversity of cynipids (Askew 1961, 1980; Cornell 1983). It is argued that by controlling cynipid numbers, parasitoids reduce the competition between cynipids for material resources, thus allowing the coexistence of many species. Because many chalcids have narrow host spectra or are polyphagous they are thought to impose intense selection for the diversification of parasitoid avoidance mechanisms. These defenses against attack by parasitoids include small or large sizes, spines and hairs, sticky secretions, hard surfaces, and tough gall tissues. Cornell (1983) suggested that the need for parasitoid-free space provides the major selective force molding complex and diverse gall structures, although Price *et al.* (1987) suggested that the evidence is not compelling, arguing that much gall variation, as in *Neuroterus,* minimizes morphological differences relevant to parasitoid attack. Certainly parasitoids can be very detrimental to populations of cynipid wasps. Rates of parasitism exceeding 80 percent are common and often all gall inducers within individual galls are killed (Askew 1975; Force 1974; Schröder 1967; Washburn and Cornell 1979, 1981). However, other factors also are responsible for mortality. Zuchlinski (1979) showed that cold winter temperatures killed many cynipids in northern regions (see below) and Taper *et al.* (1986) showed that fungus within the chambers killed many *Dryocosmus dubiosus* (Fullaway) on oak.

Alternation of Generations

Alternation of generations is a complicated type of life cycle found among many species of cynipids where an agamic generation consisting solely of females is distinct from a sexual generation of both males and females (see Chapters 5 and 6). This type of reproduction is found only among cynipids that attack trees, with distinct periods of growth for different parts of the plant such as buds, flowers, leaves, and roots (Longmann and Coutts 1974). This allows cynipids to better exploit the host plant. It may also enable the cynipids to escape from enemies by partitioning the host space (Price *et al.* 1980). For the gall inducer, the alternation of generations means an intricate life cycle, the exact genetic aspects of which are poorly understood (Folliot 1964). Whereas the males of most Hymenoptera developing from unfertilized eggs are haploid (Doncaster 1910, 1911, 1917), in some cynipids there are two types of females. In *Neuroterus lenticularis* (Olivier), for example, there are both male-producing (n) and female-producing ($2n$) females, with either a normal or nonreduced chromosome number (Patterson 1928).

Alternation of generations means a change in hosts for the entomophagous inhabitants as the primary host gall may not be available when the parasitoids emerge. An example of the complications found in the host–parasitoid relationships of four cynipids exhibiting alternation of generations is shown in Figure 16-3. Askew (1961) has found that the same parasitoid species are not usually found in galls of alternate generations of the same cynipid species.

COMMUNITIES ASSOCIATED WITH CYNIPID GALLS

Leaf Galls of *Diplolepis polita* on *Rosa acicularis*

Diplolepis polita (Ashmead) induces spherical, single-chambered galls, clothed with weak spines, on the adaxial surface of leaves of *Rosa acicularis* Lindl. (Shorthouse, 1973, 1975, 1982). Galls usually grow in clusters, several per leaflet, although single galls on a leaf can be found. Immature galls are soft, but become brittle as they mature. Galls are initiated in the early spring and reach their maximum size of about 5.0 mm in diameter by mid-July. This gall is found in Western and Central Canada and adjoining states of the United States.

Galls of *D. polita* support a community of four key parasitoids: *Eurytoma longavena* Bugbee, *Glyphomerus stigma* (F.), *Torymus bedeguaris* (L.), and *Habrocytus* sp. and an inquiline, *Periclistus pirata* (Osten Sacken). The food web of the *D. polita* community is shown in Figure 16-4. Both the larvae of *D. polita* and *P. pirata* are totally phytophagous and support the parasitoids; however, *E. longavena* and *G.*

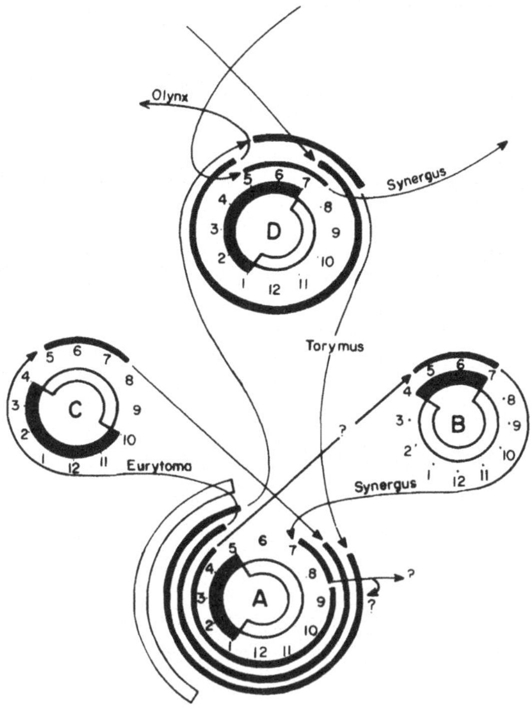

Figure 16-3. Life cycles of four galls whose inducers exhibit alternation of generations. The four galls are connected with their various parasitoids and inquilines to show some of the possible host alternations. The inner circles represent the life cycle of the gall inducer divided into a black portion (sexual generation) and a white portion (agamic generation). Around the gall inducer are drawn circles (black) of three genera of inquilines and parasitoids. The outer white semicircle is the agamic gall from which the gall inducer has already emerged, but which may still harbor developing inquilines and parasitoids. Only parts of the possible interactions are shown between the four galls; the open ended lines point to other galls. The host galls represented are A, *Cynips quercusfolii;* B, *Neuroterus quercusbaccarum* (the sexual current galls on catkins of male oak are possibly used by *Synergus*); C, *Andricus solitarius* (the agamic bottle gall on oak is host to *Eurytoma*); and D, *Biorhiza pallida* (the large oak apple gall is host to *Torymus*). All galls have other inquilines and parasitoids not mentioned in this figure.

stigma commonly consume gall tissues after they have eaten their hosts. All the parasitoids are ectoparasitoids and most will feed on each other besides the gall inducer and the inquilines (Fig. 16-4); several are cannibalistic.

The sequence of adults emerging in the spring (Fig. 16-5) corresponds closely to the trophic structure of the community. All species emerge from previous years galls and oviposit into current galls throughout the summer. Most species have a lengthy emergence period (Fig. 16-5), but it is assumed that many individuals do not find galls (and the larvae within) at a suitable size for oviposition.

D. polita emerges in late April and early May and oviposits on the surfaces of the leaflets while they are still folded in leaf buds. The first galls appear about 10 days after oviposition, soon after the leaves unfold. Galls develop rapidly and are mature by mid-July. The larvae feed within their chambers until late July and like all *Diplolepis,* overwinter within their galls in the prepupal stage. Mature galls have a thin layer of sclerenchyma cells, which makes them brittle. Galls begin abscising by mid-August, well before leaves normally abscise. Galls of *D. polita* therefore overwinter in the leaf litter and are protected by the insulating qualities of snow.

All immature galls contain a single larva of *D. polita;* however, within a week many are found by ovipositing *P. pirata.* The inquilines begin emerging in mid-May and adults are found in the field until early August (Fig. 16-5). *P. pirata* kill the larvae of *D. polita* at oviposition and as a result about 70 percent of the galls examined about 1 week after the first galls are found contain eggs of *P. pirata* along with a

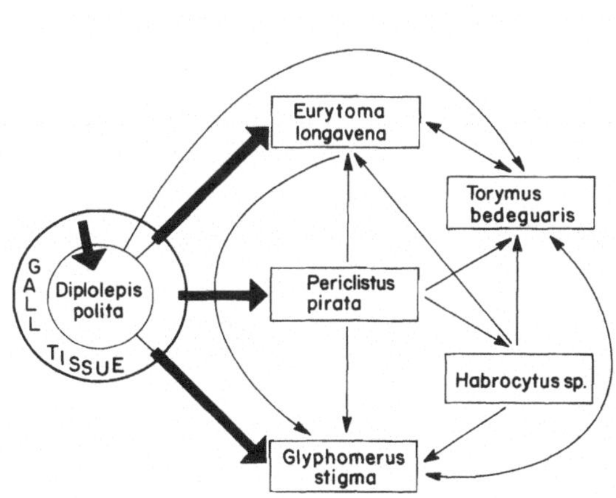

Figure 16-4. Food web of the community associated with galls of *Diplolepis polita* on *Rosa acicularis.* Galls were collected near Edmonton, Alberta.

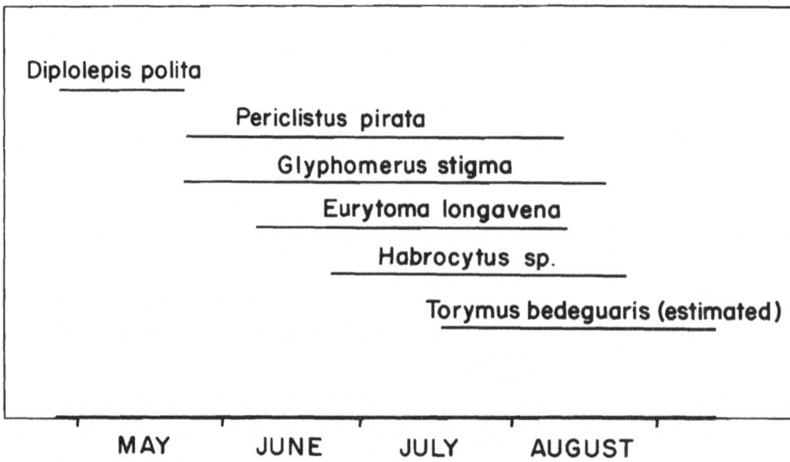

Figure 16-5. Periods of oviposition for species associated with galls of *Diplolepis polita.* Sampling was undertaken near Edmonton, Alberta.

dead *Diplolepis* larva (Table 16-3). Thus the initial decline in the *Diplolepis* population is due to *P. pirata* (Fig. 16-6). Each larva of *P. pirata* induces its own chamber within the former *D. polita* chamber and their influence more than doubles the size of normal galls (Shorthouse 1980). An average of 10.5 *P. pirata* eggs are found per gall and by the end of the season an average of 5.4 larvae are found per gall.

Eurytoma longavena is the next species associated with galls of *D. polita* (Table 16-3) even though adults of *Glyphomerus stigma* are found in the field at the same time (Fig. 16-4). Eggs of *Eurytoma longavena* are often laid among the freshly hatched larvae of *P. pirata,* which provide them with ample food as the season advances. In most cases *E. longavena* consume all larvae of *P. pirata;* however, in larger galls some *P. pirata* larvae escape. The initial reduction in galls containing *P. pirata* was due to the feeding activities of *E. longavena* (Table 16-3). Some eggs of *E. longavena* are laid on maturing larve of *D. polita. Glyphomerus stigma* has a lengthy emergence period and because its larvae consume all inhabitants of galls in which they occur, this species has a strong influence on community composition. However, *G. stigma* often lays its eggs on the larvae of *D. polita,* which in turn are frequently consumed by larvae of *Torymus bedeguaris. Habrocytus* sp. oviposit into maturing galls, but most of their larvae are found on the larvae of *P. pirata.*

Table 16-3. Seasonal Change in the Percentage of Galls Containing Members (Larvae Only) of the *Diplolepis polita* Gall Community from a Site Near Edmonton, Alberta in 1969

		Community members (% of total)				
Collection date	No. of galls	*Diplolepis polita*	*Periclistus pirata*	*Eurytoma longavena*	*Glyphomerus stigma*	*Habrocytus*
May 20	4	100.0	—	—	—	—
May 25	55	68.4	—	—	—	—
May 28	54	18.6	—	—	—	—
June 6	166	8.1	71.0	16.0	—	—
June 12	156	8.2	62.3	44.8	—	—
June 20	176	3.0	37.0	52.8	—	—
June 27	127	4.2	35.4	51.1	1.0	—
July 5	119	2.9	25.2	63.8	3.4	1.0
July 14	177	2.7	20.0	47.5	4.5	5.0
July 23	190	3.1	17.2	38.2	5.4	23.0
August 22	275	4.2	8.1	32.1	12.5	8.1

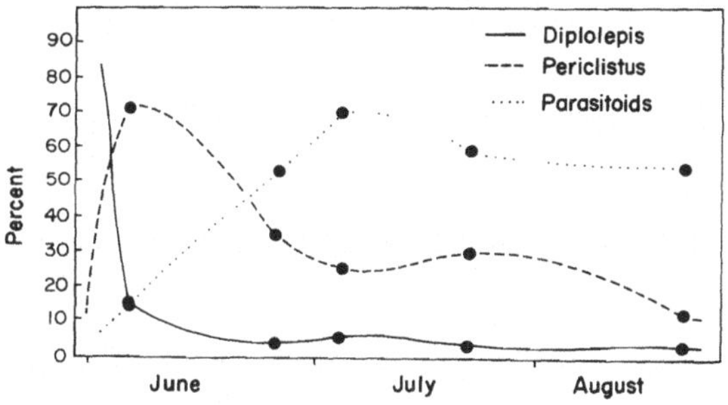

Figure 16-6. Influence of the inquiline *Periclistus pirata* and various parasitoids on the population of *Diplolepis polita.* Sampling was undertaken near Edmonton, Alberta.

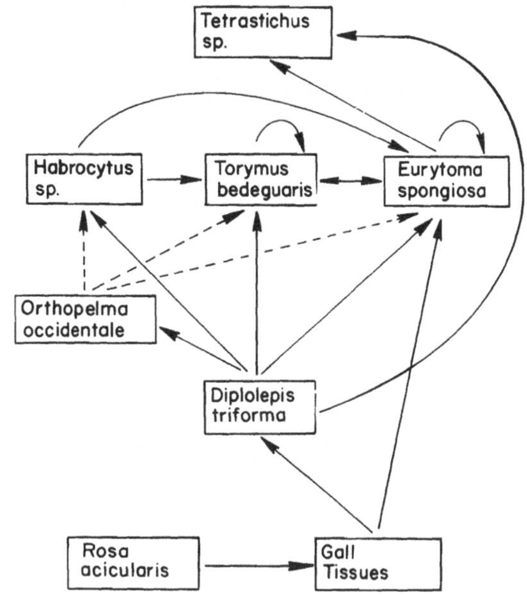

Figure 16-7. Food web of the community associated with galls of *Diplolepis triforma.* Galls were collected near Sudbury, Ontario.

Diplolepis polita persists in Canada with typically over 90 percent of the population killed annually by the combined actions of inquilines and parasitoids. The key mortality factor of *D. polita* is the inquiline *Periclistus pirata* and the composition of the community (the relative occurrence of *D. polita,* inquilines, and parasitoids) determines the composition of the community in subsequent years.

Stem Galls of *Diplolepis triforma* on *Rosa acicularis*

Diplolepis triforma Shorthouse and Ritchie induces conspicuous, usually spineless woody stem galls that are firmly attached to twigs of the previous year. The gall is multichambered and is restricted to *Rosa acicularis* (Shorthouse and Ritchie 1984). There is an average of 10 chambers per gall; however, most are occupied by parasitoids by the end of the season. Galls of *D. triforma* are found throughout the range of *R. acicularis* in Central and Western Canada (Shorthouse and Ritchie 1984) and are commonly found in patches supporting galls of *D. polita.*

Galls of *Diplolepis triforma* support a community of five key parasitoids: *Eurytoma spongiosa* Bugbee, *Orthopelma occidentale* Ashmead, *Torymus bedeguaris, Habrocytus* sp., and *Tetrastichus* sp. and the food web is shown in Figure 16-7. No *Periclistus* inquilines are found in galls of *D. triforma. D. triforma* is the only totally phytophagous species in the community and directly supports the four parasitoids; however, *E. spongiosa* commonly feeds on gall tissues once it has consumed its host. *O. occidentale* and *Tetrastichus* sp. are endoparasitoids, whereas *E. spongiosa, T. bedeguaris,* and *Habrocytus* sp. are ectoparasitoids. *O. occidentale* and *Habrocytus* sp. were found feeding only on the gall inducers whereas *T. bedeguaris* and *E. spongiosa* attacked all of the parasitoids and the larvae of *D. triforma;* they also were cannibalistic. Since *O. occidentale* is an internal parasitoid, it was indirectly consumed by other parasitoids that attacked *Orthopelma* occupied *D. triforma. Tetrastichus* sp. was an internal parasitoid of *D. triforma* and *E. spongiosa.*

The sequence of adults emerging in the spring is shown in Figure 16-8 and corresponds closely to the trophic structure of the community. *D. triforma* emerges first and the females oviposit at the base of a leaf bud depositing the eggs in a cluster near the leaf primordia. The oviposition period (the period between the first and last oviposition observed in the field) lasts approximately 15 days. Freshly initiated galls were visible about 15 days after oviposition. The larvae feed within their chambers until mid-September, and like all *Diplolepis,* over-

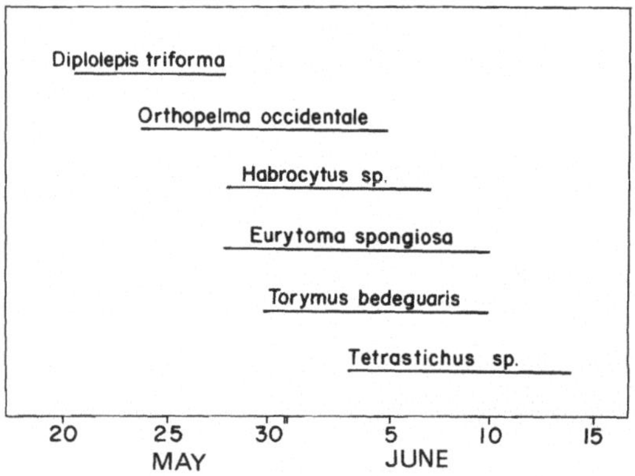

Figure 16-8. Periods of oviposition for species associated with galls of *Diplolepis triforma*. Sampling was undertaken near Sudbury, Ontario.

winter within their galls in the prepupal stage. Galls of *D. triforma* remain on the host plant throughout the winter with many galls on the upper parts of the plant remaining above the snow pack (Shorthouse *et al.* 1980). Gall size also influences the chances of *D. triforma* larvae escaping parasitoids. Larvae within galls consisting of one to four chambers had less than a 5 percent chance of escaping parasitoids, whereas those within galls consisting of 17 to 20 chambers had almost a 20 percent chance of escape.

Orthopelma occidentale is the second species to emerge, followed by *Habrocytus* sp. and *E. spongiosa* (Fig. 16-8). Galls were clearly visible when *E. spongiosa* began to emerge. *Torymus bedeguaris* and *Tetrastichus* sp. were the last species to emerge. For a short time after the first galls are initiated, *D. triforma* occupy all gall chambers. However, the community structure soon changes as the various parasitoids begin depositing their eggs into the chambers. By mid-June, larvae of *D. triforma* were found in only 76 percent of the chambers, while eurytomid larvae were found in the other 24 percent (Table 16-4). The eurytomid population increased weekly to a maximum of 74 percent by mid-July. Torymid larvae were first found on July 20 (Table 16-4) at which time they made up 10 percent of the total community composition. The larvae of *Habrocytus* sp. and *Tetrastichus* sp. appeared next but neither had a major effect on community composition. A decrease in the populations of eurytomids and torymids occurred in mid July due to some individuals being fall emergents.

The influence of all parasitoids on the number of chambers containing larvae of *D. triforma* is shown in Figure 16-9. Parasitoids that emerge in the spring begin to reduce the population of *D. triforma* within a week after the first galls are initiated and this pattern continues to the middle of July. The fall emerging eurytomids and torymids begin to emerge near mid-July and potentially can deposit eggs into chambers of the same population of galls from which they themselves emerged. Approximately 30 percent of the eurytomid population, which is approximately 30 percent of the entire parasitoid population, emerges during the latter part of the summer. Thus by the end of the summer only about 10 percent of the chambers contain a larva of *D. triforma*.

Table 16-4. Seasonal Change in the Percentage of Chambers Containing Members of the *Diplolepis triforma* Gall Community from a Site Near Sudbury, Ontario in 1976

			Community members (% of total)				
Collection date	No. of galls	Total Individuals	*Diplolepis triforma*	*Eurytoma spongiosa*	*Torymus* sp.	*Habrocytus* sp.	*Tetrastichus*
June 15	47	510	75.9	24.1	—	—	—
June 20	42	578	51.2	48.3	0.5	—	—
June 25	38	462	35.3	64.1	0.6	—	—
June 30	32	325	34.2	61.8	4.0	—	—
July 10	30	271	24.0	74.5	1.4	—	—
July 20	20	254	16.5	68.9	10.2	—	4.4
July 30	13	159	20.1	72.3	5.7	1.3	0.6
August 9	23	253	15.4	76.6	6.3	—	1.7
August 23	21	163	12.3	77.3	6.7	1.8	1.9

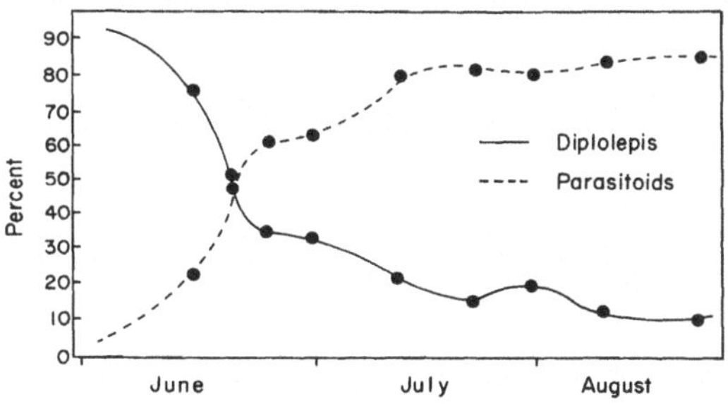

Figure 16-9. Influence of parasitoids on the population of *Diplolepis triforma.* Sampling was undertaken near Sudbury, Ontario.

Of the *D. triforma* that escape attack by parasitoids by gall maturity, many are killed during particularly cold winters with little snow. In 1977, 79 percent of the *D. triforma* that survived parasitoids were killed during the winter. *D. triforma* within galls close to the ground had a higher survival rate than those higher on the host plant. Of the *D. triforma* that overwintered in galls 25 cm or less above the ground, 33 percent were frozen, whereas 83 percent of those at 75 cm or higher were frozen. Parasitoids were less affected by cold than the gall inducers. Winter mortality of parasitoids in 1977 was 24 percent and position on the host plant also influenced winter mortality. The winter mortality of parasitoids in galls 25 cm or less above the ground was only 12 percent but rose to 44 percent for those 75 cm or more above the ground. Thus the *D. triforma* population can suffer even greater mortality from parasitoids following exceptionally cold winters.

Zuchlinski (1979) calculated the actual number of gall inducers and parasitoids available for oviposition the following spring and considered the influences of gall density on the population of *D. triforma.* In one study site with a high gall density, he found an increase in the populations of parasitoids from 1975 to 1977, which resulted in a lower number of *Diplolepis* (and thus galls), which in turn caused a considerable reduction of parasitoids in 1978. Consequently the number of emerging *Diplolepis* increased in 1979 resulting in a corresponding increase in the number of galls a year later. This in turn was followed by an increase in the population of parasitoids and thus the cycle likely continues.

Leaf Galls of *Cynips quercusfolii* on *Quercus robur*

The life cycle of *Cynips quercusfolii* is divided into two separate portions: the short, sexual, spring generation, which develops in galls in dormant buds of *Quercus robur* L., and the longer, agamic generation, which induces oak apples from summer to winter on the underside of the leaves (Fig. 16-10). All members of the community (both agamic and sexual galls) also are shown in Figure 16-10. The sexual generation emerges in late May or early June and the females oviposit in the ribs of the young leaves. The immature galls appear in late June and they grow rapidly if the inducers are not parasitized. Galls are half grown before the *Cynips* larva feeds actively in mid-July on the primary layer of nutritive tissue.

Larvae of *C. quercusfolii* mature and pupate in late September or early October. The pupal stage lasts a few weeks and the adults remain in the gall for a few days before gnawing a tunnel to the outside. Adults exit the galls from the first half of December to late January and they immediately oviposit in dormant buds in which the sexual generation will develop.

Galls of *C. quercusfolii* and their contents were studied at a site in the Netherlands by the first author over 10 consecutive years. Galls were collected periodically from the host trees and then from the ground after the trees had shed their leaves. Galls were returned to the laboratory and the contents of each determined. The bud galls were rarer and much more difficult to collect. The following discussion examines the parasitoid complexes of immature and mature oak apples and the bud galls. The sequence by which inquilines and parasitoids enter the immature and mature leaf galls is shown in Figure 16-11.

Attacks on Immature Galls

Inquilines and parasitoids are attracted to the immature oak apple galls soon after they appear on the leaf ribs (Fig. 16-11). The inquiline *Synergus nervosus* Htg. begins attacking immature galls when they are 2 mm in diameter (Wiebes Rijks 1982). One to six eggs are scattered about the inside surface of the larval

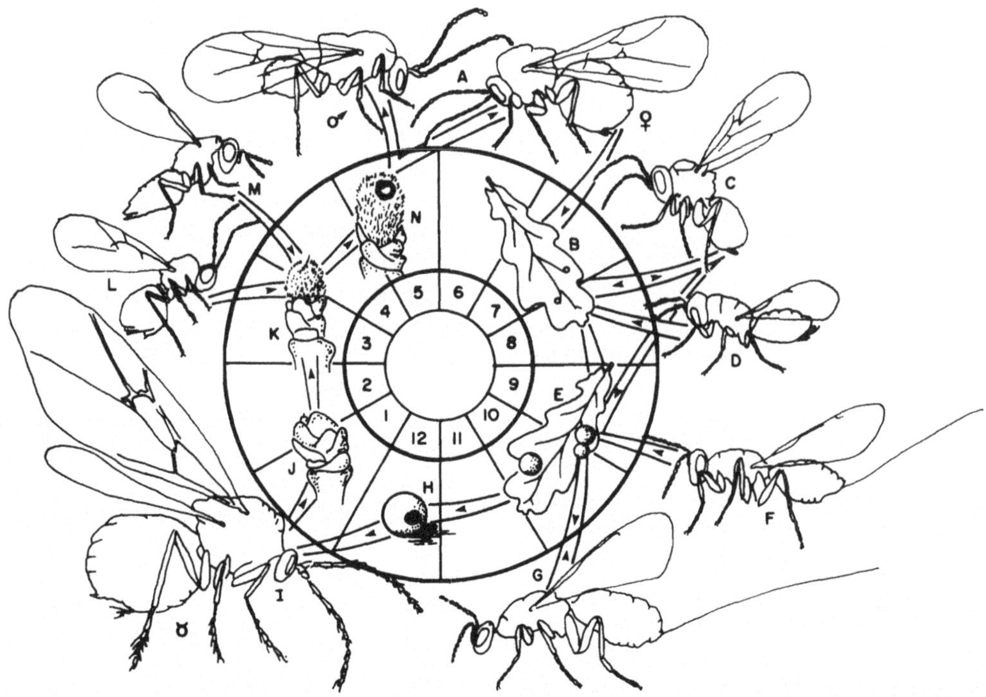

Figure 16-10. Life cycle of *Cynips quercusfolii* showing galls of both the sexual and agamic generations along with associated inquilines and key parasitoids. Months of the year are shown on the inner circle. (A) Male and female of the sexual generation. (B) Oak leaf with immature agamic galls. (C) *Synergus*. (D) *Eurytoma*. (E) Oak leaf with mature agamic galls. (F) *Torymus* first generation. (G) *Torymus* second generation. (H) Agamic gall with an emergence hole. (I) Adult of the agamic generation. (J) Oak bud at the age chosen by agamic generation for oviposition. (K) Immature sexual gall. (L) *Synergus*. (M) *Mesopolobus*. (N) Mature sexual gall with emergence hole.

chamber and when eggs are present, the regular configuration of the chamber is lost due to desiccation and a cessation of gall growth. One of the *Synergus* larvae grows faster than the others and eventually all but this one die.

Synergus nervosus is host to a parasitoid complex consisting of *Mesopolobus fasciiventris* Westw. (Table 16-2) and a primary wave of *Eurytoma brunniventris* Ratz. and *Torymus auratus* (Fourc.) may take part. Only the galls that remain small because of *S. nervosus* attack are victim to these parasitoids. The inquilines and parasitoids develop rapidly and within a month they emerge and attack other small galls. The empty oak apples are very brittle and fall from the leaves.

Attacks on Mature Galls

Immature galls that escape the early attack by *Synergus nervosus* by growing rapidly also escape the first wave of parasitoids. As occurred in the immature galls, the first insect to attack maturing galls is an inquiline (Fig. 16-11). This unidentified *Synergus* is referred to as "*Synergus* B" by Wiebes-Rijks (1979). The eggs of *Synergus* B (from one to six per gall) are not scattered about the inside surface of the larval chamber as are those of *S. nervosus*, but instead are deposited in a ring about the central chamber. The gall-inducing larva is not killed by the ovipositing female.

The larvae of *Synergus* B also have a desiccating effect on the gall tissue by which the primary nutritive layer is no longer available to *C. quercusfolii* and all but one of the *Synergus* B larvae dies. The surviving *Synergus* then represents a delayed 'superparasitism' upon its congeners after the concerted action of all to kill the *C. quercusfolii*. The impact on the *Cynips* population may be considerable causing up to 50 percent to die in some years, but not more

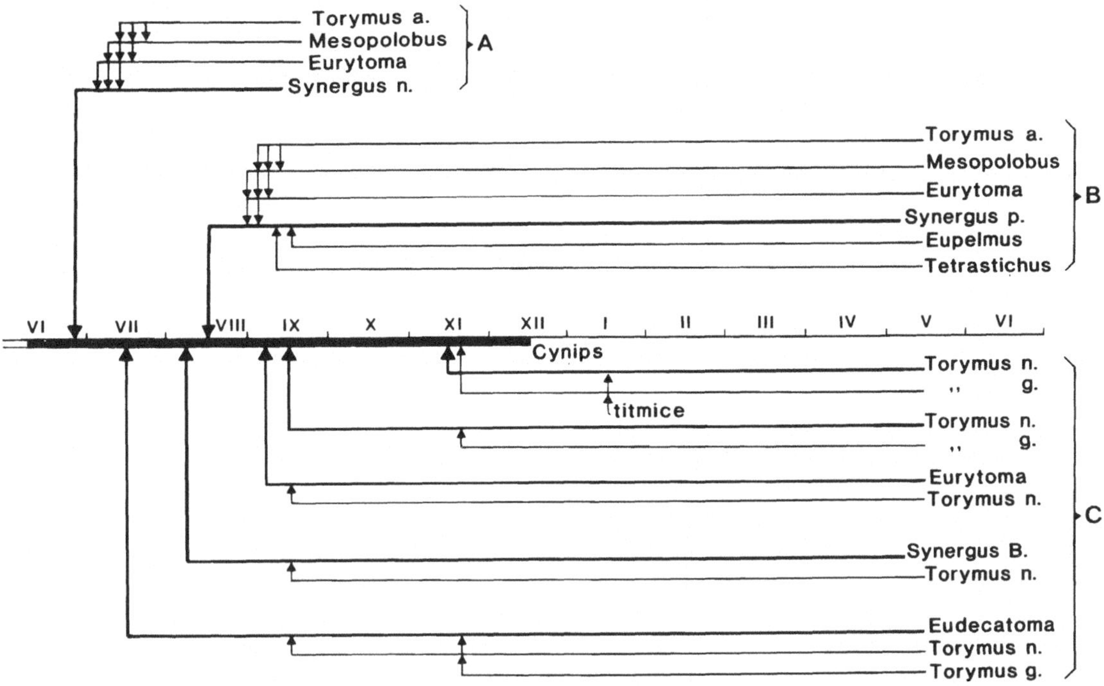

Figure 16-11. Flow chart of all inhabitants associated with galls of *Cynips quercusfolii.* Only common primary parasitoids are included. A is the first wave of parasitoids started by *Synergus nervosus;* B is the second wave of parasitoids led by *S. pallicornis;* C is the main group of parasitoids along with *Synergus* B. Some inhabitants are represented by *Torymus* a *(uratus), Torymus* n *(igricornis), Torymus* g *(eranii), Synergus n(ervosus),* and *Synergus p(allicornis).* Arrows indicate trophic relationships. For example, *Torymus* a attacks *Mesopolobus,* as well as *Eurytoma* and *Synergus* n. Arrows also indicate the period of inquiline and parasitoid attack throughout the season.

than 6 percent in other situations. A third species of inquiline, *S. pallicornis* Htg. also oviposits in galls (Fig. 16-11) and forms chambers in the peripheral part of the gall tissues without damage to the inducer.

A second wave of *Eurytoma brunniventris,* all of which had emerged from the immature galls, parasitize the larvae of both *Cynips* and *Synergus* B. *E. brunniventris* is polyphagous in that it will attack any larva found in a gall. It also attacks galls of all sizes and if it has insufficient insect tissue to complete its development, it will consume gall tissues. *E. brunniventris* usually first attack the larvae of *Synergus pallicornis,* which lie in the peripheral part of the gall. Although the *Synergus* may be a harmless commensal to the *Cynips,* they are often an additional host to its parasitoids.

Galls with mould are common in most samples. It usually follows death of the host larva or its parasitoids; however, some of the parasitoids, such as the *Torymus,* are found alive in mouldy gall chambers. A higher incidence of *Synergus pallicornis* was observed in galls in which the central chamber was filled with mould and in galls parasitized by *Torymus,* which must have occurred after the oviposition by *Synergus.* This may mean that the *Synergus* is not always harmless as it seems to be, as its numbers appear to be correlated with a higher loss of *Cynips* to slime and mould, and to *Torymus.*

The third parasitoid of mature galls is *Torymus nigricornis* Boh. In August, as many *Cynips* as *Synergus* larvae are being parasitized, but a month later *Cynips* is attacked five times as much as *Synergus.* This observation supports the assertions of Askew (1965) and Brookfield (1972) who suggested that the parasitoid fauna of the gall is determined more by the form (size) of the gall than by the species of gall inducer. It means that the second generation of

Torymus has a preference for the larger galls, by now inhabited by *Cynips* rather than by *Synergus nervosus,* the galls of which are stunted. *Torymus* with an even longer ovipositor have one more generation in oak apple galls. It appears that only *Torymus* from nearby galls of *Andricus kollari* Mayr on twigs have ovipositors long enough for oak apples over 17 mm in diameter, whereas those from nearby galls of *Biorhiza pallida* (01.) are restricted to smaller galls. The total parasitism by *Torymus nigricornis* in oak apples rises as high as 85 percent.

A second species of *Torymus, T. geranii* Walker, appears after *T. nigricornis.* Females of *T. geranii* oviposit into galls from mid-September to late October. They parasitize up to 20 percent of the *C. quercusfolii,* 2 percent of the *Synergus* and *T. nigricornis,* and are themselves parasitized by *Eurytoma brunniventris* and *Torymus nigricornis.* By now the relations in the gall become rather intricate. That is, *C. quercusfolii* is being killed and replaced by *Synergus* B and both suffer from attack by *Eurytoma* and the two species of *Torymus,* which also interact among themselves.

The larvae of *Cynips* were dissected in one season to establish parasitization by the endoparasitoid *Eudecatoma biguttata* (Swed.). By the end of the season only a few larvae issued from the host, but as early as August there occurred a parasitization of about 55 percent. When this population was followed through the winter, one sees the larvae of *Cynips* with a larva of *Eudecatoma,* which is in turn hyperparasitized by *Torymus.*The *Torymus* reduces the population of *Eudecatoma* to 5 percent by November (Fig. 16-12). Thus one has to be cautious when counting the ectoparasitoids as endoparasitoids may be present but not recognized in all instances. This brings to mind Askew's (1975) argument on the evolutionary replacement of endoparasitoids by ectoparasitoids.

Toward the end of the season, more and more galls become slimy and mouldy and their contents cannot be determined. This factor may or may not be selective. However, birds such as titmice are more selective. When titmice remove insect larvae from galls in late winter, most *Cynips* have already emerged and thus escape attack. Larvae of *Torymus,* which had attacked larger galls, are the main prey of

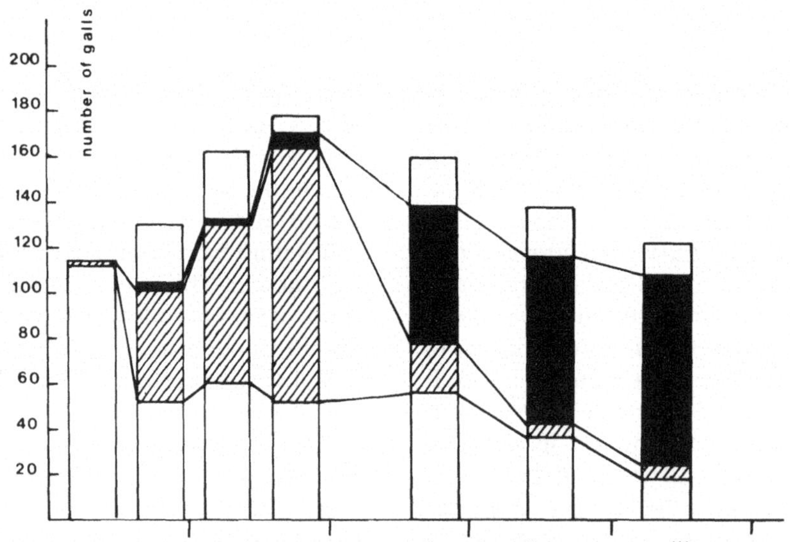

Figure 16-12. Seasonal change in the number of larvae of *Eudecatoma biguttata* (shaded) and *Torymus nigricornis* (black) in galls of *Cynips quercusfolii.* The population of *Cynips quercusfolii* larvae is along the bottom (white area) whereas the population of *Synergus* is across the top (white area). Note that by the end of August the parasitation of *Cynips* is mostly due to attack by the endoparasitoid (shaded areas). However, by November nearly all *Eudecatoma* are hyperparasitized by *Torymus nigricornis.*

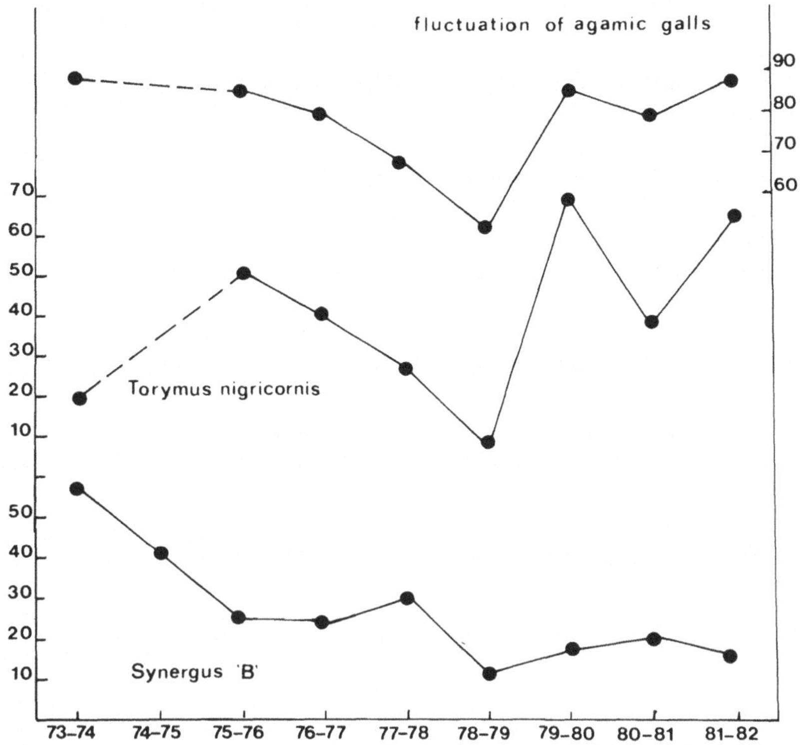

Figure 16-13. Fluctuation of agamic galls of *Cynips quercusfolii* over 9 years at a study site in the Netherlands along with percentage of galls containing *Torymus nigricornis* and *Synergus* B.

titmice and may reduce their population from 75 percent in December to 10 percent the following January.

Fluctuation of agamic galls and populations of *T. nigricornis* and *Synergus* B over several years is given in Figure 16-13. Parasitism by *Torymus nigricornis* appears to be the key factor as it is correlated with the reduction of *Cynips* over the years. In one year (1973–1974), a very high infestation by *Synergus* B was observed that resulted in death of 60 percent of the *Cynips* larvae. Also the total parasitism of the sexual galls (few data available), expressed as a percentage of the maximal number observed, seems to be correlated with the fluctuation of the agamic galls. This may mean that the number of agamic galls is directly correlated with the number of the sexual wasps emerging, which generally is the case, and that the parasitism by *Torymus nigricornis* is density dependent.

Attacks on the Sexual Generation

Bud galls of the sexual generation of *Cynips quercusfolii* appear in early May. Even though they are well concealed between the scales of the dormant buds of oak, they are attacked by *Synergus gallaepomiformis* (Boyer de Fonscolombe) whose one to six eggs lie dispersed in the gall tissue and have a desiccating effect, as mentioned above for the other species of *Synergus*. Both *Cynips* and *Synergus* are then parasitized by several *Mesopolobus* of the family Pteromalidae, mainly *M. fuscipes* (Walk.) and *tibialis* (Westw.), but also *dubius* (Walk.) and *fasciiventris* Westw. Other parasitoids are *Torymus auratus* and *T. cingulatus* Nees. The eggs of parasitoids are not only found on the larvae, but also on the nearly matures of *Cynips* and *Synergus*. Parasitism of galls of the sexual generation in many years is very high, often up to 80 percent.

DISCUSSION

Although the three species of cynipids examined in this chapter are basically different with respect to their life cycles and the anatomy of their galls, there are some common traits in the communities of inhabitants they harbor. Inquilines, for example, play an important role in the communities associated with leaf galls induced by *Diplolepis polita* and *Cynips quercusfolii*. In both cases inquilines affect the survival of the gall inducer and the complexity of the parasitoid components. Inquilines are absent in the stem gall of *Diplolepis triforma;* however, the diversity of parasitoids is about the same as in the two leaf galls.

All species of *Diplolepis*-inducing leaf galls have inquilines associated with them (Shorthouse 1975); however, the inquiline components of the oak gall communities are far more complex and intriguing. Not only do both generations of gallers in the *C. quercusfolii* complex have their specific inquilines, but the agamic generation has two specific inquilines that enter the community during an early or late phase. Other researchers have found two species of inquilines associated with one species of galler on oak (Washburn and Cornell 1981), but they did not find differences between these inquilines with respect to their resources.

Inquilines also vary in the degree of harm they cause the gall inducers. Although many examples exist whereby inquilines feed on copious amounts of gall tissues without causing harm to the gall inducer, the inquilines in the three communities examined here have negative effects on the gall inducers. These effects range from killing the gall inducer and modifying the gall as in the case of *Diplolepis polita,* to giving access to vectors such as moulds as seen with *Synergus pallicornis.* In galls of *D. polita,* of the approximate 90 percent total mortality, *P. pirata* is responsible for more than 70 percent. In galls of *C. quercusfolii,* with its more complex life cycle, inquilines usually are of secondary importance. However, the variation in importance (percentage of killed gall inducers) is enormous. Moreover, two species of *Synergus* attack early and late stages of gall development. When "top percentages" for *S. nervosus* (early, 19 percent) and *Synergus* B (late, ranging from 6 to 50 percent) are combined, the total mortality may equal that of *P. pirata.* It also is likely that inquilines will exhibit aggressive behavior when conspecific species share the same gall. Negative effects on host fitness also have been reported by Schröder (1967) for *Periclistus brandtii* Ratz. in the bedeguar galls of *Diplolepis rosae* and by Washburn and Cornell (1981) for inquilines associated with *Xanthoteras.* However, when inquilines are absent, the species of parasitoids appearing earliest are responsible for the largest mortality. *Eurytoma spongiosa,* for example, frequently kills up to 77 percent of the *D. triforma.*

Parasitoids can be subdivided into categories with respect to space and time. Traits such as endo- and ectoparasitism or host specificity deal with *where* parasitism takes place whereas synchronization with different phases of host development deal with *when.* The general scenario holds that endoparasitoids and univoltine species should be specific and highly synchronized with the host's development, whereas late parasitoids generally deposit their eggs on the host and are less host specific.

In the three communities examined here on rose and oak, egg parasitoids are absent. Unlike gall midges, for example, which lay their eggs on the host plant tissues and so are vulnerable to egg parasitoids, gall wasps drill into the tissue and so protect their eggs from this type of mortality. However, hyperparasitoids are common in many gall wasp associations. The relatively large cynipid larvae apparently allow more trophic levels than do the much smaller gall midges. Each of the three communities includes hyperparasitoids.

The endoparasitic species of *Orthopelma* in Nearctic and Palearctic *Diplolepis* galls are highly host specific as is *Eudecatoma* associated with oak galls and in all cases the endoparasitoids attack only the gall inducers. These endoparasitoids are replaced by ectoparasitoids because they are specific on early instars of the gall inducers. Such a replacement does not hold for nonspecific endoparasitoids such as the *Tetrastichus* associated with galls of *D. triforma.*

Many of the ectoparasitoids in the three communities are nonspecific. They may behave as primary and hyperparasitoids as well or may even be cannibalistic. Examples include species of *Eurytoma* and *Torymus*. However, in the studies of Schröder (1967) and Washburn and Cornell (1981), *E. rosae* seems to be host specific and *Habrocytus* seems to be host specific in all cases. Again, it attacks hosts that are active in early stages of the community development, namely the gall inducer or the inquilines.

An interesting aspect of the biology of some parasitoids is their shift from an entomophagous to phytophagous diet. Sometimes such a shift is necessary simply to exchange an emptied chamber for one with a fresh host, as often occurs when parasitoids feed on inquilines. It also occurs when early stages of host larvae are attacked and maturation of the parasitoid is not possible when the larvae are so small. In such cases the parasitoid must change its diet or starve. It also is remarkable that a change in diet may also occur in the other direction as was discussed in Chapter 3.

When one considers all mortality factors associated with various species of gall inducers, as a rule about 90 percent are eliminated by the end of the summer season. Moulds, birds, and cold winters further decrease the remaining population. Cold winters alone kill up to 62 percent of the *D. triforma* population remaining at the end of the summer season. Titmice play a similar role in galls of *C. quercusfolii.*

Studies on the long-term fluctuations in populations of gall inhabitants are uncommon, but it is an area of gall biology that warrants further attention. However, two such studies (Zuchlinski 1979; Wiebes Rijks 1982) (Fig. 16-13) indicate that gall inducers and their parasitoids "track" one another in a cyclical manner. When the population of the gall inducer is high in one year, the parasitoids reduce their numbers in the following year. A collapse of the parasitoids follows, which in turn allows the population of gall inducers to rise. What is remarkable in the 10 year study of *C. quercusfolii* (Fig. 16-13) is the high correlation of the relative frequency of the agamic wasps and that of *T. nigricornis.* It appears that in this case the *Torymus* "tracks" the gall inducer.

From the communities examined in this chapter we can see that cynipid galls are not only the result of complex interactions between insects and their host plants, they also house equally complex assemblages of phytophagous and entomophagous inhabitants. Future studies of these assemblages will likely be of interest to many biologists. Indeed, the findings should contribute to our understanding of basic principles of ecology. For example, contributions can be made with relative ease to our understanding of niche partitioning, community stability, population regulation, competition, coevolution, and host specificity. The dispersal of cynipids and the inhabitants of their galls to new patches of host plants also can be easily studied and likely would aid in understanding the distribution and abundance of many other types of insect–plant associations. Comparing the gall inhabitants of closely related cynipids may even provide clues as to the evolution of parasitoid complexes.

There are approximately 800 species of gall-inducing cynipids found across the Nearctic and Palearctic regions and nearly all support unstudied communities of parasitoids and inquilines. We hope that by highlighting the interrelationships between the inhabitants of the three galls chosen for this chapter, other researchers will be encouraged to consider cynipid gall communities as subjects for their own studies.

ACKNOWLEDGMENTS

We thank J. C. Roskam for his contribution to the discussion section of the chapter and P. W. Price for his comments on the entire chapter. We also thank C. Blomme for his help with drafting the figures and financial support from NSERC Operating Grant A0230 and the Laurentian University Research Fund.

LITERATURE CITED

Anderson, D. J., and J. Kikkawa. 1986. Development of concepts. In J. Kikkawa and D. J. Anderson (eds.), Community Ecology: Pattern and Process, pp. 3–16. Blackwell Scientific Publications, Melbourne, Australia.

Askew, R. R. 1961. On the biology of the inhabitants of oak galls of Cynipidae (Hymenoptera) in Britain. Trans. Soc. Br. Entomol. 14:237–268.

Askew, R. R. 1975. The organization of chalcid-dominated parasitoid communities centred upon endophytic host. In P. W. Price (ed.), Evolutionary Strategies of Parasitic Insects and Mites, pp. 130–153. Plenum Press, New York.

Askew, R. R. 1980. The diversity of insect communities in leaf-mines and plant galls. J. Anim. Ecol. 49:817–829.

Askew, R. R., and J. M. Ruse. 1974. The biology of some Cecidomyiidae (Diptera) galling the leaves of birch *(Betula)* with special reference to their chalcidoid (Hymenoptera) parasites. Trans. R. Entomol. Soc. London 126:129–167.

Askew, R. R., and M. R. Shaw. 1987. Parasitoid communities: Their size, structure and development. In J. Waage and D. Greathead (eds.), Insect Parasitoids, pp. 225–264. Academic Press, London.

Blair, K. G. 1944. A note on the economy of the rose bedeguar gall, *Rhodites rosae* L. Proc. Soc. London Entomol. Nat. Hist. Soc. 55–59.

Brookfield, J. F. 1972. The inhabitants (Hymenoptera: Cynipidae, Chalcidoidea) of the cynipidous galls of *Quercus borealis* in Nova Scotia. Can. Entomol. 104:1123–1133.

Caltagirone, L. E. 1964. Notes on the biology of parasites and inquilines of *Pontania pacifica.* Ann. Soc. Entomol. Am. 57:279–291.

Cornell, H. V. 1983. The secondary chemistry and complex morphology of galls formed by the Cynipidae (Hymenoptera): why and how? Am. Midl. Nat. 110:225–234.

Doncaster, L. 1910. Gametogenesis of the gall-fly *Neuroterus lenticularis* Part 1. Proc. R. Soc. London Ser. B 82:88–113.

Doncaster, L. 1911. Gametogenesis in the gall-fly *Neuroterus lenticularis* Part 2. Proc. R. Soc. London Ser. B 83:476–489.

Doncaster, L. 1917. Gametogenesis and sex-determination in the gall-fly *Neuroterus lenticularis.* Part. 3. Proc. R. Soc. London Ser. B 89:183–200.

Duffett, G. H. 1968. Some new interrelationships of Hymenoptera over-wintering within the galls of *Andricus kollari* (Hartig.). Entomol. Month. Mag. 101–105.

Dziurzynnski, A. 1961. The inhabitants of the galls of *Mikiola fagi* Hartig. Acta zool. Cracoviensia 6:10–49.

Felt, E. P. 1965. Plant Galls and Gallmakers. Hafner, New York.

Force, D. C. 1974. Ecology of insect host-parasitoid communities. Science 184:624–632.

Folliot, R. 1964. Contribution à l'étude de la biologie des cynipides gallicoles (Hymenoptères, Cynipoidea). Ann. Sci. Nat. Zool. 12 ser. 6:407–564.

Fulmek, L. 1968. Parasiteninsekten der Insektengallen Europas. Beitr. Entomol. 18:719–952.

Hawkins, B. A. and R. D. Goeden. 1984. Organization of a parasitoid community associated with a complex of galls on *Atriplex* in southern California. Ecol. Entomol. 9:271–292.

Jones, D. 1983. The influence of host density and gall shape on the survivorship of *Diastrophus kincaidii* Gill. (Hymenoptera: Cynipidae). Can. J. Zool. 61:2138–2142.

Kikkawa, J. and D. J. Anderson. 1986. Community Ecology: Pattern and Process. Blackwell Scientific Publications, Melbourne, Australia.

Longmann, K. A. and M. P. Coutts. 1974. Physiology of the oak tree. In M. G. Morris and F. H. Perring (eds). The British Oak, Its History and Natural History, pp. 194–221. Berkshire, Great Britain.

Mani, M. S. 1964. Ecology of Plant Galls. Junk, The Hague.

Moser, J. C. 1965. The interrelations of three gall-makers and their natural enemies on hackberry (*Celtis occidentalis* L.) Bull. N. Y. Mus. Sci. Serv. Bulletin No. 402.

Nordlander, G. 1973. Parasiteklar i galler av *Diplolepis rosea* (L) och *D. mayri* Schlechtd. (Hymenoptera. Cynipidae). Entomol. Tidskr. 94:149–176.

Patterson, J. T. 1928. Sexes in the Cynipidae and male-producing and female producing lines. Biol. Bull. 54:201–211.

Price, P. W. 1980. Evolutionary biology of parasites. Monographs in population biology. 15. Princeton University Press, Princeton, N. J.

Price, P. W., C. E. Bouton, P. Gross, B. A. McPherson, J. N. Thompson, and E. Weis. 1980. Interactions among three trophic levels: Influence of plants on interactions between insect herbivores and natural enemies. Annu. Rev. Ecol. Syst. 11:41–65.

Price, P. W., G. W. Fernandes, and G. L. Waring. 1987. Adaptive nature of insect galls. Environ. Entomol. 16:15–24.

Ritchie, A. J. and J. D. Shorthouse. 1987. Revision of the genus *Synophromorpha* Ashmead (Hymenoptera: Cynipidae). Can. Entomol. 119:215–230.

Rohfritsch, O., and J. D. Shorthouse. 1982, Insect galls. In G. Kahl and J. S. Schell (eds.), Molecular Biology of Plant Tumors, pp. 131–152. Academic Press, New York.

Root, R. B. 1973. Organization of a plant-arthropod association in simple and diverse habitats: The fauna of collards *(Brassica oleraceae).* Ecol. Mon. 43:95–124.

Roskam, J. C. 1979. Biosystematics of insects living in female birch catkins. II. Inquiline and predaceous gall midges belonging to various genera. Neth. J. Zool. 29:283–351.

Roskam, J. C. 1986. Biosystematics of insects living in female birch catkins. IV. Egg-larval parasitoids of the genera *Platygaster* Latreille and *Metaclisis* Forster (Hymenoptera, Platygastridae). Tijdschr. Entomol. 129:125–140.

Roskam, J. C., and G. A. van Uffelen. 1981. Biosystematics of insects living in female birch catkins. III. Plant-insect relation between white birches, *Betula* L. Section Excelsae (Koch) and gall midges of the genus *Semudobia* Kieffer (Diptera, Cecidomyiidae). Neth. J. Zool. 31:533–553.

Schröder, D. 1967. *Diplolepis* (= *Rhodites*) *rosae* (L.). (Hymenoptera: Cynipidae) and a review of its parasite complex in Europe. Commonw. Inst. Biol. Control Tech. Bull., No. 9, pp. 93–131.

Shorthouse, J. D. 1972. An emergence technique for small insects. Can. Entomol. 104:1331–1332.

Shorthouse, J. D. 1973. The insect community associated with rose galls of *Diplolepis polita* (Cynipidae, Hymenoptera). Quaest. Entomol. 9:55–98.

Shorthouse, J. D. 1975. The roles of insect inhabitants in six *Diplolepis* (Cynipidae, Hymenoptera) rose leaf galls of Western Canada. Unpublished Ph.D. Thesis, University of Saskatchewan, Saskatoon, Sask.

Shorthouse, J. D. 1980. Modification of galls of *Diplolepis polita* by the inquiline *Periclistus pirata.* Bull. Soc. Bot. Fr. (Actual. Bot.) 127:79–84.

Shorthouse, J. D. 1982. Resource exploitation by gall wasps of the genus *Diplolepsis.* In Proc. 5th Int. Symp. Insect-Plant Relationships, Wageningen, pp. 193–198. Pudoc, Wageningen.

Shorthouse, J. D., and A. J. Ritchie. 1984. Description and biology of a new species of *Diplolepis* Fourcroy (Hymenoptera: Cynipidae) inducing galls on the stems of *Rosa acicularis.* Can. Entomol. 116:1623–1636.

Stille, B. 1984. The effect of hostplant and parasitoids on the reproductive success of the parthenogenetic gall wasp *Diplolepis rosae* (Hymenoptera, Cynipidae). Oecologia 63:364–369.

Strong, D. R., J. H. Lawton, and T. R. E. Southwood. 1984. Insects on Plants: Community Patterns and Mechanisms. Blackwell Scientific Publications, London.

Taper, M. L., E. M. Zimmerman, and T. J. Case. 1986. Sources of mortality for a cynipid gall-wasp *Dryocosmus dubiosus* (Hymenoptera: Cynipidae): The importance of the Tannin-Fungus interaction. Oecologia 68:437–445.

Varley, G. C. 1947. The natural control of population balance in the knapweed gall-fly *(Urophora jaceana).* J. Anim. Ecol. 16:139–187.

Wangberg, J. K. 1976. The insect community in galls of *Diastrophus kincaidii* Gillette (Hymenoptera, Cynipidae) on Thimbleberry, Dept. Entomol. Anniv. Pub. 7:47–50.

Washburn, J. O., and H. V. Cornell. 1979. Chalcid parasitoid attack on a gall wasp population on oak. Can. Entomol. 111:391–400.

Washburn, J. O., and H. V. Cornell. 1981. Parasitoids, patches and phenology; their possible role in the local extinction of a cynipid gall wasp population. Ecology 62:1597–1607.

Weis, A. E., W. G. Abrahamson, and K. D. McCrea. 1985. Host gall size and oviposition success by the parasitoid *Eurytoma gigantea.* Ecol. Entomol. 10:341–348.

Wiebes-Rijks, A. A. 1979. A character analysis of the species of *Synergus* Hartig, Section II (Mayr. 1872) (Hymenoptera, Cynipidae). Zool. Meded. Leiden 53:297–321.

Wiebes-Rijks, A. A. 1980. The identity of the gall wasp causing stunted acorns (Hymenoptera, Cynipidae). Neth. J. Zool. 30:243–253.

Wiebes-Rijks, A. A. 1982. Early parasitation of oak-apple galls (*Cynips quercusfolii* L., Hymenoptera). Neth. J. Zool. 32:112–116.

Zuchlinski, J. A. 1979. Interactions among inhabitants of a *Diplolepis* (Hymenoptera: Cynipidae) bud gall on *Rosa acicularis* (Rosaceae) in northern Ontario. Unpublished M.Sc Thesis, Laurentian Univ., Sudbury, Ontario.

Zwölfer, H. 1979. Strategies and counterstrategies in insect population systems competing for space and food in flower heads and plant galls. Fort. Zool. 24:331–353.

17

Effects of Urbanization on the Distribution and Abundance of the Cynipid Gall Wasp, *Disholcaspis cinerosa*, on Ornamental Live Oak in Texas, U.S.A.

G. W. FRANKIE
D. L. MORGAN,
AND E. E. GRISSELL

When natural and agricultural areas are urbanized for human habitation, they become transformed into a series of new biological habitats with the introduction of various structures, paved roadways, exotic plants and animals, and human beings themselves. Generated within urban areas are accumulations of food (fresh and stored), waste, and other materials that were not present before urbanization began. Associated with the development and maintenance of these areas is a variety of unwanted animal species (pests) that find the above resources highly suitable for their food and shelter needs (Frankie and Ehler 1978). At times, conditions become so optimal that these species eventually reach intolerably high levels (NAS 1980). Although one can easily trace the obvious ways that humans establish conditions leading to pest problems (Gill and Bonnett 1973; Cornwell 1973; Ebeling 1975; Johnson and Lyons 1976; Weber 1979; NAS 1980), only recently have workers begun to examine the more subtle ways that human activities influence the development of insect populations.

Many species of phytophagous insects attack urban trees (Frankie and Ehler 1978; Frankie *et al.* 1979; Johnson and Lyons 1976) and there are numerous cases in which they cause the trees severe damage. Gall-inducing insects are also known to colonize urban trees (Hamel 1973; Ehler 1982); however, few species have been studied in this habitat. This chapter attempts to illustrate the research potential of gall-inducing insects on urban trees. Here we describe the colonization of urban live oak in Texas by the heterogonous cynipid wasp *Disholcaspis cinerosa* Bassett. We also evaluate the importance of parasitoids in influencing the abundance of the wasps and their galls in both natural and urban Texas environments. Early in the investigation, which began in 1970, we observed that asexual gall densities occasionally reached high levels in urban environments. In sharp contrast, only low densities were observed in natural or near-natural environments. Factors affecting gall density in these two environments thus formed the major focus of the investigation.

HOST TREES

Plant taxonomists agree that "live oak" exists in nature as more than one botanical species (Muller 1970, Vines 1960). A broad geographical mapping of the natural range of live oaks in Texas is presented in Figure 17-1. *Quercus vir-*

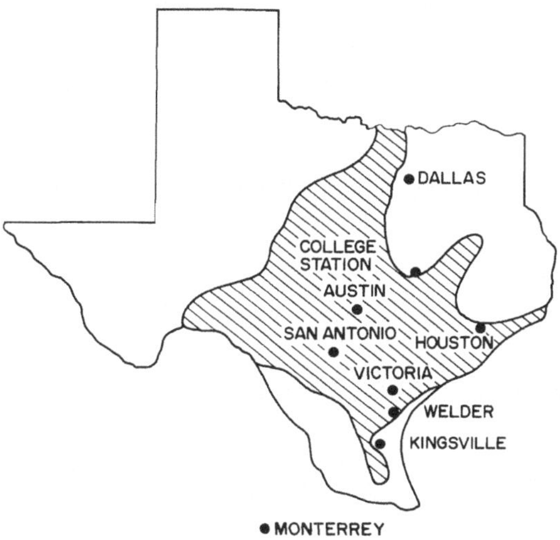

Figure 17-1. Distribution of live oak in Texas.

giniana Mill. is the "live oak" of commerce and landscape tradition due to its broad range and early use. It is found growing naturally south of San Antonio, Texas, along the Gulf Coast to the Atlantic seaboard, and northward into the state of Virginia. Northward from the east coast of Texas it occurs to about Crockett, its northernmost appearance in the state. *Quercus virginiana* is met by *Q. fusiformis* Small, which is generally found in the central and western parts of the state; here it appears on limestone outcroppings on the Coastal Plain. The range of *Q. fusiformis* extends northward up the Grand Prairie and West Cross Timbers, and into Oklahoma in the Wichita Mountains National Wildlife Refuge and Quartz Mountain State Park near Lone Wolf. Southward, it extends into the states of Coahuila, Nuevo Léon, and Tamaulipas, Mexico. *Quercus minima* (Sargent) Small, a shrubber live oak, grows typically in deep sands adjacent to beach areas along the east and central Texas Gulf Coast, and eastward to the coasts of Florida and North Carolina (Little 1971). Complete taxonomic descriptions of the native live oak species and their hybrids in Texas are found in Muller (1970).

STUDY SITES

Live oaks were surveyed and/or intensively monitored for galls of *Disholcaspis cinerosa* in several natural stands (Fig. 17-2) and urban plantings (Fig. 17-3) throughout Texas and at one natural site in Mexico.

SEASONAL HISTORY OF *DISHOLCASPIS CINEROSA*

Disholcaspis annually passes through two alternating generations on live oak (all three species) in Texas. The asexual generation develops within beige (pink initially) spherical galls (5–25 mm diameter) on branches. Development occurs centrally within the gall mantle in a small oval-shaped capsule. Generally, new asexual galls are first observed in July, and larvae within can be found from this period through most of October; pupation occurs in November. Parthenogenetic adults bore through the capsule and outer mantle and emerge from the galls in December and early January. Small cavities interspersed in the mantle are believed to assist in the emergence since females almost always begin their exit through these spaces. It is common for groups of adults to emerge synchronously on given days during this period, and these "emergence days" often correspond with relatively cold periods (daily highs of 0–4° C). Each adult lives 3–6 weeks and oviposits single or rarely two eggs within a swollen leaf or flower bud of live oak, giving rise to the sexual generation. Abandoned asexual galls may persist on trees for several years, turning gray and black as they age (Frankie *et al.* 1977).

The sexual generation remains in the egg stage until mid-March. Larval and pupal stages develop very rapidly during this period to late March, and resulting sexual galls, which resemble kernels of wheat (3–5 mm in length), develop concurrently at the bases of new leaf shoots and inflorescent stalks. Usually, one gall or rarely two form per leaf or flower bud. Sexual adults emerge from these galls around the first of April. Mated females live about a week and oviposit at the bases of leaf shoots giving rise to the asexual generation to start the cycle again (Frankie *et al.* 1977; Frankie and Morgan 1984).

First appearance of asexual and sexual galls varies according to geographic location and seasonal conditions. In south Texas, e.g., from Houston southward, asexual galls are found as

Figure 17-2. Natural live oak stand in Austin (= Austin I).
Figure 17-3. Young live oaks in parking area of NorthPark Mall, Dallas.

early as June and sexual galls first appear in mid- to late March. In north Texas, e.g., Dallas, asexual galls are usually first observed in late July/early August, and sexual galls form during the first week of April. Regardless of location, there is a significant and synchronized partial emergence of asexual females within 2 weeks after the first major cold front that sweeps through Texas in late Fall (usually from mid- to late November).

METHODS

Asexual *Disholcaspis* galls (current year and old galls) were systematically sampled from live oaks in new and older urban plantings and in natural stands (mixed ages) throughout Texas to develop information on the distribution of galls within and between stands in contrasting environments. Urban plantings (Fig. 17-3) were selected on the basis of whether they contained 100 or more live oaks, whereas natural stands (Fig. 17-2) were chosen because they represented undisturbed (or relatively undisturbed) sites where live oak was common. Twenty-five or 30 live oak trees were selected from each planting or stand.

The first systematic survey of asexual *Disholcaspis* galls from 25 randomly selected trees in Dallas and Kingsville in 1972 revealed that at a given site great intertree variation in gall densities existed, and the frequency of highly infested trees was low. To ensure that the highest infestation level would always be represented in each of the subsequent samples, 5 additional trees at each site, which were thought to support high gall numbers, were added to the 25 randomly selected trees (thus, 83 percent of the trees at a given site were randomly selected). Finally, in native stands only, a casual examination of the live oaks surrounding the block of trees designated for sampling was conducted. In all but the Welder population, this amounted to examining 300–400 extra trees per site.

Intensive sampling from several small and medium sized trees at College Station in 1971 indicated that asexual galls were generally uniformly distributed over the outer 50 percent of the crown. Based on this finding, a stratified sampling scheme was developed due to two constraints: (1) a quick and convenient method was needed to sample numerous trees of various ages and sizes, and (2) young specimen trees at urban shopping plazas had to be sampled in a manner that avoided distortion of shape through possible overcollection on one or two sides. Thus the following scheme was used: 12 0.5-m limbs (3 randomly selected per cardinal direction) were removed from mostly the outer 50 percent of the crown of each tree.

Arbitrary categories were used to sort out high, moderate, and low gall densities among the trees. Trees having <50 sampled galls were considered low producers, trees with 51–400 were viewed as moderate, and those with >400 were high producers.

Asexual and sexual galls (current year only) were also sampled annually for several years from 10 heavily infested host trees: 6 at NorthPark Mall and 4 at Samuell Grand Park, Dallas. These individuals were specifically chosen because they supported high gall densities in relatively new (NorthPark) and established (Samuell Grand) plantings. The sampling scheme that was used for the state-wide survey was also used to monitor gall populations on these trees. Twelve, 0.5-m limbs with asexual galls and 8 0.5-m limbs with sexual galls (respectively, 3 and 2 randomly selected per cardinal direction) were removed from each marked tree. Gall numbers pooled from each tree were used to compute a mean value per branch for each generation. Variations around the means were computed from several representative trees sampled in 1977/1982 (asexual generation) and 1978 (sexual generation).

Up to 150 asexual and 50 sexual galls were dissected annually from each marked tree at NorthPark and Samuell Grand Park to determine percent mortality by parasitoids and to a lesser extent by other factors. When <10 galls (pooled) were sampled per generation, a question mark follows the respective mortality value (see later figures). Numerous collections of representative parasitoids were reared in the laboratory (indoors, 80 percent; outdoors, 20 percent) by carefully extracting their larval and pupal stages (sometimes intact with the host) from the central gall area where *Disholcaspis* develops and placing them singly, or in like groups, into small, clear plastic boxes. Adults were sent to the Systematic Entomology Laboratory (USDA), Beltsville, Maryland for identification.

RESULTS

Texas Survey

Sampling asexual galls (current year and old) from live oak in urban and natural stands revealed three patterns of distribution and abundance (Fig. 17-4). First, urban stands always contained some trees that had substantially higher gall densities than oaks in natural environments. Second, the most highly infested urban trees were associated with new plantings. High densities were particularly noticeable on certain trees in new plantings at Dallas and Houston. For example, a high of 932 galls was recorded from one young tree in Dallas (Fig.

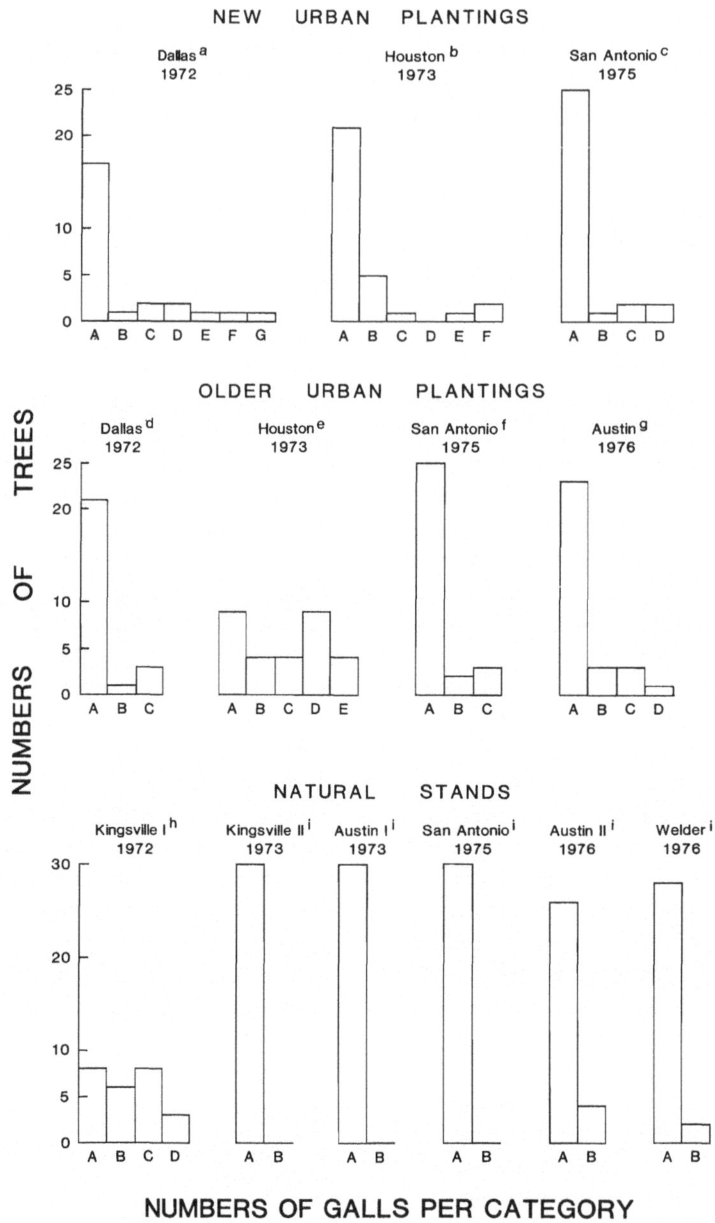

Figure 17-4. Densities of asexual *Disholcaspis* galls (current year and old) among live oaks in new and older urban plantings, and in natural stands. (a) NorthPark Mall,* (b) Galleria—Sakowitz complex,** (c) North Star Mall and vicinity,** (d) Fair Park,* (e) Hermann Park,** (f) Trinity College campus,** (g) University of Texas campus,** (h) Block of 100 trees in natural stand,* (i) Block of 100 trees in natural stand.** A = 0–25, B = 26–50, C = 51–100, D = 101–200, E = 201–400, F = 401–800, and G = 801–1600 galls. *All trees randomly selected; **83 percent of trees randomly selected.

17-4). In a later (1977) miscellaneous survey in Dallas, 1599 galls (current-year and old) were removed from one relatively young tree, using the same sampling procedure. Third, with the exception of the older Houston planting and the natural Kingsville I stand, most trees regardless of location had relatively few galls; the vast majority had less than 12 per tree. The Houston and Kingsville sites were exceptional in that each was well represented with trees having low to moderate densities. Numerous miscellaneous samples and observations in other urban (e.g., College Station and Victoria) and natural stands (scattered locations throughout the Texas Hill Country in central Texas) corroborated the above three patterns.

Within a particular live oak stand, natural or urban, a variable mixture of gall age classes (current year and old galls) was observed among the trees. However, in a few cases one gall age class predominated. For example, in the young Dallas trees (Fig. 17-4) having more than 50 sampled galls, 6 of 7 had a higher proportion (by a factor of 2 or more) of current to old galls. In the older Dallas planting (Fig. 17-4), the 4 trees with more than 25 galls each had more (by a factor of two or more) old than current year galls. At Kingsville I (Fig. 17-4), 8 of 17 trees with more than 25 galls each had approximately equal numbers of each type; the other 9 had a higher (by a factor of 2 or more) proportion of old galls.

Gall Dissections

A wide range of gall sizes was observed at all survey sites, and when current-year galls were dissected a relationship between gall size and healthy *Disholcaspis* was recorded. Galls were collected between early November and early December (prior to emergence of asexual *Disholcaspis* adults), sorted arbitrarily into three size classes, and dissected. Some collections were taken from the survey trees (Fig. 17-6) while others were taken from collections made on trees monitored over an 11-year period at NorthPark in Dallas (see below). Results of dissections are presented in Tables 17-1 and 17-2.

There was a high probability of encountering healthy *Disholcaspis* (usually a mature pupa or adult) in large-sized galls, but as gall size decreased so did the likelihood of finding healthy *Disholcaspis*. Corresponding with the decrease in gall size was a general increase in parasitism, dry remains, and empty galls. Finally, small di-

Table 17-1. Contents of Current-Year Asexual Galls from Three Size Classes at Different Urban Localities

	Galls per category (%)				
	Dallas[a]		Houston[b]		San Antonio[c]
Gall contents of diameters ≥ 18.5 mm	New ($n=643$)	Older ($n=99$)	New ($n=105$)	Older ($n=95$)	New
Healthy *Disholcaspis*	64	70.7	88.6	85.3	—
Morbid *Disholcaspis*	1.1	0	0	0	—
Parasitized	32	27.2	10.5	10.5	—
Dry remains	1.2	1	0.9	0	—
Fungal infection	0.8	1	0	3.2	—
Empty	0.6	0	0	1.1	—
Diameters 10–18.5 mm	$n=607$	$n=245$	$n=194$	$n=147$	$n=55$
Healthy *Disholcaspis*	14.2	15.1	60.5	23.8	34.5
Morbid *Disholcaspis*	0.2	0.8	0.5	0	0
Parasitized	71	73.9	30.8	33.3	60
Dry remains	10.7	7.3	3.1	10.9	5.5
Fungal infection	1.8	0.8	1	6.1	0
Empty	2.3	2	4.1	25.9	0
Diameters < 10 mm	$n=858$	$n=406$	$n=219$		$n=60$
Healthy *Disholcaspis*	0.6	0	21.9	—	20
Morbid *Disholcaspis*	0	0	0.5	—	1.7
Parasitized	71.3	67.5	64.4	—	63.3
Dry remains	20	20.2	3.2	—	13.3
Fungal infection	2.7	1.9	2.3	—	0
Empty	5.2	10.3	7.8	—	1.7

[a]Samples from young trees at NorthPark Mall (1974–1978) and older trees at Samuell Grand Park (1937–1978).
[b]Samples from young trees at Galleria–Sakowitz complex (1972–1974) and older trees at Herman Park (1973–1974).
[c]About 90 percent of sample from North Star Mall and vicinity (1975); 10 percent from an adjacent natural area (1975)

Table 17-2. Contents of Current-Year Asexual Galls from Three Size Classes at Different Natural Localities

	Galls per category (%)		
Gall contents of diameters ≥ 18.5 mm	Kingsville[a] (n=231)	Austin[b] (n=136)	Monterrey[c] (n=100)
Healthy *Disholcaspis*	87.4	45.6	84
Morbid *Disholcaspis*	0.4	0	2
Parasitized	10.8	50	13
Dry remains	0.4	2.2	1
Fungal infection	0.4	2.2	0
Empty	0.4	0	0
Diameters 10–18.5 mm	n=171	n=250	n=100
Healthy *Disholcaspis*	40.3	7.2	35
Morbid *Disholcaspis*	3.5	0.8	1
Parasitized	47.4	58.8	52
Dry remains	2.3	21.6	6
Fungal infection	0.6	8.5	3
Empty	5.8	3.2	3
Diameters < 10 mm	n=29	n=458	n=100
Healthy *Disholcaspis*	27.6	0.2	8
Morbid *Disholcaspis*	6.9	0.2	0
Parasitized	48.3	74.0	59
Dry remains	6.9	18.9	20
Fungal infection	3.4	3.1	2
Empty	6.9	3.5	11

[a]Samples from Kingsville I (1972, 1973, and 1975).
[b]Samples from Austin I (1976–1979).
[c]Samples from Monterrey, Mexico (1973 and 1974).

ameter galls had a noticeably higher frequency of fungal infections compared to the largest galls. Included in the parasitized category were very rare cases (<1 percent) of predation by an unidentified clerid larva and general predator(s) that tore apart galls. The dry remains (almost always larvae) were shriveled and amber in color. On rare occasions a thread-like attachment was observed leading from the dry remains into the gall mantle. The empty galls, which were common in the small size classes, showed no clear evidence of having been occupied by a *Disholcaspis* life stage.

A few sites had unusually high or low frequencies of healthy *Disholcaspis* and/or parasitism. For example, the middle diameter galls from new trees in Houston (Table 17-1) had an unexplained high percentage of healthy *Disholcaspis* and a low rate of parasitism (low percent parasitism also observed in the older trees at Houston). This relatively low rate of parasitism in the older Houston trees (Table 17-1) was accounted for by the exceptionally high percentage of empty galls (25.9 percent). The natural stand in Austin (Fig. 17-2) differed from the other natural stands (Table 17-2) and the urban stands in having relatively high levels (50–74 percent) of parasitism activity in all gall size classes. The high mortality levels recorded for Austin reflect levels that have been observed wherever live oaks grow in extensive forest situations (for example, the Texas Hill Country). In contrast to the natural Austin area, the oaks at Kingsville occurred in an isolated stand, and the Monterrey trees were either widely scattered or clumped in small groups.

Parasitoids of *Disholcaspis cinerosa*

It was learned early during the investigation that *Disholcaspis* was attacked by several parasitoids. Studies were subsequently conducted to assess their importance in the population regulation of this cynipid. Individually, none of these studies provided a complete picture from egg to adult for both generations (for reasons described below); therefore, a synthesis of the findings provides the assessment. It is convenient to begin by examining 4 years of survey data (from 1976–1979) on the asexual generation at a natural site in Austin (= Austin I).

Asexual Generation

Galls were sampled and dissected during November of each year when most, if not all, parasitism had already occurred; that is, just prior to adult emergence of *Disholcaspis.* Ten or 12 trees that appeared to have relatively "high" densities were selected for sampling among 300 to 500 individuals that were casually examined at Austin. Different trees were usually sampled annually since "high" gall densities on given trees regularly declined from one year to the next. Depending on the year, 24 or 36 0.5-m branches were selected per tree using the same method as described for the statewide survey. Numbers of trees and branch sample sizes varied each year according to the difficulty of finding adequate numbers of galls to sample (i.e., > 100). Pupal, adult, and to a lesser extent larval stages of *Disholcaspis* were examined for presence of parasitoids and other mortality factors. Eggs were not examined due to the difficulty of finding this stage.

The sampling at Austin produced relatively few galls each year despite extensive branch collections from individual trees (Table 17-3). Generally, the percent healthy *Disholcaspis* in the 1976–1979 sample was relatively low in all but one year (1978). When the distributions of healthy and unhealthy *Disholcaspis* were compared, a significant difference among the 4 years was observed ($\chi^2 = 16.268, p < .001$), indicating that overall mortality in natural populations may vary considerably from year to year. Other miscellaneous collections that have been examined at various times from several natural or near natural sites within 100 km of Austin also yielded low percent survival in asexual *Disholcaspis.* Dissections also indicated that parasitoids of *Disholcaspis* were extremely common; parasitism was always greater than 60 percent. Furthermore, the level of parasitism varied significantly among the years ($\chi^2 = 7.815, p < .05$) for parasitized versus other unhealthy or empty gall categories (Table 17-3). Possibly related to parasite activity was the substantial number of galls having dry remains and fungal infections. As will be discussed later, these contents may indicate previous ovipositional activity and/or host feeding by adult parasitoids.

The specific parasitoids associated with the asexual galls of Austin and other localities are listed in Tables 17-4 and and 17-5. In the Austin case, the proportions of each parasitoid taxon varied greatly from year to year.

Parasitoids from the asexual generation were reared from several natural and urban sites in Texas. Since only 20 percent of our reared parasitoids were kept outdoors, we have been conservative in providing emergence periods. Differences in emergence times of parasitoids reared indoors and outdoors have been demonstrated for other galls (Askew 1961b,c, 1965; Askew and Ruse 1974). However, dissections of periodically collected galls, including galls from

Table 17-3. Contents of Current-Year Asexual Galls Sampled at the Austin Natural Site,[a] 1976–1979

	Galls per category (%)			
Gall contents	1976[b] (*n*=175)	1977[c] (*n*=137)	1978[d] (*n*=118)	1979[d] (*n*=447)
Healthy *Disholcaspis*	2.9	13.1	22.9	7.8
Morbid *Disholcaspis*	0	2.2	0.8	0.2
Parasitized	73.1	61.3	64.4	63.5
Dry remains	11.5	19.7	4.2	22.1
Fungal infection	5.1	1.5	5.1	4.7
Empty	7.4	0.7	2.5	1.6

[a]Galls sampled at Austin I from 1976 to 1978; in 1979 the sample consisted of galls from both Austin I and Austin II.
[b]Twenty-four ½-m branches selected from 10 trees.
[c]Twenty-four ½-m branches selected from 12 trees.
[d]Thirty-six ½-m branches selected from 10 trees.

Table 17-4. Principal Parasitoids of Asexual *Disholcaspis* (Expansion of Results from Table 17-3) at Austin I Natural Site

Taxa	Parasitoids per taxa (%)		
	1976 (n = 129)	1977 (n = 84)	1978 (n = 75)
Eurytomidae			
Eurytoma sp.[a]	66.7	36.9	2.7
Sycophila disholcaspidia	1.6	9.5	17.3
Eulophidae			
Tetrastichus racemariae	4.7	3.6	10.7
Cynipidae			
Synergus sp. (possibly > 1)	13.2	23.8	34.7
Torymidae			
Torymus sp.[a](4)	6.2	2.4	28
Other parasitoids[b]	7.8	23.8	6.7

[a]See Table 17-5 for complete names.
[b]See "other parasitoids" in Table 17-5.

January to June following *Disholcaspis* emergence, provided confidence that the laboratory emergence dates were good estimations.

We made little attempt to separate parasitoids according to their role in this community (see Chapter 16). Although many of the principal parasitoid species have been observed to function as primaries, most are also known to occasionally serve as hyperparasitoids. Thus we regard most of these taxa as opportunistic.

Eurytoma *species*

These ectoparasitoids (Table 17-5) were almost always found in small diameter galls and are believed to be one of the first chalcidoid groups to enter the new galls (see Chapter 16), probably in early to mid-summer. They were also one of the first to leave the gall, usually before November of the same year. During some years at Austin (Table 17-4), they occurred in very high numbers.

Sycophila disholcaspidis (Balduf) is a suspected endoparasitoid whose period of entry into the community is unknown. The larvae, which are slow moving and also hairless, were usually associated with waste debris within the central chamber of medium-sized galls. Adults of *S.disholcaspidis* emerged from the galls during the following year from February through May.

Eulophid species

Tetrastichus racemariae Ashmead is a gregarious endoparasitoid that is thought to attack *Disholcaspis* early because of its small adult size. However, its presence was not detectable until later host development when the gall had reached a medium or sometimes large size. The occurrence of numerous small *Tetrastichus* larvae (50+) within the central gall chamber was a positive indication that this eulophid had consumed the host. Adult emergence of *T. racemariae* occurred from spring to early summer about 8–12 months after the eggs were first laid in the host cynipid.

Synergus *species*

Members of this genus have a unique relationship with *Disholcaspis.* They oviposit in galls during the summer, kill the inducer, and then stimulate the gall to continue growing, but in a slightly different and characteristic manner. The resulting larvae apparently receive nourishment by feeding directly on the inner plant tissues of this modified gall. This scenario is based on the fact that *Synergus* larvae and adults are relatively large and are usually found associated with small to medium-sized galls that are extremely dense and lack chambers. This is in sharp contrast to galls that are attacked by other parasitoids or are parasitoid free, which are noticeably less dense and commonly have small to large cavities interspersed in the mantle. Shorthouse (1973, 1980) observed a similar phenomenon with the cynipid *Periclistus pirata* (Osten Sacken), which begins its development as an inquiline in the leaf gall induced by *Diplolepsis polita* (Ashmead) on

Table 17-5. Reared Parasitoids Associated with Asexual *Disholcaspis* from Two Natural and Three Urban Areas.

	Natural		Urban		
	Austin	Kingsville	Dallas	Houston	College Station
Principal parasitoids					
Eurytomidae					
Eurytoma bugbeei			+		+
E. querciglobuli	+		+	+	+
Eurytoma sp. (1–3)	+		+	+	+
Sycophila discholcaspidis	+	+	+	+	+
Eulophidae					
Tetrastichus racemariae	+	+	+	+	+
Pteromalidae					
Acaenaclis lausus			+		+
Mesopolobus sp.	+		+		+
Cynipidae					
Synergus sp. (possibly > 1)	+	+	+	+	+
Torymidae					
Torymus warreni	+	+	+	+	+
T. racemariae	+		+	+	+
T. memnonius	+	+	+		
T. tubicola	+		+		+
Other parasitoids[a]					
Braconidae 1 sp.			+		
Eulophidae					
Tetrastichus sp. 1		+			+
Tetrastichus sp. 4			+		
Eupelmidae					
Eupelmus sp. 2		+		+	
Eupelmus sp. 3	+		+		
Eurytomidae					
Sycophila sp. 2			+	+	
Sycophila sp. 5			+		
Ormyridae					
Ormyrus sp. 1	+				
Ormyrus labotus			+		

[a]All of these parasitoids were reared infrequently.

roses. Within a short period after their arrival as eggs inside the gall the inquilines induce changes that lead to significantly larger galls (see also Blair 1945; Berland and Bernard 1951; Askew 1961d; Dailey and Campbell 1973; Askew 1975).

Emergence time of *Synergus* sp. varied greatly. Some individuals apparently emerged as adults within 3 to 4 months after oviposition. Others emerged from January to June of the year following oviposition, and a few individuals emerged up to 3.5 years after oviposition (these latter *Synergus* were kept as larvae indoors in plastic rearing boxes from 1979 to early 1983). *Synergus* occasionally were found feeding on other primary parasitoids (Table 17-5). Only rarely were they attacked themselves, and this likely was related to the hard layer that results from their gall-like feeding activity. Increased gall hardness in other galls also has been observed to deter natural enemies (Askew 1961a, 1975; Shorthouse 1973; Washburn and Cornell 1979; Weis 1982b; Cornell 1983).

Torymus *species*

Members of this genus usually entered the community late since most species require a relatively large-sized host. Consequently, they were found mostly attacking late larval, pupal, and rarely teneral adult stages. Development of these ectoparasitoids was assured on larval and perhaps pupal stages; however, it was not determined if *Torymus* could complete development on teneral adults (see also Washburn and Cornell 1979). All torymid larvae were easily recognized by their general form and body hairiness (Askew 1965). Of the four species observed throughout Texas, *T. warreni* (Cockerell) and *T. racemariae* (Ashmead) were always the most commonly encountered. On occasion, the torymids behaved as hyperparasitoids or possibly as superparasitoids in this gall community. Adult emergence of the torymids usually varied from December of the same year to June of the following year.

Principal parasitoids associated with the asexual generation were virtually the same in all areas of Texas (natural and urban) where samples had been collected (Table 17-5). Most of these were geographically widespread, and some [e.g., *Torymus racemariae* and *T. tubicola* (Osten Sacken)] are known to attack other cynipid hosts in Texas (Hamel 1973; Grissell 1979; Burks 1979). In spite of the widespread distribution of the parasitoids, the rate of parasitism and frequencies with which they were found in the non-Austin sites varied greatly (see below).

Parasitoids in the other category (Table 17-5) were only infrequently reared from the central gall area. The greater number of recorded species from Dallas reflects the more intensive rearing effort in that city. It is likely that most of these parasitoids occur in other areas of Texas as well. The roles of parasitoids in this community are unknown; however, at least some genera are known to contain facultatively hyperparasitic species (e.g., *Eupelmus* and *Sycophila*).

Sexual Generation

No effort was made to locate eggs of the sexual generation in the native stands due to the difficulty of locating this stage. In Dallas (1979–1980) and College Station (1983) several thousand buds were examined as part of a series of experimental and survey studies, respectively, and none of these contained evidence of parasitism in the egg stage (Frankie and Morgan 1984; Frankie and Vinson, unpublished data).

Numerous parasitoids were reared from the inconspicuous sexual galls at only two sites: one urban and one natural (Table 17-6). We experienced considerable difficulty in securing samples of these galls from other locations due to a critical timing problem; the trees at each site produced sexual galls during a slightly different period (a matter of days). Since a greater rearing emphasis was placed on urban Dallas, generalizations about the parasitoids of this generation come mainly from this location. Parasitoids first became associated with this generation at the specific time when galls were rapidly forming in early spring (~ 1 April), which corresponded to the new spring leaf flush. Parasitoid developmental time in the sexual galls was remarkably short with most adults having emerged from the galls by 1 May. This short period may be related to the relative ease by which these fragile galls break apart through natural causes such as wind and spring foliage growth.

Overall mortality and parasitism rates may reach high levels in this generation, especially where live oak occurs in large dense stands. For example, in one older live oak tree on the Texas A&M campus at College Station, a mortality rate of 90 percent was recorded from 143 galls that had been randomly selected in 1976. Almost 93 percent of this mortality was accounted for by parasitoids, 6.5 percent consisted of dry or morbid remains of *Disholcaspis* larvae, and 1 percent was due to fungal infection. In a Dallas stand of 1300 densely planted young oaks (11 years old), the branches of 13

Table 17-6. Reared Parasitoids of Sexual *Disholcaspis* from an Urban and Natural Area

	Urban: Dallas	Natural: Kingsville I
Eulophidae		
Tetrastichus sp. 1	+	+
Tetrastichus sp. 2	+	
Eupelmidae		
Eupelmus sp. 1	+	
Eupelmus sp. 2	+	+
Eurytomidae		
Eurytoma bugbeei	+	
Ormyridae		
Ormyrus labotus	+	
Pteromalidae		
Ormocerus n. sp.	+	+
Zatropis sp.	+	+

trees were caged with asexual females in late 1977, and a small crop of sexual galls was produced during the following spring (bags were removed from the trees long before wild parasitoids became active) (Morgan *et al.* 1983). From 5 of these trees, 249 galls were collected and dissected. An overall mortality rate of 84 percent was recorded; 47 percent was accounted for by parasites, 49 percent was due to dry remains, and 4 percent was attributed to dead or morbid *Disholcaspis*. Mortality rates varied greatly among these trees with a low of 50 percent and a high of 93 percent. Based on these urban findings we assume that natural stands have routinely high levels of parasitism in the sexual generation. With some rare exceptions (i.e., *Eurytoma bugbeei* Grissell, *Tetrastichus* sp. 1, *Eupelmus* sp. 2, and possibly *Ormyrus labotus* Walker), overlap among the parasitoid species of the two generations was insignificant (compare Tables 17-5 and 17-6) (see also Askew 1961a).

Long-Term Monitoring of Galls on Urban Trees

At NorthPark and Samuell Grand Park 10 live oaks were monitored annually for asexual and sexual gall densities from mostly 1972/1973 to 1982. Trees selected at these sites showed evidence of a high tendency to produce *Disholcaspis* galls. Monitoring coupled with gall dissections produced information on long-term population trends, percent mortality, and percent parasitism. During the first year of monitoring, some trees (nos. 67, 69, and 113, Figs. 17-7, 17-8, and 17-10) apparently had just become heavily infested as evidenced by the proportionally low numbers of old galls to new ones sampled. However, most (all Samuell Grand trees and nos. 41 and 73 at NorthPark, Figs. 17-6, 17-9, and 17-11–17-14) had substantially higher densities of old compared to new galls, suggesting earlier active periods of gall production. Tree no. 20 (Fig. 17-5) showed evidence of a previous moderate infestation (note: NorthPark trees are viewed as young on an age continuum with the Samuell Grand trees at the older end of the continuum; no. 10 is intermediate in age between the sites).

Three population trends were detected by the monitoring. The most common trend was that of substantial and rapid increases and decreases in gall density over short time periods (1 to 2 years) in half the study trees (nos. 10, 20, 69, 73, and 113, Figs. 17–5, 17–8 to 17–10, and 17–13). Second, one tree (no. 67, Fig. 17-7) had an initial period only of relatively high gall numbers (from 1 to 2 years) followed by a rapid decline in gall density. Third, one tree (no. 41, Fig. 17-6) experienced an initial period of high gall density followed by a gradual decline in gall numbers over several years.

During certain years gall densities increased noticeably on a relatively high proportion of susceptible trees in localized areas. For exam-

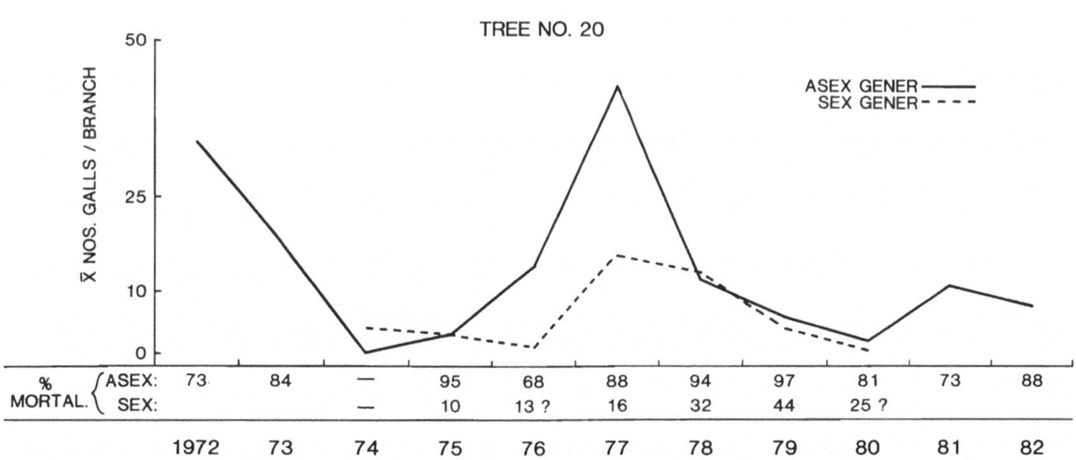

Figure 17-5. Density and mortality of wild asexual and sexual *Disholcaspis* on tree 20 and NorthPark Mall.

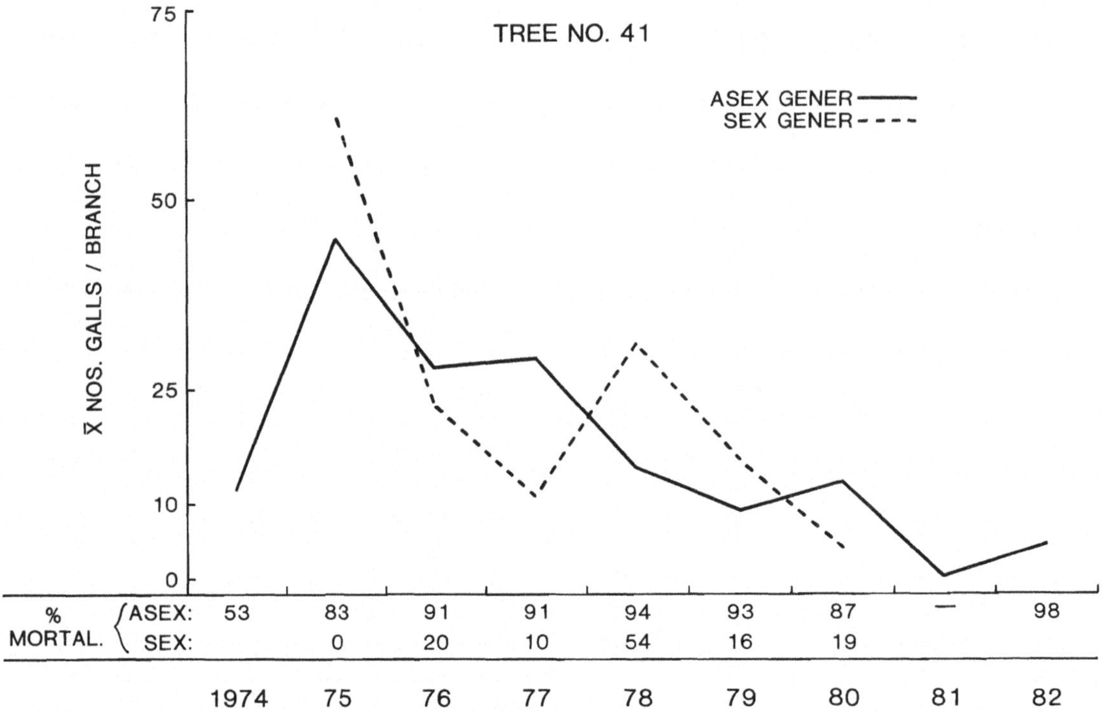

Figure 17-6. Density and mortality of wild asexual and sexual *Disholcaspis* on tree 41 at NorthPark Mall.

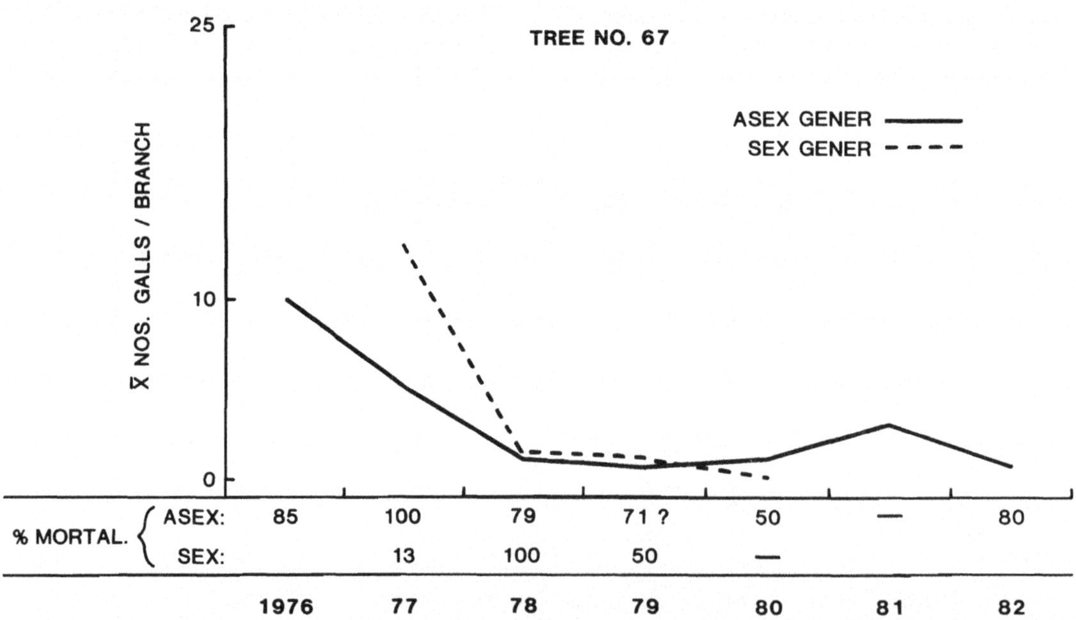

Figure 17-7. Density and mortality of wild asexual and sexual *Disholcaspis* on tree 67 at NorthPark Mall.

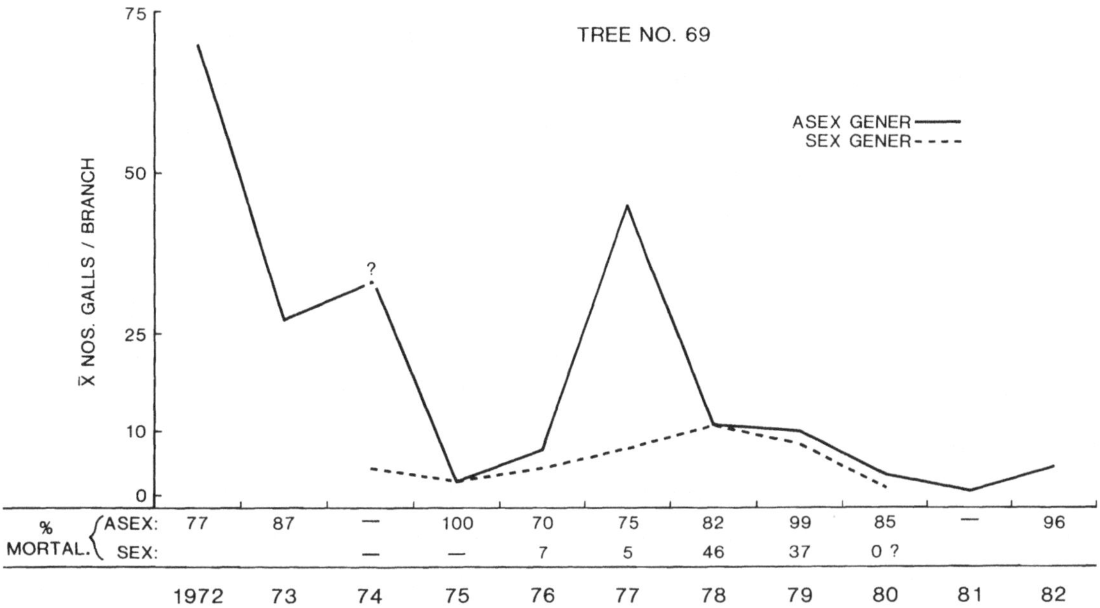

Figure 17-8. Density and mortality of wild asexual and sexual *Disholcaspis* on tree 69 at NorthPark Mall. Note in 1974, asexual galls from trees 69 and 73 (Fig. 17-9) were accidentally pooled, which resulted in estimations of asexual gall densities for both trees during that year.

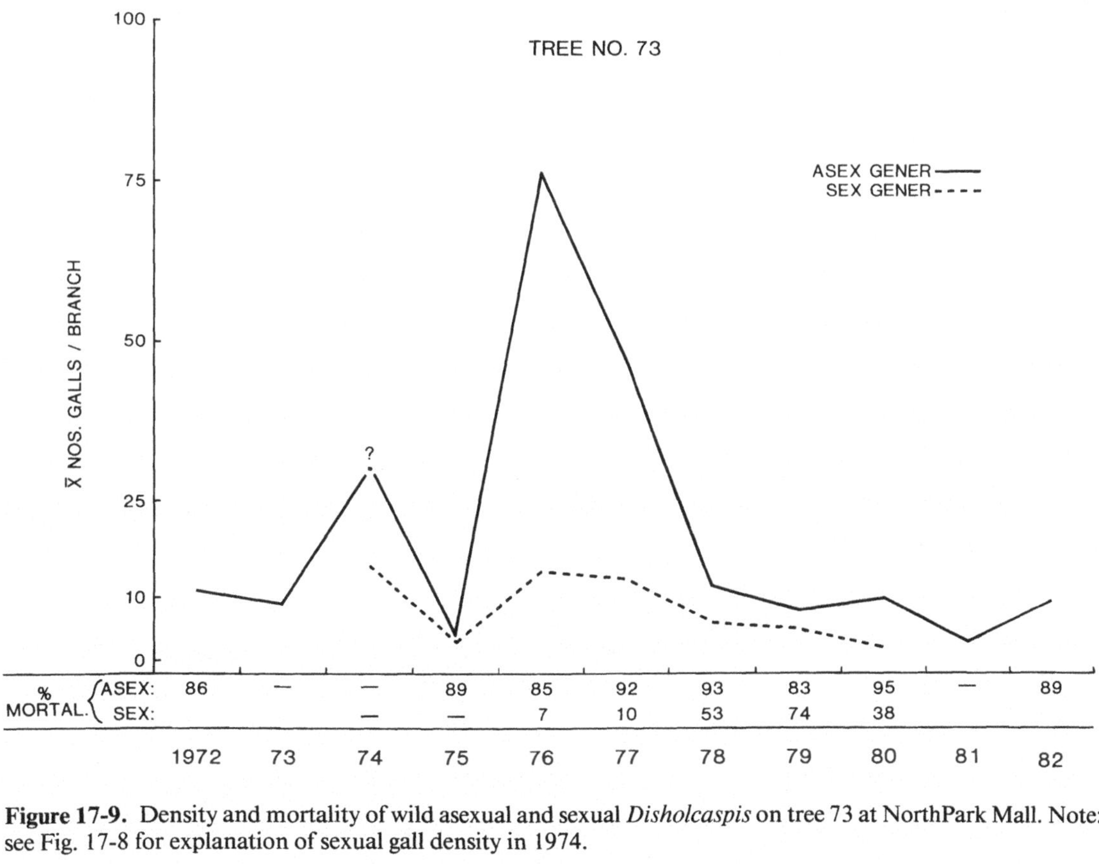

Figure 17-9. Density and mortality of wild asexual and sexual *Disholcaspis* on tree 73 at NorthPark Mall. Note: see Fig. 17-8 for explanation of sexual gall density in 1974.

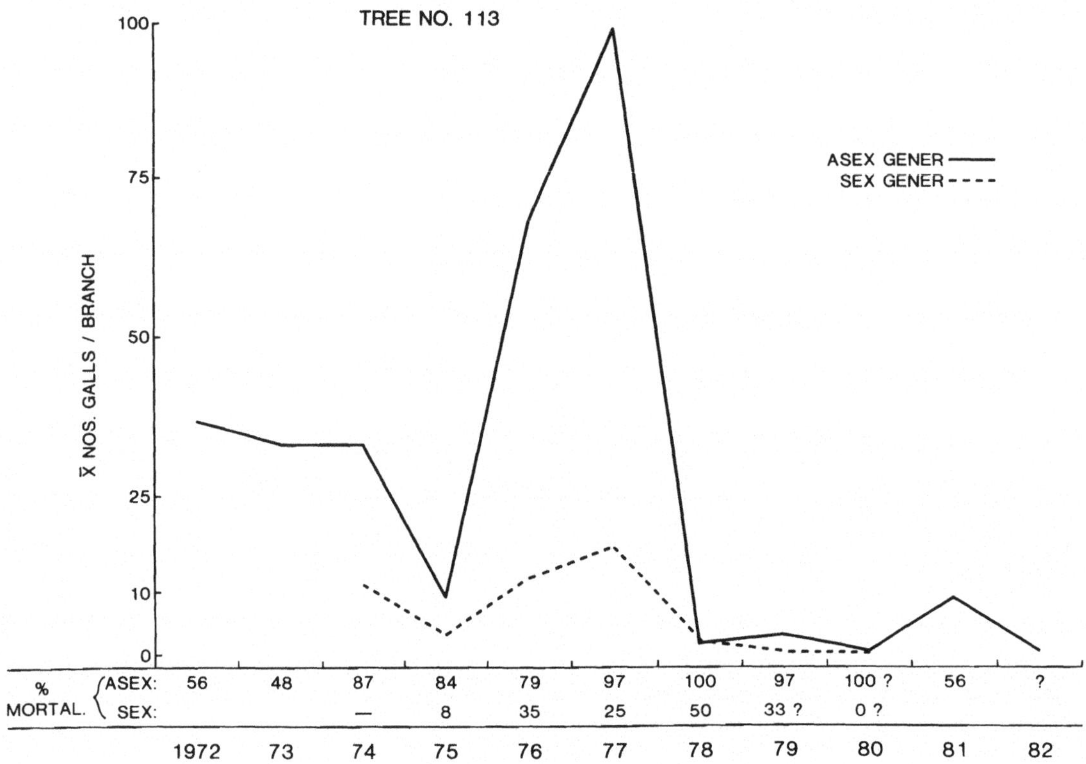

Figure 17-10. Density and mortality of wild asexual and sexual *Disholcaspis* on tree 113 at NorthPark Mall.

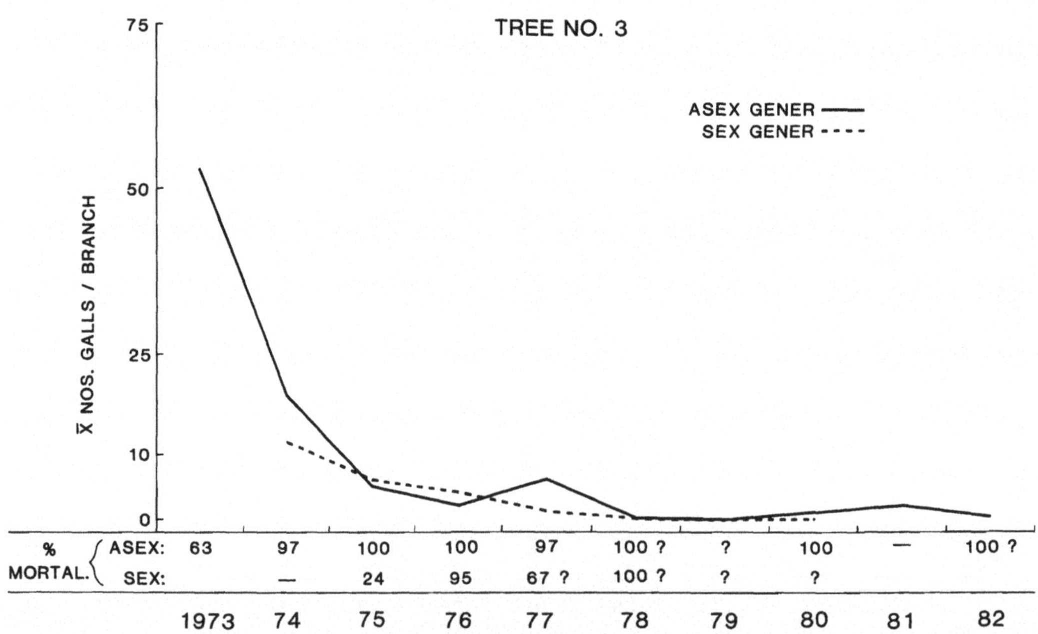

Figure 17-11. Density and mortality of wild asexual and sexual *Disholcaspis* on tree 3 at Samuell Grand Park.

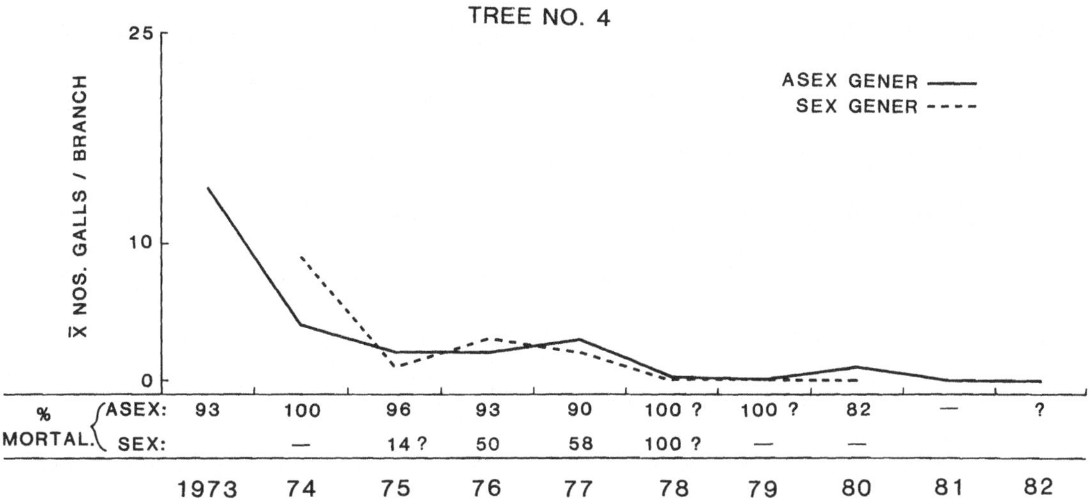

Figure 17-12. Density and mortality of wild asexual and sexual *Disholcaspis* on tree 4 at Samuell Grand Park. Note: mortality of sexual generations in 1976 differs from that reported by Frankie and Morgan (1984), which reflects new information gathered since 1984.

ple, at NorthPark and Samuell Grand Park this occurred on most marked trees in 1972/1973, in 1976/1977 (nos. 10, 20, 69, 73, 113, and possibly nos. 3 and 4, Figs. 17-5–17-9), and again in 1981 (nos. 16, 20, and 113, Figs. 17–5, 17-10, and 17-14). Casual observations on other trees in the Dallas area during these specific years also revealed noticeably higher gall numbers on a high proportion of the susceptible trees. It is important to note that as trees age (e.g., trees at Samuell Grand Park), the increase in gall density during "good" gall years was not as pronounced as in earlier years. However, as indicated by NorthPark no. 41 (Fig. 17-6) some susceptible trees operated independently of others during "good" or "poor" years.

Asexual gall densities on urban trees (Figs. 17-5–17-14) were generally much higher than those recorded for trees at the Austin natural areas (Table 17-3 and Fig. 17-4), with asexual densities on given trees during particular years almost always in excess of their sexual counter-

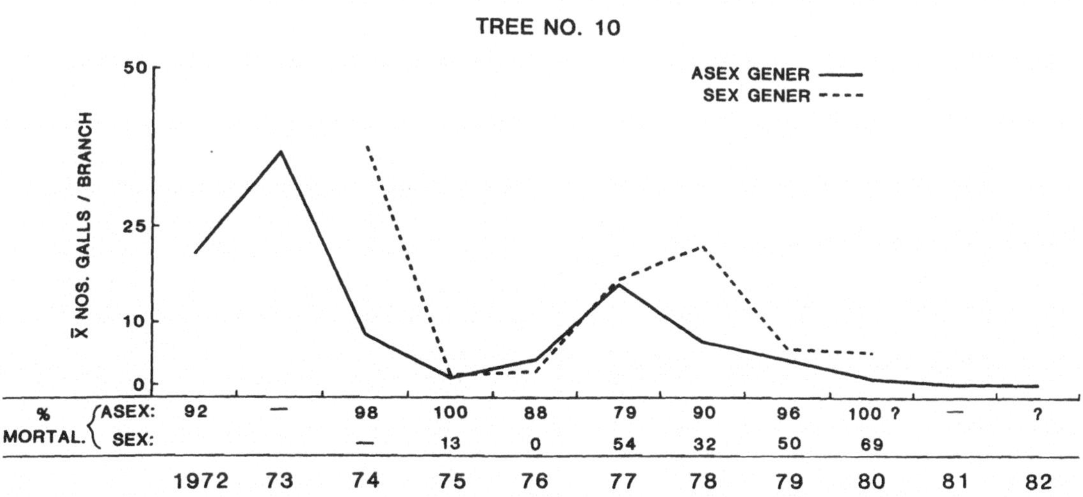

Figure 17-13. Density and mortality of wild asexual and sexual *Disholcaspis* on tree 10 at Samuell Grand Park.

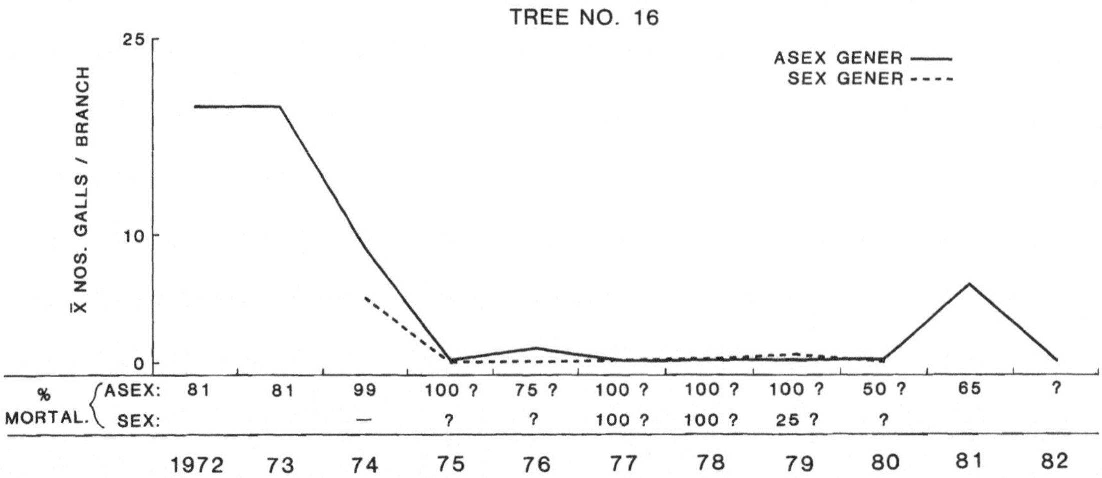

Figure 17-14. Density and mortality of wild asexual and sexual *Disholcaspis* on tree 16 at Samuell Grand Park.

parts. With a few exceptions, percent mortality in the asexual generation was consistently higher at both Dallas sites (Figs. 17-5–17-14) than at the Austin natural site (Table 17-3). Examples of the mortality factors associated with the asexual generation are presented for three representative trees at NorthPark (Table 17-7). Most of this mortality was accounted for by a complex of parasitoids. In contrast, mortality in the sexual generation was usually low at both Dallas sites; however, in some years it reached high levels among older trees at Samuell Grand Park. Most mortality observed in the sexual galls was accounted for by parasitoids (Table 17-7).

Regression analysis was used to examine the possible relationship between gall density and percent mortality for each generation of the 10 Dallas trees (Figs. 17-5–17-14). However, the regressions (not presented) failed to establish significant correlations for either generation on any marked tree, suggesting a lack of density-dependent mortality. However, significant correlations between asexual and sexual gall densities were recorded for 7 of the 10 trees (Table 17-8). A tendency toward significance was also recorded in another tree (tree 41, Table 17-8). This suggested that density in one generation was responsive to changes in the other, which in turn may be mediated by alternating high (asexual generations) and low (sexual generations) periods of parasite activity.

Table 17-7. Contents of Sexual and Asexual Galls from Four NorthPark Trees in 1977[a]

	Galls per tree (%)			
Gall contents	41	67	69	113
Sexual generation				
Healthy *Disholcaspis*	89.5	82.2	94.7	75.5
Morbid *Disholcaspis*	3.5	3.6	2.7	4.1
Parasitized	5.8	14.2	2.7	16.4
Dry remains	1.2	0	0	4.1
Asexual generation				
Healthy *Disholcaspis*	9.3	0	24.7	2.6
Morbid *Disholcaspis*	0	0	0	0.2
Parasitized	76.9	87.7	68.1	84.7
Dry remains	9.9	10.5	1.6	11.1
Fungal infection	3.2	1.8	2.8	2.1
Empty	0.9	0	2.8	0

[a]Expansion of results from Figures 17-5–17-7; see also spatial arrangement of trees in Frankie and Morgan (1984).

A year-to-year examination of percent mortality of asexual galls provided additional evidence for a relationship between gall density and percent mortality, which in turn is related to percent parasitism. Periods of low percent mortality in the asexual generation were examined for possible correlations with periods of low gall density. At NorthPark, low mortality values appeared to be associated with incipient gall infestations (see trees 20, 41, and 113) or with periods after gall densities had declined over a few years (see trees 20, 67, 69, and 113). In this regard, it is noteworthy that once the infestation began in tree 41, the percent mortality remained consistently high, i.e., 83–98 percent. Although this relationship does not generally hold for the older Samuell Grand trees where

Table 17-8. Regressions of Asexual versus Sexual Gall Densities in Dallas

Site	Tree no.	Regressions
NorthPark Mall	20	$y = 0.35x + 1.36, r = 0.81^*$
	41	$y = 1.09x - 0.93, r = 0.72$
	67	$y = 2.88x - 1.53, r = 0.99^{**}$
	69	$y = 0.06x + 4.35, r = 0.27$
	73	$y = 0.17x + 3.60, r = 0.84^*$
	113	$y = 0.17x + 1.43, r = 0.96^{**}$
Samuell Grand Park	3	$y = 0.6x + 0.44, r = 0.90^{**}$
	4	$y = 1.92x - 1.24, r = 0.85^*$
	10	$y = 1.43x + 4.71, r = 0.55$
	16	$y = 0.55x - 0.03, r = 0.99^{**}$

*$p < .05$
**$p < .01$ or better.

the park environment may be more favorable to parasites, it may apply when a lengthy period of low gall density passes (see 1980 and 1981 for tree 16).

DISCUSSION

The survey of the distribution and abundance of asexual *Disholcaspis* galls on live oak throughout Texas confirmed early observations that relatively high gall densities are to be found only on trees in urban environments. Adding 5 high gall density trees to the 25 randomly selected trees at most survey sites proved useful in only one case—the new Houston planting. One of the specially selected trees at this site fell into the highest density category (401–800 galls). Therefore, future surveys of this type would be representative with the addition of only 2–3 trees instead of 5. Other comparative studies of insect gall densities in natural and urban environments have also recorded higher densities in urban environments (Brewer 1971; Hamel 1973; Ehler 1982). The *Disholcaspis* survey also demonstrated that higher gall densities are associated with young trees, and that most trees, regardless of site, have low gall densities. Intertree differences in gall production and suspected reasons for the differences have been noted by numerous workers. An overview of these studies is presented in Morgan *et al.* (1983).

The population development of *Disholcaspis* galls occurred individually and/or synchronously among susceptible live oak trees in a particular area. Marked trees at NorthPark and Samuell Grand Park that were monitored annually for gall densities and gall contents showed distinct temporal patterns of gall development. Further, oaks surveyed throughout Texas that had varying proportions of new to old galls also provided evidence for temporal trends in gall development in successive years. In general, if a tree had a high proportion of large-sized new to old galls, there was a high probability of this population having mostly healthy *Disholcaspis.* Further, increased gall densities were likely for at least another year. At the other extreme, a high proportion of old to small-sized new galls on a tree suggested that *Disholcaspis* mortality was high and that the population was declining. Thus, by examining (1) the proportion of new to old galls and (2) the size classes of the new galls, it was possible to estimate the relative direction of population development (short-term) on given trees.

Relatively high proportions of susceptible trees in localized natural and urban areas occasionally produced sudden and synchronized increases in asexual gall densities. This phenomenon, which is little understood, is described elsewhere (Frankie and Morgan 1984). This kind of gall development may have occurred at the Houston (older trees) and Kingsville I sites (Fig. 17-4). Perhaps relevant to this phenomenon is the degree of genetic relatedness of the trees at a given site. For example, most trees at Kingsville I probably reproduced by vegetative propagation and thus had gall-producing traits in common. However, the trees in Houston represented a miscellaneous assortment of live oak stock that had been planted at various times. Due to mixed age classes and probable mixed sources and genotypes at the latter location, it is suggested that local climatic factors may be more important than relatedness and thus may operate to produce host plant conditions that are favorable for large buildups of *Disholcaspis* galls. Other examples of rapid asexual gall development on a group of live oaks may be represented by the NorthPark trees during 1972/1973, 1976/1977, and 1981 (Figs. 17-5–17-14). One result of these population increases, which occur from one year to the next, is that the gall inducer is able to escape significant parasitism for a brief period. (See Kingsville data in Table 17-2 and the NorthPark patterns in Figs. 17-5–17-10 in which percent parasitization was lower during periods of low gall densities.)

The dry remains and fungal infections recorded for galls of both generations (Tables 17-1–17-3 and 17-7) may be indications of previous parasitoid activity. That is, early adult host feeding and/or attempted oviposition probes by parasitoids (Askew 1961c; Legner 1979) may have led to dry remains and fungal-infected life stages. Direct evidence to support this hypothesis is limited and comes mostly from rare observations that have been made on thin connecting filaments between the inner asexual gall mantle and the dry remains, which may indicate a residual feeding tube from the parasite (Askew 1971; Weis 1982a) or traces of the ovipositor's route. The indirect evidence, which is stronger, consists of correlations of dry remains with small to moderate-sized asexual galls. It is in these two size categories, particularly the former, that most parasitism was observed (Tables 17-1 and 17-2). Finally, many other workers have shown that host feeding by parasitoid adults is an important mortality factor (Howard 1910; Fulton 1933; Flanders 1935, 1942, 1953; Askew and Ruse 1974; Legner 1979; Weis 1982a). DeBach (1943) suggested that host feeding may be more important as a mortality factor than parasitism itself.

It is conceivable that some dry remains may result from toxic effects produced by the plant or from some malfunction in early *Disholcaspis* development. Occasionally morbid individuals (Tables 17-1–17-3 and 17-7) which show no obvious sign of being parasitized, may represent *Disholcaspis* so affected. However, these individuals may have experienced an attempted oviposition by a parasitoid as mentioned above. Only carefully controlled caging studies in the future will reveal the factor(s) responsible for dry remains.

In the case of fungal infections, we believe that these infections may be the result of fungi that are introduced at the time of oviposition and/or host feeding by parasitoids. Future studies should examine possible fungal contaminated ovipositors, mouthparts, and bodies of the common *Disholcaspis* parasitoids. In the case of host feeding, the parasitoid taxa listed in Tables 17-5 and 17-6 would be good candidates for investigation since most belong to families or genera known for this behavior (DeBach 1943; Askew 1971).

If future studies demonstrate that most, if not all, dry remains and fungal infections are indeed associated with parasitoid activity, then calculations of mortality factors must be revised. Thus a more accurate and higher percentage of parasitism (direct and indirect) will be possible. Because we lacked these kind of data, the regressions in Table 17-8 were calculated on the basis of total percent mortality, rather than percent parasitization.

Compensatory mortality by parasitoids (sense of Huffaker and Kennett, 1966) is suspected for parasitoids that attack asexual *Disholcaspis*. At the Austin natural site (Table 17-4) it was clear that not all groups of parasitoids were equally effective or active each year. When a particular group of parasitoids was low in activity during a given year, other parasitoids appeared to perform the same mortality function and thereby proportionally reduce the number of healthy *Disholcaspis*. This may suggest that these populations of parasitoids are greatly influenced by other hosts in the area. In fact, the *Disholcaspis* food resource may be relatively unreliable from the standpoint of some of these parasitoids. Similar compensatory mortality among parasitoids has been shown for other galls (Shorthouse 1973; Force 1974; Washburn and Cornell 1981).

Parasitoids of the asexual generation appeared to cope well with the urban environment, especially *Synergus* sp. and *Torymus* sp., which are able to hold over in the galls for variable periods of time. The exception in this generation was during those periods when gall population density was extremely low for several consecutive years. However, in the case of the sexual generation parasitoids, there are probably host and shelter problems associated with certain urban habitats, especially in newly planted stands. Since parasitoids of sexual *Disholcaspis* have only a brief association with this generation, this imposes special adjustment problems on the parasitoids. For example, once parasitoids leave the galls, which occurs from April to early May, they are faced with a harsh and open environment consisting of parking lots, roadways, buildings, and widely scattered plants. This environment probably does not provide suitable protection, much less alternate food sources. These factors probably work against stabilizing populations of the parasitoids and allow *Disholcaspis* to escape high levels of parasitism during a portion of its annual cycle. However, parasitoids found in more densely vegetated areas seem able to increase to higher densities and perhaps experience greater stability due to increased plant material and the presence of alternate cynipid hosts (see also Hammel 1973). This may mean that each spring newly planted sites such as NorthPark Mall receive a substantial number of colonizing parasitoids from surrounding parks and neighborhoods having older and larger trees.

CONCLUDING REMARKS

The following is offered as a hypothesis to account for occasionally high *Disholcaspis* gall densities on live oaks in urban environments. When young oaks are planted in urban areas, they probably are first colonized by mated sexual females that are of smaller body size and that fly more readily than the heavy-bodied asexual females (Frankie and Morgan 1984). Once established on a young tree, *Disholcaspis* populations may increase to relatively high levels before parasites are able to regulate their numbers. Escape in time from parasitoids, especially in the sexual generation, is undoubtedly enhanced where trees have been planted in relatively isolated urban situations. It is believed that open, harsh conditions of isolated habitats are unfavorable to parasitoids of the short-lived sexual generation. Thus release of sexual *Disholcaspis* from heavy parasitoid loads results in production of high numbers of asexual galls for some time. However, parasitoid guilds of each generation eventually catch up with *Disholcaspis* and exert a regulatory influence on the populations. This may take many years, especially for the sexual generation. By way of contrast, parasitoid populations in natural live oak stands may be omnipresent since dense stands of oaks provide vegetative cover (Fig. 17-2) and a more diverse assortment of alternative cynipid hosts. At least one of these, *Andricus laniger* Ashmead, is known to host some of the same parasitoid species as *Disholcaspis* (Hammel 1973). Others such as *Belonocnema kinseyi* Weld and *Andricus foliatus* Ashmead are relatively common in native oak stands (Frankie and Morgan unpub.), and these may also host some of the *Disholcaspis* parasitoids. It is hypothesized that parasitoids are probably present in greater numbers in early spring in natural compared to urban habitats and are thus available to regulate *Disholcaspis* populations long before their numbers reach high levels.

Although there is good evidence to indicate that parasitoids play an important role in regulating *Disholcaspis* populations, the condition of the host plant in setting the initial level of susceptibility to gall formation must be considered. Morgan *et al.* (1983) found that the level of susceptibility changed dramatically when standard numbers of *Disholcaspis* wasps were caged on the same live oaks from 1973 to 1978 at a north Dallas site. Based on the findings of this study a subsequent and more intensive caging study was conducted from 1978 to 1980 on several of the monitored trees at NorthPark and Samuell Grand Park (Frankie and Morgan 1984). Results of the second study corroborated all of the earlier findings. It was thus hypothesized that unknown host factors operate to set the gall-carrying capacity in susceptible live oaks, regardless of site. In natural areas the full potential of a tree to produce galls is never realized since omnipresent parasitoids probably regulate both generations of *Disholcaspis* at low densities. In the urban environment, a high potential to produce galls is realized when host plant quality is optimal and parasitoid pressure is periodically relaxed (e.g., in the sexual generation). This combination is almost always observed on young trees that have some degree of isolation. After many years (40+), susceptible trees in both environments lose their capacity to produce galls through factors that are thought to be associated with host tree aging.

ACKNOWLEDGMENTS

This research was supported by the California and Texas Agricultural Experiment Stations. We thank J. Crisp, M. Gaylor, D. Hammel, J. Frankie, H. Reed, and F. Rohner for assistance in the field. J. Tucker, University of California, Davis, provided tree identifications. P. Hanson and S. Heydon provided identifications of *Ormyrus* and *Omocerus,* respectively. We benefited greatly from discussions with H. Burke, L. Caltagirone, L. Ehler, K. Hagen, C. Huffaker, W. Getz, S. Vinson, and J. Washburn. We are indebted to R. Mandel for assistance in data analysis and J. Maike and C. Tibbets for graphics. R. Askew, L. Ehler, K. Grace, M. Higgins, C. Koehler, S. Vinson, J. Washburn, and A. Weis kindly read an early draft of the manuscript. We extend our appreciation to the directors of NorthPark Shopping Mall and the Dallas Park and Recreation Department for permission to use their landscape trees for the duration of the study. The H. Heep family and the University of Texas at Austin kindly allowed us to use the live oaks in their natural stands (Heep Ranch and Breckenridge Field Laboratory, respectively).

LITERATURE CITED

Askew, R. R. 1961a. On the biology of the inhabitants of oak galls of Cynipidae (Hymenoptera) in Britain. Trans. Soc. Brit. Entomol. 14:237–268.

Askew, R. R. 1961b. Some biological notes on the pteromalid (Hym., Chalcidoidea) genera *Caenacis* Forster, *Cedidostiba* Thomson and *Hobbya* Delucchi, with descriptions of two new species. Entomophaga 6:57–67.

Askew, R. R. 1961c. A study of the biology of species of the genus *Mesopolobus* Westwood (Hymenoptera: Pteromalidae) associated with cynipid galls on oak. Trans. R. Entomol. Soc. London 113, Pt 8:155–173.

Askew, R. R. 1961d. The biology of the British species of the genus *Olynx* Forster (Hymenoptera: Eulophidae), with a note on seasonal colour forms in the Chalcicoidea. Proc. R. Entomol. Soc. London Ser. A 36:103–112.

Askew, R. R. 1965. The biology of the British species of the genus *Torymus* Dalman (Hymenoptera: Torymidae) associated with galls of Cynipidae (Hymenoptera) on oak, with special reference to alternation of forms. Trans. Soc. Br. Entomol. 16:217–232.

Askew, R. R. 1971. Parasitic Insects. American Elsevier, London.

Askew, R. R. 1975. The organization of chalcid-dominated parasitoid communities centered upon endophytic hosts. In P. W. Price (ed.), Evolutionary Strategies of Parasitic Insects and Mites, pp. 1–13. Plenum Press, New York.

Askew, R. R., and J. M. Ruse. 1974. The biology of some Cecidomyiidae (Diptera) galling the leaves of birch *(Betula)* with special reference to their chalcidoid (Hymenoptera) parasites. Trans. R. Entomol. Soc. London 126:129–167.

Berland, L., and F. Bernard. 1951. Ordre des Hyménoptères. In P. Grassé (ed.), Traité de Zoologie, No. 10, pp. 771–1276. Masson, Paris.

Blair, K. G. 1945. On the economy of the oak marble gall (*Cynips kollari* Hartig). Proc. S. Lond. Entomol. Nat. Hist. Soc. 79–83.

Brewer, J. W. 1971. Biology of the pinyon stunt needle midge. Ann. Entomol. Soc. Am. 64:1099–1102.

Burks, B. D. 1979. Eurytomidae; Eulophidae. In K. V. Krombein *et al.* (eds.), Catalog of Hymenoptera in America North of Mexico, Vol. I, Symphyta and Aprocrita (Parasitica), pp. 835–860; 967–1022. Smithsonian Institution Press, Washington, D.C.

Cornell, H. V. 1983. The secondary chemistry and complex morphology of galls formed by the Cynipinae (Hymenoptera): Why and how? Am. Midl. Nat. 110:225–234.

Cornwell, P. B. 1973. Pest Control in Buildings. Hutchinson, London.

Dailey, C. D., and L. Campbell. 1973. A new species of *Diplolepis* from California (Hymenoptera: Cynipidae). Pan-Pac. Entomol. 49:174–176.

DeBach, P. 1943. The importance of host-feeding by adult parasites in the reduction of host populations. J. Econ. Entomol. 36:647–658.

Ebeling, W. 1975. Urban Entomology. Univ. Calif. Div. Agric. Sci., Berkely, CA.

Ehler, L. E. 1982. Ecology of *Rhopalomyia californica* Felt (Diptera: Cecidomyiidae) and its parasites in an urban environment. Hilgardia 50:1–32.

Flanders, S. E. 1935. An apparent correlation between the feeding habits of certain pteromalids and the condition of their ovarian follicles. Ann. Entomol. Soc. Am. 28:438–444.

Flanders, S. E. 1942. Oosorption and ovulation in relation to oviposition in the parasitic Hymenoptera. Ann. Entomol. Soc. Am. 35:251–266.

Flanders, S. E. 1953. Predatism by the adult hymenopteran parasite and its role in biological control. J. Econ. Entomol. 46:541–544.

Force, D. C. 1974. Ecology of insect host-parasitoid communities. Science 184:624–632.

Frankie, G. W. and L. E. Ehler. 1978. Ecology of insects in urban environments. Annu. Rev. Entomol. 23:367–387.

Frankie, G. W., and D. L. Morgan. 1984. Role of the host plant and parasites in regulating insect herbivore abundance, with an emphasis on gall-inducing insects. In P. W. Price, C. N. Slobodchikoff, and W. Gaud (eds.), A New Ecology: Novel Approaches to Interactive Systems, pp. 110–140. John Wiley, New York.

Frankie, G. W., J. M. Tucker, and J. B. Welch. 1979. Arthropod fauna of live oak in urban and natural stands in Texas: I. Site descriptions, oak taxonomy/phenology and sampling methods. J. Kansas Entomol. Soc. 52:71–85.

Frankie, G. W., D. L. Morgan, M. J. Gaylor, J. G. Benskin, W. E. Clark, H. C. Reed, and P. J. Hamman. 1977. The mealy-oak gall on ornamental live oak in Texas. Texas Agr. Expt. Sta. MP-1315.

Fulton, B. B. 1933. Notes on *Habrocytus cerealellae,* parasite of the Angoumois grain moth. Ann. Entomol. Soc. Am. 26:536–553.

Gill, D., and P. Bonnett. 1973. Nature in the Urban Landscape. A Study of City Ecosystems. York Press, Baltimore.

Grissell, E. E. 1979. Torymidae. In K. V. Krombein *et al.* (eds.), Catalog of Hymenoptera in America North of Mexico, Vol. I, Symphyta and Apropita (Parasitica), pp. 748–768. Smithsonian Institution Press, Washington, D.C.

Hammel, D. R. 1973. The biology and ecology of the live oak woolly leaf gall *Andricus langier* Ashmead (Hymenoptera: Cynipidae). Unpubl. MS Thesis. Texas A&M Univ., College Station, Texas.

Howard, L. O. 1910. On the habit with certain Chalcidoidea of feeding at puncture holes made by the ovipositor. J. Econ. Ent. 3:257–260.

Huffaker, C. B., and C. E. Kennett. 1966. The biological control of *Parlartoria oleae* (Colvee) through the compensatory action of two introduced parasites. Hilgardia 37:283–334.

Johnson, W. T., and H. H. Lyons. 1976. Insects that Feed on Trees and Shrubs. Cornell Univ. Press, Ithaca, N.Y.

Legner, E. F. 1979. The relationship between host destruction and parasite reproductive potential in *Muscidifurax raptor, M. zaraptor,* and *Spalangia endius* (Chalcidoidea: Pteromalidae). Entomophaga 24:145–152.

Little, E. L., Jr. 1971. Atlas of United States trees. Vol. 1: Conifers and important hardwoods. USDA For. Serv. Mis. Pub. No. 1146.

Morgan, D. L., G. W. Frankie, and M. J. Gaylor. 1983. An evaluation of the gall-forming capacity of live oak in response to the cynipid wasp, *Disholcaspis cinerosa,* in Texas. J. Kansas Entomol. Soc. 56:100–108.

Muller, C. H. 1970. *Quercus.* In D. S. Corell and M. C. Johnson (eds.), Manual of the Vascular Plants of Texas, pp. 467–492. Texas Res. Found., Renner, TX.

NAS. 1980. Urban Pest Management. Report of committee on urban pest management. National Research Council, National Academy Press.

Shorthouse, J. D. 1973. The insect community associated with rose galls of *Diplolepis polita* (Cynipidae, Hymenoptera). Quaest. Entomol. 9:55–98.

Shorthouse, J. D. 1980. Modifications of galls of *Diplolepis polita* by the inquiline *Periclistus pirata.* Bull. Soc. Bot. Fr. (Actual Bot.) 127:79–84.

Vines, R. A. 1960. Trees, Shrubs, and Woody Vines of the Southwest. Univ. of Texas Press, Austin.

Washburn, J. O., and H. V. Cornell. 1979. Chalcid parasitoid attack on a gall wasp population on oak (*Acraspis hirta* Bassett, on *Quercus primus,* Fagaceae). Can. Entomol. 111:391–400.

Washburn, J. O., and H. V. Cornell. 1981. Parasitoids, patches and phenology: Their possible role in the local extinction of a cynipid gall wasp population. Ecology 62:1597–1607.

Weber, W. J. 1979. Health Hazards from Pigeons, Starlings and English Sparrows. Thomson Publ., Fresno, CA.

Weis, A. E. 1982a. Resource utilization patterns in a community of gall attacking parasitoids. Environ. Entomol. 11:809–815.

Weis, A. E. 1982b. Use of a symbiotic fungus by the gall maker *Asteromyia carbonifera* to inhibit attack by the parasitoid *Torymus capitae.* Ecology 63:1602–1605.

Index

The manufacturer's authorised representative in the EU for product safety is Oxford University Press España S.A. of El Parque Empresarial San Fernando de Henares, Avenida de Castilla, 2 - 28830 Madrid (www.oup.es/en or product.safety@oup.com). OUP España S.A. also acts as importer into Spain of products made by the manufacturer.
Printed and bound by CPI Group (UK) Ltd, Croydon, CR0 4YY
14/09/2026
02216286-0019